Principles of Geospatial Surveying

Principles of Geospatial Surveying

A. L. Allan

Whittles Publishing

Published by
Whittles Publishing,
Dunbeath,
Caithness KW6 6EY,
Scotland, UK
www.whittlespublishing.com

Distributed in North America by
CRC Press LLC,
Taylor and Francis Group,
6000 Broken Sound Parkway NW, Suite 300,
Boca Raton, FL 33487, USA

ISBN 978-1904445-21-0
USA ISBN 978-1420073-46-1

Typeset by Compuscript Ltd., Shannon, Ireland

Printed by Bell and Bain Ltd., Glasgow

Contents

Preface

Since the publication in 1997 of my previous textbook, *Practical Surveying and Computations*, much has changed in the profession of Geospatial Surveying, and yet nothing has changed! The same basic geometrical principles still apply – as do the need for instrumental calibration, its proper application, the suitable analysis of data, and the presentation of results to users. Although the hands-on nature of day-to-day work has almost vanished and has been replaced by rapid turnkey systems of amazing sophistication, the geospatial surveyor still has to plan and organize the work and, above all, to remain responsible to the client for its outcome and able to defend the working procedures and outcomes to a client or in court if necessary.

Because most practical work is carried out by prescribed systems, and processed by software packages, this book concentrates on those essential principles that the user needs to know, if the results are to be verified and assessed with understanding and wisdom. This change of emphasis is reflected in the new title and the order of presentation of the material. For example, the chapter on Instrumentation comes last. In any case, the proper techniques of instrument operation and handling cannot be learned from a book. The place to obtain detailed information about instrumentation is the World Wide Web, on which many thousands of relevant sites may be located. (See Appendix 10.)

The order in which the reader uses this book is very much a matter of choice and need. Many may wish to consult an appendix as the first port of call. Most of the examples have been worked in Excel spreadsheets with interim results depicted to assist the user repeat them to verify if the correct procedures are being used at each stage. The number of digits listed depends on the space available for printing and does not necessarily reflect an appropriate precision. Only at the end of a computation is the result rounded to a justifiable precision.

In compiling this new text, I have been much assisted by emeritus Professor Ian Harley, who also wrote Appendix 5, and by several former colleagues at University College London; Professor P A Cross, Dr J C Iliffe (who allowed me to publish Figure 9.11) and Mr J V Arthur. Several former students and others also gave of their time to allow me to visit their production establishments: the Ordnance Survey (M Havercroft and colleagues); the Severn Partnership (J Walton, N Glenkarn, R Otto); and the Leica Company (H Anderson and C Osborne). I am also again indebted to the editor of the Survey Review, Mr J R Smith, for allowing me free reign to publish material from several of my articles, especially the series written with the aid of Nigel Atkinson. It must be said, however, the views expressed are my own and these persons are in no way responsible for the final text.

Finally I wish to express my thanks to my wife, Daphne, for her tolerance of my neglect of family duties, and to the publisher Keith Whittles and his staff, in particular Elaine Rowan, for their understanding and support in editing and production.

A L Allan

Chapter 1
Introduction to Geospatial Surveying

This book deals with the branch of surveying concerned with the geometry of three dimensions. Three-dimensional objects, varying in size from small industrial components to the Earth itself, often need to be measured or described with respect to a coordinate system. For subsequent manipulation, the results of these measurements have to be represented in the form of a *mathematical model*. The results are prepared and presented graphically, numerically or pictorially for presentation to a client for subsequent use. What is more, the quality of the measurements, usually expressed in statistical form, is often of as much interest as the results themselves. Thus the book concerns itself not only with the measurement and modelling of three-dimensional objects but also with the quality assessment of measurements and results derived from them.

A selection of topics from a vast field is presented here, focusing on ground-based measurement techniques, including the use of photographic images. The purpose of this introductory chapter is to provide a general introduction to the technological core of the book. It therefore deals briefly with the geospatial industry, project planning, quality control, legal liability, project management, the requirements of clients and the need for archival records.

1.1 The geospatial industry

Since the beginning of civilisation, man has been concerned with the measurement and recording of spatial parameters. Historically, geospatial technology has evolved in five areas of interest: land ownership; military requirements; navigation; engineering operations and the advancement of science.

Recently, the requirement of spatial information for operations made possible by the computer, such as the location of stock items in a warehouse or the navigational guidance of motorists, has also become important. Spatial information is therefore fundamental for much of modern society. It is the job of the geospatial surveyor to provide proper geometrical data for a wide range of systems serving a vast population of users. It is instructive to identify the main areas of interest.

Military requirements

It is obvious that to conduct military operations over hostile terrain or to defend friendly territory, we need to have accurate maps depicting all features of interest, hence the need for topographical mapping. To navigate ships out of sight of land requires position fixing systems. To chart the seabed and record ocean currents, hydrographic surveying is employed. To fly modern aircraft over the surface of the Earth, accurate terrain modelling is required.

Land tenure

From the dawn of land settlement and agriculture came the need to record the limits and ownership of land plots. There arose in all countries an official mechanism to establish and maintain records of land ownership, thus various legal systems of *cadastre* were established. Some, as in the UK, were based on descriptions of the terrain itself (*descriptive cadastre*), while others sought to use spatial coordinates to establish land plot boundaries (*numerical cadastre*). Until recent times, the technological means to establish unambiguously the coordinates of land boundaries, at economical costs, has not been available. The existence of satellite-based systems has totally changed matters today. However, the reconciliation of differences between older, legally established identifiers and modern, accurate geospatial data is a matter of considerable debate and litigation.

Civil, mining and industrial engineering

Another stimulus to the development of geospatial surveying has been the construction of works such as canals, bridges, buildings and harbours, the extraction of minerals from mines and quarries, and the building of ships and aircraft. Generally such fabrications are designed on the basis of spatial information, and constructed to geometrical specifications. There is also a need to monitor movement or change of form of such constructions throughout their lifespan to ensure that they remain safe. The modern industrial practice of assembling components such as airframes and wings, which have been prefabricated on widely differing geographical locations, requires great dimensional fidelity, only established by a vigorous quality control system.

Scientific applications

Historically, geospatial surveying has played an important role in the furthering of many lines of scientific discovery such as the determination of the size and shape of the Earth itself (*geodesy*), in the prediction of earthquakes and tectonic movements and more recently in the problems of global warming.

Geographical Information Systems (GIS)

In any modern society there is a constant quest to evaluate and make the most effective use of a wide range of information related to spatial position. An immense raft of parameters has been studied statistically to establish trends and correlations to assist social scientists, commercial interests, researchers and politicians to underpin their strategies and policies. This manipulation of spatial data has gone apace often without a proper appreciation of the quality limitations of the data, even spatial data. Thus a major role of the geospatial surveyor is to ensure that all spatial data is provided with a properly assigned quality specification.

Miscellaneous users

There are many other applications for geospatial surveying, in the fields of building design, archaeology, sport, medicine and forensic matters, to name but a few. If a problem has a spatial aspect of any kind, it can be the role of a geospatial surveyor to assist with its solution.

1.2 Clients and their needs

As has been implied above, geospatial surveying and setting out procedures seek to satisfy the needs of an ever-expanding group of clients. It is imperative that the special needs of each client are viewed with sympathy so that cost effective solutions to problems can be found. Typical applications include: land and sea boundary disputes; data input to geographical information systems including underground services; the surveys of interiors and buildings; industrial applications such as ship and aircraft construction; decommissioning of nuclear reactors; measurements in sport; mining and quarry excavation; medical and surgical aspects; dimensional evidence in legal cases; map and plan making; scientific studies of ice and tectonic movements; oil platform location and construction; control for remote sensing and aerial photographic imagery; monitoring of dam and structural deformations; and the location of large engineering components and structures.

An understanding of the subsequent use of the spatial data in the computer processing chain is also needed if the full potential of the value added to the surveyor's work is to be realized. This is particularly so in land, building and engineering information systems.

1.3 Specifications

It is impossible to carry out the wishes of the client if these are not clearly understood by both parties and if solutions are not written into a sensible specification. Dangers lie in over-specifying a task just as much as under-specifying it. Issues that should be made clear include:

1. *The legal status of the client requesting the work and their ability to pay, financial arrangements and any sequential payments.*
2. *The technical specification.* This should concentrate on results and the quality assurance of these results, or on specific methods of attaining these. Too tight a technical constraint may prohibit the use of new technology or may stem from advice from another consultant whose knowledge is out of date.
3. *The timescale for the task and any penalty arrangements if it is not kept.* The likely influence of the weather in outdoor work or particular environments, such as tunnels, should not be ignored.
4. *Reasonable provision for sample inspection of the surveyor's performance by the client's expert.*
5. *Prior arrangements for settling disputes, to avoid the need for litigation.*

1.4 Planning and documentation

Once the purposes of the work have been agreed, the technical work may begin. This involves thorough planning and costing. Usually this is left to an experienced surveyor who understands the complete time sequence of events and the client's needs.

The first stage is to examine existing information about previous survey work such as old maps and documents, or to seek knowledge of prior matters such as any mine workings or previous litigation. In this respect, the internet is an invaluable source of information.

The second stage is to visit the site to see the environment, identify restrictions on access, likely problems of intervisibility, location of permanent marks, and above all to propose a detailed technical plan of execution. The degree of detail described in a report depends on the competence of the surveyors who will actually carry out the task: more detail being required for less experienced operators.

All documentation should be clearly written and in accordance with the needs of quality assurance. This will include station descriptions, a statement of methodology, field books (if any) or records of data loggers. Any changes to original specifications have to be recorded, and all documentation must be cross-referenced. It should be borne in mind that all records may have to be submitted to a court of law as evidence in the event of litigation.

In addition to technical work, especially in major tasks, administrative and logistic information should also be obtained, such as the availability of accommodation and sources of fuel, materials and possible labour.

1.5 Site organization

The organization of all operations forms a very important part of the surveyor's work. It entails the preparation of an efficient technical programme, which is both cost effective and acceptable to the workforce. Because surveying practice is still weather-dependent, work cannot always be scheduled to normal working hours.

The site environment also affects practice. Surveying near an operating railway or busy motorway is particularly hazardous; work in tropical forests is a potential danger to health; sub-zero temperatures calls for special clothing; and efficient transport requires the proper care and maintenance of vehicles, boats or aircraft including helicopters.

The provision of food, accommodation, transport and fuel has to be attended to, as has the recruitment and care of temporary staff. Technical operations are often the easiest part of any surveying task.

1.6 Computations

Previously, the surveyor was required to calculate results, such as coordinates, from the actual basic measurements. Results were combined by practical methods and the final data presented, usually in the form of maps, to the client. Today this seldom happens except in the smallest of tasks. The surveyor has 'turnkey' systems that process, statistically assess and convert data to all manner of outputs compatible with the client's needs. The client then processes these data to suit their requirements, such as road parameters or areas and volumes. Such is the sophistication of available software that the user need not know how the data are calculated. While this is very quick and convenient, it places the surveyor in a vulnerable position. There is therefore a need to interpret results independently, hence verifying them. For this reason, the surveyor needs to know, at least in principle, how the data are computed, and should calculate a few results independently by spreadsheet or other methods. The theory and worked examples in this book are therefore chosen to be the means of confirming results.

1.7 Archives

The practice of keeping old records is not always appreciated. Quite apart from the obvious advantages of leaving reference marks and clear descriptions of their whereabouts for future use, long term needs such as evidence of ground subsidence, historical information on development and evidence of land ownership can be important. With the increasing use of computer data and in some cases where no hard copy is ever produced, problems are looming for the future unless a proper archival policy is adopted.

The problem is not confined to mapping. Many large engineering structures, such as boilers and reactors, have to be assembled with the assistance of a complex sequence of surveying measurements. If records of these are lost it can be extremely difficult, if not impossible, to dismantle the structure at a later date. This is particularly true in the hostile environments of nuclear reactors.

1.8 Example of a tennis court

To help summarize the above and to focus our attention on key stages of surveying and setting-out, consider the problem of establishing a tennis court.

1. The first stage is to obtain technical information about the subject. This can easily be obtained from a website, and includes such matters as specifications for the dimensions of the court, its surrounds, drainage and other construction advice, as well as accuracy tolerances.
2. The second stage is to make an appraisal of the ownership of the site and plans of location of services, especially rainwater disposal.
3. A site visit to measure up the dimensions and shape of the existing surface, especially the exact alignment of the court within the available space, is then required.
4. Following the presentation of a layout plan, proposals are made to level the site and make good the surrounds.
5. A design plan is drawn up with volume calculations, setting out arrangements of cuttings, embankments and drainage channels.
6. The dimensions of the court and its surrounding perimeter netting have to be set out.
7. After the approved surface has been put in place, the court markings have to be positioned within tolerated specifications.
8. Finally it is usual practice to supply an as-built survey, or it may even be a legal requirement to do so, together with photographs, all digitally recorded, for possible future use and inclusion in a national database used to underpin Geographical Information Systems, and to establish the legal ownership of the

site. This implies that the new work has to be tied into the National Coordinate or Mapping system.

In the next chapter, we introduce the methodology adopted by geospatial surveyors, concentrating on the geometrical aspects of the work. These include the selection of a datum and of a working coordinate reference system; the maintenance of a length standard; the adoption of measurement systems of predictable quality; and the adoption of suitable mathematical models for computations. A method of presenting evidence of completion to the client, usually in the form of plans, elevations, cross-sections and possibly a three-dimensional visualization (which may be in the form of a video) is also introduced.

Chapter 2
Technical Procedures

This chapter introduces concepts and nomenclature to be developed in more detail in later chapters. It also discusses several miscellaneous practical matters.

The surface of an object may be thought of, and described as, consisting of a multiplicity of points. We may define the surface numerically by determining the three-dimensional coordinates of these points. The more irregular the surface and the more accurately we wish to represent it, the greater is the number of points required.

Thus the work of the geospatial surveyor can be looked at from two related points of view:

1. The creation of a mathematical model of a three-dimensional solid from various types of measurement, the assessment of the quality of this model, and means of presenting the results to a client. This is usually called *surveying*.
2. The controlling of the construction of a three-dimensional solid, described by its design model, ensuring that it meets a specified accuracy standard. This is usually called *setting out* or *building*.

For example, any housing development requires a preliminary survey of the existing terrain to be undertaken on which to design the new layout. Building then takes place under the direction of a geospatial surveyor. To check that the specification has been met, a second 'as built' survey is often made by an independent third party.

Technology now provides a range of methods and procedures with which to meet various criteria. These criteria include such things as, the size of project, the environment in which it lies, the speed with which it has to be carried out, and the documentation which has to be provided to prove that all criteria have been met, and of course the cost restraint. The basic factors will now be outlined.

2.1 Datum selection

No matter what the survey may be, a suitable reference datum is required on which to base a coordinate system that provides the mathematical foundation for all measurement work. In short, a datum must have an origin, an orientation and a scale, and its parameters should always be clearly and unambiguously defined.

At its simplest, an element of the datum might be an arbitrary starting point, physically marked on the ground. This is used as the origin of the coordinate system which is assigned arbitrary values (x, y, z). To orient the axes an arbitrary reference direction might be adopted, together with the direction of gravity (vertical).

This arbitrary reference direction will often simply be that between two points on the ground, marked in some way. In many surveys of small extent it is perfectly satisfactory to ignore the curvature of the Earth.

A more sophisticated system is to adopt the direction of north for the horizontal reference. In this case, other directions referred to north are called bearings. Three different norths can be defined—magnetic, true and projection. These three define magnetic, true and grid bearings respectively.

A magnetic bearing is that given by a magnetic compass, a true bearing is derived from star observations or satellite orbits or from a north-seeking gyroscope, and a grid bearing by calculation from either of these two, depending on the map projection used as the basis of a mapping system.

It is also usual to align one of the axes of the adopted coordinate system with the direction of gravity, the local vertical. Any time a gravity sensor such as a bubble is used to set up an instrument, the direction of gravity is controlling the geometrical calculations. Over large distances the direction of gravity varies, due to the Earth's curvature for example, for which allowances have to be made. In some special cases, such as on a floating platform, this is not practical and an arbitrary direction has to be adopted instead.

In many surveys, the absolute height of the land above sea level is of paramount importance, for example when dealing with the complex infrastructure of fresh water supply and sewage disposal in maritime countries. In countries far from the sea, it may be that irrigation and lake levels are more significant. Thus it is clear that careful attention must be given to the selection of a suitable height datum. Formerly, a long-period (19.6 years) average height of the sea at some selected point or points was adopted as the datum for heights in a national system. Today, a satellite-based height datum with its related gravity model is used.

Arbitrary datums, although sufficient for local works, are quite unacceptable on a national scale. For example, in the 20th century, it was necessary to adopt local datums for the title surveys of land plots (cadastral surveys) to enable a register of ownership to be established. These local surveys were not totally uncorrelated, however, as they were oriented by true bearings from north obtained by observations made of the Sun. This practice, governed by the lack of technology at the time, sometimes gave rise to overlapping claims to land, and land disputes. On an international scale, the inconsistency of national datums has given rise to many disputes over international boundaries and to litigation concerning the boundaries of concessions assigned to oil companies extracting oil offshore.

It might be thought that, with the introduction of a worldwide system of reference coordinates based on satellite technology, these datum problems would have vanished. This is only partly true however, as satellite technology cannot provide accurate enough datums for all applications, for example in navigation. Arising from the various mathematical models adopted and from the differing levels of accuracy obtained from instrumentation, great care is needed to ensure that consistent datums are used.

As the results of surveys depend on the adopted datum, it is vital that they are only used and interpreted within the proper context. Great care must be taken when the results of two or more surveys are combined. Such a combination usually means that some transformation of coordinates is required.

2.2 Scale and length standards

Another vital factor in surveying is the maintenance of a standard of length. A fundamental part of any object model is its size or scale. Clearly this requires that any measurement of length is traceable to an acceptable length standard, usually the international metre. This metre has to be defined in an unambiguous way and by a process which is reasonably practicable to be repeated anywhere in the world. Today, the metre is defined in terms of time and the speed of light, to an accuracy of at least one part in ten million as: *the distance travelled by a light beam in vacuo during one* 299 792 458*th of a second of time.*

A most important implication of this fact is that since most electronic distance measurers actually measure time intervals, an instrumental timing accuracy of at least 1 in 10^{12} is required to achieve millimetre accuracy.

This definition of the metre requires advice from a suitable national metrology institute, such as the National Institute of Standards and Technology (NIST) in the USA, Physikalisch-Technische Bundesanstalt (PTB) in Germany or the National Physical Laboratory (NPL) in the UK, which also maintains a service to enable surveyors to trace the scale of their instruments back to a standard. Most countries provide a similar service, which many now extend to the calibration of satellite-based systems.

In many countries, units of length other than the metre are in current use, or were used in the past. The US foot is one such unit. Rather surprisingly, there were also many different versions of the metre in common use. It is vital that the surveyor uses the correct conversion factor when relating old to new work.

Of equal importance is the need for the surveyor to check, at periodic intervals, that any distance measuring equipment is calibrated against a traceable length standard.

2.3 Angles and directions

The circle

The circle is surprisingly important and useful in practical life. Just look around your house to see how many circular objects there are in it: plates, knobs, water pipes, bottles, tin cans, buttons, wheels and so on. One reason for this popularity is that a circle is easy to make, say on a lathe, or draw with a pair of compasses. In science too, the circle is important. It is the basis of goniometers (devices for measuring angles), it is used to describe sections through spheres and cones, and is the basis for much theory about the shape of curves. In surveying and mapping the circle is second in importance only to the straight line.

Angles

Figure 2.1 shows an angle AOB subtended by an arc AB of length s at the centre of a circle, radius r, whose centre is at O. Angles are measured in a variety of units.

First of all, one complete revolution of the radius may be called one cycle. Again, the whole circle may be divided equally into four parts, called right angles, which are themselves further divided in different ways.

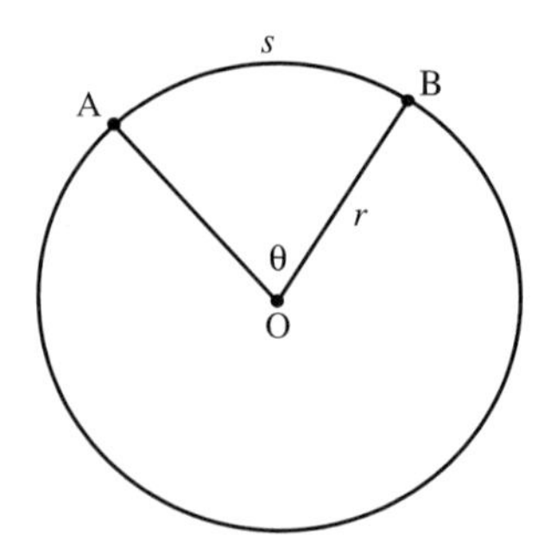

Figure 2.1

The Sexagesimal angle system

A common method is to divide a right angle into 90 parts or degrees, written 90°. Each degree is further divided into 60 parts, or minutes of arc, written 60', and finally each minute of arc is divided into 60 parts or seconds of arc, written as 60".

The Sexagesimal time system

You will have noticed that we used the phrase 'seconds of arc' in the above explanation. This is because of another way to divide up a whole circle into units of time. The average clock face is divided into twelve parts, one for each hour. Thus a right angle consists of three hours. Each hour is divided into 60 minutes of time, written 60^m, and further divided into 60 seconds of time, written 60^s.

These minutes and seconds are not the same size as their arc counterparts. A minute of time is 15 times larger than a minute of arc. This follows from the fact that 360/24 = 15. A second of time is 15 times larger than a second of arc. In surveying, both units are used.

The Centesimal system

Another way to divide up a right angle is into 100 parts, called *gons,* written ^{g}, which is now the standard method on the continent of Europe and elsewhere. Each gon is divided into 100 centigons, written ^{c}, and each centigon is divided into 100 parts, written ^{cc}. This decimal system greatly simplifies arithmetic. (Note: another name for gon is grad.)

$$\text{An angle of } 47.2245^g = 47^g\ 22^c\ 45^{cc}$$

Thus no arithmetic is needed in a conversion. Compare this to the following algorithm to convert an angle of 47° 22' 45" in the sexagesimal system to decimals of a degree:

$$47 + 22/60 + 45/3600 = 47.379\ 167°$$

The Radian system

In mathematics, the unit of angle employed is the *radian* instead of these arbitrary systems. This quite simple concept can be explained as follows, referring to Figure 2.1. If the angle θ is such that the arc of the circle AB = s is equal in length to the radius r, then θ is defined to be one radian. The reason for adopting this system of angular units is to allow us to relate the angle at the centre of a circle to the length of arc it subtends in a simple way. *When angles are referred to in mathematical formulae, the unit of measurement is always the radian unless otherwise stated.* The degree and gon systems are used for instruments, such as theodolites or protractors, and the radian system in mathematics.

There are approximately 57.3°, 3438' or 206 265" in a radian.

The length of arc subtended by an angle of one sexagesimal second of arc (1") on the surface of the Earth, whose radius is 6 378 140 m, is given by

$$s = r\theta = \frac{6\ 378140 \times 1 \times \pi}{180 \times 60 \times 60} = 30.922\,\text{m} \quad (\text{approximately}\ \ 100\,\text{ft})$$

Example In seconds of arc, what angle is subtended at the centre of the Earth by a distance of 3 mm?

The angle subtended by 30.922 m (= 30922 mm) is 1". Therefore 3 mm subtends an angle of 3/30922" = 0.0001" (approximately).

The implication of this for the cartographer is that very small angles are involved in maps and map projections. They require special care in calculations.

2.4 Time, frequency and wavelength

Intervals of time can be measured to accuracies of better than one part in a hundred million by electronic methods and atomic clocks. Traditional time standards, such as the rotating Earth, are not regular to such accuracy. All time-keepers are calibrated by the International Time Service available from National Physical Laboratories or other sources, such as the Bureau International de l'Heure (BIH), Paris.

There are approximately 366.2422 sidereal days in the solar year, which contains 365.2422 solar days. The one-day difference is due to the fact that the Earth rotates once around the Sun in a year. Thus each sidereal day is about 4 minutes shorter than the solar day.

The period T of an oscillation is related to the frequency f in units of cycles per second, or hertz (Hz), by

$$T = \frac{1}{f}$$

The speed of light c is related to the frequency and wavelength λ by the relationship

$$\mathrm{c} = f\lambda$$

where c ≈ 300 × 10^6 m sec^{-1}.

The distance s travelled in time t by light on a forward and return trip (double transit time) from one end of a line to another is given by

$$s = 0.5\,\mathrm{c}t$$

Because many distance measurement systems are really timing systems, the following approximate relationships are useful in mental work:

1. A frequency of 1500 MHz (megahertz) implies a wavelength of $\lambda = 0.2$ m.
2. The length equivalent of a double transit time of 10^{-9} sec (a nanosecond) is 300 mm.

2.5 Coordinate systems

The common coordinate systems used to model geometrical shapes are (see Figure 2.2):

1. Cartesian coordinates (x, y, z);
2. Polar coordinates (U, V, R);
3. Cylindrical coordinates (U, d, z);
4. The two-plus-one hybrid on a plane e.g. Easting, Northing and height (E, N, h);
5. The two-plus-one hybrid on the spheroid e.g. latitude, longitude and height (ϕ, λ, h).

2.6 Cartesian coordinates

Cartesian coordinates employing a set of orthogonal axes i.e. at right angles to each other are most common. Not only do such systems generate obviously convenient mathematical forms, but they are easy to create in practice. Computer monitors and TV screens are undoubtedly the most common applications of a grid and coordinate systems in modern life.

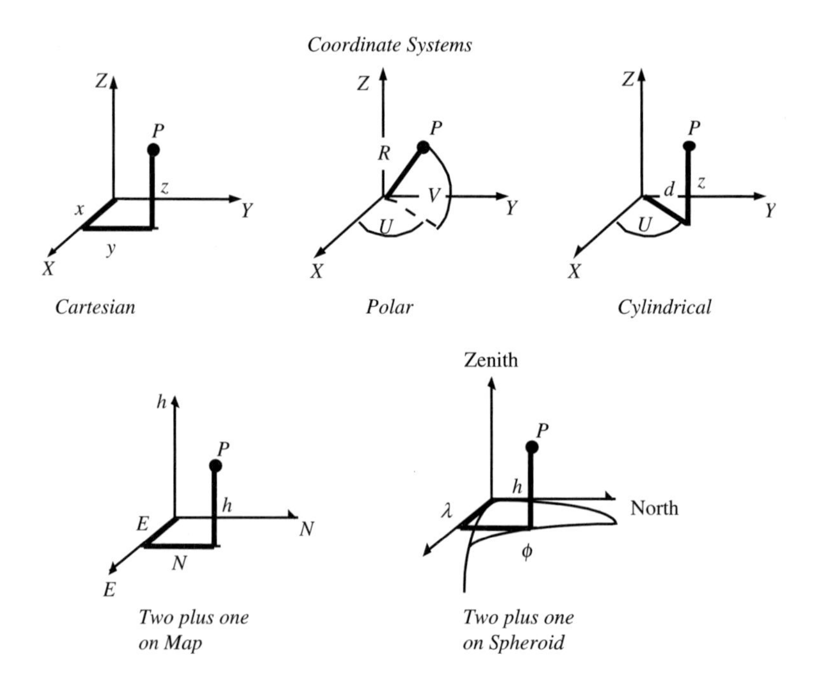

Figure 2.2

Euclid described how to set out or draw a right angle 2300 years ago (see Section 2.15), how to set up a grid of squares, and how to take measurements in such a system. It is equally practicable to construct rectangular plotting devices such as coordinatographs and digitising tables using such methods.

Because a grid is so fundamental to practical work it is advisable to check it for size and orthogonality. This can be done by a reversal procedure, such as drawing two copies of a grid and checking if they match when placing one on top of the other after rotating one through a right angle.

Three-axis coordinate measuring machines (CMMs) are a mechanical way of establishing three-dimensional grids in the industrial sector. Other systems create the necessary right angles optically via pentagonal prisms or mechanically.

The map grid, not only valuable as a reference system and a necessity for plotting, also provides a key to any distortion of the paper on which the map is drawn. It is often vital to allow for this distortion when working with old maps, as in legal disputes or in the construction of a Geographical Information System (GIS) for the computer handling of map and other data.

Current computational strategy is tending towards a rigorous Cartesian three-dimensional approach in preference to the older two-plus-one hybrids, even if this procedure requires more complex mathematical models.

2.7 Polar coordinates

Both the modern theodolite with its integral distance measurement capability (the total station) and the traditional instrument using optical distance measurement are based on the polar coordinates system. In two dimensions, the drafting protractor still has some advantages over Cartesian methods. Mechanical polar coordinate measurement systems still have some advantages in that they are more compact than Cartesian systems, as seen for example in an industrial robot and laser tracker.

Polar coordinates are basic to many mathematical models such as vector geometry and spherical trigonometry, and the total station (theodolite plus distance measurer) is a mechanical analogue.

2.8 Cylindrical coordinates

Although not as widely used as the other two systems, the cylindrical system has advantages when measuring cylindrical and rectangular objects, such as ships, large pipes, tunnels and oil storage tanks.

2.9 The two-plus-one system

Since the Earth's surface is comparatively flat, it can be modelled conveniently in two separate coordinate systems: plan and height. The coordinate system for plan is reduced to two dimensions, Easting and Northing (E, N) or latitude and longitude (ϕ, λ), on the spheroid from which the z coordinate or height h is separate. This is often described as a 'two plus one' coordinate system. This system is necessary when a three-dimensional surface is represented on a map or plan. In this case the heights are represented in a conventional way as isolines of equal height (contours), or by spot heights written on the map.

It should be noted that the same separation occurs in most survey methods; differences in height are often measured independently of measurements in plan. The most important reason for this, however, is that the Earth's atmosphere generally causes much more serious curvature of light rays in a vertical plane than in a horizontal plane. For practical reasons, height differences are usually required with greater accuracy than plan positions.

2.10 Fixation of points in space: geometrical principles

In three-dimensional space, a point can be described uniquely by three independent pieces of information. This information consists of some form of measurement such as length, length difference, length sum, direction, direction difference or combinations of these. The position in space is ultimately determined by the intersection of at least two vectors (lines), and the accuracy of this fixation depends on two factors:

1. Their angle of intersection—a feature of the network design,
2. Their likely lateral shift—a feature of measurement accuracy.

The former depends on the design of the measurement system and the latter on the design and quality of the measuring instruments. Quality may be improved by taking more measurements, by changing to a better instrument, or by changing to a better-designed measurement system using the same instruments.

A graphical treatment of a problem can often yield a good idea of the potential accuracy of a system and especially when presenting results to a client. For example, lines meeting at right angles are preferable to those intersecting at an acute angle.

Fundamental to all surveying networks is that they must cover the whole area to be surveyed, so that all subdivision will be interpolated within a strong framework, and not extrapolated outside it. This is often described as working from the whole to the part. With computer software, networks design can be tested without making any observations at all (see Chapters 5 and 6). Sometimes this principle has to be abandoned as in tunnelling.

For simplicity we show the various geometrical methods in two dimensions (see Figure 2.3).

The two most common methods of fixation are:

1. *Radiation* as from a total station and
2. *Lateration* as from artificial Earth satellites (GPS, GLONASS etc.).

Radiation

Radiation, for example by total station, is by far the most common ground method to capture the position of the point in space. It involves the measurement of two angles

Principal Methods of Fixation in Two Dimensions

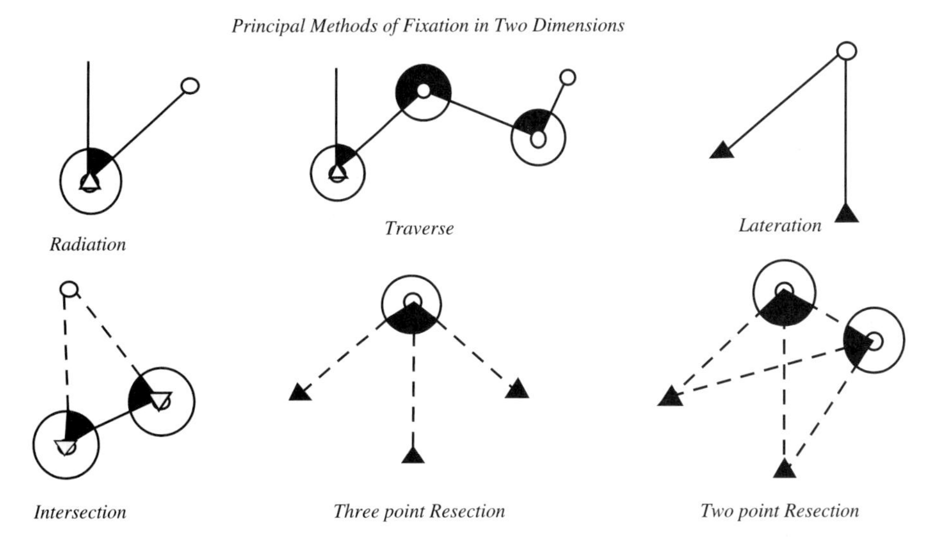

Figure 2.3

in mutually orthogonal planes and a distance from the instrument to the new point: typically a *horizontal* and a *vertical* angle and a *slant distance*.

A series of radiations joining up a chain of points is called a *traverse*. This is often the only way to survey points within an enclosed space such as in a tunnel.

Point fixations

The coordinates of points can be derived by astronomical or satellite methods using some of the fundamental geometrical methods described above (see Appendix 7 and Chapter 7).

Bearings and azimuths

From astronomical measurements, the direction of a line on the surface of the Earth can be determined (see Appendix 7). This direction, referred to as *true north*, is called an astronomical *azimuth*. If a north seeking gyroscopic theodolite is used, a similar azimuth is obtained. By contrast, if a magnetic compass is used, a magnetic azimuth is obtained which changes with time because of the drift of the Earth's magnetic pole; this is not used often in precise surveying. Also, when converted to a projection system via the grid convergence, a *grid bearing* is obtained from an azimuth (see Appendix 8).

Azimuth or bearing measurements are treated in Least Squares analysis in a similar way to a direction but with no added orientation parameter (see Equation (6.26)).

Direction vectors

The output from GPS software is often a vector (direction and length) which has been obtained from differential lateration to fix two points. Such a vector is treated in network analysis as two correlated input measurements: bearing and distance.

Intersection

A new point can be located by the intersection of two rays from two points of known location, without the measurement of distances. In photogrammetry, most models are formed by considering the intersection of lines in space.

Lateration

In a similar manner, two measured distances from two known points will locate a new point in a plane (three distances are required in three dimensions).

Resection from three points

In two dimensions it is possible to fix a new point from two angle measurements to three given locations only. This was a favoured method used by mariners from angles measured by a sextant aboard ship. The principle is also fundamental to the creation of photogrammetric models.

Resection from two points

It is also possible to fix a new point from angle measurements at two connected stations to only two known points. This was a method favoured by topographic surveyors and also has applications in photogrammetry.

2.11 Mathematical modelling

It is not practical to determine the position of every point on an object, nor is it necessary. For example, a straight line is described fully by two points in theory, one at each end; a circle by three suitable points and so on. It is more economical to model a surface or line with the minimum of data. This modelling may be done by the surveyor by selecting discrete points at changes of direction or slopes on the ground, or by the cartographer when digitising a map. The process of tracing lines in a photogrammetric plotter is, by contrast, non-selective. Even in this process, some data condensation has to take place if, as is usual, a record of the trace is kept in digital form.

Clearly, a limit to the accuracy is implied by the modelling process. In the traditional line-map a plotting accuracy of 0.2 mm was the accepted error bound. Hence the real accuracy depends on the map scale. For example at a scale of 1 to 500 the accuracy limit is 0.1 m. With the increased use of computer drafting, such conventional concepts of accuracy have been ignored—with serious results.

Again, it is reasonable to collect more than the minimum amount of data in order to provide a means of checking against mistakes, or to control the quality of the work. A redundancy of data imposes the need to reconcile inconsistencies by a simple method or by the method of Least Squares, which in turn allows a statistical estimate of precision to be evaluated and affords an opportunity to detect blunders.

2.12 Algebraic and algorithmic methods

Many fundamental geometrical concepts can be described in simple graphical terms, and many results can be depicted in graphical form. On the other hand, algebraic methods are required if complex problems are to be treated. Although conventional, matrix and vector algebras are each of value in expressing concepts, and in the increasing use of spreadsheet computer systems, many numerical problems have to be recast in suitable algorithmic form to avoid errors arising from the number crunching itself. These numerical methods need to be robust. Hence it is vital to check any computer software against standard sets of data to ensure that there are no 'bugs' in the system and that proper theory has been used. In the first analysis, it is the professional surveyor who is liable for his results, not the software company who wrote the program.

In the last analysis, there is no substitute for the use of common sense methods to check all work, be it fieldwork, computations or plotting and setting out. Often a simple graphical treatment will be the basis of this common sense.

The straight line

Mathematically, a straight line can be defined with reference to a plane by an equation such as

$$y = mx + c$$

This is an ideal or theoretical line which we can say is a *model* of the real thing, which is generally assumed to be imperfect. Actual straight lines will always be imperfect and depart from straightness. The accuracy required varies with circumstances. In photogrammetry, errors of microns are significant, while in road construction a centimetre is sufficient (see Section 6.7 for fitting a line to points).

To draw a straight line on a piece of paper all we need to do is to copy the side of a straight edge in a consistent manner, using a sharp pencil, say. (A straight edge is a piece of steel like a ruler but without any markings.) This begs the question: how is the straight edge initially manufactured? To check for straightness it is helpful to draw a line, then reverse the straight edge and hold it against the line already drawn. With certain limitations, lack of straightness will show.

Line of sight and laser beams

Another way to establish straightness is by line of sight. How straight then is a line of sight? Considering that the human eye can resolve an angle of about 20 sexagesimal seconds of arc, this means we can see departures from a line of sight to a tolerance of 20/206 265 or about 1 part in 10 000. Over a distance of 100 m, this amounts to 100/10 000, or 10 mm. Telescopic viewing is capable of much greater precision.

However, the line of sight may bend due to the effect of refraction on light, thus the line need not be straight at all. In a tunnel, for example, where there are temperature gradients close to the tunnel walls, this effect can be considerable (as much as several centimetres). Inside buildings, temperature gradients such as those found close to heating ducts, should be considered. In the open air, if there is gentle air circulation, the refracting effect is less. However, in very hot weather, the air becomes so turbulent that practical lines of sight have to be restricted to about 70 m due to light shimmer. These statements are equally true of a laser beam.

Stretched string

Another way to establish a straight line in two dimensions is to pull a fine string or wire very tightly. Although the string will sag in the middle, it will not deviate from side to side, unless there is a side wind present. A way to avoid such an effect over a short distance is to shield the string from the wind. Often a long tube is used for this. It has been found that nylon thread is very effective for many precise engineering applications.

Strings are often used in construction works to mark a chalk line on a hard surface, such as a road. Chalk is run along the string, which after being pulled tight in the correct location is flicked to impart a chalk line to the surface.

Sloping and vertical lines

It should be stressed at this point that much of what has been said above also applies to sloping or vertical lines, also used in measurement and surveying.

The geodesic line

The geodesic line is the shortest distance between two points (see Appendix 4.6). For example if A and B are two points on opposite sides of a room, there are two possible geodesics:

1. down one wall across the floor and up to the second point
2. up to the ceiling which it crosses, and down the wall,

as shown in Figure 2.4. The way to establish the route is to develop (open out) the three dimensional model of the walls and floor into a plane, and draw the straight line between A and B on this plane. This defines the geodesic. On the surface of an Earth ellipsoid, the geodesic is a curved line lying between the two normal sections.

Orthogonality or normality

Second in importance to the straight line are lines which are orthogonal or normal to each other, usually forming the basic geometrical shapes such as rectangles, squares etc. Since this orthogonality is vital to many operations in drafting and construction work, some method of checking for right angles is required. In map making and precise engineering, mechanical draughting machines and other mechanical (analogue) devices such as optical prisms are used. From time to time, these devices should be checked against some standard. For example, to check a grid plotted by machine, two copies should be drawn and compared with each other with one rotated through a right angle relative to the other.

Alternatively, another mechanical device such as a set square might be used which can easily be checked by reversal as shown in Figure 2.5. Graphical and field methods of setting out right angles are described in the next section.

2.13 Geometrical construction

Figures 2.6(a) and (b) demonstrate how right angles can easily be established geometrically, either by drawing or, at a larger scale, by sweeping arcs with a tape or string or rope.

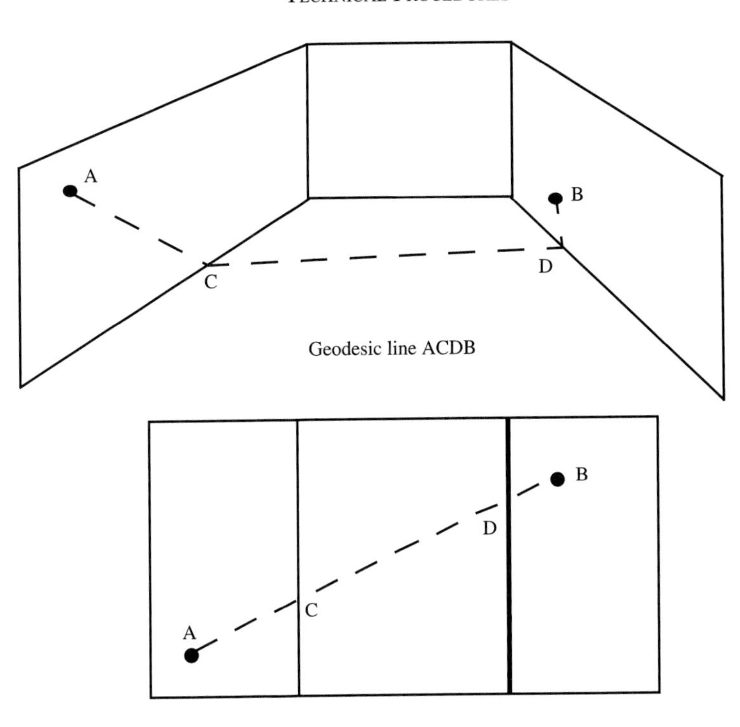

Figure 2.4

In Figure 2.6(a), the line CD has to be constructed at right angles to the given line AB through the point P located on AB. From P, points A and B are set out such that AP = BP. Then arcs of equal circles are drawn from centres A and B to establish points C and D, giving CPD perpendicular to AB.

In a similar way Figure 2.6(b) shows how to construct the orthogonal line CD through a point E which does not lie on AB. In this case, the points A and B are established by sweeping a circular arc to cut the given line at A and B. (Chapter 11 includes more practical details on setting out works.)

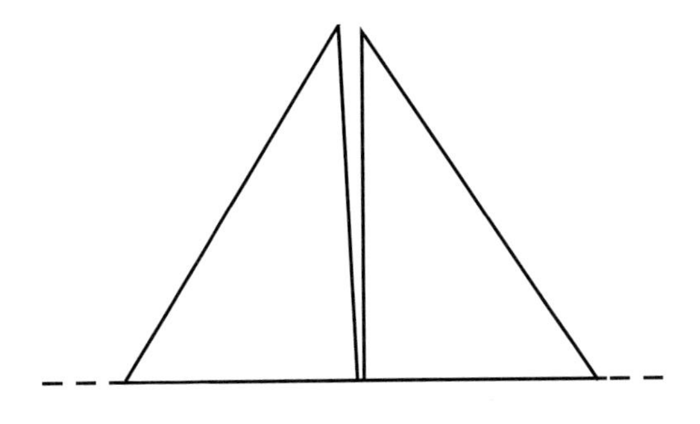

Figure 2.5

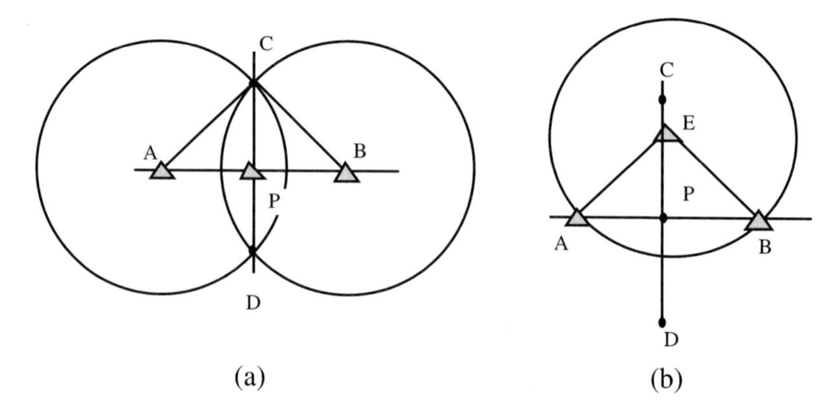

Figure 2.6

Pythagoras' method

A quick way to establish a right angle is to use a tape to set out a Pythagorean triangle, the most common one being with sides in the ratio of 3: 4: 5, illustrated by Figure 2.7. Lengths in the ratios 3: 4: 5 units are set out to produce the right angle at point A.

By hand

A crude method in the field, accurate to about 5°, is as follows. Stand on the line with the hands outstretched and also on line. Then bring the hands rapidly together at an approximately orthogonal direction. This is quite useful in preliminary location to be followed up by a more accurate technique.

By instrument

The pentagonal prism has the useful property of establishing two mutually orthogonal parts of a line of sight which passes through the prism. Field applications normally use a theodolite or total station to establish right angles. Care is required with centring and observing techniques. These topics are dealt with more fully in Chapter 13.

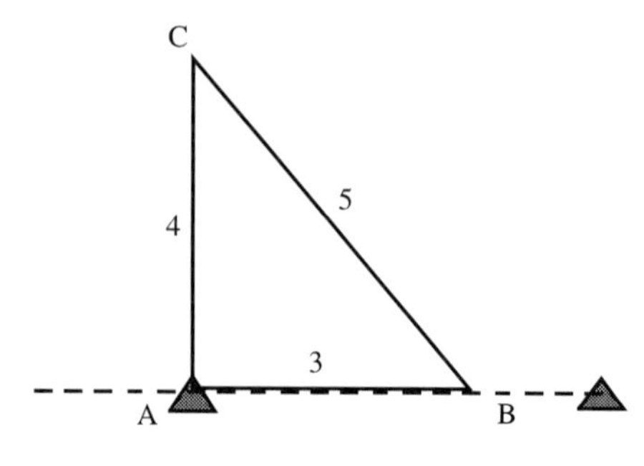

Figure 2.7

2.14 Paper sizes

Consider the sheets of paper on which maps are drawn and printed. For this purpose a sheet of paper is laid flat on a table or drawing board. Thus we can say that the paper lies in a *plane*. A sheet of paper can be rolled into other shapes such as a *cylinder* and *cone,* folded into boxes and complex shapes in the art of origami. For mapping purposes we deal with the paper lying in a plane.

Previously, many different sizes of paper were used for drawing and printing maps, including *quarto*, *foolscap* and *elephant*. However, there has to be some agreement about paper sizes so that maps fit together and that printing machines can be constructed to suit them. The accepted international measure for paper sizes is referred to as the *A series*.

Most people are familiar with the A4 size of paper used for everyday office work and student note pads, etc. The first thing to note about any paper sheet is that it forms a *rectangle.* A rectangle has parallel and equal opposite sides, and equal diagonals. When the sheet is oriented as in Figure 2.8(a) it is in *portrait* position, and Figure 2.8(b) represents *landscape* orientation.

Example Measure the sides and diagonals of an A4 sheet of paper, ABCD, as in Figure 2.8(a). The results (mm) may be something like:

Short sides (mm)	Long sides (mm)	Diagonals (mm)
AB = 211	BC = 296	AC = 362.8
CD = 210	AD = 297	BD = 363.5

The reason that we do not have a perfect rectangle is due to errors in the measurements; see below for the theoretically exact dimensions.

Example Show that the ratio AB/BC = 0.713 is approximately equal to $0.5 \times BC/AB = 0.701$.

This means that, if the paper is folded in two, the rectangle formed is almost the same shape as the original but half its size. The smaller paper size is A5. In fact, it

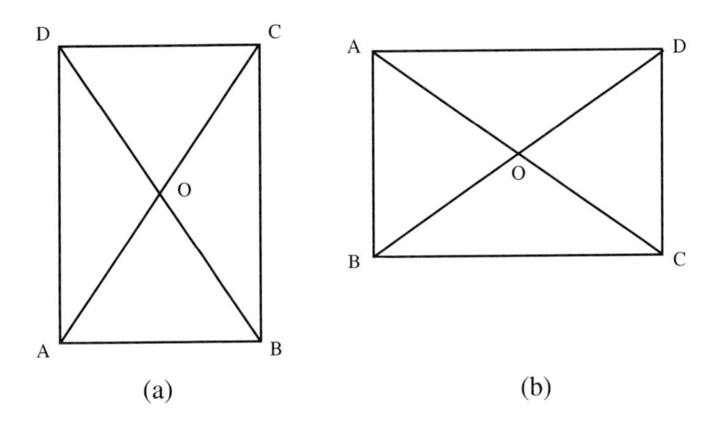

Figure 2.8

should be exactly the same shape. The way the A paper sizes are devised follows a pattern. If the paper proportions are chosen so that

$$\frac{AB}{BC}=\frac{0.5\times BC}{AB} \quad \text{or} \quad AB^2=0.5\times BC^2 \quad \text{or} \quad AB=\frac{BC}{\sqrt{2}}$$

when folded over, the sides will be in exactly the same ratios. The A system of paper sizes begins with a sheet exactly one square metre in area with sides of these proportions i.e. such that

$$AB\times BC=1 \quad \text{and} \quad AB=\frac{BC}{\sqrt{2}}$$

Thus

$$\frac{BC^2}{\sqrt{2}}=1$$

therefore

$$BC=\sqrt{\sqrt{2}}=1.1892 \quad \text{and} \quad AB=0.8409$$

The sheet of paper with these dimensions is referred to as A0 size. Halving and halving in sequence gives all the other A paper sizes. Thus, A4 is

$$1.1892/4 = 0.2973 \text{ m by } 0.8409/4 = 0.2102 \text{ m}$$

$$\text{or } 297.3 \text{ mm by } 210.2 \text{ mm}$$

2.15 Reference grid

A network of two sets of regularly spaced parallel lines orthogonal to each other forms a *grid.* In surveying and mapping the accuracy of the grid is vital to most operations. It is therefore important to consider this in some detail both for its own sake and to introduce some basic geometrical concepts.

It has always been possible to create a right angle by simple methods and therefore to construct a grid. To capture the spirit of this truth, the reader is invited to perform the task of drawing a grid as outlined here using only a ruler, pencil and sheet of A4 paper.

Start by drawing two straight lines roughly diagonally across the paper. Let them intersect at point O. From O mark off four equal lines OA, OB, OC and OD as shown in Figure 2.9(a), and join across the sides AB, BC etc. You have now drawn a rectangle.

Check that opposite sides are equal (within a drawing tolerance of 0.2 mm) and that the diagonals AC and BD are also equal. If not, repeat the task and try to see why the error has crept in. The opposite sides should also be parallel to each other. Test this by sliding a set-square against a ruler, as shown in Figure 2.9(b).

Mathematically, ABCD is defined as a perfect rectangle. However, in reality it will only be a close approximation to one. We say that the perfect rectangle ABCD is a

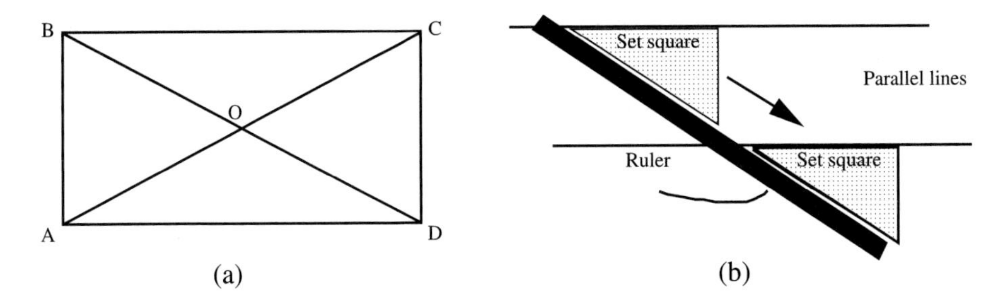

Figure 2.9

mathematical model of the real thing. The differences between the two will be the subject of statistical and error analysis. In practical surveying, engineering, woodworking etc. we have to decide on an acceptable tolerance for the practical creation of the rectangle. For example, a wooden door is probably good to about 5 mm, a tennis court to about 10 mm and the grid of a map to about 0.2 mm.

The rectangle ABCD is now used to construct a grid as in Figure 2.10. Point A is chosen as the *origin* of the grid and AD and AB its *axes*. Along AD and BC mark off points, a, b,...p, q etc. at multiples of 40 mm apart, say. (Do this first by counting along at 40 mm intervals and observe how the error accumulates!) The correct method is to use the ruler to mark off all distances from the origin i.e. 40 mm, 80 mm and so on.

The lines ap, bq etc. are all parallel and equally spaced. Again from the base line AD mark off the points g, h ...k, l etc. and join the lines across to form the grid. Except by accident, or prior calculation, the original construction rectangle ABCD does not fit the edges of the grid exactly.

On a building or archaeological site, where the grid is used to set out or measure the footings of the walls, the grid intersections are marked by pegs or steel plates. Sometimes, for example in archaeology or botanical sampling, the grid is physically marked by wires or strings. Computer and cartographic plotters use a mechanical

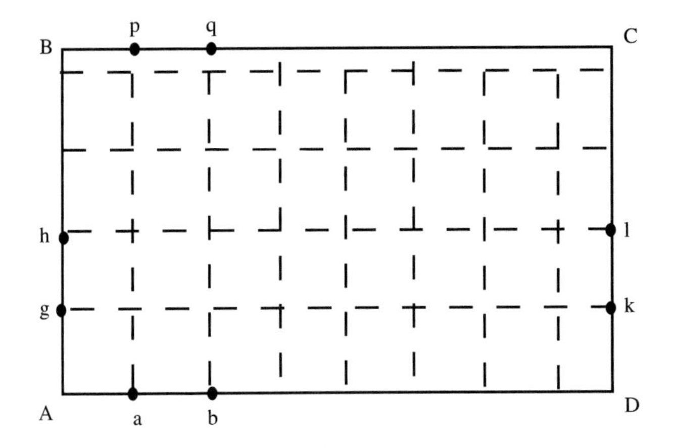

Figure 2.10

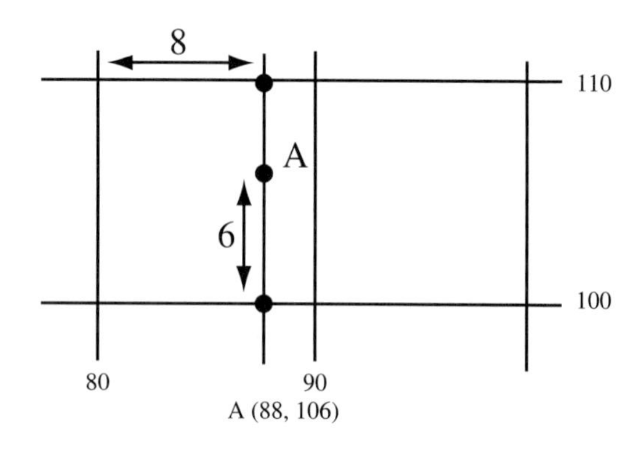

Figure 2.11

system of orthogonal rails to establish the grid quickly. Sometimes these rails are incorrectly aligned and need to be checked.

When setting out a grid by total station, the grid points can be positioned directly from their coordinates using any point as the origin. The bearings and distance to all points are calculated and set out directly. Chapter 11 provides more details on this subject.

Plotting control points

The control points of the survey have to be plotted within the relevant grid square by measuring fractional distances along pairs of grid lines, and joining opposite pairs of lines.

For example, in Figure 2.11 we show the relevant square within which to plot a point A. The fractional part of its Easting is measured along both grid lines above and below the point, and a north–south line drawn to join them. On this line, the Northing of the point is plotted. This procedure is often adopted using strings stretched between grid pegs in setting out. Of course the alternative of first using two north–south lines is equally acceptable.

Checking computer plotted grids

It is not unknown for an automated plotting system to plot incorrectly. The method of checking a grid for orthogonality is to make two plots and see if they fit on overlay after a rotation of 90° with respect to each other. Size has to be verified by a calibrated scale.

Position and orientation

Often the grid is required to be located in a special position. For example, a map may have to be positioned central to a page which also contains other items such as a title, a map key, a coordinate list and so on. The map may even have two different grids on it. In engineering, the site grid will need to be placed correctly in relation to the ground. In these cases, a coordinate transformation may be required (see Chapter 3). Often

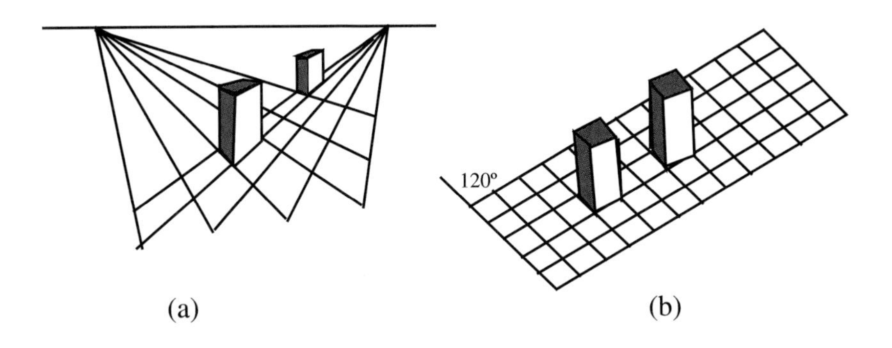

Figure 2.12

some simple off-sets from known points, such as page corners (or existing buildings) will suffice to locate the grid. Clearly these must be checked most carefully.

2.16 Perspective and isometric views

Two non-rectangular 'grids' are often used to depict a solid on a two dimensional page. These are perspective and isometric grids, depicted in Figure 2.12.

The *perspective* grid (used by artists, Figure 2.12(a)) is that seen when looking at a grid from an oblique position. Sets of parallel lines converge to vanishing points on the horizon as shown in the diagram, which might have been a photograph. Although it is possible to make measurements on such a grid using the technique of *anharmonic ratios*, it is not popular.

An *isometric* grid (Figure 2.12(b)) is created by drawing sets of oblique parallel lines (i.e. which do not converge to a point). Although disturbing to the eye, it is easy to make measurements along the parallel lines on such a grid if the sets are drawn at 120° to each other.

Chapter 3
Coordinate Systems

3.1 Cartesian coordinates

Fundamental to surveying are rectangular Cartesian coordinates: describing the position of points in a plane in two dimensions, or in space in three dimensions. Problems arise with the conventions in use. In a two dimensional system (X, Y) or (N, E), it has become customary to direct the X-axis to the north and the Y-axis to the east, because a bearing U is reckoned clockwise from north in surveying. This convention accords with mathematics, as long as we are working in a plane. In three dimensions great care is necessary in defining positive directions of axes and rotations (see Figure 3.1). We have

$$\Delta X = S\cos U \quad \text{and} \quad \Delta Y = S\sin U$$

$$\Delta x = R\cos\theta \quad \text{and} \quad \Delta y = R\sin\theta$$

The order x, y is, however, in direct contrast to the conventions of map reading and cartography, where it is customary to quote Easting E before Northing N (see Figure 3.1(c)). Here the *bearing* is commonly denoted by T (especially in German texts), giving

$$\Delta E = S\sin T \quad \text{and} \quad \Delta N = S\cos T$$

However, the convention to quote geographical coordinates latitude ϕ before longitude λ once more accords with x and y respectively of our convention, so we have the *approximate* transformations

$$\Delta\phi = S\cos\alpha \quad \text{and} \quad \Delta\lambda = S\sin\alpha$$

where α is the *azimuth* of the line as shown in Figure 3.1(d).

To avoid potential confusion, all listed coordinates should clearly indicate which definition is being used, as well as the units of measurement.

Example A point may have coordinates:

$$x = 2.5,\ y = 5.5$$

$$E = 5.5,\ N = 2.5$$

which correspond to values (S, U) in a polar system, of

$$S = 6.041$$

$$U = 65.556°$$

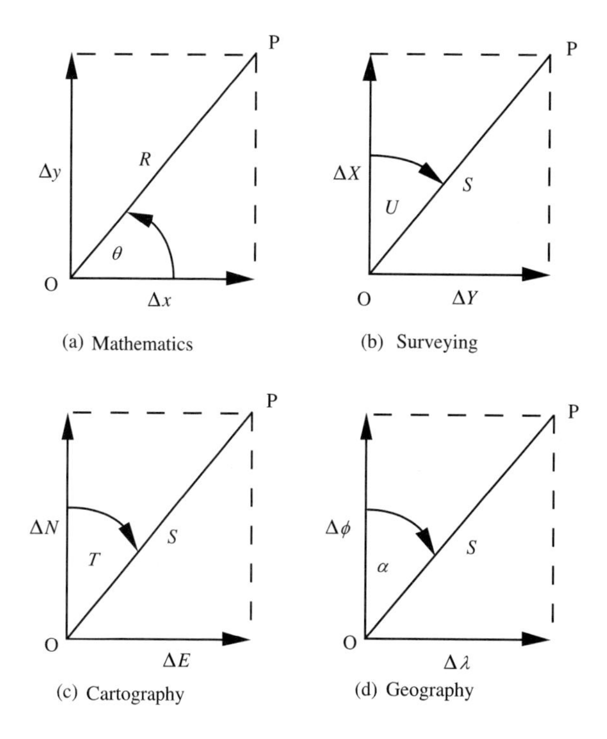

(a) Mathematics (b) Surveying

(c) Cartography (d) Geography

Figure 3.1

The bearings U are reckoned in sexagesimal degrees.

Example The reverse problem of converting Cartesian to polar coordinates is very important in surveying. Some care is needed to cater for all quadrants. For example, converting the point with survey coordinates (–2.5, –5.5) from a calculator with a *rectangular-to-polar* function key, we obtain the values

$$S = 6.041$$

$$U = -114.443°$$

Because all survey bearings are reckoned clockwise from north, this particular value of U, which was provided by a calculator, is added to 360° to give the survey bearing of

$$U = 245.556°$$

The mathematical functions employed are

$$S^2 = x^2 + y^2$$

$$\arctan U = \frac{y}{x}$$

The arctan function can accommodate the negative signs of both x and y, ensuring that an erroneous first quadrant angle is not returned as

–5.5/–2.5 = 2.2, which is the tangent of 65.556°.

A better algorithm function is listed as *ATAN2* in contrast with *ATAN* which does not deal with all quadrants properly. If no *ATAN2* function is available, the following algorithm may be used:

$$S = \mathrm{SQRT}(x^2 + y^2)$$

$$\mathrm{IF} \quad S + x = 0 \quad \mathrm{THEN} \quad U = 180$$

$$\mathrm{ELSE} \quad U = 2 \times ATAN\left(\frac{y}{S+x}\right)$$

adding 360° if necessary.

Since not all algorithms deal properly with the various quadrants when computing reverse trigonometrical functions, a check should be made when using a calculator or computer algorithm for the first time. Also, a check should be made to see if a correct result is given for the limiting cases of points on the coordinate axes. In other words, we could check that the correct related data are

x	6	0	–6	0
y	0	6	0	–6
U	0	90	180	–90

The last result has to be added to 360° to give the whole circle bearing of 270°.

Again, most computer algorithms require angles to be converted to radians before calling a trigonometric function, and in the reverse mode, provide results in radians. Thus a suitable algorithm in Excel to calculate the sine of 12° 34' 56" is

SIN(PI()*(12+(34+56/60)/60)/180) = 0.217 840 422

and to convert 12.582 222 22° to traditional form (degrees, minutes and seconds of arc) might be

x = 12.582 222 22

INT(x) = 12°

y = (x – INT (x)) * 60 = 34.933 333

INT(y) = 34'

INT ((y – INT (y)) * 60 + 0.5) = 56"

If no function for PI is available, a convenient algorithm is

PI = 4 * ATAN(1)

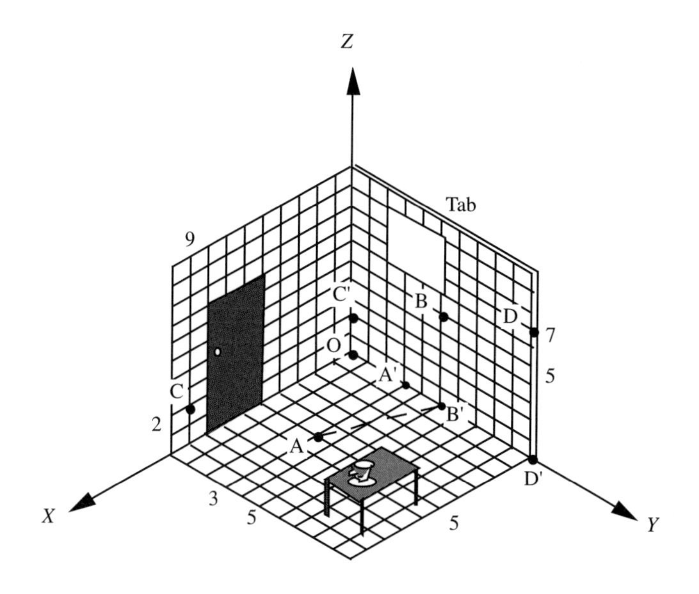

Figure 3.2

Precision of plane coordinates

The precision of computed results is examined throughout this book. The treatment of coordinate conversions in geodesy requires special care because of the very large numbers and high precision required. Again, any computer system should be tested to ensure that the required precision is being achieved.

3.2 Cartesian coordinates in three dimensions

Figure 3.2 shows the corner of a room illustrating a system of Cartesian coordinates in three dimensions (X, Y, Z). The articles of furniture are placed to keep the origin O appearing away from and not towards the viewer.

The system is called *right-handed* since if the right hand is held as shown in Figure 3.3 with the index finger aligned in the direction of the X-axis and the second finger in the direction of the Y-axis, then the thumb will automatically be aligned with the Z-axis.

In the *isometric* view of the orthogonal axes, the X-axis is inclined to OY by 120°, and unit vectors along each of the three axes appear to be of equal length. Thus it is possible to read off or measure the Cartesian coordinates from the diagram.

Example The coordinates of the points of Figure 3.2 are listed in Table 3.1 below.

Although lines parallel to the axes are true to scale, measurements cannot be made in other directions. The table is 2.5 units long and 1.8 units wide. Its diagonals have to be calculated.

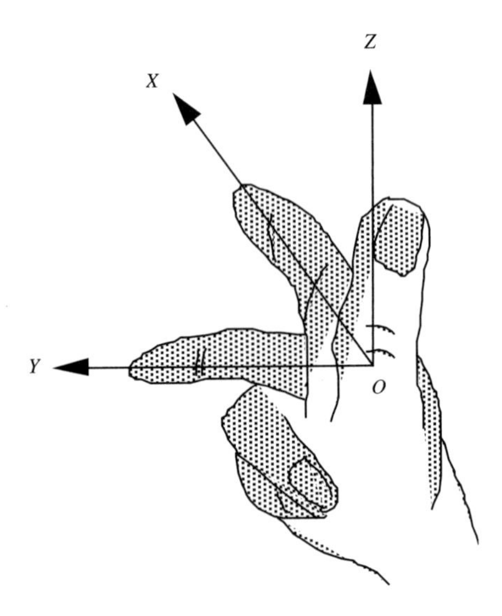

Figure 3.3

Table 3.1

Point	*A*	*A'*	*B*	*B'*	*C*	*C'*	*D*	*D'*	*O*
X	5	0	0	0	9	0	0	0	0
Y	3	3	5	5	0	0	10	10	0
Z	0	0	5	0	2	2	7	0	0

Example

Distance CD is obtained from

$$\text{CD}^2 = \Delta X^2 + \Delta Y^2 + \Delta Z^2 = (0-9)^2 + (10-0)^2 + (7-2)^2 = 206$$

$$\text{CD} = 14.35$$

3.3 Polar to Cartesian coordinates

The most commonly used field surveying instrument is the *total station*, in which a horizontal angle U, a vertical angle V and a distance R as shown in Figure 3.4 are measured.

To obtain Cartesian coordinates (X, Y, Z) from the polar system (R, U, V) we have the following transformations

$$x = R\cos V\cos U; \quad y = R\cos V\sin U; \quad z = r\sin V$$

$$R = \sqrt{x^2 + y^2 + z^2}$$

$$\tan U = \frac{y}{x}$$

$$\tan V = \frac{z}{\sqrt{x^2 + y^2}}$$

A useful check is

$$R = (x \cos U + y \sin U) \cos V + z \sin V$$

Example Given that R = 63.64 mm, U = 65.556° and V = 18.423°, calculate the Cartesian coordinates (x, y, z).

$$x = 63.640 \times \cos 18.423^\circ \times \cos 65.556^\circ = 24.98 \text{ mm}$$

$$y = 63.640 \times \cos 18.423^\circ \times \sin 65.556^\circ = 54.97 \text{ mm}$$

$$z = 63.640 \times \sin 18.423^\circ = 20.11 \text{ mm}$$

Example The coordinates of P in mm are (25, 55, 20). Calculate R, U and V.

$$\tan U = \frac{55}{25} = 2.2 \Rightarrow U = 65.556^\circ$$

$$\tan V = \frac{20}{60.04} = 0.331 \Rightarrow V = 18.423^\circ$$

$$R = \sqrt{25^2 + 55^2 + 20^2} = 63.640 \text{ mm}$$

Check

$$R = (25 \cos 65.556^\circ + 55 \sin 65.556^\circ) \cos 18.423^\circ + 20 \sin 18.423^\circ = 63.640 \text{ mm}$$

The slight inconsistencies in results arise from limiting the number of decimal points listed.

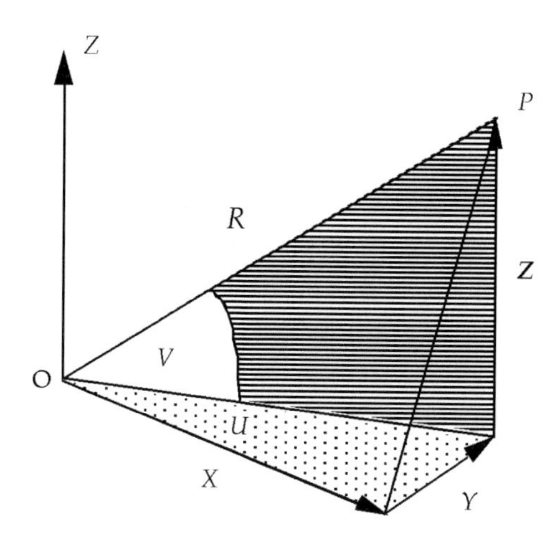

Figure 3.4

3.4 Coordinate differences

In practice, coordinates are seldom referred directly to the origin O. Usually connections between two points, such as A and B, are required. In this case the above formulae need only a little modification to put

$$\Delta x = x_B - x_A; \quad \Delta y = y_B - y_A; \quad \Delta z = z_B - z_A$$

The transformations become

$$\Delta x = r\cos V\cos U; \quad \Delta y = r\cos V\sin U; \quad \Delta z = \sin V$$

$$R = \sqrt{\Delta x^2 + \Delta y^2 + \Delta z^2}$$

$$\tan U = \frac{\Delta y}{\Delta x}; \quad \tan V = \frac{\Delta z}{\sqrt{\Delta x^2 + \Delta y^2}}$$

A useful check is

$$R = (\Delta x\cos U + \Delta y\sin U)\cos V + \Delta z\sin V$$

Example The coordinates of A and B in mm are (50, 70, 20) and (75, 125, 40) respectively. Calculate R, U and V.

$$\Delta x = 75 - 50 = 25; \quad \Delta y = 125 - 70 = 55; \quad \Delta z = 40 - 20 = 20$$

$$\tan U = \frac{55}{25} = 2.2 \Rightarrow U = 65.556^\circ$$

$$\tan V = \frac{20}{\sqrt{25^2 + 55^2}} = 0.333 \Rightarrow V = 18.423^\circ$$

$$R = \sqrt{25^2 + 55^2 + 20^2} = 63.640 \text{ mm}$$

Check

$$R = (25\cos 65.556^\circ + 55\sin 65.556^\circ)\cos 18.423^\circ + 20\sin 18.423^\circ = 63.640 \text{ mm}$$

3.5 Geographical spherical coordinates

At its simplest, the Earth can be considered as a sphere of radius 6 378 000 m, and points on its surface described by latitude, longitude and height above or below the sphere. This simple spherical model is most valuable as a teaching aid, giving insight into geodetic principles and the theory of map projections. It has practical uses for approximate calculations such as finding the direction of Mecca or the distance flown by carrier pigeons. The various calculations involve only the basic formulae of spherical trigonometry. A more accurate model of the Earth uses an oblate spheroid, discussed later.

Latitude and longitude

In Figure 3.5 the Earth is represented by a sphere, centre O, whose *Z*-axis coincides with the Earth's spin axis passing through the North and South poles. (In practice this

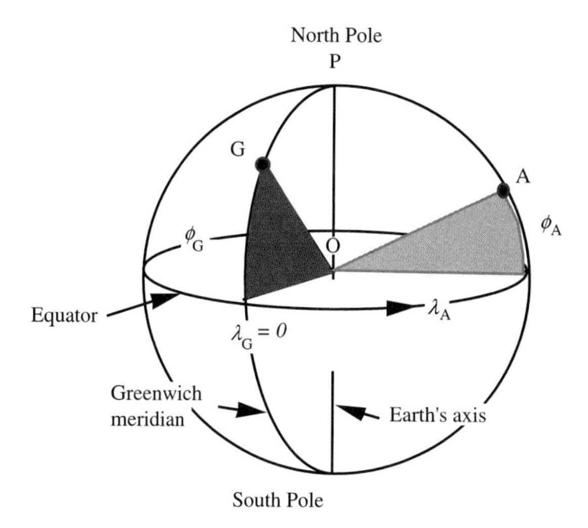

Figure 3.5

is determined from astronomical observations.) The plane of the equator generates a great circle of the sphere perpendicular to the Z-axis.

A great circle containing the axis and any point A is called the meridian of A. The angular distance in the meridian plane from the equator to this point is its latitude. It is sometimes necessary, as in astronomy, to define north latitude as positive and south latitude as negative. Latitude is usually denoted by the Greek letter phi ϕ.

Longitude is the angular distance round the equator between any two meridians: positive eastwards and negative westwards. Since there is no natural starting reference line, an arbitrary system for worldwide purposes was adopted based on the meridian of Greenwich in England. For many purposes, however, only longitude differences are used. Longitude is usually denoted by the Greek letter lambda λ.

It is important to grasp the dimensions involved. One second of latitude subtends an arc length of about 30 metres on the surface of the Earth, so millimetre precision demands angular work to 0.0001 second.

Rectangular Cartesian coordinates

Most computations are made today using not the geographical system (ϕ, λ) but a rectangular system oriented to suit the purposes in hand (Figure 3.6). For a worldwide system we direct the OX axis along the Greenwich meridian, the OZ axis north–south and the OY axis towards the east. Longitudes then become positive east and negative west. The formulae to transform the geographical coordinates of any point A into Cartesian values are therefore:

$$\begin{aligned} X &= (R+h)\cos\phi\cos\lambda \\ Y &= (R+h)\cos\phi\sin\lambda \\ Z &= (R+h)\sin\phi \end{aligned} \tag{3.1}$$

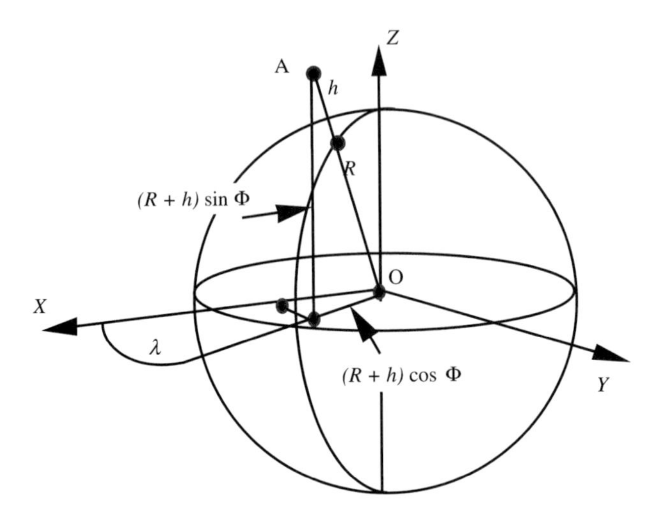

Figure 3.6

Example If $R = 6\ 378\ 000$ m, calculate the Cartesian coordinates of the point B from the following data:

$\phi = 15°\ 39'\ 22.5964''$
$\lambda = 35°\ 23'\ 37.454''$
$h = 500.00$ m

The reverse transformation is achieved by:

$$\tan\lambda = \frac{Y}{X}$$
$$\tan\phi = \frac{Z}{\sqrt{X^2 + Y^2}} \qquad (3.2)$$
$$h = Z\sec\phi - R$$

$$\Rightarrow (X, Y, Z) = (5\ 006\ 777.17, 3\ 557\ 307.73, 1\ 721\ 338.52)$$

As will be seen later, the reverse problem when dealing with ellipsoidal coordinates is less simple.

3.6 Ellipsoidal coordinates

In worldwide geospatial surveying (*geodesy*), a spheroid or ellipsoid of reference has to be used to model the size and shape of the Earth. The ellipsoid of geodesy is the figure described by the rotation of an ellipse about its minor axis, an oblate spheroid. (The ellipsoid generated by rotating the ellipse about the major axis is a *prolate spheroid*.)

The worked examples in the following section will use the data listed in Tables 3.2 and 3.3.

Table 3.2 *Clarke 1880 Spheroid*

semi-major (equatorial) axis a	*flattening f*	$e^2 = f(2-f)$
6 378 249.145	$\frac{1}{293.5}$	0.006 802 701 3

Table 3.3 *Data points*

	Longitude	*Latitude*	*Height (m)*
A	35° 18' 16.7559"	15° 52' 29.9096"	600
B	35° 23' 37.4540"	15° 39' 22.5964"	500
C	35° 28' 22.7070"	15° 44' 56.7567"	550

3.7 The meridian ellipse

The ellipse that defines an ellipsoid is called the meridian ellipse. The parallels of latitude are small circles in planes parallel to the equator, which is a great circle. An ellipse is defined in many ways and has a multiplicity of geometrical properties. We shall consider it defined with respect to its semi-major axis a and semi-minor axis b by the equation

$$\frac{x^2}{a^2}+\frac{y^2}{b^2}=1 \tag{3.3}$$

The coordinates of a point on the ellipse with respect to the origin at its centre are (x, y). The following properties will also be employed:

$$e^2 = f(2-f) \tag{3.4}$$

$$b^2 = a^2(1-e^2) \tag{3.5}$$

where e is the eccentricity of the ellipse, which is about 1/12 for a terrestrial ellipsoid. The flattening given by

$$f = \frac{a-b}{a} \tag{3.6}$$

is about 1/300 for the Earth.

Example From a and f (Table 3.2) we calculate the eccentricity e and the semi-minor axis b

e = SQRT(1/293.5*(2–1/293.5)) = 0.006 802 701 3

b = SQRT(6 378 249^2*(1–0.006 802 701 3)) = 6 356 517.32

Given either of the size parameters a or b, and a shape parameter e or f, the others may be derived. It is usual to define a meridian ellipse, and therefore an ellipsoid, in terms

Table 3.4

Ellipsoid	*Year*	*a*	*1/f*
Everest	1830	6 377 304	300.9
Bessel	1841	6 377 397	299.2
Clarke	1858	6 378 293	294.3
Clarke	1866	6 378 206	295.0
Clarke	1880	6 378 249	293.5
Helmert	1906	6 378 200	298.3
Hayford	1910	6 378 388	297.0
WGS	1984	6 378 137	298.3

of a and f. For historical reasons, there are many reference ellipsoids recommended for use in different parts of the world, because all previous surveys and maps were based on them.

The parameters of early ellipsoids were calculated from terrestrial arc measurements in specific parts of the Earth. With the advent of artificial Earth satellites, geodesists have been able to use the Earth as a whole to define an ellipsoid. The basic parameters, obtained from orbital analyses, are the semi-major axis a and the first harmonic of gravitational potential J_2. The latter can be used to derive an equivalent flattening according to a selected mathematical model. Datums, such as the world geodetic reference system WGS 84, are used for satellite work. The subject is one of much complexity, and influences position fixing. It is especially important in oil exploration, where geodetic cadastral problems arise. Table 3.4 gives approximate values of a few of these system parameters.

It will be clear that, because the flattening is not listed to the same precision as the semi-major axis, the semi-minor axis cannot be calculated to the same precision. Early measurements were unable to establish ellipsoidal shape with sufficient accuracy. However, the WGS 1984 flattening reciprocal has been quoted with sufficient precision as

$$298.257\ 222\ 101$$

Before embarking on important computations involving ellipsoids and datums, geodetic sources should be consulted to obtain the latest information on parameters.

3.8 Geodetic latitude and longitude

Geodetic latitude

Geodetic latitude ϕ_G is the angle between the normal to the ellipsoid at a point and the plane of the equator (see Figure 3.7). The plane of the equator is perpendicular to the spin axis of the Earth.

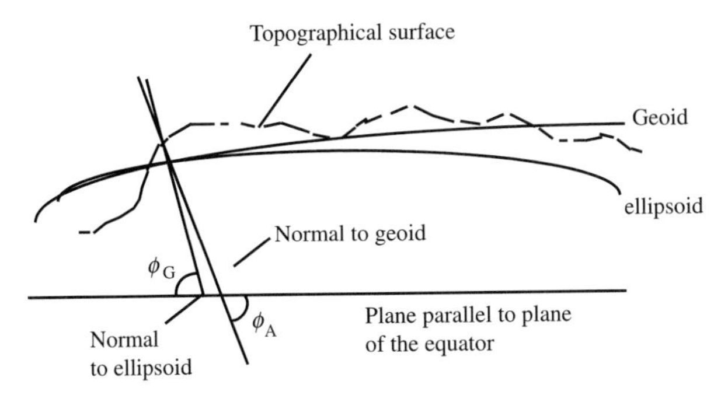

Figure 3.7

Note: *Astronomical latitude* ϕ_A is the angle between the normal to the geoid (see Chapter 9) at a point and the plane of the equator (see Figure 3.7). It is not used in ellipsoidal computations.

Geodetic longitude

Geodetic longitude λ_G is the angle measured along the equator between the meridian of Greenwich and the meridian ellipse of the location. Longitude is reckoned positive east of Greenwich.

3.9 Radii of curvature of the ellipsoid

The double curvature of the surface of the ellipsoid is usually resolved into two components: ρ in the plane of the meridian ellipse and ν in a plane at right angles to the meridian i.e. the plane of the *prime vertical*. The respective radii of curvature in these two directions, denoted by the Greek letters ρ and ν, are given by

$$\rho = \frac{a^2b^2}{p^3} = \frac{2a(1-e^2)}{3(1-e^2\sin^2 f)} \tag{3.7}$$

$$\nu = \frac{a^2}{p} = \frac{2a}{1-e^2\sin^2 f} \tag{3.8}$$

where $p^2 = a^2\cos^2\phi + b^2\sin^2\phi$ and ϕ is latitude. For any given ellipsoid for which the values of a and e are defined, both radii are functions of latitude alone.

For an oblate ellipsoid, ρ is always less than or equal to ν, except at the poles where they are equal. For more details of the geometry of the ellipsoid see Appendix 4.

Example From the data of Tables 3.2 and 3.3, we find at the point B

$$\rho = 6\ 339\ 570.25 \text{ and } \nu = 6\ 379\ 829.56$$

Examples We also calculate ν for points A and C, giving

$$\nu_A = 6\ 379\ 872.89$$

$$\nu_C = 6\ 379\ 847.88$$

3.10 Ellipsoidal Cartesian coordinates

In Figure 3.8, the meridian ellipse shows two points P_1 and P_2 at heights h_1 and h_2 above the ellipsoid. Since the length PS in each case is $(h + \nu)$ and the distance WS is $e^2\nu$ (see Appendix 4) we have the transformations

$$z = [(1 - e^2)\nu + h]\sin\phi \tag{3.9}$$

$$u = (\nu + h)\cos\phi \tag{3.10}$$

where u is a line.

Therefore

$$\begin{aligned} x &= u\cos\lambda \\ y &= u\sin\lambda \end{aligned} \tag{3.11}$$

These equations are easy to solve in the forward case, i.e. given ϕ and h, as ν is itself a function of ϕ.

Example Computing the Cartesian coordinates of A, B and C (Table 3.3) we find the following results:

Point	x	y	z
A	5 008 373.63	3 546 737.05	1 733 437.32
B	5 008 213.28	3 558 328.08	1 710 120.06
C	5 001 023.61	3 563 637.54	1 720 021.60

The reverse computation is not as simple because ν is a function of the unknown ϕ. An iterative approach may be adopted using a first approximation

$$\tan\phi \approx \frac{z}{U} \tag{3.12}$$

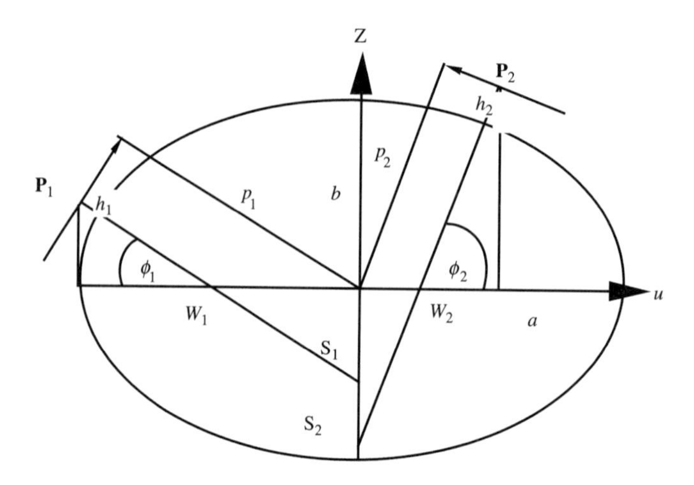

Figure 3.8

to give a first value of ϕ and ν from Equation (3.8) and then h from Equation (3.10). These parameters in turn use the equation

$$\tan\phi = \frac{z(\nu+h)}{u[(1-e^2)\nu+h]} \tag{3.13}$$

derived from Equations (3.9) and (3.10) to calculate an improved value of ϕ, and so on until the solution converges with successive unchanged values for ϕ. However, Bowring and Vicenty (1978) have shown that the following nearly closed formula provides adequate accuracy. By setting $u^2 = x^2 + y^2$ and $\tan k = \frac{z}{u}\frac{a}{b}$, we have the working formula

$$\tan\phi = \frac{z+\varepsilon b\sin^3 k}{u-ae^2\cos^3 k} \tag{3.14}$$

where the second eccentricity of the ellipse ε is given by

$$\varepsilon = \frac{a^2-b^2}{b^2} \tag{3.15}$$

Once ϕ has been determined, the height h is calculated directly from

$$h = u\sec\phi - \nu \tag{3.16}$$

Having found u and z from ϕ and h, the coordinates with respect to another meridian at an angle of longitude λ are given by Equation (3.11). To find λ, the reverse calculation poses no problems as

$$\tan\lambda = \frac{y}{x} \tag{3.17}$$

Example The reverse calculation for point B is as follows:

e = 0.006 849 339 834

$$\tan\lambda = \frac{3558328.08}{5008213.28} = 0.71049851 \Rightarrow \lambda = 35.3937372 = 35°23'37''.454$$

$$u = \sqrt{5008213.28^2 + 3558328.08^2} = 6143606.35$$

$$\tan k = \frac{1710120.06}{6143606.35}\times\frac{6378249.145}{6356517.32} = 0.27930934 \Rightarrow k = 15.6055448° = 0.27236814 \text{ rads}$$

According to Equation (3.12),

ϕ = 0.27325358 rads ≡ 15.6562768° = 15° 39' 22.5964"

According to Equation (3.8),

ν = 6 379 829.56

and h = 500.

Computation of the surface of the ellipsoid

In the pre-satellite era, networks of points were computed for the surface of the ellipsoid by complex formulae (see Allan 2004, Torge 2001). Only in a few special cases would this now be done. See also Appendix 4.

Chapter 4
Coordinate Transformations

4.1 Introduction

A major problem of geospatial surveying is the establishment of relationships between different sets of coordinates. A total station, laser scanner or pair of stereoscopic photographs derive three-dimensional coordinates in a local system based on the instrument. These have to be transformed in some way into the general coordinate system used to model the space in question. In a similar way, GPS coordinates have to be linked with local systems, or vice versa, and in the field of cadastre, new surveys have to be reconciled with old maps and surveys. The reader is therefore encouraged to study the elements of matrix algebra before attempting to master the subject matter of this chapter.

The problem concerns the linking of datums together. This process involves three basic factors: the origins of the systems, the relative alignments of their axes, and a possible change of scale between them.

There are two cases to be considered:

(a) When the origin shifts and the rotation of axes and the scale changes are known.

(b) When these parameters are unknown and have to be calculated from two sets of points whose coordinates in both systems are known. We refer to this as the 'Reverse Transformation Problem'.

Often stage (b) is used to determine the parameters from a few selected points, and used to transform most other points by (a), as in the case of digital mapping.

The reverse problem can be treated in a similar way to that of straight line fitting (see Chapter 6). There are two cases:

1. When *only one set* of coordinates is assumed to be observed, and the other set fixed. This model gives useful simple results. In this method, the treatment of redundant observations by Least Squares uses the simple form of equations:

$$Ax - L = v$$

   We treat most examples by this method for its simplicity and merit.
2. When *both sets of coordinates* obtained from observations have associated dispersion matrices. This occurs, for example, when data from GPS observations are to be matched to a classical triangulation network. This

model, requiring the application of the general Least Squares process, uses the more complicated equation of the form:

$$Ax + Cv = CL$$

We give one example of the latter form with respect to a two-dimensional case, but first we deal with the straightforward direct cases.

4.2 Direct transformations in two dimensions

The most common case of the direct transformation in two dimensions will be considered first. The coordinates used by surveyors are generally orthogonal and conformal. That is, the grid lines form squares and there is no scale change over the area of the survey. (Note that if a map projection is used, the scale changes differentially.)

We consider the points ABC, the origins o (O) and the centroids, sometimes referred to as centres of gravity, of the points G, all shown in Figure 4.1 whose various coordinate values are given in Table 4.1.

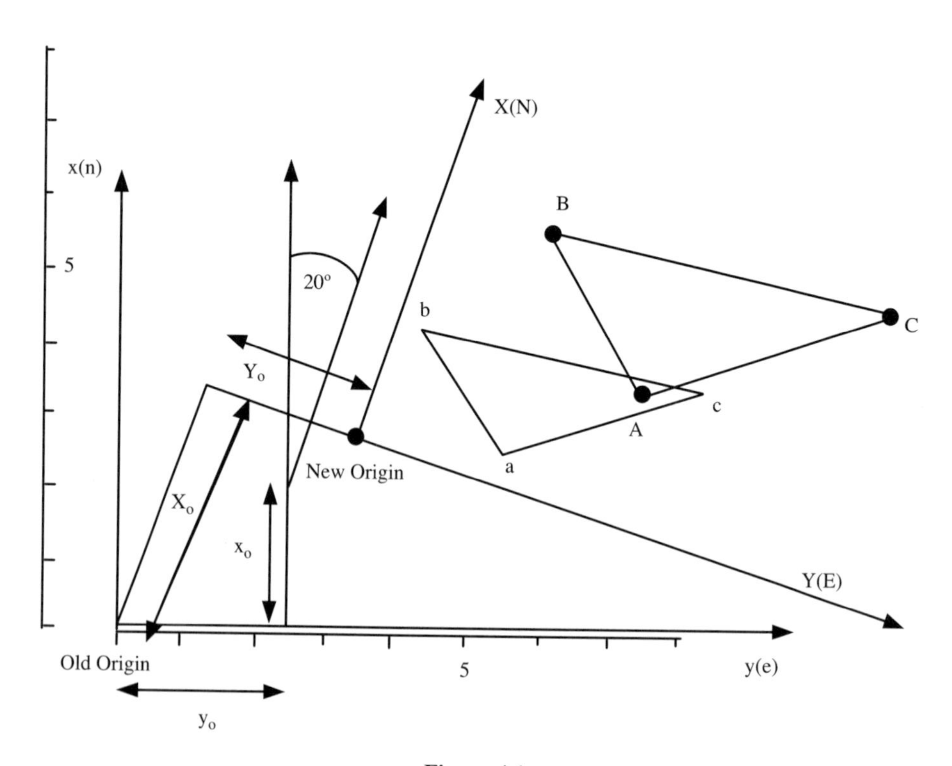

Figure 4.1

Table 4.1

		A	B	C	O	G
(x, y) system	x	2.5	4.5	3.5	2.0	3.5
	y	5.5	4.5	8.5	2.5	6.17
Origin change	x'	0.5	2.5	1.5	–2.0	1.5
	y'	3.0	2.0	6.0	–2.5	3.7
Rotated system	X'	1.5	3.03	3.46	-2.73	2.67
	Y'	2.65	1.02	5.13	-1.67	2.93
Scaled by $K = 1.3$	X	1.95	3.939	4.498	-3.549	3.46
	Y	3.445	1.326	6.669	-2.171	3.81

The original values (x, y) are changed to a new origin at (2.0, 2.5) giving the (x', y') set. Then a forward rotation of 20° is applied to give the (X', Y') set (not shown in the Figure), which is finally scaled by 1.3 to yield the (X, Y) set.

A common practical application of this type of change arises when a survey is first computed on an assumed position and an assumed bearing, both later to be corrected, and a foot–metre conversion has to be applied. We have chosen simple numbers by way of easy illustration.

The various operations are illustrated in Table 4.1.

The change of origin is given by

$$\begin{bmatrix} x_0 \\ y_0 \end{bmatrix} = \begin{bmatrix} 2.0 \\ 2.5 \end{bmatrix}$$

Thus we obtain a new set of coordinates relative to the new origin, but with the same orientation. This operation can be expressed as

$$\begin{bmatrix} x'_i \\ y'_i \end{bmatrix} = \begin{bmatrix} x_i \\ y_i \end{bmatrix} - \begin{bmatrix} x_0 \\ y_0 \end{bmatrix} \tag{4.1}$$

or in matrix algebra as

$$\mathbf{x}' = \mathbf{x} - \mathbf{x}_0$$

Notice that the vector **x**, when expressed in bold type, represents both the x and the y coordinates. The subscript i refers to the points A, B, C and G. The coordinates of the previous origin o with respect to the new origin O should be noted. This point is treated just as any other in the new system.

We now rotate the axes through an angle $\theta = 20°$ to give coordinates in the (X', Y') system. The relationship between the two systems is written

$$\mathbf{X}' = \mathbf{R}\,\mathbf{x}'$$

or in full

$$\begin{bmatrix} X' \\ Y' \end{bmatrix} = \begin{bmatrix} \cos\theta & \sin\theta \\ -\sin\theta & \cos\theta \end{bmatrix} \begin{bmatrix} x' \\ y' \end{bmatrix} \tag{4.2}$$

$$\begin{bmatrix} X'_A \\ Y'_A \end{bmatrix} = \begin{bmatrix} 0.9397 & 0.3420 \\ -0.3420 & 0.9397 \end{bmatrix} \begin{bmatrix} 0.5 \\ 3.0 \end{bmatrix} = \begin{bmatrix} 1.5 \\ 3.0 \end{bmatrix}$$

Finally the scale change $K = 1.3$ is applied, giving

$$\begin{bmatrix} X_A \\ Y_A \end{bmatrix} = \begin{bmatrix} 1.95 \\ 3.445 \end{bmatrix}$$

These three operations—translation, rotation and scaling—could be written simultaneously as

$$\mathbf{X} = K\,\mathbf{R}(\mathbf{x} - \mathbf{x}_0) = K\,\mathbf{R}\,\mathbf{x} - K\,\mathbf{R}\,\mathbf{x}_0$$

But

$$K\,\mathbf{R}\,\mathbf{x}_0 = \mathbf{X}_0$$

so we may write

$$\mathbf{X} = K\,\mathbf{R}\,\mathbf{x} - \mathbf{X}_0$$

or, in full

$$\begin{bmatrix} X \\ Y \end{bmatrix} = \begin{bmatrix} K\cos\theta & K\sin\theta \\ -K\sin\theta & K\cos\theta \end{bmatrix} \begin{bmatrix} x \\ y \end{bmatrix} - \begin{bmatrix} X_0 \\ Y_0 \end{bmatrix} \tag{4.3}$$

The transformation for point A is

$$\begin{bmatrix} X_A \\ Y_A \end{bmatrix} = \begin{bmatrix} 1.2216 & 0.4446 \\ -0.4446 & 1.2216 \end{bmatrix} \begin{bmatrix} 2.5 \\ 5.5 \end{bmatrix} - \begin{bmatrix} -3.54 \\ -2.17 \end{bmatrix} = \begin{bmatrix} 1.94 \\ 3.44 \end{bmatrix}$$

Note that exact agreement of these calculations is only achieved if all figures are retained consistently in the spreadsheets throughout. We have rounded all values in the tables.

4.3 Transformation by polar coordinates

Another way to transform the original (x, y) system into the new (X, Y) system is through the polar coordinates, by directly altering the survey bearings and recalculating new coordinates. For example, beginning with the original coordinates (x, y) we could obtain the distances and bearings (r, U) then change all bearings using

$$U'' = U - \theta$$

Thus we can convert all the bearings to the new system. Distances could also be rescaled if necessary. The coordinates are then moved to the origin at O via (x_0, y_0). Although it is always possible to treat these problems by polar coordinates instead of Cartesian coordinates, the method is tedious.

4.4 Note on change of scale

When a change of scale is required, a scale factor k is introduced in such a way that all lengths, including coordinates, are altered in proportion. For example, if we wish to plot a map at a scale of 1:500, all sizes are scaled down by this amount. Formally, we state

$$r' = Kr; \quad x' = Kx; \quad y' = Ky$$

where $K = 1/500$, referred to as the *nominal scale* of the map.

A change of units from metres to feet is another example of coordinate change, with a scale factor of $K = 1/0.3048$.

An additional way in which the scale may alter involves only a small change. This occurs when a reduced distance is plotted on a map projection (see Appendix 8).

However a scale change can arise from the use of incorrect lengths in the survey itself. The use of a non-standard tape, or the wrong refractive index in electromagnetic distance measurement (EDM) will cause such a scale change. In such a case, the scale factor k is approximately equal to one, and we set

$$k = 1 + e$$

where e is called the *scale error*. This is an unfortunate term because it can be confused with random errors of observation. It is in fact a deliberate systematic error. A scale factor of this type is often found when an old survey is repeated with modern instruments capable of better accuracy and the two have to be combined into a homogeneous or consistent system.

4.5 Reverse transformations

The reverse problem is to find the parameters (X_0, Y_0), θ and K from the two sets of coordinates $\mathbf{x}$ and $\mathbf{X}$. To do so we must rearrange Equations (4.3). Multiplying out,

$$\begin{aligned} X &= xK\cos\theta + yK\sin\theta - X_0 \\ Y &= -xK\sin\theta + yK\cos\theta - Y_0 \end{aligned} \tag{4.5}$$

Setting $a = K\cos\theta$ and $b = K\sin\theta$ we have

$$\begin{aligned} X &= xa + yb - X_0 \\ Y &= -xb + ya - Y_0 \end{aligned}$$

or, in matrix form

$$\begin{bmatrix} X \\ Y \end{bmatrix} = \begin{bmatrix} x & y & -1 & 0 \\ y & -x & 0 & -1 \end{bmatrix} \begin{bmatrix} a \\ b \\ X_0 \\ Y_0 \end{bmatrix}$$

Clearly we need data from at least one other point for a solution to be obtained. Further, we have K and θ from

$$K^2 = a^2 + b^2; \quad \tan\theta = \frac{b}{a}$$

Example Using the values of the original and new coordinates for points A and B (see Table 4.1) we have

$$\begin{bmatrix} X_A \\ Y_A \\ X_B \\ Y_B \end{bmatrix} = \begin{bmatrix} x_A & y_A & -1 & 0 \\ y_A & -x_A & 0 & 1 \\ x_B & y_B & -1 & 0 \\ y_B & -x_B & 0 & 1 \end{bmatrix} \begin{bmatrix} a \\ b \\ X_0 \\ Y_0 \end{bmatrix}$$

$$\begin{bmatrix} 1.94467876 \\ 3.44248813 \\ 3.94325339 \\ 1.33163535 \end{bmatrix} = \begin{bmatrix} 2.5 & 5.5 & -1 & 0 \\ 5.5 & -2.5 & 0 & 1 \\ 4.5 & 4.5 & -1 & 0 \\ 4.5 & -4.5 & 0 & 1 \end{bmatrix} \begin{bmatrix} a \\ b \\ X_0 \\ Y_0 \end{bmatrix}$$

Therefore

$$\begin{bmatrix} a \\ b \\ X_0 \\ Y_0 \end{bmatrix} = \begin{bmatrix} 2.5 & 5.5 & -1 & 0 \\ 5.5 & -2.5 & 0 & 1 \\ 4.5 & 4.5 & -1 & 0 \\ 4.5 & -4.5 & 0 & -1 \end{bmatrix}^{-1} \begin{bmatrix} 1.94467876 \\ 3.44248813 \\ 3.94325339 \\ 1.33163535 \end{bmatrix} = \begin{bmatrix} 1.2216 \\ 0.44463 \\ 3.55477 \\ 2.16475 \end{bmatrix}$$

We can therefore calculate K and θ as:

$$K = \sqrt{1.2216^2 + 0.44463^2} = 1.30; \quad \tan\theta = \frac{0.44463}{1.2216} \Rightarrow \theta = 20.0^\circ$$

In a practical case, the two sets of coordinates are not usually consistent with each other and so slightly different results will be obtained depending on the points used for the calculation. To get a best estimate, we use the method of Least Squares incorporating data from a number of points. This gives the set of equations

$$\mathbf{A}\mathbf{X} = \mathbf{L}$$

$$\mathbf{X} = (\mathbf{A}^T\mathbf{A})^{-1}\mathbf{A}^T\mathbf{L}$$

See Section 4.7 for an example, where we treat the Method of Centroids. However, the more correct model is to use the general Least Squares version also in Section 4.7.

4.6 Transformations in three dimensions

The transformation of coordinates in three dimensions is similar to that for the two-dimensional problem, with an **R** matrix of dimensions (3 × 3).
We shall consider two applications:

1. Where it is known that there is only one rotation about the vertical (z) axis.
2. When likely rotations about all three axes are small.

Case (1)
When the problem arises in connection with a total station or a laser scanner application, the unknown orientation angle (i.e. in the horizontal plane) is the only

one likely to be of some size. If the instruments are levelled properly the z-axis is assumed untilted. Thus the transformation is of the form

$$\begin{bmatrix} X \\ Y \\ Z \end{bmatrix} = \begin{bmatrix} K\cos\theta & K\sin\theta & 0 \\ -K\sin\theta & K\cos\theta & 0 \\ 0 & 0 & K \end{bmatrix} \begin{bmatrix} x \\ y \\ z \end{bmatrix} - \begin{bmatrix} X_0 \\ Y_0 \\ Z_0 \end{bmatrix} \tag{4.6}$$

Expanding, rearranging and substituting a = K cosθ; b = K sinθ as before implies we have three equations for each of two points, giving the minimum six required for a solution.

$$\begin{bmatrix} X_1 \\ Y_1 \\ Z_1 \\ X_2 \\ Y_2 \\ Z_2 \end{bmatrix} = \begin{bmatrix} x & y & 0 & -1 & 0 & 0 \\ -y & x & 0 & 0 & -1 & 0 \\ 0 & 0 & 1 & 0 & 0 & 1 \\ x & y & 0 & -1 & 0 & 0 \\ -y & x & 0 & 0 & -1 & 0 \\ 0 & 0 & 1 & 0 & 0 & 1 \end{bmatrix} \begin{bmatrix} a \\ b \\ K \\ X_0 \\ Y_0 \\ Z_0 \end{bmatrix}$$

Example Refer to Figure 4.2. This shows a wall to be surveyed by a laser scanner. To incorporate the scan points (of which there can be thousands) into the site coordinate system, four reference points (A, B, C and D) are surveyed by total station to give Coordinate Set 1, as listed in Table 4.2.

The scanner is then set up at the same station, in the same approximate plan position but at a different height. We have offset it by 4 mm in x, 6 mm in y and 120 mm in z. The orientation is also only approximate; we have offset this by 2°.

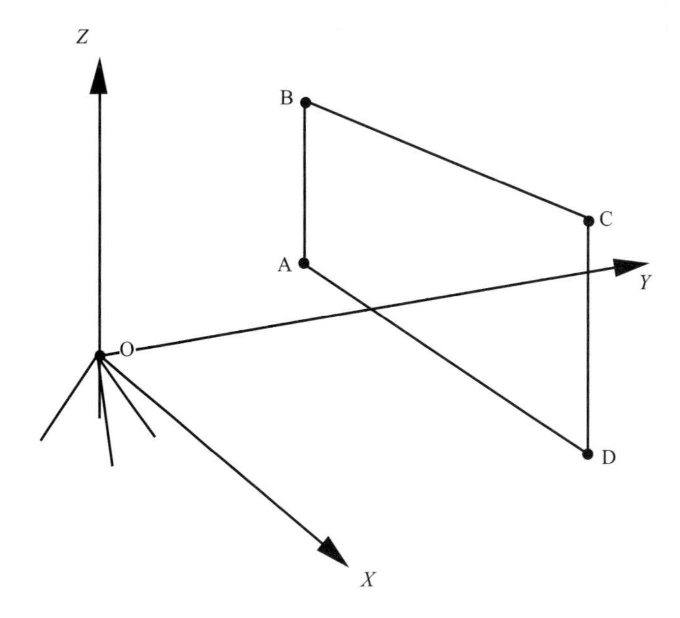

Figure 4.2

Table 4.2 *Set 1: Site system coordinates*

Point	*x*	*y*	*z*
P	0.000	0.000	0.000
A	–5.000	10.000	0.000
B	–5.000	10.000	10.000
C	5.000	10.000	10.000
D	5.000	10.000	0.000

The scanner coordinates are given in Set 2 (see Table 4.3). Before the scanner coordinates can be incorporated into the site system, the orientation and centring errors have to be identified and allowed for by applying the forward transformation formula, Equation (4.6).

Equation (4.6) for points B and D are then:

$$\begin{bmatrix} -4.6539592 \\ 10.1724058 \\ 10.12 \\ 5.3399491 \\ 9.82341079 \\ 0.12 \end{bmatrix} = \begin{bmatrix} -5 & 10 & 0 & -1 & 0 & 0 \\ 10 & 5 & 0 & 0 & -1 & 0 \\ 0 & 0 & 10 & 0 & 0 & -1 \\ 5 & 10 & 0 & -1 & 0 & 0 \\ 10 & -5 & 0 & 0 & -1 & 0 \\ 0 & 0 & 0 & 0 & 0 & -1 \end{bmatrix} \begin{bmatrix} a \\ b \\ K \\ X_0 \\ Y_0 \\ Z_0 \end{bmatrix}$$

$$a = K\cos\theta = 0.99939083$$

$$b = K\sin\theta = 0.0348995$$

which implies $\theta = 2°$ and $(X_0, Y_0, Z_0) = (0.006, -0.004, -0.12)$

Check

Having found the necessary parameters, they should be checked back on the original data. It is usual to make this check using a point which was not involved in their derivation.

Table 4.3 *Set 2: Scanner coordinates*

Point	*X*	*Y*	*Z*
A	–4.654	10.172	0.120
B	–4.654	10.172	10.120
C	5.340	9.823	10.120
D	5.340	9.823	0.120

We obtain the result for point A from the equations

$$\begin{bmatrix} x \\ y \\ z \end{bmatrix} = \begin{bmatrix} 0.99939 & 0.03489 & 0 \\ -0.03489 & 0.99939 & 0 \\ 0 & 0 & 1 \end{bmatrix} \begin{bmatrix} -4.65395 \\ 10.17240 \\ 0.12 \end{bmatrix} - \begin{bmatrix} 0.006 \\ -0.004 \\ -0.12 \end{bmatrix} = \begin{bmatrix} -5 \\ 10 \\ 0 \end{bmatrix} \quad (4.7)$$

Case (2)

In geodesy problems, when establishing a connection between two datums, the orientation angles are likely to be small. Therefore the assumption is made that $\cos\theta = 1$ and $\sin\theta = \theta$ for each of the three rotations, θ_x, θ_y, θ_z giving a simple rotation matrix of the form

$$\mathbf{R}_x\mathbf{R}_y\mathbf{R}_z = \begin{bmatrix} 1 & -\theta_z & \theta_y \\ \theta_z & 1 & -\theta_x \\ -\theta_y & \theta_x & 1 \end{bmatrix} = \mathbf{I} + \varepsilon_x + \varepsilon_y + \varepsilon_z$$

This can be observed if we multiply the following three matrices and neglect powers of small angles greater than the first.

$$\mathbf{R}_x = \begin{bmatrix} 1 & 0 & 0 \\ 0 & 1 & -\theta_x \\ 0 & \theta_x & 1 \end{bmatrix} = \mathbf{I} + \varepsilon_x$$

$$\mathbf{R}_y = \begin{bmatrix} 1 & 0 & \theta_y \\ 0 & 1 & 0 \\ -\theta_y & 0 & 1 \end{bmatrix} = \mathbf{I} + \varepsilon_y$$

$$\mathbf{R}_z = \begin{bmatrix} 1 & -\theta_z & 0 \\ \theta_z & 1 & 0 \\ 0 & 0 & 1 \end{bmatrix} = \mathbf{I} + \varepsilon_z$$

Hence the direct transformation equations for small angles in three dimensions become

$$\begin{bmatrix} X \\ Y \\ Z \end{bmatrix} = K \begin{bmatrix} 1 & -\theta_z & \theta_y \\ \theta_z & 1 & -\theta_x \\ -\theta_y & \theta_x & 1 \end{bmatrix} \begin{bmatrix} x \\ y \\ z \end{bmatrix} - \begin{bmatrix} X_0 \\ Y_0 \\ Z_0 \end{bmatrix} \quad (4.8)$$

Once again, to derive the parameters from two sets of coordinates we have to solve the equations in seven parameters. This requires data from two points and one relevant piece of data from a third. We say relevant, because not every point will do and a singular matrix may result. This occurs if we use the X value of point A.

$$\begin{bmatrix} X_1 \\ Y_1 \\ Z_1 \\ X_2 \\ Y_2 \\ Z_2 \\ Y_3 \end{bmatrix} = \begin{bmatrix} -y & z & 0 & x & -1 & 0 & 0 \\ x & 0 & -z & y & 0 & -1 & 0 \\ 0 & -x & y & z & 0 & 0 & 1 \\ -y & z & 0 & x & -1 & 0 & 0 \\ x & 0 & -z & y & 0 & -1 & 0 \\ 0 & -x & y & z & 0 & 0 & 1 \\ x & 0 & -z & y & 0 & -1 & 0 \end{bmatrix} \begin{bmatrix} K\theta_z \\ K\theta_y \\ K\theta_x \\ K \\ X_0 \\ Y_0 \\ Z_0 \end{bmatrix}$$

Applying this method to the data in Tables 4.2 and 4.3 yields the solutions:

$$K\theta_z = -1.90\ (-2.0);\ K\theta_y = 0.01\ (0);\ K\theta_x = -0.097\ (0);\ X_0 = -0.009\ (-0.006);$$
$$Y_0 = 0.0075\ (0.004);\ Z_0 = -0.138\ (-0.12).$$

Although the model is inappropriate, the solution is quite good when compared with the correct results in brackets.

4.7 Method of centroids

Although there are other methods of transformation which will give better results than the above, we move straight away to the best general method, involving the centroids. The concept is to move all scaling and rotations to the middle of the areas concerned. The method has two stages: a translation, and a combined rotation and scaling.

First the coordinate system is moved to the centroid in each case, giving new coordinates expressed by

$$\begin{bmatrix} x_i^G \\ y_i^G \end{bmatrix} = \begin{bmatrix} x_i \\ y_i \end{bmatrix} - \begin{bmatrix} x_G \\ y_G \end{bmatrix}$$
$$\begin{bmatrix} X_i^G \\ Y_i^G \end{bmatrix} = \begin{bmatrix} X_i \\ Y_i \end{bmatrix} - \begin{bmatrix} X_G \\ Y_G \end{bmatrix} \tag{4.9}$$

The notation implies the *abscissa of point i* with the centroid X_G as origin. For example, for point A, = $X_A^G = 1.89 - 2.67 = -0.78$. See Tables 4.4 and 4.5 for an example.

Table 4.4

Point	*A*	*B*	*C*	*G*
x	2.50	4.50	3.50	3.50
y	5.50	4.50	8.50	6.17
x_G	–1.00	+1.00	0.00	0.00
y_G	–0.67	–1.67	2.33	0.00

Table 4.5

Point	*A'*	*B'*	*C'*	*G'*
X	1.89	2.91	3.2	2.67
Y	2.74	1.66	4.39	2.93
X_G	–0.78	0.24	0.53	0.00
Y_G	–0.19	–1.27	1.46	0.00

These coordinates, reduced to the centroid, are next transformed by the formula

$$\mathbf{X}^{G} = \mathbf{R}\,\mathbf{x}^{G}$$

We now introduce the general notation for the elements of the scaled rotation matrix $\mathbf{R}$ i.e. r_{11}, r_{12}, etc. where $r_{11} = r_{22} = K\cos\theta$ and $-r_{21} = r_{12} = K\sin\theta$. In terms of $\mathbf{R}$,

$$\begin{bmatrix} X_1^G \\ Y_1^G \\ X_2^G \\ Y_2^G \end{bmatrix} = \begin{bmatrix} x_1^G & y_1^G & 0 & 0 \\ 0 & 0 & x_1^G & y_1^G \\ x_1^G & y_1^G & 0 & 0 \\ 0 & 0 & x_1^G & y_1^G \end{bmatrix} \begin{bmatrix} r_{11} \\ r_{12} \\ r_{21} \\ r_{22} \end{bmatrix}$$

For points A and B,

$$\begin{bmatrix} -0.78 \\ -0.19 \\ 0.24 \\ -1.27 \end{bmatrix} = \begin{bmatrix} -1.0 & -0.67 & 0 & 0 \\ 0 & 0 & -1.0 & -0.67 \\ 1.0 & -1.67 & 0 & 0 \\ 0 & 0 & 1.0 & -1.67 \end{bmatrix} \begin{bmatrix} r_{11} \\ r_{12} \\ r_{21} \\ r_{22} \end{bmatrix}$$

Therefore,

$$\mathbf{R} = \begin{bmatrix} r_{11} & r_{12} \\ r_{21} & r_{22} \end{bmatrix} = \begin{bmatrix} 0.6243 & 0.2286 \\ -0.2271 & 0.6257 \end{bmatrix}$$

The complete transformation can then be written as:

$$\mathbf{X} = \mathbf{X}_G + \mathbf{X}^G = \mathbf{X}_G + \mathbf{R}\,\mathbf{x}^G = \mathbf{X}_G + \mathbf{R}(\mathbf{x} - \mathbf{x}_G) \tag{4.10}$$

$$\begin{bmatrix} X \\ Y \end{bmatrix} = \begin{bmatrix} X_G \\ Y_G \end{bmatrix} + \begin{bmatrix} 0.6243 & 0.2286 \\ -0.2271 & 0.6257 \end{bmatrix} \begin{bmatrix} x - x_G \\ y - y_G \end{bmatrix}$$
$$= \begin{bmatrix} 2.67 \\ 2.93 \end{bmatrix} + \begin{bmatrix} 0.6243 & 0.2286 \\ -0.2271 & 0.6257 \end{bmatrix} \begin{bmatrix} x - 3.5 \\ y - 6.17 \end{bmatrix}$$

Note: The application of this formula to points A, B and C gives perfect agreement with the original values of Tables 4.4 and 4.5 as expected. The agreement of point C comes about because the two sets of values are entirely consistent. In practice, the two surveys of the points of the triangle would contain errors which would give rise to slight inconsistencies. In this case, a Least Squares approach is used.

4.8 General case of centroids with Least Squares

We shall now deal with the problem of inconsistent data, and where both sets of coordinates are associated with known dispersion matrices. The reader who is unfamiliar with the Least Squares process should read Chapter 6 and/or Appendix 3, which explain these matters and the notation being used. Although the example is for a two-dimensional coordinate problem, the method is identical in three dimensions.

The situation is that the triangles ABC and A'B'C' are not exactly similar in shape as well as being of different sizes. In addition, the coordinates in both systems have associated dispersion matrices $\mathbf{D}_\mathbf{x}$ and $\mathbf{D}_\mathbf{X}$. The method follows similar lines to the centroid method, except that each centroid is obtained as a weighted operation. We first set up the observation equations for each set of coordinates separately, in the form

$$\mathbf{Ax} = \mathbf{L} + \mathbf{v}$$

For the (x, y) system:

$$\begin{bmatrix} 1 & 0 \\ 0 & 1 \\ 1 & 0 \\ 0 & 1 \\ 1 & 0 \\ 0 & 1 \end{bmatrix} \begin{bmatrix} x_G \\ y_G \end{bmatrix} = \begin{bmatrix} 2.5 \\ 5.5 \\ 4.5 \\ 4.5 \\ 3.5 \\ 8.5 \end{bmatrix} + \mathbf{v}$$

The dispersion matrix $\mathbf{D}_\mathbf{x}$ could contain many nonzero covariances. However, to simplify the calculations we assume the upper triangle of this symmetric matrix to be

$$\begin{bmatrix} 0.1 & 0.05 & 0 & 0 & 0 & 0 \\ & 0.1 & 0 & 0 & 0 & 0 \\ & & 0.1 & 0.05 & 0 & 0 \\ & & & 0.1 & 0 & 0 \\ \text{symmetric} & & & & 0.1 & 0.05 \\ & & & & & 0.1 \end{bmatrix}$$

The coordinates of the centroid are then given by

$$\mathbf{G}_\mathbf{x} = (\mathbf{A}^\mathrm{T}\mathbf{W}\mathbf{A})^{-1}\mathbf{A}^\mathrm{T}\mathbf{W}\,\mathbf{x} \tag{4.11}$$

where

$$\mathbf{W} = (\mathbf{D}_\mathbf{x})^{-1}$$

Thus the coordinates of the centroid are

$$\begin{bmatrix} G_x \\ G_y \end{bmatrix} = \begin{bmatrix} 3.5 \\ 6.17 \end{bmatrix}$$

In the same way, using the coordinates from the (X, Y) system where it should be noted a change has been made to the coordinate of point C to ensure that the two triangles are no longer similar, we have the equations

$$\begin{bmatrix} 1 & 0 \\ 0 & 1 \\ 1 & 0 \\ 0 & 1 \\ 1 & 0 \\ 0 & 1 \end{bmatrix} \begin{bmatrix} G_X \\ G_Y \end{bmatrix} = \begin{bmatrix} 1.89 \\ 2.74 \\ 2.91 \\ 1.66 \\ 3.20 \\ 4.00 \end{bmatrix} + \mathbf{v}$$

and a dispersion matrix which, we assume for simplicity, is the same as for the (x, y) system, that is $\mathbf{D_x} = \mathbf{D_X}$.

We obtain the centroid coordinates of

$$\begin{bmatrix} G_X \\ G_Y \end{bmatrix} = \begin{bmatrix} 2.67 \\ 2.80 \end{bmatrix}$$

New coordinate systems $\mathbf{x}_G$ and $\mathbf{X}_G$ with respect to these centroids give respective values for the points in columns two and three of Table 4.6. As before, we wish to find the matrix $\mathbf{R}$ which affects the transformation

$$\mathbf{X}_G = \mathbf{R}\,\mathbf{x}_G$$

4.9 Simple Least Squares model

A simple solution can be found for $\mathbf{R}$ if we assume it is a scaled orthogonal matrix and only the (x, y) coordinates are treated as observed. We have for the three points the equations

$$\mathbf{A}\,\mathbf{X} = \mathbf{L}$$

Table 4.6

Point	x_G	*Observed* X_G	*Provisional* X_G	L
x_A	–1.00	–0.777	–0.777	0.00
y_A	–0.67	–0.060	–0.190	0.13
x_B	1.00	0.243	0.243	0.00
y_B	–1.67	–1.140	–1.270	0.13
x_C	0.00	0.530	0.530	0.00
y_C	2.33	1.200	1.460	–0.26

where

$$\mathbf{A} = \begin{bmatrix} -1 & -0.6666667 \\ -0.6666667 & 1 \\ 1 & -1.6666667 \\ -1.6666667 & -1 \\ 0 & 2.3333333 \\ 2.3333333 & 0 \end{bmatrix} \quad \text{and} \quad \mathbf{L} = \begin{bmatrix} -0.7766667 \\ -0.19 \\ 0.24333333 \\ -1.27 \\ 0.53333333 \\ 1.46 \end{bmatrix}$$

Here we have

$$\mathbf{A}^{\mathrm{T}}\mathbf{A}\,\mathbf{X} = \mathbf{A}^{\mathrm{T}}\mathbf{L}$$

$$\begin{bmatrix} 10.667 & 0 \\ 0 & 10.667 \end{bmatrix}\mathbf{X} = \begin{bmatrix} 6.667 \\ 2.437 \end{bmatrix}$$

Therefore

$$\mathbf{X} = (\mathbf{A}^{\mathrm{T}}\mathbf{A})^{-1}\mathbf{A}^{\mathrm{T}}\mathbf{L}$$

and the solution is:

$a = 0.6253125$; $b = 0.2284375$; $K = 0.66573224$; $\cos\theta = 0.93928529 \Rightarrow$
$\theta = 0.35025486$ rad or $\theta = 20.0681253$ deg

4.10 General Least Squares model

Because both sets of coordinates are observed and have dispersion matrices, and the equations to be solved are not linear in the unknowns, we proceed by the combined Least Squares method. This assumes an initial value for the **R** matrix, and for one of the sets of coordinates. We assume that

$$\mathbf{R} = \mathbf{R}^*$$

For this assumed value, we use the **R** matrix found previously by the direct method, that is

$$\mathbf{R}^* = \begin{bmatrix} r_{11} & r_{12} \\ r_{21} & r_{22} \end{bmatrix} = \begin{bmatrix} 0.6243 & 0.2286 \\ -0.2271 & 0.6257 \end{bmatrix}$$

From the x values in Table 4.6 we compute the provisional values of $\mathbf{X}_G$ in column four of the same table. On subtraction of these from the observed values in column three we obtain the **L** vector in column five.

The simple Least Squares model can be used. This assumes that only the first set of coordinates (x, y) is observed and subject to error.

Next, we derive the combined Least Squares model of the form

$$\mathbf{Ax} + \mathbf{Cv} = \mathbf{CL}$$

In this case the **x** vector consists of the four small corrections to the elements of the provisional **R** matrix used to calculate the **L** vector.

The mathematical model is as before, and for simplicity we have dropped the superscripts G. Following the usual procedure we have

$$\begin{bmatrix} X \\ Y \end{bmatrix} = \begin{bmatrix} r_{11} & r_{12} \\ r_{21} & r_{22} \end{bmatrix} \begin{bmatrix} x \\ y \end{bmatrix}$$

$$\begin{bmatrix} \mathrm{d}X \\ \mathrm{d}Y \end{bmatrix} = \begin{bmatrix} \mathrm{d}r_{11} & \mathrm{d}r_{12} \\ \mathrm{d}r_{21} & \mathrm{d}r_{22} \end{bmatrix} \begin{bmatrix} x \\ y \end{bmatrix} + \begin{bmatrix} r_{11} & r_{12} \\ r_{21} & r_{22} \end{bmatrix} \begin{bmatrix} \mathrm{d}x \\ \mathrm{d}y \end{bmatrix}$$

or recasting we have

$$\mathbf{0} = \begin{bmatrix} x & y & 0 & 0 \\ 0 & 0 & x & y \end{bmatrix} \begin{bmatrix} \mathrm{d}r_{11} \\ \mathrm{d}r_{12} \\ \mathrm{d}r_{21} \\ \mathrm{d}r_{22} \end{bmatrix} + \begin{bmatrix} r_{11} & r_{12} & -1 & 0 \\ r_{21} & r_{22} & 0 & -1 \end{bmatrix} \begin{bmatrix} \mathrm{d}x \\ \mathrm{d}y \\ \mathrm{d}X \\ \mathrm{d}Y \end{bmatrix}$$

which is of the form

$$\mathbf{0} = \mathbf{A}\,\mathbf{x} + \mathbf{C}\,\mathbf{dx}$$

This is the data available from each point. The full matrix for all three points of the example with numerical values is therefore

$$\mathbf{A}\,\mathbf{x} = \begin{bmatrix} -1 & -0.667 & 0 & 0 \\ 0 & 0 & -1 & -0.667 \\ 1 & -1.667 & 0 & 0 \\ 0 & 0 & 1 & -1.667 \\ 0 & 2.333 & 0 & 0 \\ 0 & 0 & 0 & 2.333 \end{bmatrix} \begin{bmatrix} \mathrm{d}r_{11} \\ \mathrm{d}r_{12} \\ \mathrm{d}r_{21} \\ \mathrm{d}r_{22} \end{bmatrix}$$

The vector **dx** is split into **L** + **v** where the **L** vector consists of the observed-minus-provisional values of the observed parameters. Because we selected the observed values of the **x** system to be their provisional values, and from them computed the provisional values of the **X** system, the first part of the column vector **L** will be zero. In full, the transposed vector is

$$\mathbf{L}^{\mathrm{T}} = (0, 0, 0, 0, 0, 0, 0, 0.13, 0, 0.13, 0, -0.26)$$

The **C** matrix has a particular structure for three points as follows

$$\mathbf{C} = \begin{bmatrix} \mathbf{R} & 0 & 0 & -\mathbf{I} & 0 & 0 \\ 0 & \mathbf{R} & 0 & 0 & -\mathbf{I} & 0 \\ 0 & 0 & \mathbf{R} & 0 & 0 & -\mathbf{I} \end{bmatrix}$$

The sub-matrices **R** and **I** are of dimensions (2×2) in a two-dimensional coordinate system (x, y). In our example, **R** is the provisional transformation matrix defined previously.

The full (12×12) dispersion matrix of the two observed sets of two-dimensional coordinates for the three points has the structure

$$\mathbf{W}^{-1} = \begin{bmatrix} \mathbf{D} & 0 & 0 & 0 & 0 & 0 \\ & \mathbf{D} & 0 & 0 & 0 & 0 \\ & & \mathbf{D} & 0 & 0 & 0 \\ & & & \mathbf{D} & 0 & 0 \\ & & & & \mathbf{D} & 0 \\ & \text{symmetric} & & & & \mathbf{D} \end{bmatrix}$$

Where each submatrix **D** is of the form

$$\mathbf{D} = \begin{bmatrix} 0.10 & 0.05 \\ 0.05 & 0.10 \end{bmatrix}$$

This gives $\mathbf{C}\,\mathbf{W}^{-1}\mathbf{C}^{\mathrm{T}}$ with the structure

$$\begin{bmatrix} \mathbf{P} & 0 & 0 \\ 0 & \mathbf{P} & 0 \\ 0 & 0 & \mathbf{P} \end{bmatrix}$$

where each submatrix **P** is of the form

$$\mathbf{P} = \begin{bmatrix} 17.32 & -8.91 \\ -8.91 & 21.14 \end{bmatrix}$$

and finally

$$\mathbf{A}^{\mathrm{T}}(\mathbf{C}\mathbf{W}^{-1}\mathbf{C}^{\mathrm{T}})^{-1}\mathbf{A}\,\mathbf{x} = \mathbf{A}^{\mathrm{T}}(\mathbf{C}\mathbf{W}^{-1}\mathbf{C}^{\mathrm{T}})^{-1}\mathbf{C}\mathbf{L}$$

$$\begin{bmatrix} 0.1474 & -0.0737 & 0.0621 & -0.0310 \\ & 0.6387 & -0.0311 & 0.2692 \\ & & 0.1208 & -0.0604 \\ \text{symmetric} & & & 0.5235 \end{bmatrix} \mathbf{x} = \begin{bmatrix} 3.106 \times 10^{-6} \\ -5.177 \times 10^{-6} \\ 6.040 \times 10^{-6} \\ -1.007 \times 10^{-5} \end{bmatrix} \qquad (4.12)$$

which gives the solution for **x** and the final values of the **R** matrix as shown in Table 4.7.

Table 4.7

x	*Provisional* **R**	*Final* **R**
1.5272×10^{-6}	0.6243	0.62430153
-6.433×10^{-6}	0.2286	0.22859357
1.6916×10^{-6}	−0.2271	−0.2270983
-7.125×10^{-6}	0.6257	0.62569287

4.11 Added constraints

In the above transformation it has not been assumed that the matrix **R** consists of a scaled orthogonal matrix. If we wish this to be so, for example in photogrammetry or cartography, we can do so by imposing constraints on the solution, such as

$$\check{r}_{11} = \check{r}_{22} \quad \text{and} \quad \check{r}_{12} = -\check{r}_{21}$$

therefore

$$\begin{aligned} \dot{r}_{11} + \mathrm{d}r_{11} &= \dot{r}_{22} + \mathrm{d}r_{22} \\ \dot{r}_{12} + \mathrm{d}r_{12} &= -\dot{r}_{21} - \mathrm{d}r_{21} \end{aligned}$$

giving the constraint equations

$$\begin{aligned} \mathrm{d}r_{11} - \mathrm{d}r_{22} &= \dot{r}_{22} - \dot{r}_{11} = 0.0014 \\ \mathrm{d}r_{12} + \mathrm{d}r_{22} &= \dot{r}_{12} - \dot{r}_{21} = -0.0015 \end{aligned}$$

When these equations are added to the normal equations (Equations 4.12) we obtain the hypermatrix, where **k** is the (1×2) Lagrangian column vector (see Appendix 3).

$$\begin{bmatrix} 0.1474 & -0.0737 & 0.0629 & -0.0310 & 1 & 0 \\ -0.0737 & 0.6387 & -0.0310 & 0.2691 & 0 & 1 \\ 0.0621 & -0.0310 & 0.1208 & -0.0604 & 0 & 1 \\ -0.0310 & 0.2691 & -0.0604 & 0.5235 & -1 & 0 \\ 1 & 0 & 0 & -1 & 0 & 0 \\ 0 & 1 & 1 & 0 & 0 & 0 \end{bmatrix} \begin{bmatrix} \mathbf{x} \\ \mathbf{k} \end{bmatrix} = \begin{bmatrix} 3.106 \times 10^{-6} \\ -5.177 \times 10^{-6} \\ 6.040 \times 10^{-6} \\ -1.007 \times 10^{-6} \\ -0.0014 \\ 0.0015 \end{bmatrix}$$

The solution and final values of the coefficients of **R** are listed in Table 4.8.

The matrix coefficients are then split into the scale and rotation elements K and θ from

$$K = \sqrt{0.6254^2 + 0.2286^2} = 0.665$$

$$\theta = \arctan\left(\frac{0.2286}{0.6254)}\right) = 20.07^\circ$$

Table 4.8

x	*Provisional* **R**	*Final* **R**
0.0011	0.6243	0.625433
2.841×10^{-5}	0.2286	0.228621
–0.0015	–0.2271	–0.228622
–0.0002	0.6257	0.625433

Example

We repeat the above transformation using a scaled orthogonal matrix at the very outset, instead of the empirical method with a later constraint. Applying a direct solution without any provisional values and iterations, we have

$$\mathbf{A} = \begin{bmatrix} -1 & -0.6666667 \\ -0.6666667 & 1 \\ 1 & -1.6666667 \\ -1.6666667 & -1 \\ 0 & 2.3333333 \\ 2.3333333 & 0 \end{bmatrix}; \quad \mathbf{L} = \begin{bmatrix} -0.776667 \\ -0.19 \\ 0.2433333 \\ -1.27 \\ 0.5333333 \\ 1.46 \end{bmatrix}$$

$a = 0.6253125$; $b = 0.2284375$; $K = 0.66573224$; $\cos\theta = 0.93928529 \Rightarrow$
$\theta = 0.35025486$ rad or $\theta = 20.0681253$ deg

As expected in this simple case, the results are identical.

4.12 Summary

The various methods of coordinate transformation are all useful, depending on the accuracy required and the distortions that exist between the two sets of coordinates.

The full Least Squares method is worth programming because of its general application. It is also possible to obtain error estimates of the parameters, and to carry out tests for the location of possible blunders in the data.

These matters are discussed in Chapter 5 and Appendix 6. To reduce the working, the data of the example was kept to a minimum and therefore will not yield useful statistical data.

Other methods of transforming coordinates have roles to play, such as linear interpolation within pre-calculated values (see Chapter 10), and polynomial modelling (Iliffe 2000).

Chapter 5
Theory of Errors and Quality Control

Chapter Summary

In this chapter we discuss how to assess the quality of measurements and results derived from them. This involves an understanding of the theory of errors of measurement, statistical methods of selecting the best estimates of derived parameters such as coordinates, ways in which to combine extra observations of various kinds into a consistent set of data, and whether or not to accept or reject values on the basis of statistical hypotheses.

Clients now readily demand such information as a means of assessing the quality of the product presented to them by the surveyor. *Quality control* forms part of a wider management system called *quality assurance.*

Initially, we establish mathematical ways to describe measurements and results derived from them. We then examine some basic statistical procedures used to handle data in a consistent manner. Finally, we look closely at the major methods of reconciling mixed and inconsistent observational data into consistent figures for use in dimension control and for other purposes.

5.1 Introduction to errors of measurement

At school we were once asked to draw a straight line 300 mm long and then to measure it, using a 100 mm rule for both processes. We were surprised to discover that the exercise was not as ridiculous as it first appeared, and that it embodied many basic concepts to do with the errors of measurement.

Some of the discoveries made were:

(a) Most lines drawn by pupils were of different lengths, as could easily be seen by direct comparison without the use of rulers. This direct comparison is an example of an independent checking procedure, something always to be sought after in surveying, which gives a measure of *reliability*.

(b) When we measured with a ruler graduated in centimetres only, the same result was obtained each time, but when we used a ruler with precise millimetre divisions, the results varied within small limits. This illustrates the need for sufficient *precision* in measurement.

(c) Different precise answers were obtained with different rulers, illustrating the need to calibrate the rulers by comparison with a known standard length. The calibration process reduces the *systematic* error of the rulers to within known limits.

(d) One pupil had miscounted, and gave the length as 400 mm. Mistakes are the most serious of all the errors.

(e) In the end we were uncertain as to the exact lengths of all the lines, which were each supposed to be 300 mm. This is an example of uncertainty in meeting the defined specification of 300 mm. No measurement *tolerance* was asked for nor was there a *statistical criterion* for this tolerance. Arguments about specifications are often the subject of legal disputes and often arise because these matters were not defined beforehand.

Thus an apparently simple exercise was full of problems: problems which arise in all attempts to set out measurements according to a specification, and to measure the exact value of a quantity. In addition, it raised the issue of what is meant by 'the exact value'.

Since land and geodetic surveying is almost wholly concerned with measurements and setting out to specifications, these problems are of paramount importance. Indeed, in many cases, it is almost as important to know the accuracy of a result as to know the result itself. Clearly there is a need to define the meaning of 'quality' in unambiguous ways.

5.2 Definition of concepts and terms

The following definitions and symbols are essential for an understanding of subsequent sections. While there is unfortunately no universally accepted form of notation for the various quantities used in error theory, we shall adopt a notation in agreement with most of the main texts and publications. However, readers should be warned of the general lack of consistency in the literature.

Mathematical and statistical formulae are defined such that problems can be treated consistently and with computer. Most of the problems ultimately end up as the near intersection of two lines in space, or combinations of such intersections.

True value T

Just as the concept of 'truth' is an abstract idea, so also in most cases is the concept of true value. In general, the true value of a quantity will never be found, or, if it is found, we will never know that we have found it. For example, we can only know the length of a ruler within specific limits set by calibration. This lack of knowledge of the truth applies to a single observed quantity only, for we often know the true value of a combination of quantities that are observed singly. For example, the sum of the three angles of a plane triangle is known to be 180°.

Population and sample means μ and $\bar{x}$

By *population* we mean all the possible values that could exist of a particular kind, such as a measurement of the 300 mm line we described above, made by every 100 mm ruler in the world. This is a very large population indeed.

Our particular small set of measurements drawn from this population is called a *sample*. It is also a sample of another population consisting of all the measurements that could be made of the line using one ruler alone.

The mean or average of all possible values is the population mean usually denoted by the Greek letter μ; the mean or average of the sample is denoted by $\bar{x}$ Clearly these two means will not generally be equal. However, it would be reasonable to think that if the number of samples increased, the sample average would tend towards the population mean. Thus we introduce the concept of mathematical expectation.

Expectation E

The *mathematical expectation* of a quantity is the value to which the average of such quantities will tend as the number of measured values increases to infinity. Thus the expectation of the sample mean $\bar{x}$, written as $E(\bar{x})$, is the population mean μ. If we were able to remove all systematic errors (see below) from the measurements, this *population mean* would be the *true value* T. Some authors define the true value in this sense. However, the assumption that all systematic errors have been removed is implausible, and the two concepts should be kept separate.

Observed value $\overset{o}{x}$

The observed value is the first numerical information you obtain about a result, for example: a length from a tape, an angle from a theodolite or a reading from a GPS receiver. In fact these numerical values are very often derived within an instrument from a series of other measurements, such as time in an EDM instrument or satellite receiver, which the user generally does not see. Thus we must settle for the first set of numerical values as the 'observations'.

The superscript o is used *over* a parameter, say x, to indicate that it is the observed value of x.

True error Δ

Like true value, the true error Δ of a single observed quantity can never be found. It is merely an abstract idea defined to be the difference between the true value and the observed value i.e.

$$\Delta = T - \overset{o}{x}$$

The best estimate $\hat{x}$

The best estimate derived from observations is that value which is more likely than any other to be the true value, judged on the evidence available. We shall accept that the best estimate which can be derived from a series of observations is the arithmetic mean of the set, provided that the observations are independent of each other and that they are of the same quality. Observations of differing quality will be treated later.

The superscript ^ is used over a parameter x to indicate that it is the best estimate. It is widely accepted in literature that using the symbol ^ to indicate best estimate implies it has been derived from the principle of Least Squares or (the same thing) minimum variance (see below).

Sample Residual v

A residual v is defined to be the difference between the best estimate and the observed value i.e.

$$v = \hat{x} - \overset{o}{x}$$

Note that in statistical literature, deviations from the mean, or residuals, are defined in the opposite sense, i.e.

$$v = \overset{o}{x} - \hat{x}$$

This makes no difference to any results, provided we are consistent. We use the definiton common to survey.

Population Residual V

In a similar manner, we define the population residual V to be the difference between the population mean and the observed value i.e.

$$V = \mu - \overset{o}{x}$$

Corrections

It should be noted that in deriving the best estimates from observed values, in no sense are these observations 'corrected'. The observations are simply used to derive best estimates. A change made to an arbitrary provisional value to give the best estimate is called an *increment* and the word 'correction' is avoided altogether.

Weight w

The weight w of an observation is a measure of its quality relative to other observations, usually expressed as a number proportional to the inverse of the variance (see later). Weights are needed to reduce mixed mode types of observations e.g. angles and sides, to a dimensionless number for combined work, and to allow for the differing quality of various instruments and observers.

Precision and accuracy

If a quantity is measured several times, the degree of agreement between the measures is the *precision* of the set. Thus if the residuals are small the observations are precise, and *vice versa*. The *accuracy* of the set is the difference between the best estimate and the true value. A high degree of precision is no indication of great accuracy, however. For example, an expensive watch may be precise to a second, but could easily be set at the wrong time.

The efforts to obtain accuracy usually involve a high degree of precision but, more importantly, they require much effort to remove systematic errors by *calibration* and mistakes by a *quality control* system. For example the watch should be compared with radio time signals at regular intervals.

Systematic and random errors

In the classification of errors, two main types are distinguished according to the way they affect a result:

(a) Errors which have a cumulative or constant effect are *systematic errors*.
(b) Errors which have a tendency to compensate one another are *random* errors, sometimes called *accidental* errors.

In our example of length measurement, if the ruler is either too long or too short then clearly it will always give an incorrect answer. This information will have been found by *testing* it against a standard. The way to solve the problem is to *calibrate* the ruler, that is compare it with a known standard of length and apply a calibration correction to all readings.

Even after this calibration has been applied, there will be small differences each time anyone uses the ruler, provided it is precisely graduated. If the chances are even that these small errors are either too large or too small their effects tend to compensate and are therefore called random errors. Notice that the calibration process itself is subject to these random errors, and the calibration figure applied to the measurements is also affected by random error.

Mistakes or blunders

A third type of error is sometimes distinguished on the basis of the mere size. This is the *gross error*, *mistake* or *blunder*. Since mistakes may be systematic or random this is not a genuine third type of error. It is however true to say that they are treated in a different manner by most surveyors simply because they may be detected by a self-checking procedure built into the particular survey process involved.

Mistakes are the most serious of all the errors. This fact can easily be forgotten in the light of the attention paid to small errors by the method of Least Squares, and to the space taken up by the latter in survey literature, including in this book. The importance of a topic is not proportional to the number of words written on the subject nor to the effort required to understand it. Quality assurance techniques are therefore of paramount importance if reliable work is to be performed.

Mistakes are very serious simply because of their size, and care must be taken to avoid them. With the advent of automated production line work, for example when observations made from total stations are recorded by data loggers which give direct output for processing, there is less likelihood of mistakes in booking, but at the same time, there are fewer opportunities to check work by common sense methods. It is therefore necessary to incorporate into computer software various statistical tests to locate blunders and to incorporate self-checking observational procedures to locate any mishaps. This means that statistical theory has to be used more than ever.

5.3 Systematic errors and calibration

Systematic errors arise from some physical phenomenon or psychological tendency on the part of the observer, and may only be eradicated if the laws governing these contributing factors are known, albeit empirically.

A troublesome source of systematic error in levelling and distance measurement is atmospheric refraction, and an example of a personal tendency is visual bias when observing a vertical angle. Systematic error is often difficult to assess and equally troublesome to remove. However, it is often practicable to reduce its effect considerably by calibration or by balancing observations. Systematic error can even be completely removed by the application of the important principle of instrument reversal such as changing face in theodolite work.

In dealing with an over-determined set of observations and parameters by the method of Least Squares, efforts can be made to locate systematic errors by incorporating a suitable mathematical model, such as a scale factor in an EDM instrument. In this case, the systematic effect becomes another unknown parameter. Sometimes, but not always, this error can then be determined and eliminated.

Processes most likely to be significantly affected are those involving lengthy repetitive techniques, such as levelling and traversing, where a chain of small errors can accumulate.

In some cases, although the magnitude of a systematic error may be assessed, its sign may be unknown. For example, the standardization of a tape is itself subject to error. If the length of the tape is quoted as 100.025 m, all field measurements will be corrected by the 0.025/100 × length measured. The standardization itself could be in error by, say, ± 0.0005 i.e. either plus or minus. Whatever its sign it will affect the measurements in a systematic manner. Generally, this sign is unknown. If ± 0.0005 is the *standard error* of the calibration measurements, it is a *standard systematic error* when affecting the field measurement. Of course, the standard error of the calibration could be worse than 0.0005. This figure is taken only if no other information is available.

The propagation of systematic errors

As an example of the propagation of systematic errors, consider a line of n tape lengths each with a systematic error $e_1, e_2, \ldots, e_n$. Then the resultant systematic error of the whole base is given by

$$E_S = \sum e_i$$

If these values are standard systematic errors, they are either *all positive* or *all negative*.

If the e_i are all numerically equal to σ, then we have

$$E_S = n\sigma$$

It will be seen later that this contrasts with the propagation law for random errors which gives $\sigma\sqrt{n}$ for n equal standard errors σ.

The proportional error for the base as a whole does not improve with its length but remains σ / L where L is the tape length. Again this contrasts with the effect of accidental error, where the proportional error of the base improves with distance measured.

5.4 Random or accidental errors

The theory of random errors is now considered at length. However, attention is redrawn to the importance of mistakes and systematic error, neither of which conform to the same laws. They must be studied in each particular case, and the treatment of discrepancies applied accordingly. In theory, random errors remain after mistakes and systematic errors have been removed. In practice, some systematic errors will remain and, for want of an alternative, are treated with random errors.

The concepts of error theory are quite simple; it is the notation and algebra necessary to develop these ideas which can initially be confusing. Although it is quite possible to develop these ideas without employing matrix algebra, it is very inconvenient not to use this very effective tool of mathematics. The reader is therefore encouraged to study the elements of matrix algebra before attempting to master the subject matter of this chapter.

However, we will develop many basic concepts through the medium of simple examples, both to assist the beginner and to define terms. A more direct treatment is given in Appendix 3.

5.5 Statistical analysis of results

The study of random errors is concerned with: probability, presentation of results as histograms and probability density functions, calculation of sample statistics and population parameters, and the detection of outliers which do not seem to accord with the theoretical distributions expected. Typical problems to be handled are:

(a) When should we reject what appears to be a bad observation or, more likely, what rejection criteria do we write into our computer software?
(b) How can we tell if an instrument is performing to the manufacturer's specification?
(c) How can we assess the reliability and precision of work computed by the method of Least Squares, so that we may inform the client of this and if necessary defend our results in a court of law?

Probability

Most of our findings are based on probabilities, and very few results are certain. Because we shall be dealing with levels of statistical significance and probabilities which are themselves subject to error, it is essential to have a good grasp of some fundamental concepts. Thus we shall deal with the idea of probability in basic terms.

If there are two balls in a hat, one white and one red, the probability that we shall draw out a red ball is one in two, or 1/2. This does not mean that we shall draw out a red ball in one of two chances. We might pick a white ball in the first six attempts. It means that if we make a very large number of attempts, say 1000, 50 % of the time we shall draw out a red ball, and 50 % a white ball. Notice that the probability of drawing any colour of ball is a certainty of 100 %, and that the sum of the probability of drawing a red plus the probability of drawing a white is also 100 %.

If there are two white and ten red balls in the hat, the probability of drawing out a white ball is two in twelve or 2/12, and that of drawing out a red ball is ten in twelve, where it is again understood that a very large number of chances are involved. Hence we obtain the rule

$$P(x) = \frac{\text{the number of ways of obtaining a result}}{\text{the total number of possible results}}$$

In this way, the probability of the occurrence of an error is related to the number of times this particular error can occur and the total possible number of all errors that could occur. This involves analysing and counting errors of various magnitudes. For such an analysis to be valid, a very large number of errors should be *sampled.* A sample of more than thirty observations can often be taken as large when it comes to statistical testing.

However, it is seldom practicable to make a very large number of observations, nor is it necessary. Usually we make samples of ten or so unless the measurement process is highly automatic.

The theory below will therefore deal mainly with small samples and their *statistics,* the mean and the standard deviation, and estimates from these samples of the *population parameters* which they represent.

Histograms

The histogram is a diagrammatic way of displaying sets of data which have a statistical or *stochastic* distribution about a central value.

This method of analysing results will be explained in terms of the game of 'shove halfpenny'. This traditional game, once played in English country pubs with a small coin, is a useful way to obtain statistical data. Figure 5.1 shows a smooth board on which a number of parallel lines are ruled at a distance of just over an inch i.e. slightly larger than the diameter of the coin. The player places his coin at a position just protruding over the edge of the board. The aim is to strike the coin with the wrist, getting it to land as close to the middle as possible. The centre of the coin is taken as the reference point.

The figure shows the results of two players A and B, each of whom took 157 shots. It is obvious that A was a much better player than B, because his observations are clustered closer to the mean, or we say they show less *dispersion* than Player B's.

The bands between the lines are called the *class intervals* of the analysis. Their width has to be carefully chosen with respect to the skill of the players if a difference between them is to be detected. For example, if the class interval had been the complete depth of the board no difference in the respective skills would have been found. Again, if the intervals had been very small, an abnormally large number of shots would have to be played to land more than one shot in each space. The selection of class intervals always needs care and experiment.

The number of times that a shot lands in each space, or class, is plotted on a block diagram called a *histogram* (see Figure 5.2). The histogram for A is bounded by the solid line and that for B by the broken line. The area of each histogram is the same since it represents the same number of shots. For a fair comparison to be made this should always be the case.

It is possible to analyse observational data in the manner of the game. A decision has to be made about the size of the class interval, selecting about one fifth of the

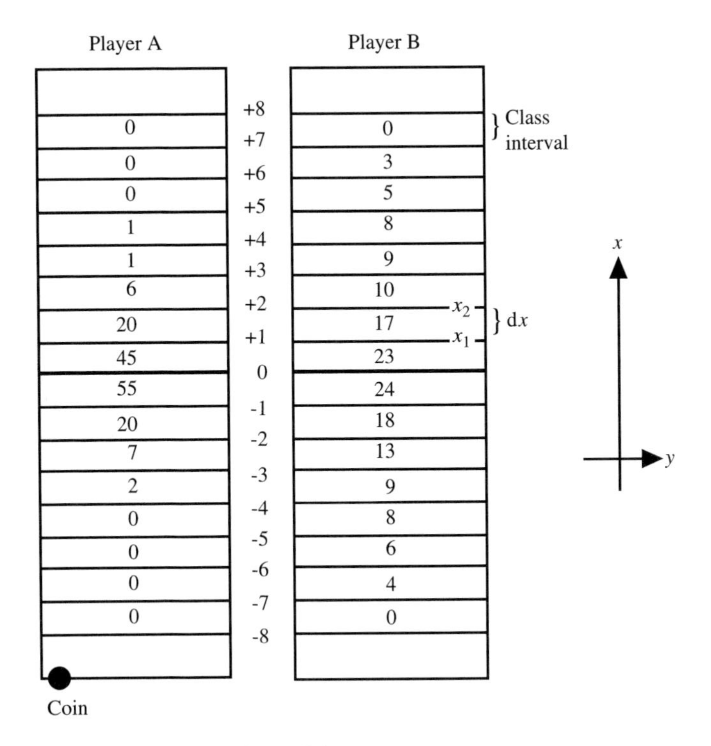

Figure 5.1

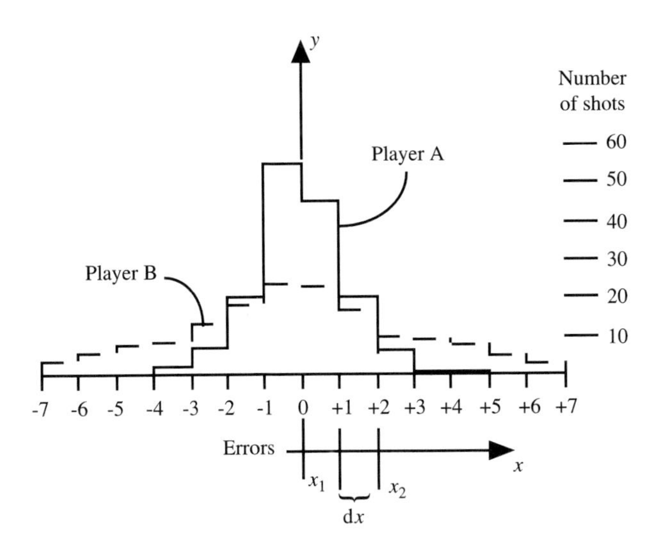

Figure 5.2

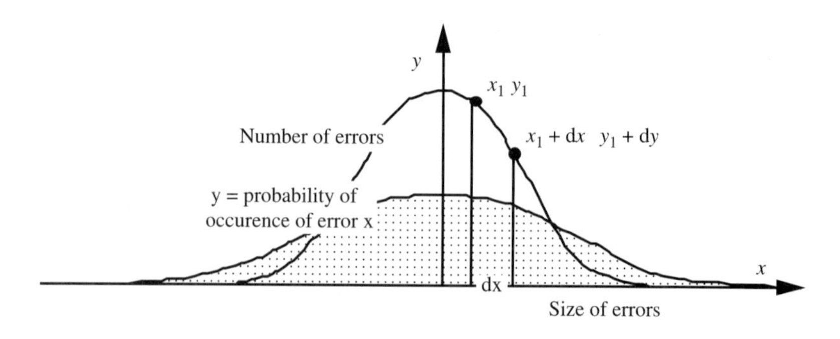

Figure 5.3

range of measurements for this. If the interval proves either too large or too small, it has to be reselected. For example, the data of Figure 5.3 could have been the results of measurement with two different EDM instruments A and B.

Mathematical treatment

The statistical information may be expressed mathematically:

(a) An error is denoted by x_i and the class interval by dx. Hence $x_2 = x_1 + \mathrm{d}x$ and so on.
(b) The height of the column of the histogram can be expressed as y_i to correspond to x_i.

Thus we convert the numbers in each class interval into probabilities by dividing each by the total number of observations made e.g. 55/157 = 0.35.

Notice that the probability of any result at all is 157/157 = 1 i.e. a certainty. The graph of all these probabilities is called a probability density function graph, which usually has a bell shape as shown in Figure 5.3. The equations of these bell-shaped curves are considered later.

Random error and central tendency

Continuing with the shove halfpenny example, if the number of shots is increased to a very great number i.e. towards infinity, while at the same time the class interval is decreased, there is a continuous curve towards which the histograms will tend.

The respective curves representing A and B's performances are given in Figure 5.3. The area contained between each curve and the x-axis represents the total number of all possible errors in the system, an infinite number in theory. This is true for both curves: neither curve will meet the x-axis but approach it asymptotically. For any point on the curve, the ordinate y can be considered as either the number of errors of abscissa x that occur, or the probability of the occurrence of an error of magnitude x. Hence the area under the curve represents either (a) the total number of errors n or (b) the probability of all errors occurring with a probability of 1 or 100 %.

From the shape of the curve, which shows *central tendency* about the mean, the distribution of random errors is described as follows:

(a) Small errors are more frequent than large ones.
(b) Positive and negative errors are equally likely to occur.
(c) Very large errors seldom occur.

These statements are reasonable if systematic errors have been removed and the observer, or player of the game, has a good degree of skill.

It can be established theoretically by the *central limit theorem* that the combination of four or more error distributions approximates to the normal. Since most survey 'observations' are the result of at least four error sources, the normal distribution is a reasonable statistical model to accept.

Sample deviation curve

More usually than in the case say of triangle closures, where we are dealing with true errors, we obtain a set of observations distributed about a mean value. We can plot a histogram of deviations (residuals) about this mean value in just the same way, and also show their distribution on a smooth curve, which illustrates the degree of dispersion of the observations.

In this case the abscissae x_i become the residuals v_i, and the ordinates y_i the number of such residuals in each class.

Such a curve is defined by its mean $\bar{x}$ and general shape, which shows the *dispersion* about the mean. A good indicator of this dispersion is the *variance* s^2. This is defined to be the average of the squares of the residuals about the mean. For n observations, we have

$$s^2 = \frac{\sum v_i^2}{n} = \frac{v_1^2 + v_2^2 + \ldots + v_n^2}{n}$$

or, in matrix terminology,

$$s^2 = \frac{\mathbf{v}^{\mathrm{T}}\mathbf{v}}{n}$$

where

$$\mathbf{v}^{\mathrm{T}}\mathbf{v} = \begin{bmatrix} v_1 & v_2 & \ldots & v_n \end{bmatrix} \begin{bmatrix} v_1 \\ v_2 \\ \vdots \\ v_n \end{bmatrix}$$

It is stressed that the definition of variance implies nothing at all about the nature of the distribution of the observations. Variance is a convenient statistic which gives a measure of the spread or dispersion of results about a mean value. Provided there is more than one observation, a variance is calculable.

Example An angle was observed nine times obtaining the results tabulated in column one of Table 5.1. The average is calculated as

$$\bar{x} = \frac{\sum x_i}{n} = 109^\circ 25' 07.7''$$

Subtracting each observed value $\overset{o}{x}$ from $\bar{x}$ gives the positive and negative residuals v_i in columns two and three of Table 5.1. The sum of the residuals should be zero or approximately zero due to rounding error. The sum of squares of these residuals calculated from column four is found to be 30.64. The variance s^2, defined to be the average of the squares of these residuals, is therefore given by 30.64/9. The standard deviation s is therefore $\sqrt{3.4} = 1.8$.

This *sample statistic* can easily be calculated using a direct function key on a hand-held calculator. However, it is wise to check which formula is being used by the calculator. As we shall see later another similar formula with a denominator of $n-1$ is used to calculate an unbiased estimate of the *population variance.*

We shall return to this example later and calculate the result a different way to illustrate the general method of observation equations used in the Least Squares process.

5.6 Statistical models

We now give a formal treatment which will incorporate all the above ideas and summarize them in mathematical form. We begin with the Normal or Gaussian distribution, most generally applied to randomly observed data. This distribution, which applies to a population, will be treated in rather more detail than the others

Table 5.1

Observed	$+v$	$-v$	v^2
6.3	1.4		1.96
7.2	0.5		0.25
10.4		–2.7	7.29
4.3	3.4		11.56
9.6		–1.9	3.61
5.8	1.9		3.61
7.9		–0.2	0.04
8.3		–0.6	0.36
9.1		–1.4	1.96
Sum = 68.9;	+7.2	–6.8	30.64
$\mu = 7.7$			
	remainder = 0.4		$s^2 = 30.64/9$
			$\Rightarrow s = 1.8$

which follow, and which apply to samples: namely the Student t (see Secton 5.10), the Fisher ratio, and the χ^2 (Chi squared) distributions.

The normal distribution function

Although many of our results and techniques—such as calculation of variance—have nothing to do with the Normal distribution, this distribution is widely used to represent the kind of error dispersion met with in practice. This is because most errors result from the combination of others, which are centrally distributed. It can be shown by the Central Limit theorem that the combination of only a few such errors leads to a set of normally distributed errors. The bell-shaped curve, which represents the frequency of residuals about a mean, derived from randomly observed variables, may be expressed by the *probability density function* or PDF given by the formula

$$\phi(x) = y = \frac{h}{\sqrt{\pi}} \exp(-h^2 x^2) \tag{5.1}$$

To accord with previous definitions and notation, the variable x in this formula is equivalent to the residual v defined by

$$x = v = \mu - \overset{\mathrm{o}}{x}$$

Such statistical (or stochastic) models are derived from probability theory in advanced statistical textbooks such as *Advanced Theory of Statistics* (Kendal and Stuart 1967) to which the reader may refer. However, to give understanding of this most important distribution, we will trace this function and show that a bell shaped curve results.

$$\text{When } x = 0, \quad y = \frac{h}{\sqrt{\pi}}$$

giving the maximum height of the curve. A glance at Figures 5.2 and 5.3 will show that the less dispersed observations made by observer A have a greater height than those of B. The quantity h is therefore known as the *index of precision*.
Differentiating, we obtain

$$\frac{\mathrm{d}y}{\mathrm{d}x} = -2h^2 xy = 0$$

when $y = 0$ and $x = \pm\infty$.

Therefore

$$\frac{\mathrm{d}^2 y}{\mathrm{d}x^2} = -2h^2 y(1 - 2h^2 x^2) = 0$$

when $x = \pm\infty$, or at the points of inflexion, when

$$x = \pm\frac{1}{h\sqrt{2}}$$

A rough sketch of the curve can now be made, which clearly follows the general pattern shown in Figure 5.4. We will now show that the index of precision h is related to the standard error σ by the expression

$$\sigma = \pm \frac{1}{h\sqrt{2}}$$

Since $x = v$, the variance is defined as the *expectation* of v^2 or

$$E(v^2) = \sigma^2$$

Thus the population variance

$$\sigma^2 = \frac{\text{sum of squares of residuals}}{\text{total number of residuals}}$$

or in mathematical terms

$$\sigma^2 = \frac{\int_{-\infty}^{+\infty} x^2 y \, \mathrm{d}x}{\int_{-\infty}^{+\infty} y \, \mathrm{d}x}$$

(Note that there are y values of each residual x.)

But the value of the denominator is the probability of all residuals being selected, namely a certainty, therefore

$$\int_{-\infty}^{+\infty} y \, \mathrm{d}x = 1$$

However,

$$\frac{\mathrm{d}^2 y}{\mathrm{d}x^2} = -2h^2 y(1 - 2h^2 x^2)$$

and therefore

$$\int_{-\infty}^{+\infty} \frac{\mathrm{d}^2 y}{\mathrm{d}x^2} \mathrm{d}x = 4h^4 \int_{-\infty}^{+\infty} x^2 y \, \mathrm{d}x - 2 \int_{-\infty}^{+\infty} h^2 y \, \mathrm{d}x$$

implying

$$0 = 4h^4 \sigma^2 - 2h^2$$

hence the result

$$\sigma = \pm \frac{1}{h\sqrt{2}}$$

We may now express the normal probability distribution function $\phi(x)$ in terms of the standard error σ, the population mean μ and the observed parameters $\overset{o}{x}$ as follows

$$\phi(x) = y = \frac{h}{\sqrt{\pi}} \exp(-h^2 x^2) = \frac{1}{\sigma\sqrt{2\pi}} \exp\left(-\frac{v^2}{2\sigma^2} \right) = \frac{1}{\sigma\sqrt{2\pi}} \exp\left(\frac{-(\mu - \overset{o}{x})^2}{2\sigma^2} \right) \qquad (5.2)$$

This function is completely defined if the population mean μ and the standard error σ are known. However it is convenient to standardize this formula by dividing x and y by σ and moving the origin to the mean μ. The new simpler equation is then of the form

$$\phi(t) = \frac{1}{\sqrt{2\pi}} \exp\left(-\frac{t^2}{2}\right) \tag{5.3}$$

where $t = v / \sigma = x / \sigma$. Hence the variance of t is 1. This reduction to a standard form has two distinct advantages:

(a) It enables observed quantities of all different kinds to be treated in a general way, reduced to a common weight.
(b) It allows standard sets of tables to be used for all problems.

For example, a residual of 15" (seconds of arc) belonging to a distribution whose standard error is 5" is converted to a dimensionless number t = 15"/5" = 3, which is conveniently tabulated. A residual of 15 mm and standard error of 5 mm is identically tabulated as 3. A special case is a residual equal to the standard error giving a ratio of 1.

Although the population standard errors σ can be used to bring about this reduction to a common dimensionless standard, in practice we must allow for errors in estimating the standard errors themselves, and if necessary reconsider our estimates in the light of more data becoming available.

5.7 Statistical tests

In statistical literature the standardized normal distribution function is written as N (0, 1), meaning ‘a normal distribution with zero mean and variance 1’. The distribution in its original general form of Equation (5.2) would be written

$$N\ (\mu, \sigma^2)$$

This and other probability distribution functions are used to help decide if we have made a mistake, if a systematic error has gone undetected or to give statistical limits within which to expect a result to lie.

Any decision-making is based on the probability of obtaining a result within selected bounds or limits. The area of the figure bounded by the curve and two ordinates, with selected abscissae $x = a$ and $x = b$, or in the standardized table $t = t_a$ and $t = t_b$, gives the probability that a value will lie in this region. This information is represented by the equation

$$P(a < x < b) = \int_a^b y\,\mathrm{d}x = \int_a^b \phi(x)\,\mathrm{d}x = \int_{a/\sigma}^{b/\sigma} \phi(t)\,\sigma\,\mathrm{d}x$$

or

$$P(t_a < t < t_b) = \frac{1}{\sqrt{2\pi}} \int_{t_a}^{t_b} \exp\left(-\frac{t^2}{2}\right) \mathrm{d}t \tag{5.4}$$

These probabilities will always be < 1 (or 100 %).

Equation (5.4) is used in two ways:

(a) From given values of a and b, we have to find $P(x)$. For example, the probability that a value will lie within selected limits, such as $\pm\sigma$ about the mean.

(b) Or, we are given $P(x)$ and have to find a and b. For example we select a probability such as 95 % as acceptable for the retention of an observed value and thus find the range of acceptable limits within which to accept or reject values.

Although it is quite practicable to evaluate the integrals by computer, it is traditional to tabulate the area under the curve, the *cumulative probability function,* from minus infinity to successive values of x, and to interpolate subsequently within these tables. A required integral is obtained as the difference of two as follows:

Figure 5.4 shows $P(a < x < b)$ in graphical terms, where

$$\text{area C} = \text{area B} - \text{area A}$$

or

$$P(a < x < b) = \int_{a}^{b} y\,\mathrm{d}x = \int_{-\infty}^{b} y\,\mathrm{d}x - \int_{-\infty}^{b} y\,\mathrm{d}x$$

$$P(t_a < t < t_b) = \frac{1}{\sqrt{2\pi}} \int_{-\infty}^{t_b} \exp\left(-\frac{t^2}{2}\right)\mathrm{d}t - \frac{1}{\sqrt{2\pi}} \int_{-\infty}^{t_a} \exp\left({}^{2}-\frac{t}{2}\right)\mathrm{d}t$$

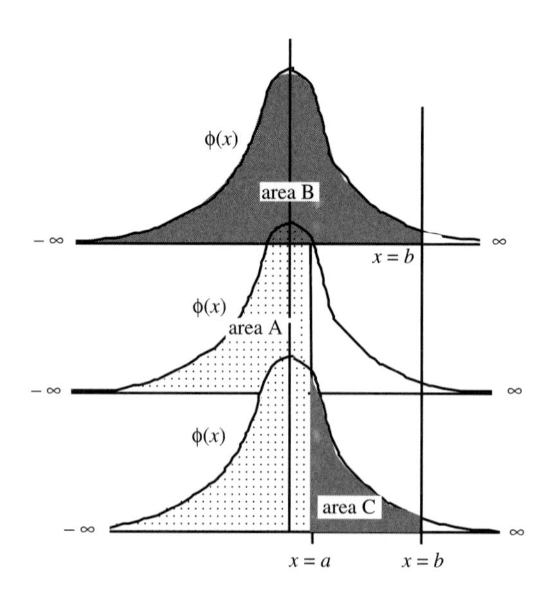

Figure 5.4

Table 5.2 *Table of cumulative normal probability*

t	*0*	*1*	*2*	*3*	*4*	*5*	*6*	*7*	*8*	*9*
0.0	.500	.504	.508	.512	.516	.520	.524	.528	.532	.536
0.6								.750		
1.0	.841	.844	.846	.848	.851	.853	.855	.858	.860	.862
1.6					.950					
1.7						.960				
1.9							.975			
2.0	.977	.978	.978	.979	.979	.980	.981	.981	.981	.982
3.0	.999	.999	.999	.999	.999	.999	.999	.999	.999	.999

For numerical work, these areas or integrals are tabulated in a standardized cumulative probability table as shown in Table 5.2. The specific entries of 0.750, 0.950, and 0.975 have been added for use in the examples which follow.

It will be noticed that when $x = 0$ the table gives half the total area covered by the integral from $-\infty$ to $+\infty$, or a cumulative probability of 0.5 (or 50%). When x is less than 0 the required cumulative probability is 1 – the tabular entry.

Thus, if $x = -2.05$, the cumulative probability from $-\infty$ to -2.05 is $1-0.980 = 0.02$ (or 2 %). This is shown by the dark area in Figure 5.5(a).

We interpret this result in this way: the probability of x being *less* than -2.05 is 2 %. Likewise, we can also say that the probability of x being *greater* than $+2.05$ is also 2 %, because the function is symmetric. We write these ideas down more formally as

$$P(x < -2.05) = 2\% \quad \text{or} \quad P(x > 2.05) = 2\%$$

One-sided statements or tests are called *one-tailed tests.* They arise in dealing with dimensional problems. For example, the length of a competition swimming pool must never be shorter than a given length if swimming records are to be recognized. If the pool is too *long* however, it is a disadvantage to the competitors but does not invalidate a record time.

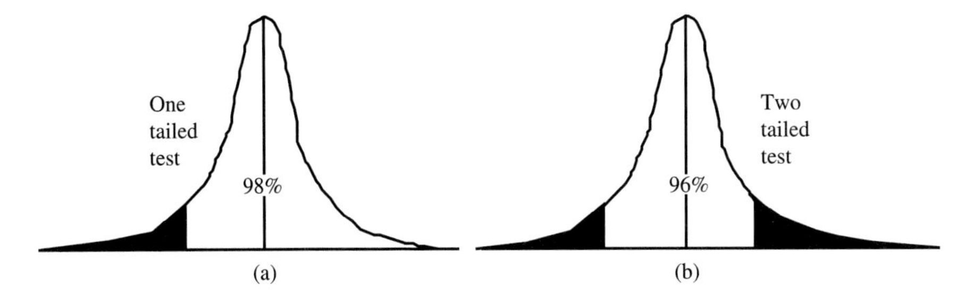

Figure 5.5

On the other hand we can say that the probability of x being *both* less than –2.05 *and* greater than + 2.05 is 4 %. This is a *two-tailed test*, as shown by the two dark areas in Figure 5.5(b).

Another way of expressing this last result is to say that there is a 96 % probability of x falling within in the range –2.05 to +2.05, which is written formally as

$$P(-2.05 < x < 2.05) = 96\,\%$$

This two-tailed test is common when a dimension, such as steelwork, has to be neither longer nor shorter than a given specification, to within acceptable statistical limits.

Critical tables

Another way to express these probabilities is by critical values only. These critical tables are much shorter and often give sufficient information by themselves. Table 5.3 gives similar information to Table 5.2 for selected critical values only. The most common values to be used are the 5 % or 2.5 % cumulative probabilities.

The argument $1 - f(x)$ of the second row is used in one-tailed tests: for example, to answer such questions as: 'What is the value of x such that there is only a 5 % probability that it will be *exceeded*?' Look up the cumulative probability table (Table 5.2), choose the entry at 0.950 and note that its argument is 1.64. Alternatively, from the critical percentage table (Table 5.3), select the 5.0 % entry from row two for a one-tailed test, and read off the critical percentage point $x = 1.645$.

A traditional percentage point in a two-tailed test is 50 % i.e. tabular entry of 75 %, giving $x = 0.6745$. This relates a 'probable' or 50 % error e to the standard error σ by $e = 0.6745\ \sigma$. The 50 % probability is of value when considering the rounding errors in arithmetic processes.

Example A typical example of the application of some of these ideas and tables is as follows. An angle of a primary triangulation is 63"47'12.6". During a third order extension, it is re-measured with a standard error of 5", obtaining a value of 63°47'22". Is this value significantly different from the original?

There are two populations to consider.

(a) The primary angle with its given population mean, say μ_0.
(b) The observed third order population, say with unknown mean μ, from which the known sample mean is drawn. The standard error σ of this population is known or it can be estimated from the sample itself.

Table 5.3 *Critical percentage points for normal probability*

1	*f(x)*	*95.0*	*97.5*	*99.0*	*99.5*	*99.9*	*99.95*
2	$1-f(x)$	5.0	2.5	1.0	0.5	0.1	0.05
3	$2(1-f(x))$	10.0	5.0	2.0	1.0	0.2	0.1
4	x	1.6459	1.960	2.3263	2.5758	3.0902	3.2905

We must now select a *level of significance* for the test.

A level of 5 % is considered *significant.*
A 2.5 % level is considered *very significant.*
A 1 % level is considered *highly significant.*

It is common practice to deal with the 5 % significance level or the 95 % *confidence level*. We must also decide on the nature of the test: one-tailed or two? The question posed is 'Is this value significantly different from the original?' In other words we wish to know whether the two means are likely to be equal.

Because we are not asking whether one is bigger or smaller, we will apply a two-tailed test with confidence limits of ±1.96 derived from Table 5.3, selecting 5 % from row three and the limit of 1.96 from row four. The test statistic is

$$\frac{\hat{x}-\mu_0}{\sigma}=\frac{9.4}{5}=1.88$$

which falls inside the rejection limit of 1.96. We therefore accept that this sample could belong to a population which has the same mean as the primary population.

Scaling back to the dimensioned data gives the limits for the mean of

$$\mu=\hat{x}\pm(5\times1.96)=\hat{x}\pm9.8$$

However, because the test statistic is only just within the acceptance criterion we are concerned that we may have made a mistake in accepting the result of the test, and that this sample mean really belongs to another population with a different mean, say m_1. If we know what this other mean is we can make a similar, alternative, test on it.

If we assume that the alternative population mean is 5 $\times$ σ greater than the actual value i.e. that it is 63°47'37.6", the test statistic is now

$$\frac{\bar{x}-\mu_1}{\sigma}=\frac{15.6}{5}=3.1$$

which is well outside the 1.96 confidence limit and would be rejected from the alternative hypothesis.

Type I and Type II errors

Remembering that acceptance or rejection depends upon arbitrary probability limits, it is possible that a decision is wrong. Suppose that in the first test, the test statistic

$$t=\frac{\bar{x}-\mu_0}{\sigma}$$

had been 2.0 i.e. > 1.96; we would have rejected the value of x. But there is a 2.5 % probability that it really does belong to this population. Thus in rejecting it when it was actually correct, we would have made what is called a Type I error with a known probability of 2.5 %. Conversely, if we accept a value that actually belongs to another population, we have made a Type II error.

The probability of making this error depends on the value of the alternative population mean chosen for the alternative hypothesis. Suppose the two means are separated by an amount given by

$$\Delta = \mu_1 - \mu_0 = (t + t_1)\sigma$$

The probabilities associated with t from the first mean and t_1 from the second are α and β. In Figure 5.6, the probability of making a Type II error is β, which is equivalent to the shaded area to the left of the line marked A.

The alternative hypothesis is often selected, as in our example, to differ from the sample mean by five times the standard error of the sample i.e.

$$\mu_1 = \mu_0 + \Delta = \mu_0 + 5\sigma$$

The line marked A in Figure 5.6 is selected by the choice of 0.5 α in a two-tailed test on the null hypothesis. Thus the probability of accepting a value to the left of A i.e. outside the limit of t_1 that should rightly belong to the alternative sample is known to be β. It will be seen that the values of (α, t), (β, t_1) and Δ are linked together. Given any two of these groups, is possible to calculate the remainder.

In accepting values within the range 0 to t we could be wrong by β. Thus we can say that we are β % sure that no value greater than 5 σ from the mean μ_0 has been accepted.

Identical tests can be applied to a residual, scaled by its standard error, to decide whether to accept or reject the observation from which it was derived, or to assess the reliability of an observation against an alternative hypothesis. The variance of a residual is obtained from the dispersion matrix $\mathbf{D}_v$, calculated as part of the least squares process (see Appendix 3).

5.8 Formal statement of ideas

Most of the concepts and ideas used in the above arguments have been formalized into statistical notation which at first sight is rather formidable.

Refer to Figure 5.6 in the following discussion. The test statistic t is first standardized as described above, to enable us to use normal tables. The outcome of the test is called a *null hypothesis,* written H_0. In this case the null hypothesis queries

$$\text{does } \mu = \mu_0?$$

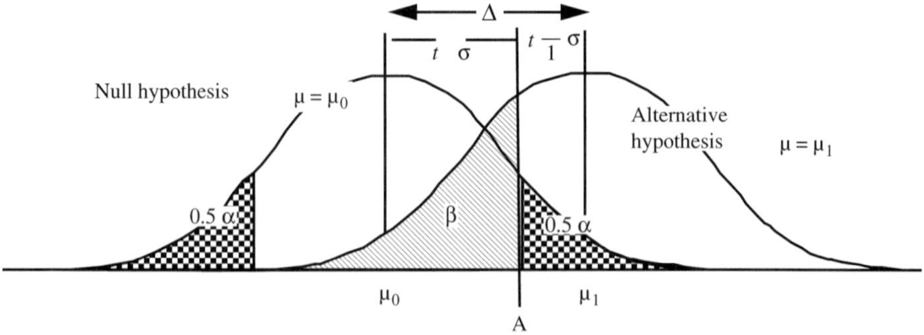

Figure 5.6

The *alternative hypothesis*, referred to as H_1 in this case, is

$$\mu \neq \mu_0?$$

The significance level, such as 5 %, is denoted by α in one-tailed tests and by 0.5 α in two-tailed tests. The corresponding confidence limits are denoted by t_α in one-tailed tests and by $t_{0.5\alpha}$ in two-tailed tests. In this case, $t_{0.5\alpha} = 1.96$.

All this information is brought together into one statement:

$$H_0\text{: } \mu = \mu_0\text{; } H_1\text{: } \mu \neq \mu_0$$

$$\text{accept } H_0 \text{ if } -t_{0.5\alpha} < t < t_{0.5\alpha}$$

$$\text{because } P(-t_{0.5\alpha} < t < t_{0.5\alpha}) = 1 - \alpha$$

$$\text{where in this case, } t = \frac{\breve{x} - \mu_0}{\sigma} \text{ and } 1 - \alpha = 95\ \%$$

This is the general form used to express most statistical tests although the distributions and tests statistics vary according to the nature of the problem.

Degrees of freedom

In a mathematical problem, the essential amount of information required for a solution is called the *necessary information.* For example, two distinct equations are needed to solve for two unknowns. If more than sufficient information is available, say we have five distinct equations in these two variables, we have a *degree of freedom* to select the two necessary equations.

If we have m equations in n unknowns ($m > n$) there are $m - n$ degrees of freedom in the problem.

To obtain a best estimate of one observed quantity, we obviously need at least one observation. With only one observation there is no redundant information to yield residuals or statistical analysis. Therefore the necessary number of observations is one. Hence if there is more than one observation, say n, there will be $n - 1$ degrees of freedom. In statistics, the number of degrees of freedom in a problem enhances the validity of predictions and calculations. It is usually denoted by the greek letter ν.

5.9 Unbiased estimators of σ

We have already calculated a sample variance s^2 from nine observations. The question now arises whether this sample statistic can be used to estimate the population parameter σ. Is, for example, the expectation, or average of more and more sample variances likely to be σ^2? The straight answer is no, because each sample is biased by its mean. However, as will be shown later, the statistic

$$\frac{n}{n-1}s^2$$

does tend to equal the population variance σ^2. Or, put another way, we say the expectation, written

$$\text{Exp}\left[\frac{n}{n-1}s^2\right]$$

is σ^2. It is common to use the symbol $\hat{\sigma}^2$ for this expected value.

In statistical language, we say that $\hat{\sigma}^2$ is an *unbiased estimator* of the population variance σ^2. Substituting for s^2 leads to the well-known formula

$$\hat{\sigma}^2 = \frac{n}{n-1}s^2 = \frac{\sum v^2}{n-1} \tag{5.5}$$

If the population variance is unknown, we can use this formula to estimate it. Hence, in the case of 9 samples from a calculated sample variance of 3.4, we obtain

$$\hat{\sigma}^2 = \frac{9}{8} \times 3.4 = 3.825 \quad \text{and} \quad \hat{\sigma} = 1.96$$

5.10 Pooled variance

If we have several samples and their respective variances, such as from rounds of angles observed under similar conditions at several stations, is it possible to pool this information to provide a better estimate of $\hat{\sigma}^2$? The answer is a qualified 'yes'.

The formula for the *pooled variance* from two samples of n_1 and n_2 observations, with variances s_1^2 and s_2^2 respectively, illustrates the general form to be adopted for any number of samples:

$$\hat{\sigma}^2 = \frac{n_1 s_1^2 + n_2 s_2^2}{n_1 + n_2 - 2} \tag{5.6}$$

Before using this formula for the pooled variance, we should test that each of the samples could have belonged to populations with equal variances. The Fisher F test on the variances is applied. Figure 5.7 shows the approximate shape of the F statistic for degrees of freedom greater than three. The actual shape varies for each degree of freedom. It is also tabulated in Table 5.4.

Suppose we calculated the sample variance of 9 observations to be 3.4, and that we have a similar set of results from another station giving a variance of 5.6 from 6 observations. To test whether it is legitimate to derive a pooled variance to estimate the population variance, we test the ratio of the larger variance to the smaller. The test statistic is therefore:

$$t = \frac{s_2^2}{s_1^2} = \frac{5.6}{3.4} = 1.64$$

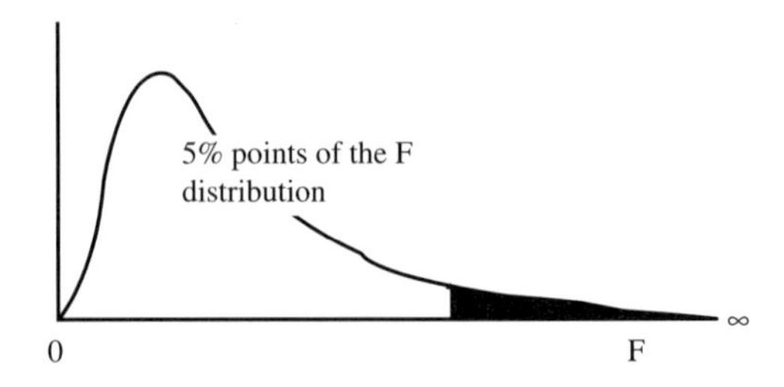

Figure 5.7

Table 5.4 *5% points of Fisher Table*

	$n_1 - 1$			
$n_2 - 1$	5	6	7	8
7	3.97	3.87	3.79	
8	3.69	3.58	3.50	
9	3.48	3.37	3.29	
10	3.33	3.22	3.14	3.07
...	...	...	...	...
∞	2.21	2.10	2.01	1.94

The respective degrees of freedom are 5 and 8, giving a rejection limit of 3.69 from the Fisher Table (Table 5.4) at the 5 % confidence level. Since 1.64 is well within the rejection limit of 3.69, we can accept these samples and pool them, giving the pooled unbiased estimate of the population variance

$$\hat{\sigma}^2 = \frac{(6 \times 5.6) + (9 \times 3.4)}{9 + 6 - 2} = 4.9$$

Testing of sample values

When dealing with less than thirty samples, as is generally the case with surveying, the normal distribution is not strictly applicable.

Each value in a sample affects its mean and therefore the residuals derived from it. At least one value is needed, but any subsequent values are not essential. The larger the sample, the better we are able to estimate the variance derived from it.

Instead of the normal distribution, we use a separate distribution for each sample size. This distribution is called the Student or '*t*' distribution. Although cumulative probability tables could be used, there would need to be one for each sample size. Instead, it is perfectly adequate, and a lot simpler, to use critical percentage point tables, such as Table 5.3. A sample of these points is given in Table 5.5. It will be noted that when the sample size is very large, the Student distribution is the same as the normal distribution and gives identical results for the critical values.

This table is used in the same way as for the normal distribution, entering the row with the appropriate *degree of freedom*, or $n = \nu - 1$, where n is the size of the sample.

Using the data from the earlier example, in which the mean is

$$\bar{x} = 7.7$$

and the sample variance is

$$s^2 = 3.4 \Rightarrow s = 1.84$$

we can check the largest residual (3.4) for acceptability at the 5 % level in a two-tailed test. The test statistic is

$$\frac{3.4}{1.84} = 1.85$$

Table 5.5 *Critical points in the Student's 't' distribution*

	Probability	
two-tail test	*10 %*	*5 %*
one-tail test	*5 %*	*2.5 %*
1	6.31	12.7
9	1.83	2.26
10	1.81	2.23
20	1.72	2.09
30	1.70	2.04
∞	1.64	1.96

The degrees of freedom are

$$\nu = n - 1 = 9$$

This gives a rejection limit of 2.26 from Table 5.5. Clearly this residual is quite acceptable. A residual of $2.26 \times 1.84 = 4.2$ would be on the rejection limit.

5.11 Sample variance: χ^2 distribution

Just as we wish to know the quality of a value extracted from a sample, we also wish to know the quality of the sample variance estimate. There is no point in quoting a standard error to a second of arc if it is uncertain to five seconds. To reach a conclusion about the quality of an estimated variance, we use the probability distribution of the standardized residuals i.e. the distribution of the function

$$\sum\left(\frac{v}{\sigma}\right)^2$$

One example of this distribution is illustrated in Figure 5.8, which, being a function of squares, has no negative values.

If the statistic v / σ has a normal distribution about a zero mean i.e. is N(0,1), then the statistic $\Sigma(v / \sigma)^2$ has a χ^2 (chi-squared) distribution, which takes the approximate shape shown in Figure 5.8 for samples with more than three degrees of freedom. Recalling that the sample variance is defined as

$$s^2 = \frac{\sum v^2}{n}$$

it follows that

$$\frac{ns^2}{\sigma^2}$$

has a χ^2 distribution. This χ^2 probability distribution varies for each value of n, and is tabulated for each degree of freedom ν in the problem. Since we use the sample

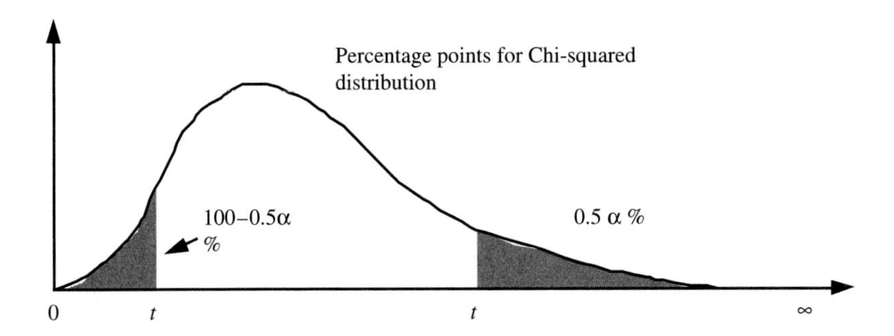

Figure 5.8

mean to calculate the residuals v, there are only $n - 1$ degrees of freedom in the simple problem. It is usual to tabulate only the critical percentage points for confidence limits set by a 95 % probability limit for one or two-tailed tests, as in Table 5.6.

Suppose we wish to know the confidence limits for σ for the example of Section 5.1, where

$$s^2 = 3.4; \quad n = 9; \quad \nu = 8$$

Applying a two-tailed test at 97.5 % and 2.5 % probabilities, confidence limits of 2.18 and 17.53 can be extracted from Table 5.6, equivalent to limits of 3.74 and 1.32 for the standard error.

This can be seen from:

$$2.18 < \frac{ns^2}{\sigma^2} < 17.58$$

$$2.18 < \frac{9 \times 3.4}{\sigma^2} < 17.58$$

$$\frac{9 \times 3.4}{2.18} > \sigma^2 > \frac{9 \times 3.4}{17.58}$$

$$14.0 > \sigma^2 > 1.74$$

$$3.74 > \sigma > 1.32$$

Table 5.6 χ^2 *critical percentage points*

	Probability			
n – 1	*97.5 %*	*95 %*	*5 %*	*2.5 %*
5	0.83	1.15	11.07	12.83
8	2.18	2.73	15.15	17.53
9	2.70	3.33	16.92	19.02
10	3.25	3.94	18.31	20.48
20	9.59	10.85	31.41	34.17
60	40.48	43.19	79.08	83.30

The calculated standard deviation of the sample is 1.8" or the estimated population standard error is

$$\sqrt{\frac{n}{n-1}} \times 1.8 = 1.9''$$

From the above analysis, this value of 1.9 could be in error by 3.74 – 1.9 = +1.84" or by 1.90 – 1.32 = –0.68" at the 95 % confidence level. There is therefore no point in quoting a figure of better than 2" for the standard error, since the extra decimal would be misleading.

5.12 Creating a random sample from a normal distribution

Suppose we wish to simulate a sample of ten residuals extracted from a normal distribution. From a random number table, we selected ten values, entering the table anywhere at random. These two-digit random numbers are listed in column 1 of Table 5.7.

To convert these two-digit random numbers to a symmetrical set, add 0.5 to all and multiply by 0.01, thus reducing them to random fractions of 1.00. We then select values of residuals corresponding to these fractional probabilities taken from a standardized normal distribution table. These values are given in column 3. Since the distribution set is chosen to have a standard deviation of 5 mm, these residuals are scaled up to give the final rounded values in column 5. The set of residuals is slightly adjusted to make their sum exactly zero, as theory indicates it should be.

Table 5.7 *Random sample of residuals*

Random number	*To 1.000 base*	*Value from N(0,1)*	*Scaled by SD of 5 mm*	*Rounded residuals*
89	0.895	1.25	6.25	6.2
90	0.905	1.31	6.55	6.6
26	0.265	–0.63	–3.15	–3.2
36	0.365	–0.35	–1.75	–1.8
22	0.225	–0.75	–3.75	–3.7
74	0.745	0.66	3.30	3.3
71	0.715	0.57	2.85	2.8
13	0.135	–1.10	–5.50	–5.5
74	0.745	0.66	3.30	3.3
05	0.055	–1.60	–8.00	–8.0
			sum = 0.1	sum = 0.0

Readers may care to verify that the standard deviation computed from the created residuals is 5.1 mm, i.e. close to the expected 5 mm.

It is not always necessary to adopt this manual procedure, because there are many software packages which carry out the operation on demand.

5.13 Errors in derived quantities

The real benefits from error analysis arise in the assessment of the likely quality of results derived from measurements whose errors are known or which are postulated in design problems.

Consider Figure 5.9 which shows a point P(x, y). The abscissa x and the ordinate y are measured and plotted at right angles. For the moment we assume that a perfect right angle is possible. Suppose that these measurements are in error by small amounts δx and δy respectively. If we plot the new position of P as P' we see that there are corresponding changes to the angle α, the distance s = OP and the area of the rectangle $A = xy$. Hence if we know the values of the errors δx and δy we can calculate their effects on the derived quantities α, s and A. Let these effects be δα, δs and δA.

A simple way of finding these changes is to recalculate them using new values derived from the originals altered by the known changes. Although inelegant, this procedure should be considered as it can provide quick rough estimates. Consider the following example:

$$\text{Let } x = 300 \text{ mm}, y = 200 \text{ mm}, \delta x = 0.1 \text{ mm and } \delta y = -0.1 \text{ mm.}$$

$$\text{Thus } s = 360.555 \text{ mm}, \alpha = 33°41'24'' \text{ and } A = 6 \times 10^4 \text{ mm}^2$$

$$\text{Recalculation with } x = 300.1 \text{ mm and } y = 199.9 \text{ mm gives}$$

$$s = 360.583 \text{ mm}, \alpha' = 33°40'05'' \text{ and } A' = 59\ 989.99 \text{ mm}^2.$$

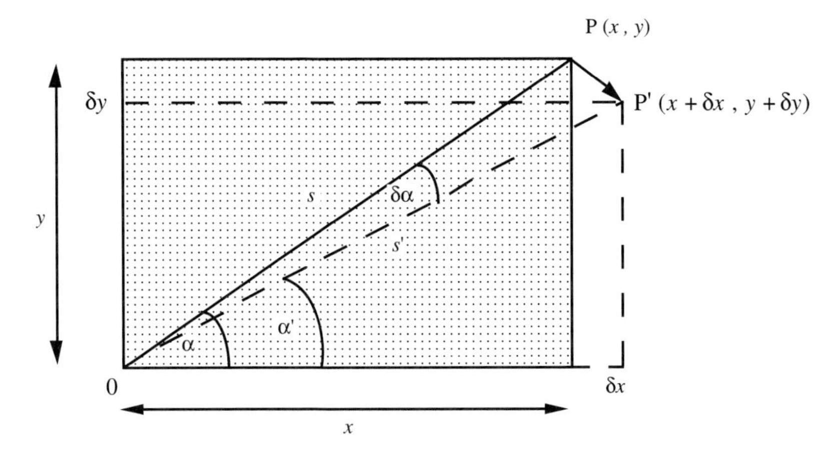

Figure 5.9

Thus, by subtraction, we have the changes:

$$\delta s = 0.028 \text{ mm}, \delta\alpha = -01'19'' \text{ and } \delta A = -10.01 \text{ mm}^2.$$

However, this procedure does not lend itself to the treatment of random errors in complex problems nor the statistical assessment of results. To do this, we need a different approach using differential calculus and matrix algebra, putting this simple procedure on a much more flexible footing.

The three formulae used in these calculations are

$$s^2 - x^2 - y^2 = 0 \quad \text{or} \quad F_1(s:x,y) = 0$$

$$\tan\alpha - \frac{y}{x} = 0 \quad \text{or} \quad F_2(\alpha:x,y) = 0$$

$$A - xy = 0 \quad \text{or} \quad F_3(A:x,y) = 0$$

A more general statement is to relate any derived parameter p to the measured variables x and y by the expression $F(p: x, y) = 0$.

We find it is helpful to separate the parameters from the observed quantities by a colon, retain the full notation of the original problem, carry out its partial differentiation, and convert the expression to the general form suited to matrices. For example, the partial differentiation of this general expression is written

$$\left(\frac{\partial F}{\partial p}\right)\delta p + \left(\frac{\partial F}{\partial x}\right)\delta x + \left(\frac{\partial F}{\partial y}\right)\delta y = 0$$

which, when applied to these three specific functions, gives

$$2s\,\delta s - 2x\,\delta x - 2y\,\delta y = 0$$

$$\sec^2\alpha\,\delta\alpha + \frac{y}{x^2}\,\delta x - \frac{1}{x}\,\delta y = 0$$

$$\delta A - y\,\delta x - x\,\delta y = 0$$

Putting $x = 300$ mm and $y = 200$ mm, and ignoring the fact that the coefficients can be simplified, the equations become

$$721 \times \delta s - 600 \times \delta x - 400 \times \delta y = 0$$

$$1.444 \times \delta\alpha + 0.00222 \times \delta x - 0.00333 \times \delta y = 0$$

$$\delta A - 200 \times \delta x - 300 \times \delta y = 0$$

On setting $\delta x = 0.1$ mm and $\delta y = -0.1$ mm, we obtain the results $\delta s = -0.028$ mm, $\delta A = 3.8435 \times 10^{-4}$ radians $= 01'19''$ and $\delta A = -10.00$ mm^2. The very slight disagreement in δA is due to the approximation in taking first differentials only.

The above equations may be written in matrix form as follows:

$$\begin{bmatrix} 721\,\delta s \\ 1.444\,\delta\alpha \\ \delta A \end{bmatrix} + \begin{bmatrix} -600 & -400 \\ 0.00222 & -0.00333 \\ -200 & -300 \end{bmatrix} \begin{bmatrix} \delta x \\ \delta y \end{bmatrix} = \begin{bmatrix} 0 \\ 0 \\ 0 \end{bmatrix}$$

or

$$\begin{bmatrix} 721 & 0 & 0 \\ 0 & 1.444 & 0 \\ 0 & 0 & 1 \end{bmatrix} \begin{bmatrix} \delta s \\ \delta\alpha \\ \delta A \end{bmatrix} + \begin{bmatrix} -600 & -400 \\ 0.00222 & -0.00333 \\ -200 & -300 \end{bmatrix} \begin{bmatrix} \delta x \\ \delta y \end{bmatrix} = \begin{bmatrix} 0 \\ 0 \\ 0 \end{bmatrix}$$

which is of the form

$$\mathbf{A\,x} + \mathbf{C\,s} = \mathbf{0}$$

It should be remembered that the coefficients of the variables are the partial differentials of the various functions, assembled above as a matrix, for example

$$\mathbf{C} = \begin{bmatrix} \dfrac{\partial F_1}{\partial x} & \dfrac{\partial F_1}{\partial y} \\ \dfrac{\partial F_2}{\partial x} & \dfrac{\partial F_2}{\partial y} \\ \dfrac{\partial F_3}{\partial x} & \dfrac{\partial F_3}{\partial y} \end{bmatrix} \tag{5.7}$$

It is defined as the *Jacobian matrix* of the vector $\mathbf{F} = (F_1, F_2, F_3)^{\mathrm{T}}$ with respect to the vector $\mathbf{x} = (x, y)$.

Very often the coefficients in this matrix can be obtained by a semi-graphic method with little difficulty. Reconsider the problem illustrated in Figure 5.9. The error figure for this is given in Figure 5.10. We see at once the effects of the changes δx and δy in producing s, δa and δs which can obtained by inspection as

$$\delta s = \cos\alpha\,\delta x - \sin\alpha\,\delta y$$
$$s\,\delta\alpha = \sin\alpha\,\delta x + \cos\alpha\,\delta y$$

These are the same first two equations as before, reduced to their simplest forms. It can also be observed that these equations are derived by a rotation of the error coordinates (δx, δy) by an angle α to give the new coordinates (δs, $s\ \delta\alpha$). This illustrates the power of the semi-graphic approach to simple geometrical problems. Complex mathematical models are best treated by conventional differentiation.

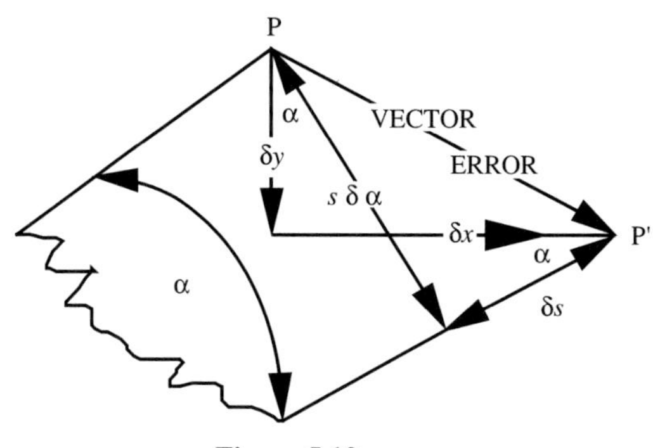

Figure 5.10

Chapter 6
Least Squares Estimation

6.1 The theory of Least Squares

Since the nineteenth century, applied scientists have adopted the principle of Least Squares to select the best estimates of parameters from a series of observations. In fact, when we accept the arithmetic mean of a number of observations as the best estimate of an observed parameter, we are making use of this principle.

Most of the theory and procedures developed by Legendre and Gauss have been applied for over a century, but it has not been until the development of computers that Least Squares has become commonplace in everyday applications.

The principle provides a method of making the best use of observations when more than the minimum number has been observed, and also a way of combining different types of observations to yield best estimates of parameters.

It should be noted that the application of Least Squares does not imply any statistical distribution of observed variables. However, when an attempt is made to infer quality from them and derive results, it is usual to assume that a Normal distribution is involved.

Most of the procedures used to analyse observed data can also be used without data as a pre-analysis tool when designing control networks, before making any observations, thus saving much time and money.

Note: Results were calculated using spreadsheets using maximum precision data in order to avoid rounding errors leading to inconsistencies. It is important, however, that final results are quoted to a realistic precision. For example, the coordinates of the point G referred to in Section 6.13 can only be quoted to the nearest mm.

6.2 Application to a single observed parameter

To define terms and explain basic ideas, consider the case of the nine observations of the angle whose sample variance was computed in Table 5.1. Firstly we relate each observation, denoted by $\overset{o}{x}_i$, to the best estimate of the angle $\hat{x}$, by a typical *observation equation*

$$\hat{x} = \overset{o}{x}_i + v_i$$

The equation for the first observation of Table 5.1 is

$$\hat{x} = 109^\circ 25' \, 06.3'' + v_1$$

This is only a statement of how much the observed value 109°25' 06.3" differs from the final best estimate $\hat{x}$ by the residual v_1. The procedure of Least Squares is to find this best estimate $\hat{x}$. Additionally, if we wish some information about the quality of the work, we can substitute the best estimate in the observation equations to obtain the residuals. From these residuals, statistical information such as the variance can be derived.

Returning to the example, we now introduce a device to reduce the size of the figures in calculations as follows. Let

$$\overset{o}{x}_i = x^* + L_i \quad \text{and} \quad \hat{x} = x^* + \delta$$

where x^* is some provisional, or base value, close to $\hat{x}$ and L_i is described as the absolute term. Here the value of x^* is chosen to be 109°25' simply for convenience (it is usual, however, to adopt the first observation as the base value).

In linear problems, that is if the mathematical model only involves linear functions of parameters such as in levelling, there is no *need* to adopt this device of a provisional value. However it is always a good idea to do so, because the sizes of the numbers used in calculation are thereby greatly reduced. In non-linear problems, as in distance measurement, it is *essential* to adopt this procedure, so the Newton's method of solution can be used.

Substitution for $\hat{x}$ reduces the observation equation to

$$\delta = \overset{o}{x}_i - x^* + v_i = L_i + v_i$$

e.g. for the first observation

$$\delta = 109°25'\,06.3'' - 109°25' + v_1 = 6.3 + v_1$$

The absolute term L is always expressed as 'observed minus provisional' provided it is written to the right of the equality sign in the equations.

Casting all nine observations of Table 5.1 in this way gives equations in matrix form

$$\begin{bmatrix}1\\1\\1\\1\\1\\1\\1\\1\\1\end{bmatrix}\delta = \begin{bmatrix}6.3\\7.2\\10.4\\4.3\\9.6\\5.8\\7.9\\8.3\\9.1\end{bmatrix} + \begin{bmatrix}v_1\\v_2\\v_3\\v_4\\v_5\\v_6\\v_7\\v_8\\v_9\end{bmatrix}$$

These equations are of the form

$$\mathbf{A\,x} = \mathbf{L} + \mathbf{v} \tag{6.1}$$

which is the standard form adopted for this type of Least Squares problem. In this simple case there is only one unknown, therefore

$$\mathbf{x} = \delta$$

Because there are more equations than unknowns (in this example, nine equations to find one parameter) we have to adopt a principle to give a unique result. The most common principle adopted is the Least Squares, or minimum variance principle. This principle finds that value of δ which makes the sum of the squares of the residuals a minimum. Let the sum of squares of residuals be Ω. Then we have to make

$$\Omega = \sum v_i^2$$

a minimum. This will be the case when

$$\frac{\partial \Omega}{\partial \delta} = 0$$

The mechanism of applying the Least Squares principle to this simple problem, using two mathematical methods, is explained below. The reader should follow the left hand treatment to its conclusion, before considering the matrix equivalent on the right, which is a restatement of the results derived by ordinary algebra on the left. There are m observations of one parameter.

Ordinary Algebra	Matrix Algebra
$\Omega = \sum v_i^2$	$\Omega = \mathbf{v}^{\mathrm{T}}\mathbf{v}$
$= v_1^2 + v_2^2 + \ldots + v_m^2$	
$= (\delta - L_1)^2 + (\delta - L_2)^2 + \ldots + (\delta - L_m)^2$	
$\frac{\partial \Omega}{\partial \delta} = 2(\delta - L_1) + 2(\delta - L_2) + \ldots + 2(\delta - L_m)$	
$= 2\sum v_i$	$\frac{\partial \Omega}{\partial \mathbf{x}} = 2\mathbf{v}^{\mathrm{T}}\mathbf{A}$
But	Therefore
$\frac{\partial \Omega}{\partial \delta} = 0 \Rightarrow \sum v_i = 0$	$\mathbf{v}^{\mathrm{T}}\mathbf{A} = \mathbf{0}$ (6.2)
	(Note also that
	$\mathbf{A}^{\mathrm{T}}\mathbf{v} = \mathbf{0}$)

The very important result at Equation (6.2) will be derived for the more general case later. We now proceed to the solution, using the same two parallel mathematical models.

Adding up the columns of the nine equations we have

Ordinary Algebra	Matrix Algebra
$m\delta = \sum L + \sum v = \sum L$	$\mathbf{A}^{\mathrm{T}}\mathbf{A}\mathbf{x} = \mathbf{A}^{\mathrm{T}}\mathbf{L} + \mathbf{A}^{\mathrm{T}}\mathbf{v} = \mathbf{A}^{\mathrm{T}}\mathbf{L}$
$\delta = \frac{\sum L}{m}$	$\mathbf{x} = (\mathbf{A}^{\mathrm{T}}\mathbf{A})^{-1}\mathbf{A}^{\mathrm{T}}\mathbf{L}$

The inverse $(\mathbf{A}^T\mathbf{A})^{-1}$ is merely the reciprocal of m, and $\mathbf{A}^T\mathbf{L}$ is ΣL. Finally, we obtain the best estimate of x to be

$$\begin{aligned}\hat{x} &= x^* + \delta \\ &= x^* + \sum Lm \\ &= x^* + \frac{\sum \overset{o}{x}_i - \sum x^*}{m} \\ &= x^* + \frac{\sum \overset{o}{x}_i}{m} - x^* \\ &= \bar{x}\end{aligned}$$

Thus we have proved that the *best estimate* is the *arithmetic mean* of the observations. Completing the arithmetic as in Table 5.1, we find again that δ = 07.7", giving the final result

$$\hat{x} = \bar{x} = 109^\circ 25' 07.7''$$

Next we obtain the same residuals as in Table 5.1 and the variance as before.

The reader should be completely satisfied that only the arithmetic mean gives the minimum sum of squares of residuals. A simple exercise to demonstrate this is to select values slightly greater and slightly less than the mean, find the residuals from these means, and their sums of squares which both prove to be greater than Ω.

6.3 Weighted observations

Suppose that the observed angles of Table 5.1 differ in quality. This could arise for several reasons: they may have been observed with different instruments, by different people or under different conditions. We could carry out test measurements under these different circumstances to find the variances of each angle and thus have an estimate of their relative quality. The idea is to make allowances for the different quality of the residuals in a Least Squares process by scaling each residual by its standard error, thus producing new residuals, each of the same quality. Let the new residuals $\mathbf{u}$ be given by

$$u = \frac{v}{\sigma} \tag{6.3}$$

Clearly the better observations have smaller standard errors and thus their scaled residuals are given more weight in the solution. To allow for the fact that we can only *estimate* the standard errors σ we introduce an unknown constant σ_0 into the equation:

$$\frac{u}{\sigma_0} = \frac{v}{\sigma} \quad \text{or} \quad u = \frac{\sigma_0 v}{\sigma}$$

If the standard errors have been accurately estimated, we expect this constant σ_0 (usually known as the *standard error of unit weight*) to be equal to 1. However, in practice, it is evaluated as a routine part of each problem and acts as an indicator of how well the observational variances were initially estimated, or *a priori.*

The effect of this modification on the Least Squares procedure is that we now minimize the sum of squares of the reduced residuals u; that is we minimize

$$\Omega = \mathbf{u}^{\mathrm{T}}\mathbf{u}$$

where

$$\mathbf{u}^{\mathrm{T}} = \sigma_0\left(\frac{v_1}{\sigma_1}, \frac{v_2}{\sigma_2}, \ldots, \frac{v_9}{\sigma_9}\right)$$

A neater way to introduce these standard errors and the unknown constant is to create a special matrix containing their squares or variances. This is called the weight matrix **W**. It is a diagonal matrix containing the reciprocals of the respective variances, scaled by the variance of unit weight. By the introduction of this weight matrix we convert

$$\Omega = \mathbf{u}^{\mathrm{T}}\mathbf{u}$$

into the equivalent expression

$$\Omega = \mathbf{v}^{\mathrm{T}}\mathbf{W}\,\mathbf{v} \tag{6.4}$$

where the weight matrix **W** is a diagonal matrix of dimensions $m \times m$ where m is the number of observations. A typical 3×3 weight matrix has the following structure:

$$\mathbf{W} = \sigma_0^2\begin{bmatrix} 1/\sigma_1^2 & 0 & 0 \\ 0 & 1/\sigma_2^2 & 0 \\ 0 & 0 & 1/\sigma_3^2 \end{bmatrix}$$

Note that this weight matrix is the scaled inverse of the dispersion matrix. Also, for independent observations with zero covariances,

$$\mathbf{W}^{-1} = \frac{1}{\sigma_0^2}\mathbf{D}$$

because

$$\mathbf{W}^{-1} = \frac{1}{\sigma_0^2}\begin{bmatrix} \sigma_1^2 & 0 & 0 \\ 0 & \sigma_2^2 & 0 \\ 0 & 0 & \sigma_3^2 \end{bmatrix} = \frac{1}{\sigma_0^2}\mathbf{D}$$

6.4 Example of weighted observations

Reconsider the example of Table 5.1. We now assign different weights to the twelve observations. Let us assume that each of the first six observations was observed with a standard error of 2", and the last three, with standard errors of 5". This could happen if two different theodolites had been used.

The diagonal dispersion matrix **D** of the observations is:

$$\begin{bmatrix} 4 &&&&&&&& \\ & 4 &&&&&&& \\ && 4 &&&&&& \\ &&& 4 &&&&& \\ &&&& 4 &&&& \\ &&&&& 4 &&& \\ &&&&&& 25 && \\ &&&&&&& 25 & \\ &&&&&&&& 25 \end{bmatrix}$$

The scaled inverse of **D** is the weight matrix **W** i.e.

$$\mathbf{W} = \sigma_0^2 \mathbf{D}^{-1} \tag{6.5}$$

$$\mathbf{W} = \sigma_0^2 \begin{bmatrix} 1/4 &&&&&&&& \\ & 1/4 &&&&&&& \\ && 1/4 &&&&&& \\ &&& 1/4 &&&&& \\ &&&& 1/4 &&&& \\ &&&&& 1/4 &&& \\ &&&&&& 1/25 && \\ &&&&&&& 1/25 & \\ &&&&&&&& 1/25 \end{bmatrix}$$

(Note: All off-diagonal terms are zero.)

For correlated observations, the off-diagonal terms (covariances) are generally non-zero. In this case the inverse **W**, though not so easily found, is treated in just the same way. It is difficult to call such a matrix a 'weight' matrix. Generally the term is retained for a diagonal matrix of uncorrelated observations, and we use the term 'inverse of the dispersion matrix' for the general case. In many texts, this matrix **W** is denoted by **P**.

Example In practice we start by making $\sigma_0^2 = 1$. In our simple problem converting to ordinary algebra we have

$$\mathbf{A}^{\mathrm{T}}\mathbf{W}\,\mathbf{A} = \sum w \quad \text{and} \quad \mathbf{A}^{\mathrm{T}} wL = \sum wL$$

giving the result

$$\begin{aligned}\hat{x} &= \frac{\sum wL}{\sum w} \\ &= \frac{\frac{1}{4}(6.3+7.2+10.4+4.3+9.6+5.8)+\frac{1}{25}(7.9+8.3+9.1)}{\frac{6}{4}+\frac{3}{25}} \qquad (6.6) \\ &= 7.35''\end{aligned}$$

This result is called the *weighted mean*. The residuals formed from the weighted mean are: 1.05, 0.15, 3.05, –3.05, 2.25, –1.55, 0.55, 0.95, 1.75, giving

$$\mathbf{v}^{\mathrm{T}}\mathbf{W}\,\mathbf{v} = \frac{1}{4}(1.05^2+\ldots+1.55^2)+\frac{1}{25}(0.55^2+\ldots+1.75^2) = 6.8+0.17 = 6.97$$

An estimate of the standard error of unit weight σ_0 is obtained from the *weighted residuals* using the formula

$$\begin{aligned}\sigma_0^2 &= \frac{\mathbf{v}^{\mathrm{T}}\mathbf{W}\,\mathbf{v}}{m-1} = \frac{6.97}{8} = 0.87 \\ &\Rightarrow \sigma_0 = 0.93\end{aligned}$$

That σ_0 is close to 1 indicates that the original weights were well estimated. This very important concept of a weight matrix also enables observations of different types (e.g. directions and lengths) to be combined in a Least Squares estimate of parameters.

Summary

The treatment of observations of differing quality is brought about by the use of a matrix **W**. It is formed from the inverse of the dispersion matrix of the observed parameters, scaled by the variance of an observation of unit weight, initially assumed equal to 1, and later estimated as part of the computation process. The expression to be minimized is: $\mathbf{\Omega} = \mathbf{v}^{\mathrm{T}}\mathbf{W}\mathbf{v}$ (Equation (6.4)).

The variance of an observation of unit weight is given by:

$$\sigma_0^2 = \frac{\mathbf{v}^{\mathrm{T}}\mathbf{W}\,\mathbf{v}}{m-n}$$

where there are m equations in n variables, $m \geq n$.

6.5 Observations to estimate more than one parameter

The extension of the above theory to deal with more than one parameter follows similar lines except that **A** has one column for every parameter to be estimated. Consider the small levelling net shown in Figure 6.1. There are six lines of levels connecting four points A, B, C and D. Since only three such lines of levels would be sufficient to give relative heights between the points, three lines are extra to requirements. Therefore there are *three degrees of freedom* in the problem.

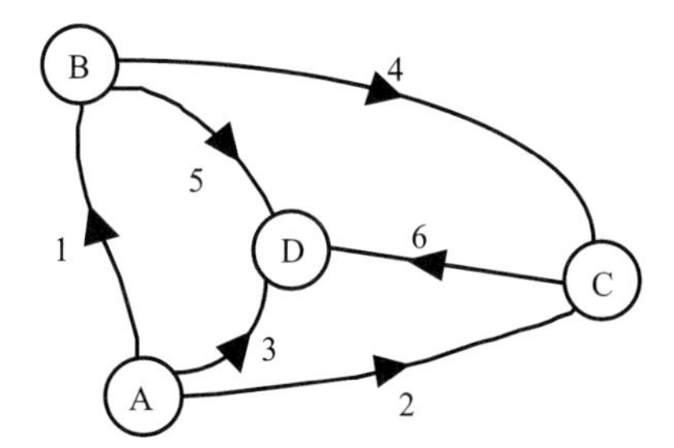

Figure 6.1

Since we only know *differences* in level, we can only find the relative heights of the points. We may select any one to be a datum. In this case the height of A is fixed, hence $\delta h_A = 0$ leaving the three parameters h_B, h_C and h_D to be estimated.

Consider Table 6.1 in which all the necessary data are listed. Although any three height differences could be selected as the provisional values, with the proviso that they are close to the observed values, it is more efficient to select the first three observed values thus making the first three values of the **L** vector zero.

In general, when there are $m - n$ degrees of freedom in the problem, we select the first n observed parameters to be the n provisional values, thus reducing the first n values of the **L** vector to zero. The remaining provisional values are derived from these first n. In this case,

$$h_C^* - h_B^* = h_C^* - h_A^* - (h_B^* - h_A^*) = +69.36 - 12.00 = 57.36$$

and so on for the other lines.

In this case of observation equations which are already linear, it is not essential but however convenient to use the method of approximations. For example, the observation for the line CD is

$$h_D^* - h_C^* = -70.87 = -70.91 + 0.04$$

Table 6.1

Line	*Observation*	*Observed*	*Provisional*	***L***
AB	$h_B - h_A$	+12.00	+12.00	0.00
AC	$h_C - h_A$	+69.36	+69.36	0.00
AD	$h_D - h_A$	–1.55	–1.55	0.00
BC	$h_C - h_B$	+57.28	+57.36	–0.08
BD	$h_D - h_B$	–13.66	–13.55	–0.11
CD	$h_D - h_C$	–70.87	–70.91	+0.04

thus we have

$$\delta h_D - \delta h_C = +0.04$$

thereby reducing the sizes of the numbers in the problem. Care must be taken with signs. These follow strictly from the order in which we write down the parameters in the equations, as indicated by the arrows on the lines of Figure 6.1. The observation equations for this problem can then be assembled in matrix form:

$$\begin{bmatrix} 1 & 0 & 0 \\ 0 & 1 & 0 \\ 0 & 0 & 1 \\ -1 & 1 & 0 \\ -1 & 0 & 1 \\ 0 & -1 & 1 \end{bmatrix} \begin{bmatrix} \delta h_B \\ \delta h_C \\ \delta h_D \end{bmatrix} = \begin{bmatrix} 0 \\ 0 \\ 0 \\ -0.08 \\ -0.11 \\ 0.04 \end{bmatrix} + \begin{bmatrix} v_1 \\ v_2 \\ v_3 \\ v_4 \\ v_5 \\ v_6 \end{bmatrix}$$

These equations are in the standard form of Equation (6.1): $\mathbf{Ax} = \mathbf{L} + \mathbf{v}$.

Minimizing the sum of squares of weighted residuals i.e. Equation (6.4):

$$\Omega = \mathbf{v}^T \mathbf{W} \mathbf{v}$$

means that

$$\frac{d\Omega}{dx} = 0$$

Differentiating (see Appendix 3),

$$\mathbf{v}^T \mathbf{W} \mathbf{A} = \mathbf{0}$$

and also

$$\mathbf{A}^T \mathbf{W} \mathbf{v} = \mathbf{0}$$

as the weight matrix $\mathbf{W}$ is symmetric.

Multiplying each side of Equation (6.1) by $\mathbf{A}^T\mathbf{W}$ gives

$$\mathbf{A}^T \mathbf{W} \mathbf{A} \mathbf{x} = \mathbf{A}^T \mathbf{W} \mathbf{L} + \mathbf{A}^T \mathbf{W} \mathbf{v}$$
$$= \mathbf{A}^T \mathbf{W} \mathbf{L}$$

which can be written

$$\mathbf{N} \hat{\mathbf{x}} = \mathbf{b} \tag{6.7}$$

This is a series of n linear equations in n unknowns. In older literature these n equations are known as the *Normal equations.* As the name is useful it is retained in this book.

Here we take the weight matrix to be the unit matrix assuming the observations are all of the same weight. The normal equations become:

$$\mathbf{N} = \mathbf{A}^T \mathbf{A}$$

and the absolute term is given by

$$\mathbf{b} = \mathbf{A}^{\mathrm{T}}\mathbf{L}$$

We now return to the example of the level net. The normal equations are formed to give

$$\begin{bmatrix} 3 & -1 & -1 \\ -1 & 3 & -1 \\ -1 & -1 & 3 \end{bmatrix} \begin{bmatrix} \delta h_{\mathrm{B}} \\ \delta h_{\mathrm{C}} \\ \delta h_{\mathrm{D}} \end{bmatrix} = \begin{bmatrix} 0.19 \\ -0.12 \\ -0.07 \end{bmatrix}$$

giving the solution

$$\hat{\mathbf{x}} = \begin{bmatrix} \delta h_{\mathrm{B}} \\ \delta h_{\mathrm{C}} \\ \delta h_{\mathrm{D}} \end{bmatrix} = \begin{bmatrix} 0.048 \\ -0.030 \\ -0.017 \end{bmatrix}$$

Substituting the solution in the original observation equations gives the residuals

$$\mathbf{v} = \begin{bmatrix} 0.048 \\ -0.030 \\ -0.017 \\ 0.002 \\ 0.045 \\ -0.027 \end{bmatrix}$$

The arithmetic check on the work is that

$$\mathbf{A}^{\mathrm{T}}\mathbf{v} = \mathbf{0}$$

In this case we find

$$\mathbf{A}^{\mathrm{T}}\mathbf{v} = \begin{bmatrix} 0.001 \\ -0.001 \\ 0.001 \end{bmatrix}$$

The slight disagreement is due to rounding errors in the arithmetic processes. We can also compute the variance from

$$\frac{\mathbf{v}^{\mathrm{T}}\mathbf{v}}{m-n} = \frac{0.006251}{3} = 0.002084$$

and the standard error

$$\sigma_0 = 0.0456$$

From this result we see that it would have been better to assign weights of 1/0.002084 to each of the observation equations. Although this would not have changed the solution (as the weight matrix appears on each side of the equations) it would have altered the estimate of the variance. This would have been calculated from

$$\frac{\mathbf{v}^{\mathrm{T}}\mathbf{W}\,\mathbf{v}}{m-n}$$

which would just have been

$$\frac{\mathbf{v}^{\mathrm{T}}\mathbf{v}}{0.002084(m-n)}$$

which is clearly 1. Hence in problems, when we assign realistic weights to each observation, we expect that the estimated variance will be 1.

6.6 Mathematical models with more than one observed parameter

In the two simple problems considered above, there was only one observed parameter in each equation. To deal with problems with more than one observed parameter per equation, we introduce equations of the form

$$\mathbf{A\,x}+\mathbf{C\,v}+\mathbf{C\,L}=\mathbf{0}$$

These equations ultimately lead to the solution of a set of symmetric equations of the same form as before i.e.

$$\mathbf{N}_1\mathbf{x}=\mathbf{b}_1 \tag{6.8}$$

where

$$\begin{aligned}\mathbf{N}_1 &= \mathbf{A}^{\mathrm{T}}\mathbf{N}^{-1}\mathbf{A}\\ \mathbf{b}_1 &= \mathbf{A}^{\mathrm{T}}\mathbf{N}^{-1}\mathbf{b}\\ \mathbf{N} &= \mathbf{C\,W}^{-1}\mathbf{C}^{\mathrm{T}}\\ \mathbf{b} &= \mathbf{CL}\end{aligned}$$

$\mathbf{W}$ is the weight matrix of the observations.

We will derive these equations in the course of fitting a straight line to observed data points. (See Appendix 3 for a purely formal derivation.)

6.7 Example of fitting a straight line by Least Squares

In the following sections we use the example of fitting a straight line to data as a means of explaining the general principles connected with the Least Squares estimation of parameters. Not only does this particular problem occur in several fields such as surveying, cartography, engineering surveying and photogrammetry, but the method of treatment is the same for all manner of curve fitting problems. Thus the following treatment of the straight line problem is typical of many.

Only two points are needed to define a straight line. For simplicity, we confine our attention to the two-dimensional problem of fitting a straight line in a plane. Three cases of fitting the corresponding values of x and y, listed in the first two columns of Table 6.2, will be treated.

Table 6.2 *Line fitting data*

x	y	*Provisional* y^*	$o-c=L$	v	$\hat{y}$
1	2.812	2.5	0.312	–0.023	2.789
2	3.066	3.0	0.066	–0.035	3.030
3	3.218	3.5	–0.282	0.053	3.271
4	3.482	4.0	–0.518	0.030	3.513
5	3.713	4.5	–0.787	0.041	3.754
6	4.033	5.0	–0.967	–0.037	3.996
7	4.278	5.5	–1.222	–0.040	4237
8	4.445	6.0	–1.555	0.034	4.478
9	4.783	6.5	–1.717	–0.063	4.720
10	4.920	7.0	–2.080	0.041	4.961

In two dimensions, the equation of a straight line is

$$y = mx + c \tag{6.9}$$

Generally the problem is to find values for the gradient m and the intercept c on the y-axis, from given values of x and y. In the formal functional statement, we adopt the convention of placing the unobserved parameters first, separated from the observed parameters by a colon.

$$F(m, c: x, y) = 0$$

There are two main cases of the line-fitting problem:

1. when only one of the parameters x or y is observed, and
2. when both parameters x and y are observed.

In the first case, the Least Squares model follows the method already considered above. We shall deal with this briefly to set the scene for the more complex treatment which follows.

In both cases y is an observed quantity burdened by observational error. If a large-scale graph is drawn of the values of y and x it will be seen that a straight line cannot be found to fit the data exactly. This graph is illustrated in Figure 6.2, showing the eye-balled line 1–10, from which the parameters can be measured as a check to give

$$c = 2.5; \quad m = \arctan(13^\circ)$$

The Least Squares process is an analytical method of drawing this graph. It produces unique results and will not depend on the skill of the person drawing it. The criterion used is that the sum of squares of the observation residuals is minimum.

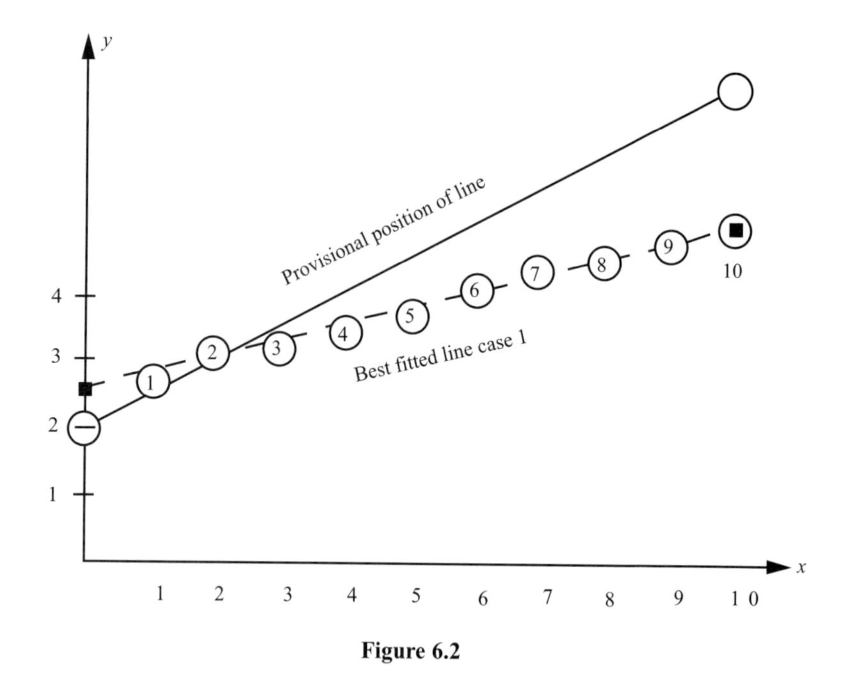

Figure 6.2

The first case is similar to a regression analysis of two variables, in which the small residuals are considered to be only in the direction of the y-axis. The second requires that the residuals lie perpendicular to the line and thus need to be treated by the general Least Squares method. In a later section we will deal with another case in which the solution is constrained in some way, for example to force the line to pass through a given fixed point. These examples illustrate the basic mechanisms for tackling a wide range of such problems.

The figures of the example have been selected from an initial line whose equation is

$$y - 0.5x - 2.0 = 0 \tag{6.10}$$

to which random residuals have been added.

6.8 Case 1: One observable per equation

It must be stressed that because this mathematical equation is non-linear (because of the product mx), we must use Newton's method to linearize the equations involved.

For each observed value, $\overset{o}{y}$, there is an observation equation. Following the usual method we select provisional values of the parameters to be

$$m^* = 0.5; \quad c^* = 2.0$$

Ten values of y^*, listed in column three of Table 6.2, are calculated using Equation (6.10) and values of $x = 1$ to 10.

The objective is to find the best estimates of the unobserved parameters $\hat{m}$, $\hat{c}$ and of the observed parameters $\hat{y}$. As usual, the various quantities are related by the equations:

$$\hat{m} = m^* + \delta m$$
$$\hat{c} = c^* + \delta c$$
$$\hat{y} = y^* + \delta y$$

Since there is only one observed parameter $\overset{o}{y}$ in each equation the residuals v are written

$$v = \hat{y} - \overset{o}{y} = y^* - \overset{o}{y} + \delta y$$

therefore

$$\delta y = \overset{o}{y} - y^* + v = L + v$$

Because the best estimates and the provisional values satisfy Equation (6.10) we may write:

$$\begin{aligned} F(\hat{m}, \hat{c} : x, \hat{y}) &= 0 \\ F(m^*, c^* : x, y^*) &= 0 \end{aligned} \tag{6.11}$$

The connection between these equations and the small correction is given by the linear part of the expansion applied to Equation (6.11) as follows

$$F(\hat{m}, \hat{c} : \hat{y}) = F(m^*, c^* : y^*) + \frac{\partial F}{\partial m^*}\delta m + \frac{\partial F}{\partial c^*}\delta c + \frac{\partial F}{\partial y^*}\delta y$$

Hence from Equations (6.11) we have

$$\frac{\partial F}{\partial m^*}\delta m + \frac{\partial F}{\partial c^*}\delta c + \frac{\partial F}{\partial y^*}\delta y = 0 \tag{6.12}$$

Applying (6.12), we obtain

$$x \times \delta m + 1 \times \delta c = L + v \tag{6.13}$$

It will be remembered that x is considered error free because it is not an observed parameter. Evaluating Equation (6.13) for each of the ten observation equations gives:

$$\mathbf{A\,x} = \mathbf{L} + \mathbf{v}$$

$$\begin{bmatrix} 1 & 1 \\ 2 & 1 \\ 3 & 1 \\ 4 & 1 \\ 5 & 1 \\ 6 & 1 \\ 7 & 1 \\ 8 & 1 \\ 9 & 1 \\ 10 & 1 \end{bmatrix} \begin{bmatrix} \delta m \\ \delta c \end{bmatrix} = \begin{bmatrix} 0.312 \\ 0.066 \\ -0.282 \\ -0.518 \\ -0.787 \\ -0.967 \\ -1.222 \\ -1.555 \\ -1.717 \\ -2.080 \end{bmatrix} + \mathbf{v}$$

The normal equations, on the assumption that observations are all of equal weight, are:

$$\mathbf{A}^{\mathrm{T}}\mathbf{A}\,\mathbf{x}=\mathbf{b} \quad \text{or} \quad \mathbf{N}\,\mathbf{x}=\mathbf{b}$$

$$\begin{bmatrix} 385 & 55 \\ 55 & 10 \end{bmatrix}\begin{bmatrix} \delta m \\ \delta c \end{bmatrix}=\begin{bmatrix} -69.458 \\ -8.75 \end{bmatrix}$$

Giving the solution

$$\begin{bmatrix} \delta m \\ \delta c \end{bmatrix}=\begin{bmatrix} 0.25858 \\ 0.5472 \end{bmatrix}$$

Back substitution in the observation equations provides the residuals of the fifth column in Table 6.2. Thus we can calculate the variance factor σ_0^2 from

$$\sigma_0^2=\frac{\mathbf{v}^{\mathrm{T}}\mathbf{v}}{m-n}=\frac{0.017\,228}{8}=0.0021535$$

$$\Rightarrow \sigma_0=0.044$$

Finally, the best estimates of the derived parameters are obtained from

$$\hat{m}=\overset{*}{m}+\mathrm{d}m=0.2414; \quad \hat{\alpha}=13.5716°$$

$$\hat{c}=\overset{*}{c}+\mathrm{d}c=2.5472$$

and the best estimates of the observed parameters from

$$\hat{y}=\overset{\mathrm{o}}{y}+v_y$$

These are listed in the final column of Table 6.2. The beginner is encouraged to work through this problem both by the graphical and computational methods before proceeding to the next section.

6.9 Case 2: Both x and y observed parameters

We now extend the treatment to the more general case in which both x and y are observed. The treatment is as before, but we also have to estimate

$$\hat{x}=\overset{\mathrm{o}}{x}+v_x$$

The functional models to be satisfied are

$$F(\hat{m},\hat{c}:\hat{x},\hat{y})=0 \quad \text{and} \quad F(\overset{*}{m},\overset{*}{c}:\overset{*}{x},\overset{*}{y})=0$$

$$\frac{\partial F}{\partial \overset{*}{m}}\delta m+\frac{\partial F}{\partial \overset{*}{c}}\delta c+\frac{\partial F}{\partial \overset{*}{x}}\delta x+\frac{\partial F}{\partial \overset{*}{y}}\delta y=0$$

$$\overset{*}{x}\times\delta m+1\times\delta c+\overset{*}{m}\times\delta x-1\times\delta y=0$$

The notation indicates that the coefficients of the variables are evaluated using the provisional values adopted for the problem. Less formally, the equation may be written without the superscript dots as:

$$x \times \delta m + 1 \times \delta c + m \times \delta x - 1 \times \delta y = 0 \tag{6.14}$$

Separating the unobserved from the observed parameters we have:

$$\begin{bmatrix} x & 1 \end{bmatrix} \begin{bmatrix} \delta m \\ \delta c \end{bmatrix} + \begin{bmatrix} m & -1 \end{bmatrix} \begin{bmatrix} \delta x \\ \delta y \end{bmatrix} = 0$$

These equations can be written

$$\mathbf{A\,x} + \mathbf{C\,s} = \mathbf{0}$$

For the line-fitting example, the matrices **A** and **C** have the dimensions (10×2) and (10×20) respectively, and the vectors **x** and **s** have the dimensions (2×1) and (20×1) respectively.

Forming ten equations for every one of the ten observed points gives

$$\begin{bmatrix} 1 & 1 \\ 2 & 1 \\ \vdots & \vdots \\ 10 & 1 \end{bmatrix} \begin{bmatrix} \delta m \\ \delta c \end{bmatrix} + \begin{bmatrix} 0.5 & -1 & 0 & 0 & & & \\ 0 & 0 & 0.5 & -1 & & & \\ & & & & \ddots & & \\ & & & & & \ddots & \\ & & & & & 0.5 & -1 \end{bmatrix} \begin{bmatrix} \delta x_1 \\ \delta y_1 \\ \delta x_2 \\ \delta y_2 \\ \vdots \\ \delta x_{10} \\ \delta y_{10} \end{bmatrix}$$

In general there are r equations in m observed parameters to estimate n derived parameters.

Notice the change in the notation from the ordinary algebra of the problem (in this case m, c, x and y) to that of matrix notation in which we always write the derived vector as **x** and the observed vector **s**.

As usual the small changes to the observed parameters are split into two parts: the known part **L,** and the unknown part **v.** Thus for example

$$\delta x_1 = L_{x_1} + v_{x_1}$$

The final equations are of the form

$$\mathbf{A\,x} + \mathbf{C\,v} = -\mathbf{C\,L}$$

$$\mathbf{A\,x} + \mathbf{C\,v} = \mathbf{b}$$

The reader should now refer to Appendix 3 where it is shown that by minimizing the sum of squares of weighted residuals, we obtain the normal equations

$$\mathbf{N_1\,x} = \mathbf{b}_1$$

where

$$\mathbf{N_1} = \mathbf{A^T N^{-1} A}; \quad \mathbf{b_1} = \mathbf{A^T N^{-1} b}$$

$$\mathbf{N} = \mathbf{C\,W^{-1} C^T}; \quad \mathbf{b} = \mathbf{C\,L}$$

W is the weight matrix.

Key information is presented only.

6.10 Approximate solution with weights

We assign weights of 1 to x and 2 to y giving a (20 × 20) weight matrix with these values alternating down the diagonal.

In this example, because the numbers have been chosen to simplify the arithmetic, the matrix **N** (10 × 10) is diagonal with the value 0.75 for all of its diagonal terms. Its inverse is simply:

$$\mathbf{N}^{-1} = \frac{4}{3}\mathbf{I}$$

where **I** is the unit matrix.

L = 0 as the provisional values were chosen to be the same as the observed values. It must be remembered that the provisional values have to be consistent and

$$F(m^*, c^* : x, y^*) = 0$$

The **L** values for the y's are the same as before.

The problem finally reduces to a solution of the normal equations

$$\mathbf{N}_1 x = \mathbf{b}_1$$

i.e.

$$\begin{bmatrix} 513.33 & 73.33 \\ 73.33 & 13.33 \end{bmatrix} \begin{bmatrix} \delta m \\ \delta c \end{bmatrix} = \begin{bmatrix} -77.138 \\ -9.086 \end{bmatrix}$$

$$\begin{bmatrix} \delta m \\ \delta c \end{bmatrix} = \begin{bmatrix} -0.24686 \\ 0.676 \end{bmatrix}$$

As would be expected, this solution is not very different from the previous method. In this case the residuals are perpendicular distances from the observed points to the line.

6.11 Solution with additional constraint

Sometimes new measurements have to be constrained to fit previous work upon which maps have been plotted, or building construction begun.

Suppose that the line just fitted to the data has to pass through a fixed point (x', y'). This means that there is an equation which must be satisfied exactly:

$$y' - mx' - c = 0$$

This is a constraint equation which the estimated parameters must also satisfy exactly i.e. we have

$$y' - \hat{m}x' - \hat{c} = 0$$

One practical way to treat the problem is to assign a very high weight to the fixed-point coordinates and treat them as observations in the usual way. If we hold the tenth point nearly fixed by assigning it a weight of 100, we obtain the solution:

$$\begin{bmatrix} \delta m \\ \delta c \end{bmatrix} = \begin{bmatrix} -0.2901362 \\ 0.8348 \end{bmatrix}$$

Although this procedure is often acceptable in practice, it is theoretically incorrect because an infinite weight cannot be handled computationally. However, the solution is seen to be acceptable when compared with the exact solution which follows.

6.12 Additional constraints

Since theoretically correct treatment is not difficult, it should certainly be employed in scientific studies. Two methods are available.

A conceptually simple method is to eliminate one parameter from the problem by expressing it in terms of the other. Although this would be the simplest way to treat the particular problem of our example, which only has two parameters, it is not convenient in complex cases. For this reason we present a general alternative method of treatment.

Constraints by Lagrangian multipliers

As explained in Appendix 3, the use of Lagrange's method yields equations of the form:

$$\mathbf{N}_1\mathbf{x} + \mathbf{E}^{\mathrm{T}}\mathbf{k} = \mathbf{b}_1 \tag{6.15}$$

$$\mathbf{E}\,\mathbf{x} - \mathbf{d} = \mathbf{0} \tag{6.16}$$

These can be combined into one hyper-matrix as follows

$$\begin{bmatrix} \mathbf{N}_1 & \mathbf{E}^{\mathrm{T}} \\ \mathbf{E} & 0 \end{bmatrix} \begin{bmatrix} x \\ k \end{bmatrix} = \begin{bmatrix} \mathbf{b}_1 \\ \mathbf{d} \end{bmatrix}$$

If we hold point 10 fixed, its observation equation becomes the constraint equation E, and the hyper-matrix is:

$$\begin{bmatrix} 513.33 & 73.33 & 1 \\ 73.33 & 13.33 & 10 \\ 1 & 10 & 0 \end{bmatrix} \begin{bmatrix} \delta m \\ \delta c \\ k \end{bmatrix} = \begin{bmatrix} -77.138 \\ -9.086 \\ -2.080 \end{bmatrix}$$

the solution of which is

$$\begin{bmatrix} \delta m \\ \delta c \end{bmatrix} = \begin{bmatrix} -0.2922 \\ 0.8426 \end{bmatrix}$$

This compares very well with the previous weighted solution.

6.13 Example of combined network

We shall now consider an example of a network combining GPS, angle and distance measurements and make further analyses of results.

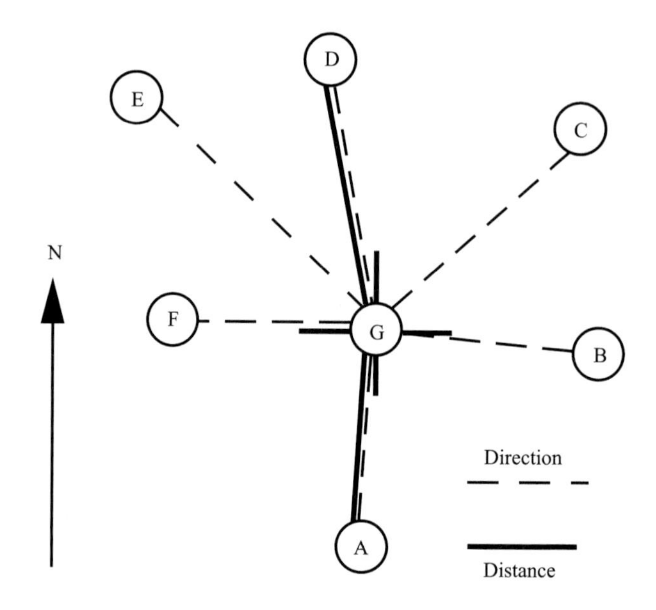

Figure 6.3

Figure 6.3 shows a typical surveying network involving seven points A to G located on a map grid. The position of G was determined from GPS observations. Points A to F were derived from a traverse. The distances AG and DG were measured by EDM. Point G was also fixed by angles observed from the traverse (see Chapter 8). For simplicity we consider that the traverse points A to F are fixed, and dwell only on estimating the best values for the coordinates of G.

Table 6.3 lists the values in metres of the coordinates of points A to F, the bearings U (in sexagesimal degrees), and their tangents.

The provisional coordinates of G obtained from the rays BG and CG are (516.330, 448.990).

To calculate coordinates of G, any two rays that intersect will suffice. We can select two rays from six in any of 15 ways which will give 15 different versions of the

Table 6.3

Point	*Easting E*	*Northing N*	*Ray*	*U = OBS Bg to G (°)*	*tan U*
A	512.402	386.280	AG	3.5588	0.0621928
B	567.895	443.275	BG	276.3248	–9.0220816
C	564.439	487.776	CG	231.1238	1.2403675
D	500.458	507.498	DG	164.8525	–0.2707103
E	457.825	503.912	EG	133.2231	–1.0640318
F	474.929	454.756	FG	98.0069	–7.1091575

Table 6.4

	Best Bearing $\hat{U}$ *(°)*	*Obs Bearing* *U (°)*	*Residual* $v = \hat{U} - U° - U$ *(°)*	*Residual v* *(sec of arc)*
AG	3.558302	3.558756	–0.0004535	–1.6
BG	276.287579	276.324829	–0.0372503	–134.1
CG	231.119795	231.123833	–0.0040380	–14.5
DG	164.857584	164.852482	0.0051026	18.3
EG	133.221335	133.223144	–0.0018087	–6.5
FG	97.974722	98.006931	–0.0322095	–115.9

position of G. Using the data from Table 6.4, Figure 6.4 shows a scale diagram of how the various rays intersect in the vicinity of G. (The intersection of BG with FG cannot be shown at this scale.) Clearly we can say that there is no unique solution to the problem as it stands, or that the position of G is *overdetermined.* This redundancy of information can be put another way. We say that, because there are six equations, only two of which are necessary and sufficient, there are 6 – 2 = 4 *degrees of freedom* in the problem. In general if we have m equations in n variables, $m > n$, the number of degrees of freedom is $m - n$.

On Figure 6.4, the reader can verify that rays AG and BG meet at the point whose coordinates are (516.302, 448.993), and rays BG and CG meet at (516.330, 448.990). The final objective of our analysis is to select one unique position for G which uses all the rays in some regular systematic way. The finally accepted position of G, resulting from the Least Squares process, is

$$E_G = 516.300 \text{ and } N_G = 448.960$$

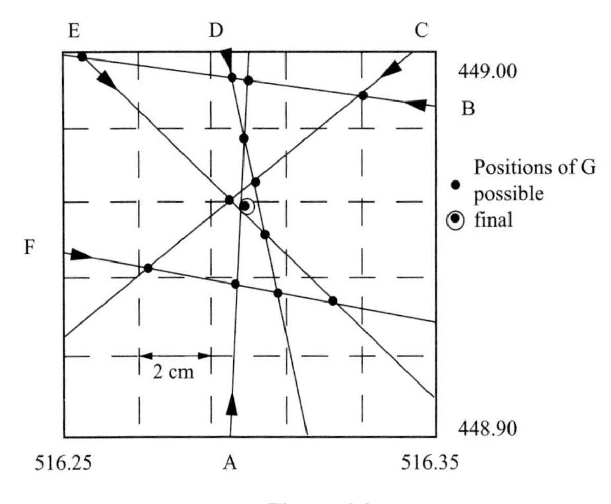

Figure 6.4

Bearing Equations

A full description of Least Squares theory is included in Appendix 3. Note that since we are considering the points A to F as fixed, the inward bearings are known. In the computation of a traverse in Section 8.4, allowance has to be made for the unknown orientation of directions at each station thus introducing an additional orientation parameter Z for each observation station.

Consider first the functional model linking the coordinates of two points A and G and its bearing U. In its general form, the connection is expressed as

$$F(\hat{E}_{\mathrm{A}}, \hat{N}_{\mathrm{A}}, \hat{E}_{\mathrm{G}}, \hat{N}_{\mathrm{G}}, \hat{U}) = 0$$

This equation is exactly satisfied by the Least Squares estimates indicated by the hat ^. It is also exactly satisfied by a set selected or provisional consistent values identified by the asterisk *, i.e.

$$F(\overset{*}{E}_{\mathrm{A}}, \overset{*}{N}_{\mathrm{A}}, \overset{*}{E}_{\mathrm{G}}, \overset{*}{N}_{\mathrm{G}}, U^{*}) = 0$$

Specifically, these equations for a bearing are

$$(\hat{N}_{\mathrm{G}} - \hat{N}_{\mathrm{A}}) \tan \hat{U} - \hat{E}_{\mathrm{G}} + \hat{E}_{\mathrm{A}} = 0 \tag{6.17}$$

and

$$(\overset{*}{N}_{\mathrm{G}} - \overset{*}{N}_{\mathrm{A}}) \tan U^{*} - \overset{*}{E}_{\mathrm{G}} + \overset{*}{E}_{\mathrm{A}} = 0 \tag{6.18}$$

There are two ways of dealing with the expression linking the provisional values:

1. We can calculate a provisional value of the 'observable' U so that $F(\overset{*}{E}_{\mathrm{A}}, \overset{*}{N}_{\mathrm{A}}, \overset{*}{E}_{\mathrm{G}}, \overset{*}{N}_{\mathrm{G}}, U^{*}) = 0$. Then obtain K from

$$K = U - U^{*}$$

2. Alternatively, we use the provisional values along with the observed bearing $\overset{o}{U}$ to find K more directly

$$F(\overset{*}{E}_{\mathrm{A}}, \overset{*}{N}_{\mathrm{A}}, \overset{*}{E}_{\mathrm{G}}, \overset{*}{N}_{\mathrm{G}}, \overset{o}{U}) = K$$

We shall use the second method for the moment because it is slightly the easier to calculate. Normally in this calculation we use the given coordinates of A, the provisional coordinates of G and the observed value of U. However, in this case, since we used the observed value of U to obtain the provisional coordinates of G, K will be zero (the reader should verify this). To illustrate the general method, we will select other provisional values close to the computed ones. We select

$$\overset{*}{E}_{\mathrm{G}} = 516.302 - 0.102 = 516.200 \quad \text{and} \quad \overset{*}{N}_{\mathrm{G}} = 448.993 - 0.093 = 448.900$$

Then

$$F(\overset{*}{E}_{\mathrm{A}}, \overset{*}{N}_{\mathrm{A}}, \overset{*}{E}_{\mathrm{G}}, \overset{*}{N}_{\mathrm{G}}, \overset{o}{U}_{\mathrm{AG}}) = (448.900 - 386.28) \times 0.0621928 - 516.200 + 512.402 = 0.09651 = K_{\mathrm{AG}}$$

Similarly, for the ray BG,

$$F(\overset{*}{E}_{\mathrm{B}}, \overset{*}{N}_{\mathrm{B}}, \overset{*}{E}_{\mathrm{G}}, \overset{*}{N}_{\mathrm{G}}, \overset{o}{U}_{\mathrm{BG}}) = (\overset{*}{N}_{G} - \overset{*}{N}_{B}) \tan U^{*} - \overset{*}{E}_{G} + \overset{*}{E}_{B} = 0.94579 = K_{\mathrm{BG}}$$

The K values for the other four rays are

$$K_{CG} = 0.01847307;\ K_{DG} = 0.12108216;\ K_{EG} = 0.15951738;\ K_{FG} = 0.36022632$$

All that remains is to evaluate the partial differentials. We proceed as follows by partially differentiating the equation for the ray AG i.e. the equation

$$(N_G - N_A)\tan \overset{o}{U}_{AG} - E_G + E_A = 0 \tag{6.19}$$

Thus we obtain

$$\frac{\partial F}{\partial E_A} = 1; \quad \frac{\partial F}{\partial N_A} = -\tan \overset{o}{U}_{AG}; \quad \frac{\partial F}{\partial E_G} = -1$$

$$\frac{\partial F}{\partial N_G} = \tan \overset{o}{U}_{AG}; \quad \frac{\partial F}{\partial U} = (N_G - N_A)\sec^2 \overset{o}{U}_{AG}$$

giving the Equation

$$\delta E_A - \delta E_G - \delta N_A \tan \overset{o}{U}_{AG} + \delta N_G \tan \overset{o}{U}_{AG} + (N_G - N_A)\sec^2 \overset{o}{U}_{AG}\, \delta U_{AG} + K_{AG} = 0 \tag{6.20}$$

The similar equation for the ray BG is

$$\delta E_B - \delta E_G - \delta N_B \tan \overset{o}{U}_{BG} + \delta N_G \tan \overset{o}{U}_{BG} + (N_G - N_B)\sec^2 \overset{o}{U}_{BG}\, \delta U_{BG} + K_{BG} = 0 \tag{6.21}$$

Equations (6.20) and (6.21) are the general expressions for the bearings AG and BG in terms of all the parameters. In the case of simple intersection, the points A and B are kept fixed so

$$\delta E_A = 0; \quad \delta N_A = 0; \quad \delta E_B = 0; \quad \delta N_B = 0$$

and we have

$$-\delta E_G + \delta N_G \tan U_{AG} + (N_G - N_A)\sec^2 U_{AG}\, \delta U_{AG} + K_{AG} = 0 \tag{6.22}$$

$$-\delta E_G + \delta N_G \tan U_{BG} + (N_G - N_B)\sec^2 U_{BG}\, \delta U_{BG} + K_{BG} = 0 \tag{6.23}$$

Where we have only two rays necessary to fix the point G, the bearings U will not be changed so

$$\delta U_{AG} = 0 \quad \text{and} \quad \delta U_{BG} = 0$$

hence Equations (6.22) and (6.23) simplify greatly to

$$-\delta E_G + \delta N_G \tan U_{AG} + K_{AG} = 0 \tag{6.24}$$

$$-\delta E_G + \delta N_G \tan U_{BG} + K_{BG} = 0 \tag{6.25}$$

The numerical versions of these are

$$-\delta E_G + 0.0621928 \times \delta N_G + 0.09651 = 0$$

$$-\delta E_G - 9.0220816 \times \delta N_G + 0.94579 = 0$$

The solution is

$$\delta E_G = 0.102 \quad \text{and} \quad \delta N_G = 0.093$$

And the final coordinate values are

$$\hat{E}_G = E_G^* + \delta E_G = 516.200 + 0.102 = 516.302$$
$$\hat{N}_G = N_G^* + \delta N_G = 448.900 + 0.093 = 448.993$$

which, as expected, are the same as those obtained by the direct solution.

Least Squares application

We form observation equations like (6.20) for all the observed bearings. Although these coefficients can be recast in simpler forms for computation, and we will do so later in Chapter 8, they can be expressed in algebraic form as

$$a_{11}\delta E_G + a_{12}\delta N_G + L_1 = v_1 \tag{6.26}$$

where for AG

$$a_{11} = \frac{1}{(N_G - N_A)\sec^2 U_{AG}}; \quad a_{12} = -\frac{\tan U_{AG}}{(N_G - N_A)\sec^2 U_{AG}};$$
$$L_1 = -\frac{K_{AG}}{(N_G - N_A)\sec^2 U_{AG}}$$

In the same way we can form equations for the other rays and obtain the full array of six equations as follows:

$$\begin{aligned} a_{11}\delta E_G + a_{12}\delta E_G + L_1 &= v_1 \\ a_{21}\delta E_G + a_{22}\delta E_G + L_2 &= v_2 \\ a_{31}\delta E_G + a_{32}\delta E_G + L_3 &= v_3 \\ a_{41}\delta E_G + a_{42}\delta E_G + L_4 &= v_4 \\ a_{51}\delta E_G + a_{52}\delta E_G + L_5 &= v_5 \\ a_{61}\delta E_G + a_{62}\delta E_G + L_6 &= v_6 \end{aligned} \tag{6.27}$$

which are neatly expressed in matrix form as

$$\mathbf{Ax} + \mathbf{L} = \mathbf{v}$$

i.e. Equation (6.1). (Note that this equation can be written either in the above form or as $\mathbf{Ax} = \mathbf{L} + \mathbf{v}$ where $\mathbf{A}$ is the (6×2) matrix of coefficients, $\mathbf{x}$ is the (2×1) column vector of variables $(\delta E_G, \delta N_G)^T$ and $\mathbf{v}$ is the (6×1) column vector of residuals $(v_1, v_2, v_3, v_4, v_5, v_6)^T$.)

Table 6.5 shows the calculations of the coefficients for all six equations.

Normal equations

The observation equations (6.27) can be written in full:

$$\begin{bmatrix} a_{11} & a_{12} \\ a_{21} & a_{22} \\ a_{31} & a_{32} \\ a_{41} & a_{42} \\ a_{51} & a_{52} \\ a_{61} & a_{62} \end{bmatrix} \begin{bmatrix} x_1 \\ x_2 \end{bmatrix} = \begin{bmatrix} L_1 \\ L_2 \\ L_3 \\ L_4 \\ L_5 \\ L_6 \end{bmatrix} + \begin{bmatrix} v_1 \\ v_2 \\ v_3 \\ v_4 \\ v_5 \\ v_6 \end{bmatrix}$$

where $m = 6$ and $n = 2$, and the variables $x_1 = \delta E_G$ and $x_2 = \delta N_G$. The solutions of these normal equations give the required parameters, and are of the form

$$\mathbf{A}^T\mathbf{A}\,\mathbf{x} + \mathbf{A}^T\mathbf{L} = \mathbf{0} \tag{6.28}$$

$\mathbf{A}^T\mathbf{A}$ has dimensions (2×2) and $\mathbf{A}^T\mathbf{L}$ (2×1). The equations in full are

$$\begin{aligned} 0.000\,697\,87\,\delta E_G + 0.000\,122\,8\,\delta N_G &= 7.6929 \times 10^{-5} \\ 0.000\,122\,8\,\delta E_G + 0.001\,189\,12\,\delta N_G &= 8.3487 \times 10^{-5} \end{aligned} \tag{6.29}$$

with solution $\delta E_G = 0.100$ and $\delta N_G = 0.060$ and as before

$$\hat{E}_G = E_G^* + \delta E_G = 516.2 + 0.100 = 516.300$$

$$\hat{N}_G = N_G^* + \delta N_G = 448.9 + 0.060 = 448.960$$

Calculation of the residuals

Once the variables δE_G and δN_G are defined we substitute them in the observation equations (6.27) to obtain the residuals. As a numerical check, we show that $\mathbf{A}^T\mathbf{v} = \mathbf{0}$. Table 6.6 shows these residuals in radians and seconds of arc obtained by back substitution in the observation equations of Table 6.5.

Distance and point measurements

We now extend the analysis to include two further situations:

1. when distances have been measured;
2. when coordinates have been measured directly.

Table 6.5

i	*Ray*	δE	δN	*Dist*	K_i	a_{i1}	a_{i2}	L_i
1	AG	3.798	62.620	62.735	0.09651314	0.015911	–0.000965	0.001535
2	BG	–51.695	5.625	51.600	0.94579100	0.002080	0.019118	0.002005
3	CG	–48.239	–38.876	61.954	0.01847307	–0.010128	0.012567	–0.000187
4	DG	15.742	–58.598	60.675	0.12108216	–0.015917	–0.004276	–0.001933
5	EG	58.375	–55.012	80.212	0.15951738	–0.008550	–0.009073	–0.001364
6	FG	41.271	–5.856	41.685	0.36022632	–0.003370	–0.023752	–0.001193

Table 6.6 *Residuals*

Radians	*Seconds of arc*
-7.915×10^{-6}	–1.6
-6.501×10^{-4}	–134.1
-7.048×10^{-5}	–14.5
8.9057×10^{-5}	18.3
-3.157×10^{-5}	–6.5
-5.622×10^{-4}	–115.9

First, we derive the equation to be solved if the length of AG has been measured to fix the position of G. The functional model is

$$S_{AG}^2 - (E_G - E_A)^2 - (N_G - N_A)^2 = 0 \tag{6.30}$$

In the statement of Newton's method of solution we obtain the general functional equation for a distance measurement as

$$\frac{\partial F}{\partial E_A}\delta E_A + \frac{\partial F}{\partial N_A}\delta N_A + \frac{\partial F}{\partial E_G}\delta E_G + \frac{\partial F}{\partial N_G}\delta N_G + \frac{\partial F}{\partial S}\delta S + F(E_A^*, N_A^*, E_G^*, N_G^*, S^*) = 0 \tag{6.31}$$

Since

$$\begin{aligned} &\frac{\partial F}{\partial E_A} = 2(E_G - E_A); \quad \frac{\partial F}{\partial N_A} = 2(N_G - N_A); \quad \frac{\partial F}{\partial E_G} = -2(E_G - E_A); \\ &\frac{\partial F}{\partial N_G} = -2(N_G - N_A); \quad \frac{\partial F}{\partial S} = 2S_{AG}; \quad F(E_A^*, N_A^*, E_G^*, N_G^*, S^*) = K_{AG} \end{aligned} \tag{6.32}$$

the explicit observation equation can be written

$$\begin{aligned} 2(E_G - E_A)\delta E_A + 2(N_G - N_A)\delta N_A - 2(E_G - E_A)\delta E_G - 2(N_G - N_A)\delta N_G \\ + 2S_{AG}v_{AG} + K_{AG} = 0 \end{aligned} \tag{6.33}$$

Dividing by $2S_{AG}$ gives the final form as

$$\begin{aligned} \frac{(E_G - E_A)}{S_{AG}}\delta E_A + \frac{(N_G - N_A)}{S_{AG}}\delta N_A - \frac{(E_G - E_A)}{S_{AG}}\delta E_G - \frac{(N_G - N_A)}{S_{AG}}\delta N_G \\ + v_{AG} + L_{AG} = 0 \end{aligned} \tag{6.34}$$

i.e. of the form. $\mathbf{A}\,\mathbf{x} + \mathbf{L} = \mathbf{v}$

Evaluating the coefficients of Equations (6.34), the distance observation equations become

$$\begin{aligned} 0.0605403\,\delta E_G + 0.99816575\,\delta N_G - 0.085 = v_{AG} \\ 0.25944504\,\delta E_G + 0.96575787\,\delta N_G + 0.070 = v_{DG} \end{aligned} \tag{6.35}$$

We derive now the observation equations representing the position of G measured by a GPS satellite receiver. In this simple case, there is no functional model, but merely an observational model. The residuals are defined as

$$v = \hat{E} - \overset{\text{o}}{E}$$

We let

$$\hat{E} = E^* + \delta E$$

therefore

$$\delta E + E^* - \overset{\text{o}}{E} = v$$

Thus the observation equations are of the form

$$\begin{aligned} \delta E_{\text{G}} + L_{E_{\text{G}}} &= v_{E_{\text{G}}} \\ \delta N_{\text{G}} + L_{N_{\text{G}}} &= v_{N_{\text{G}}} \end{aligned} \tag{6.36}$$

If the measured GPS values are

$$E_{\text{G}}^{\text{o}} = 516.285; \quad N_{\text{G}}^{\text{o}} = 448.988$$

then

$$\begin{aligned} \delta E_{\text{G}} = L_{E_{\text{G}}} + v_{E_{\text{G}}} &= 0.085 + v_{E_{\text{G}}} \\ \delta N_{\text{G}} = L_{N_{\text{G}}} + v_{N_{\text{G}}} &= 0.088 + v_{N_{\text{G}}} \end{aligned} \tag{6.37}$$

6.14 Combined Least Squares estimation

We now have the theory to tackle the problem of combining observed parameters of different quality and of a different type, and of estimating the quality of the results. We shall employ the data in Table 6.3, incorporating distance and GPS measurements into the original intersection problem. These extra equations are Equations (6.35) and (6.37).

One might argue that the GPS positions are not 'observations' since measurements are derived from a great many observations which have been digitally processed. The same thing can also be said of distances and angles. However, we refer to the numerical output of the systems as 'observations', although many authors refer to them instead as 'observables'.

We have already dealt with the six direction equations, and there are also two distance and two position equations giving ten equations in two variables i.e. $m = 12$ and $n = 2$ as shown in Table 6.7.

We must now select suitable elements for the dispersion matrix. Assume that the directions are uncorrelated, each with a standard error of 20 seconds of arc. Also assume that the distances are uncorrelated with a standard error of 0.01 m and that the position of point G is obtained by differential GPS relative to point A with a standard error of 0.014 m in Easting and a more accurate 0.0014 in Northing.

The value of σ_0^2 will be investigated. However we must also include a value for the covariance of the GPS results, say –0.000 006. Also, since we worked the directions in radians, their standard error of 20 seconds has to be converted to radians.

Therefore

$$\frac{20 \times \pi}{180 \times 60 \times 60} = 9.70 \times 10^{-5} = S_\theta, \text{ say}$$

And the variance

$$S_\theta^2 = 9.4012 \times 10^{-9} = P, \text{ say}$$

The dispersion matrix, $\mathbf{D}_o$, is therefore defined as

$$\mathbf{D}_o = \begin{bmatrix} P & 0 & 0 & 0 & 0 & 0 & 0 & 0 & 0 & 0 \\ 0 & P & 0 & 0 & 0 & 0 & 0 & 0 & 0 & 0 \\ 0 & 0 & P & 0 & 0 & 0 & 0 & 0 & 0 & 0 \\ 0 & 0 & 0 & P & 0 & 0 & 0 & 0 & 0 & 0 \\ 0 & 0 & 0 & 0 & P & 0 & 0 & 0 & 0 & 0 \\ 0 & 0 & 0 & 0 & 0 & P & 0 & 0 & 0 & 0 \\ 0 & 0 & 0 & 0 & 0 & 0 & 0.0001 & 0 & 0 & 0 \\ 0 & 0 & 0 & 0 & 0 & 0 & 0 & 0.0001 & 0 & 0 \\ 0 & 0 & 0 & 0 & 0 & 0 & 0 & 0 & 0.0002 & -0.000006 \\ 0 & 0 & 0 & 0 & 0 & 0 & 0 & 0 & -0.000006 & -0.000006 \end{bmatrix}$$

where P is as defined as above.

The remainder of the process is purely arithmetical. The various stages are shown in the following equations:

The weight matrix **W** is defined as the inverse of the dispersion matrix i.e.

$$\mathbf{W} = \mathbf{D}_o^{-1}$$

Table 6.7

Obs	a_{i1}	a_{i2}	*L*	*Residuals** v	*Shifts* Sv
AG	0.015911	–0.000965	0.001535	-9.485×10^{-5}	–0.0059502
BG	0.00208	0.019118	0.002005	-2.247×10^{-4}	–0.0116832
CG	–0.010128	0.012567	-1.87×10^{-4}	2.5913×10^{-4}	0.01605399
DG	–0.015917	–0.004276	–0.001933	5.8688×10^{-5}	0.00356095
EG	–0.00855	–0.009073	–0.001364	-2.034×10^{-4}	–0.0163144
FG	–0.00337	–0.023752	–0.001193	–0.0010939	–0.0455997
Dist AG	0.0605403	0.99816575	0.085	0.00336109	
Dist DG	0.25944504	–0.9657579	–0.070	0.0148884	
Easting	1	0	0.085	0.01053062	
Northing	0	1	0.088	–0.0052706	

* The residuals are computed later from the solution.

$$= \begin{bmatrix} 1/P & 0 & 0 & 0 & 0 & 0 & 0 & 0 & 0 & 0 \\ 0 & 1/P & 0 & 0 & 0 & 0 & 0 & 0 & 0 & 0 \\ 0 & 0 & 1/P & 0 & 0 & 0 & 0 & 0 & 0 & 0 \\ 0 & 0 & 0 & 1/P & 0 & 0 & 0 & 0 & 0 & 0 \\ 0 & 0 & 0 & 0 & 1/P & 0 & 0 & 0 & 0 & 0 \\ 0 & 0 & 0 & 0 & 0 & 1/P & 0 & 0 & 0 & 0 \\ 0 & 0 & 0 & 0 & 0 & 0 & 10000 & 0 & 0 & 0 \\ 0 & 0 & 0 & 0 & 0 & 0 & 0 & 10000 & 0 & 0 \\ 0 & 0 & 0 & 0 & 0 & 0 & 0 & 0 & 5494.5 & 16483.51 \\ 0 & 0 & 0 & 0 & 0 & 0 & 0 & 0 & 16483.51 & 549450.54 \end{bmatrix}$$

The normal equations are

$$\mathbf{Nx} = \mathbf{b}$$

$$\begin{bmatrix} 80436.277 & 27645.6036 \\ 27645.6036 & 695226.164 \end{bmatrix} \mathbf{x} = \begin{bmatrix} 9971.23131 \\ 60156.6311 \end{bmatrix}$$

where

$$\mathbf{N}^{-1} = \begin{bmatrix} 1.2604 \times 10^{-5} & -5.012 \times 10^{-7} \\ -5.012 \times 10^{-7} & 1.4583 \times 10^{-6} \end{bmatrix}$$

giving the solution

$$\mathbf{x} = \mathbf{N}^{-1}\mathbf{b} = \begin{bmatrix} 0.09553062 \\ 0.08272938 \end{bmatrix}$$

The final position of point G is $(E_G, N_G) = (516.296, 448.983)$.

Note: The diagonal elements of the inverse N^{-1} are the variances of the coordinates of G, and the off-diagonal term is their covariance. Hence the coordinate variances are in Easting 1.2604×10^{-5} and in Northing 1.4583×10^{-6}, giving corresponding standard errors 0.0355 m and 0.0012 m. See Appendix 6 for a discussion on error ellipses.

We calculate the residuals from $\mathbf{Ax} + \mathbf{L} = \mathbf{v}$ which are placed in Table 6.7. From these residuals we calculate s_0^2, an estimate of σ_0^2, from the formula

$$s_0^2 = \frac{\mathbf{v}^{\mathrm{T}}\mathbf{W}\,\mathbf{v}}{m-n} \tag{6.38}$$

In this case, $s_0^2 = 20.237100$.

The fact that this is not close to unity indicates either that the mathematical model is imperfect or that the dispersion matrix has not been modelled correctly.

6.15 Statistical tests for outliers

As discussed in Chapter 5, the huge benefit arising from a Least Squares estimation of results is the statistical information available as a by-product. Some of these will now be

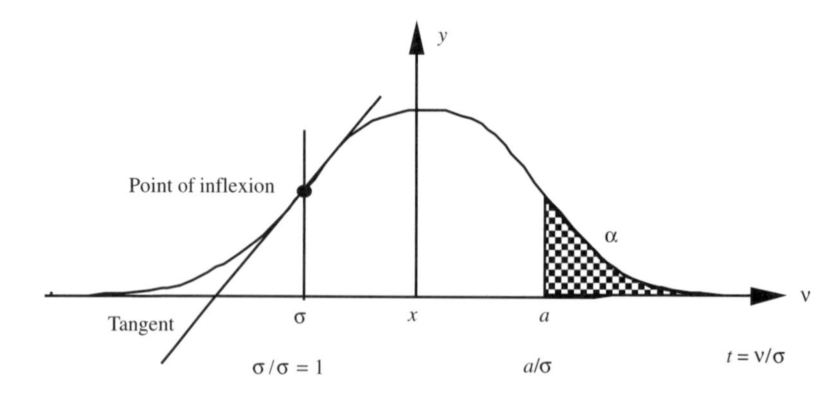

Figure 6.5

described. We start by assuming that residuals have a Normal distribution. It is stressed that for most tests to be valid, the sample must contain more than approximately 30 individual measurements. In our simple example we only have ten observations, in which case we should be discussing the Student's *t* distribution. However, since most Least Squares problems involve much more than 30 variables, our assumption is valid.

From the Least Squares process we obtain the residuals v. We assume these are normally distributed about an origin x. A theoretical graph of the residuals about an origin at x is shown in Figure 6.5.

In order to calculate the probability of the existence of a residual with value greater than the standard error σ, we look up the tables with $t = v / \sigma = 1$. In this case, P = 0.8413. Complete Normal distribution tables are readily available in many textbooks on surveying or elsewhere. Table 6.8 shows selected values from such a table, where e.g. the probability of $t = 1.02$ is 0.846.

Table 6.8 *Cumulative Normal probability*

t	*0.00*	*0.01*	*0.02*	*0.03*	*0.04*	*0.05*	*0.06*	*0.07*	*0.08*	*0.09*
0.0	.500	.504	.508	.512	.516	.520	.524	.528	.532	.536
0.5	.674									
0.6								.750		
1.0	.841	.844	.846	.848	.851	.853	.855	.858	.860	.862
1.6					.950					
1.7						.960				
1.9							.975			
2.0	.977	.978	.978	.979	.979	.980	.981	.981	.981	.982
3.0	.999	.999	.999	.999	.999	.999	.999	.999	.999	.999

The chance of a residual being > σ is therefore

$$1 - P = 1 - 0.8413 = 0.16$$

or 16 %. The probability of the residual either having a value > +σ or < –σ is 32 %. Consideration of the Normal distribution is used to detect *outliers* in the data i.e. a suspect observation or result.

Not all of the directions in Figure 6.4 appear to be correct. A non-arbitrary test is required which can be applied automatically, to decide whether or not to reject an observation. Although we can use a simple test such as: reject residuals greater or less than three times the standard error, such tests have limited applications. It is almost as easy to test the probability of a residual exceeding a particular value, or the value outside a given probability. Suppose, for example, that we decide to reject all residuals with probability of occurrence ≥5 %. This means that a two-sided test for 2.5 % is applied. The table value corresponding to 97.5 % is 1.96. We then apply the test:

$$\text{is } t = v / \sigma \geq 1.96?$$

If so, we reject the observation.

To apply this test to our example we first have to calculate the standard error of each residual. The variance of the residuals are readily obtained as the diagonal terms of the matrix

$$\mathbf{D}_v = \mathbf{D}_o - \mathbf{D}_s \tag{6.39}$$

where

$$\mathbf{D}_s = \mathbf{A}\,\mathbf{D}_x\mathbf{A}^{\mathrm{T}} \tag{6.40}$$

$\mathbf{D}_s$ is the dispersion matrix of the observables, and $\mathbf{D}_o$ the originally estimated dispersion matrix of the observations. The square roots of the diagonals of $\mathbf{D}_v$ are the required standard errors. (See Appendix 3 for derivation of this formula.)

The definitions of these matrices and parameters are listed in Table 6.9.

In calculating

$$\mathbf{D}_x = \sigma_o^2 \mathbf{N}^{-1}$$

we will use $\sigma_o^2 = 1$ instead of the clearly incorrect value of s_o^2.

Table 6.9

Parameter	*Matrix*	*Variance (matrix diagonal)*	*Standard error (square root of variance)*
observed, s_o	$\mathbf{D}_o$	$\sigma_{s_o}^2$	σ_{s_o}
best, s	$\mathbf{D}_s$	σ_s^2	σ_s
unobserved, x	$\mathbf{D}_x$	σ_x^2	σ_x
residuals, v	$\mathbf{D}_v$	σ_v^2	σ_v

The dispersion matrix of observables is defined as

$$\mathbf{D}_s = \mathbf{A}\,\mathbf{D}_x\mathbf{A}^{\mathrm{T}}$$

$$= \begin{bmatrix} 3.21 & 0.238 & -2.15 & -3.16 & -1.63 & -0.454 & 2.80 & 61.2 & 201 & -9.38 \\ 0.238 & 0.547 & 0.168 & -0.379 & -0.385 & -0.693 & 27.7 & -21.6 & 16.6 & 26.8 \\ -2.15 & 0.168 & 1.65 & 2.03 & 0.933 & -0.104 & 15.2 & -57.3 & -134 & 23.4 \\ -3.16 & -0.37 & 2.03 & 3.15 & 1.68 & 0.627 & -10.2 & -53.1 & -198 & 1.74 \\ -1.63 & -0.385 & 0.933 & 1.68 & 0.963 & 0.560 & -15.1 & -18.1 & -103 & -8.94 \\ -0.454 & -0.693 & -0.104 & 0.627 & 0.560 & 0.885 & -34.7 & 23.8 & -30.5 & -32.9 \\ 2.8 & 27.7 & 15.2 & -10.2 & -15.1 & -34.7 & 1430 & -1300 & 262 & 1420 \\ 61.2 & -21.6 & -57.3 & -53.1 & -18.1 & 23.8 & -1300 & 2450 & 3750 & -1530 \\ 201 & 16.6 & -134 & -198 & -103 & -30.5 & 262 & 3750 & 12600 & -501 \\ -9.38 & 26.8 & 23.4 & 1.74 & -8.94 & -32.9 & 1420 & -1530 & -501 & 1450 \end{bmatrix} \times 10^{-9}$$

To calculate σ_v, hence t, we extract the diagonals of $\mathbf{D}_o$ and $\mathbf{D}_s$, as displayed in Table 6.10.

These ratios t will be used to detect blunders in the observations, assuming that the observations are Normally distributed. We select a suitable rejection level of significance, usually 0.01 %, and test to see if any observations lie outside this level. From a Normal distribution table, we see that this critical value is 2.57, exceeded by three values in Table 6.10. We do not reject them all.

We reject the observation with the largest rejection ratio (t = 11.8), the direction from the sixth point F, and repeat the analysis. This is a simple task if using a spreadsheet, since all that is required is to set the coefficients of equation six to 0 while retaining the original weight matrix.

After recalculating, the value for the variance of unit weight is an improved 2.67. Table 6.11 displays the test ratio calculations for the new situation, highlighting how two directions fail to meet the test.

Table 6.10 *Diagonal elements*

$\mathbf{D}_o$	$\mathbf{D}_s$	$\mathbf{D}_v = \mathbf{D}_o - \mathbf{D}_s$	$\sigma_v = \sqrt{\mathbf{D}_v}$	v	$t = v/\sigma_v$
9.4×10^{-9}	3.21×10^{-9}	6.19×10^{-9}	7.87×10^{-5}	9.48×10^{-5}	1.21
9.4×10^{-9}	5.48×10^{-10}	8.85×10^{-9}	9.41×10^{-5}	2.25×10^{-4}	2.39
9.4×10^{-9}	1.65×10^{-9}	7.75×10^{-9}	8.80×10^{-5}	2.59×10^{-4}	2.94
9.4×10^{-9}	3.15×10^{-9}	6.25×10^{-9}	7.91×10^{-5}	5.87×10^{-5}	0.74
9.4×10^{-9}	9.64×10^{-10}	8.44×10^{-9}	9.19×10^{-5}	2.03×10^{-4}	2.21
9.4×10^{-9}	8.86×10^{-10}	8.52×10^{-9}	9.23×10^{-5}	1.09×10^{-3}	11.85
1.0×10^{-4}	1.44×10^{-6}	9.86×10^{-5}	9.93×10^{-3}	3.36×10^{-3}	0.34
1.0×10^{-4}	2.46×10^{-6}	9.75×10^{-5}	9.88×10^{-3}	1.49×10^{-2}	1.51
2.0×10^{-4}	-1.26×10^{-5}	2.01×10^{-4}	1.42×10^{-2}	1.05×10^{-2}	0.77
2.0×10^{-6}	1.46×10^{-6}	5.42×10^{-7}	7.36×10^{-4}	5.27×10^{-3}	7.17

Table 6.11 *Diagonal elements for calculation with direction from point F removed*

$\mathbf{D}_o$	$\mathbf{D}_s$	$\mathbf{D}_v = \mathbf{D}_o - \mathbf{D}_s$	$\sigma_v = \sqrt{\mathbf{D}_v}$	v	$t = v/\sigma_v$
9.4×10^{-9}	3.23×10^{-9}	6.17×10^{-9}	7.85×10^{-5}	3.64×10^{-5}	0.46
9.4×10^{-9}	6.04×10^{-10}	8.80×10^{-9}	9.38×10^{-5}	1.36×10^{-4}	1.45
9.4×10^{-9}	1.65×10^{-9}	7.75×10^{-9}	8.80×10^{-5}	2.73×10^{-4}	3.10
9.4×10^{-9}	3.20×10^{-9}	6.20×10^{-9}	7.88×10^{-5}	2.19×10^{-5}	0.28
9.4×10^{-9}	1.00×10^{-10}	8.40×10^{-9}	9.17×10^{-5}	2.75×10^{-4}	3.00
9.4×10^{-9}	0.00	9.40×10^{-9}	9.70×10^{-5}	0.00	0.00
1.0×10^{-4}	1.58×10^{-6}	9.84×10^{-5}	9.92×10^{-3}	7.82×10^{-3}	0.79
1.0×10^{-4}	2.53×10^{-6}	9.75×10^{-5}	9.87×10^{-3}	1.18×10^{-2}	1.20
2.0×10^{-4}	-3.83×10^{-7}	2.00×10^{-4}	1.42×10^{-2}	1.45×10^{-2}	1.02
2.0×10^{-6}	1.59×10^{-6}	4.14×10^{-7}	6.44×10^{-4}	1.04×10^{-3}	1.61

There are now two options: to remove the three directions from F, C and E, or investigate the circumstances of the observations to see if there may be some explanation for these large residuals.

Adopting the latter approach, it is discovered that at the time of the GPS fix, the EDM distances to A and D were measured, so there is no likelihood of mis-centring at G (see Figure 6.6). However, for use as a reference object later, a flagpole was erected at G and kept in place by guy wires. The inward angle observations were taken at different times, those at A and D in the same field visit as the EDM and GPS. The observations from B, C, E and F were acquired at a different time when running the loop surround traverse. The centring of the pole was not checked. (The field books should have been closely inspected at the outset.) However, we could still go

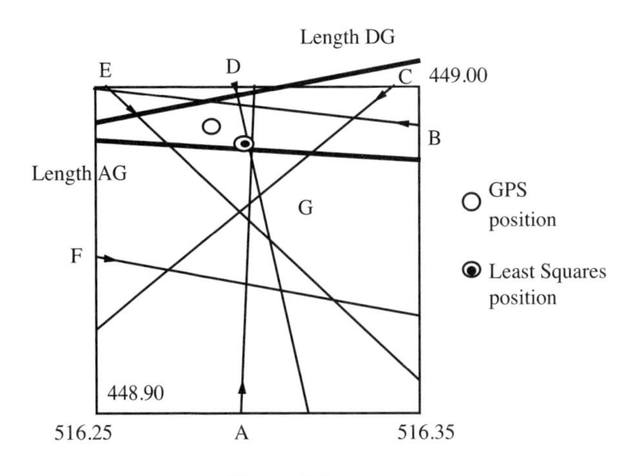

Figure 6.6

into the field and inspect the flagpole. In this particular case it was found that the pole could wobble by about 2–3 cm about the vertical, so we should assign standard errors to the four poor observations in inverse proportion to the lengths of lines, say by 0.025/distance^2.

The respective variances for directions BG, CG, EG and FG are

	BG	CG	EG	FG
Distances	52.000	61.954	80.212	41.684
Variances	2.311×10^{-7}	1.628×10^{-7}	9.714×10^{-8}	3.597×10^{-7}

Repeating the calculations provides the test results as listed in Table 6.12, and $s_o^2 = 1.07$, an acceptable value. All the values of the t test ratios are also acceptable. Therefore, in practice, when the statistical ratio shows a problem in some of the residuals (and therefore in the original observations), it should be investigated whether the problem is due to a blunder or an incorrect estimation of weights.

Common blunders include mixing up the order of the stations for an angle, or misidentifying a station. Correcting this often makes the network solution converge. Over- or under-estimating the observation weights (in this case due to faulty equipment) can similarly cause major problems. The statistical test outlined here will point us towards the cause of the error, and we can make a better estimate of relevant weights.

6.16 Reliability testing

Reliability concerns the ability to check work. For example the *reliability* of a position fix is a measure of the ease with which gross errors may be detected: the

Table 6.12 *Diagonal elements for calculation when standard errors assigned to poor observations*

$\mathbf{D}_o$	$\mathbf{D}_s$	$\mathbf{D}_v = \mathbf{D}_o - \mathbf{D}_s$	$\sigma_v = \sqrt{\mathbf{D}_v}$	v	$t = v/\sigma_v$
9.40×10^{-9}	4.19×10^{-9}	5.22×10^{-9}	7.22×10^{-5}	6.28×10^{-5}	0.87
2.31×10^{-7}	6.67×10^{-10}	2.30×10^{-7}	4.80×10^{-4}	1.30×10^{-4}	0.27
1.63×10^{-7}	2.12×10^{-9}	1.61×10^{-7}	4.01×10^{-4}	2.95×10^{-4}	0.74
9.40×10^{-9}	4.12×10^{-9}	5.28×10^{-9}	7.27×10^{-5}	2.01×10^{-6}	0.03
9.71×10^{-8}	1.26×10^{-9}	9.59×10^{-8}	3.10×10^{-4}	2.66×10^{-4}	0.86
3.60×10^{-7}	1.08×10^{-9}	3.59×10^{-7}	5.99×10^{-4}	1.21×10^{-3}	2.03
1.00×10^{-4}	1.74×10^{-6}	9.83×10^{-5}	9.91×10^{-3}	8.20×10^{-3}	0.83
1.00×10^{-4}	3.04×10^{-6}	9.70×10^{-5}	9.85×10^{-3}	1.09×10^{-2}	1.11
2.00×10^{-4}	-5.89×10^{-7}	2.01×10^{-4}	1.42×10^{-2}	1.28×10^{-2}	0.91
2.00×10^{-6}	1.76×10^{-6}	2.42×10^{-7}	4.92×10^{-4}	5.46×10^{-4}	1.15

greater the ease of detection, the more reliable is the system. Reliability also depends on the amount of redundancy in the problem. Imagine a point fixed by only two measured distances from two known points. There are no redundant measurements, so we have no residuals. Thus the test $t = v / \sigma_v$ is indeterminate and a check for gross errors is not possible. We consider such a fix as *unreliable.* It could in fact be very good, but there is also the chance that it could be very wrong! We need as much redundancy as is practicable. In our example, we have eight degrees of freedom so the fix should be reliable. With this example, we can make some other useful reliability statements to assess quality in comparative terms.

In Section 6.15 we described how outliers are detected, based on a boundary statistic of $2.57 \times \sigma$ with a confidence limit of 99 % (α = 1 % significance level) obtained from a Normal Distribution Table. We now extend this idea to consider the marginal error with respect to two hypotheses: (1) that there is no gross error in a measurement (the null hypothesis) and (2) that there is a gross error (the alternative hypothesis), assuming there is only one outlier in the data. See Figure 6.7.

Illustrated below are these two hypotheses. Residuals are distributed normally about a particular observation, referred to as the *i*th observation. The preferred hypotheses (called the Null Hypothesis) shows the dispersion of residuals in this *i*th observation about a zero mean, and the other hypothesis (called the Alternative Hypothesis) about a mean δ_i^u. The latter distribution is believed to arise from the existence of a maximum gross error in the *i*th observation of Δ_i^u, where

$$\delta_i^u = \frac{\Delta_i^u}{\sigma_i} \tag{6.41}$$

and σ_i is the standard error of the *i*th residual. The *u* indicates the upper limit. Considering the null hypothesis, rejecting residuals outside the bound (defined by α) includes good data and a Type I error is said to have been made. According to statistical theory, very small or very large residuals are possible, but unlikely. A Type I error is a small price to pay for the detection of bad data.

However, outliers may sometimes be so small that the test is passed and data containing outliers are accepted. When this occurs, we say a Type II error has

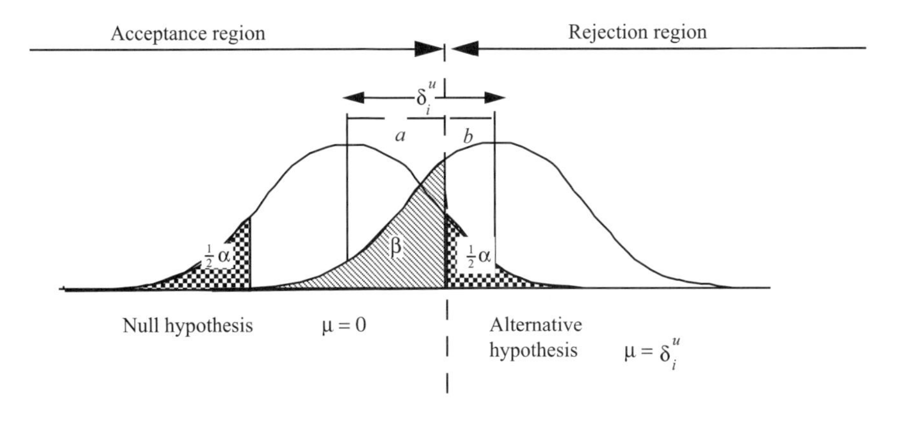

Figure 6.7

occurred. Considering the alternative hypothesis, the probability of such an event is β, or the probability of detecting the outlier is $1-\beta$. This probability is referred to as the *Power of the Test.* To quantify the extent to which outliers can be detected, the power of the test is selected (usually 80 %) and we calculate the magnitude of the outlier that may be detected with this probability. This magnitude is called the Marginal Detectable Error (MDE). The 80 % probability is selected so that we can say the magnitude of the outlier can be found with reasonable certainty.

The choice of α determines the rejection limit and so affects our actions. The choice of β has no real effect on the rejection process; it merely affects any statement about the data quality. To make comparisons between different sets of data, it is usual to accept these probability levels for both distributions.

For example, in Figure 6.7 we select both α and β, find a and b from Normal Tables and calculate the MDE from

$$\text{MDE} = \Delta_i^u = \delta_i^u \sigma_i = (a+b)\sigma_i \tag{6.42}$$

where σ_i is the standard error of the residual v_i.

6.17 Internal reliablity

The internal reliability is expressed in terms of MDEs. It is common to test the hypothesis that only one observation has a gross error and we calculate the MDE for each observation.

Columns 4, 5, and 6 of Table 6.12 list values of σ, v and t. We have already shown that since no value of t exceeds the test statistic 2.57, no further data are rejected. From Normal Distribution Tables $\alpha = 0.01$, the entry of 99.5 % (two-tailed) yields $a = 2.57$ and 80 % (one-tailed) $b = 0.84$. Hence, we obtain the MDEs for each observation. The first MDE is given by Equation (6.42) i.e.

$$\text{MDE for AG } = (a+b)\sigma_i = (2.57+0.84)\times 7.22\times 10^{-5} = 2.47\times 10^{-4}$$

All the MDEs are listed in Table 6.13.

The quality is an assessment of the size and nature of undetected errors that remain in the solution. In this case, the largest MDE is 4.84×10^{-2} or approximately 5 mm in the GPS Easting. Such a low figure indicates a high internal reliability. We can say of the quality when outlier detection is carried out with a level of significance of 1 %, there is an 80 % chance an outlier of 5 mm will be detected.

6.18 External reliability

External reliability is a more useful concept because a large undetected outlier may have little effect on the solution. External reliability is assessed by the largest effect of an observational MDE on the solution, in this case on the coordinates of G. We compute the separate effect of each MDE on the solution, and quote the largest to describe the quality of the fix. We make a series of computations to find the contribution of each MDE in turn, using the formula

$$\mathbf{dx} = \mathbf{N}^{-1}\mathbf{A}^{\mathrm{T}}\mathbf{W}\,\mathbf{dL} \tag{6.43}$$

Table 6.13

	MDE	d*L*	d*E*	d*N*
AG	2.47×10^{-4}	2.47×10^{-4}	6.89×10^{-3}	-2.90×10
BG	1.64×10^{-3}	0.00×10^{-0}	1.63×10^{-4}	2.30×10^{-4}
CG	1.37×10^{-3}	0.00×10^{-0}	-1.46×10^{-3}	2.36×10^{-4}
DG	2.48×10^{-4}	0.00×10^{-0}	-6.85×10^{-3}	4.89×10^{-5}
EG	1.06×10^{-3}	0.00×10^{-0}	-1.47×10^{-3}	-1.19×10^{-4}
FG	2.05×10^{-3}	0.00×10^{-0}	-2.36×10^{-4}	-2.26×10^{-4}
Dist AG	3.39×10^{-2}	0.00×10^{-0}	1.38×10^{-4}	5.82×10^{-4}
Dist DG	3.36×10^{-2}	0.00×10^{-0}	1.63×10^{-3}	-6.22×10^{-4}
Easting	4.84×10^{-2}	0.00×10^{-0}	3.90×10^{-3}	1.25×10^{-3}
Northing	1.68×10^{-3}	0.00×10^{-0}	-8.78×10^{-5}	1.61×10^{-3}

Note: Note the third column is typical only for the calculation of the effect of d*L* from AG. The values in columns four and five are from successive calculations for each individual contribution d*L* in turn

where **dL** = MDE (column 2, Table 6.13) is zero except in the case of observation AG. **dx** is calculated:

$$\mathbf{dx} = \begin{bmatrix} dE \\ dN \end{bmatrix} = \begin{bmatrix} 1.6454\times10^{-5} & -5.887\times10^{-7} \\ -5.887\times10^{-7} & 1.7579\times10^{-6} \end{bmatrix} \mathbf{A}^{\mathrm{T}}\mathbf{W}\,\mathbf{dL} = \begin{bmatrix} 6.89\times10^{-3} \\ -2.90\times10^{-4} \end{bmatrix}$$

Thus the effects of the MDE in the first observation on the coordinates of G are approximately 7 mm in Easting and 0.3 mm in Northing. Columns three and four of Table 6.13 show the MDEs in position caused by the MDE of each observation in turn. It can therefore be seen that the external reliability is also good, as the largest MDEs in position are 6 mm in Easting and 2 mm in Northing.

6.19 Variance (Fisher F) test

Although this test has a wide application, it is used mainly in Least Squares problems to investigate if the value calculated for the unit variance is close to the expected value of unity.

In a Least Squares problem, the sample statistic is the unit variance calculated from the weighted residuals i.e. from

$$\frac{\mathbf{v}^{\mathrm{T}}\mathbf{W}\,\mathbf{v}}{m-n}$$

This is compared with the theoretical value for the population variance. Thus the test statistic is given by

$$t = \frac{s_o^2}{\sigma_o^2}$$

Since $\sigma_o^2 = 1$, in theory the test is

$$t = s_o^2$$

To check if the value for the unit variance of 1.07 is acceptable, we compare the value from $r_1 = 8$ degrees of freedom with the theoretical value of 1.0 from an infinite number of values i.e. $r_2 = \infty$. From Table 5.4 we have the bound of 1.94. Thus the tested value is acceptable.

6.20 Test for normality: χ^2 test

In addition to or as an alternative to the F test, most software packages incorporate the χ^2 test on the residuals to determine if the error distribution of the observations is Normal. The test statistic is

$$t = \frac{(m-n)\, s_0^2}{\sigma_0^2}$$

Since $\sigma_0^2 = 1$ in theory, the test becomes

$$t = (m-n)\, s_0^2$$

Since $r_1 = 8$ degrees of freedom from Table 5.6, the acceptable bounds are 2.18 and 17.53. In this case $t = 8.56$ and so passes the test.

In practice this test can be misused to authenticate the quality of work, by removing observations with large residuals from the data until the test is passed. Many surveyors prefer to treat the test with a degree of suspicion.

6.21 Dispersion matrices of derived quantities

So far we have assumed or obtained measures of *precision*: variances found as the diagonal terms of the dispersion matrices and co-variances as their off-diagonal terms. It is also possible to extend the analysis to obtain measures of precision of any parameter derived from the final coordinates of a network. This parameter could be an important dimension such as the location of bridge piers, or a bearing on which to base a tunnel drive.

Suppose we want to calculate the standard error in the length of the side BG which has not been directly measured. Its length is calculated from the coordinates in the usual way. We can also obtain its standard error by setting up the coefficients of an observation equation for this length BG. The elements of the **A** matrix are

$$\frac{(E_G - E_B)}{BG} \quad \text{and} \quad \frac{(N_G - N_B)}{BG}$$

We then apply the formulae

$$\mathbf{D}_s = \mathbf{A}\,\mathbf{D}_x\,\mathbf{A}^T \tag{6.44}$$

Table 6.14

Point	*E*	*N*
B	567.895	443.275
G	516.2	448.9
Δ	–51.695	5.625
BG	52.000	
A	–0.9941321	0.1081728

Consider the following example, referring to Table 6.14.

The dispersion matrix of the calculated coordinates derived by the Least Squares method is:

$$\mathbf{D}_x = \sigma_0^2 \mathbf{N}^{-1} = 1 \times \begin{bmatrix} 1.6454 \times 10^{-5} & -5.887 \times 10^{-7} \\ -5.887 \times 10^{-7} & 1.7579 \times 10^{-6} \end{bmatrix}$$

The variance of the length of BG is then obtained from (6.39)

$$\sigma_x^2 = \begin{bmatrix} -0.9941328 & 0.1081728 \end{bmatrix} \begin{bmatrix} 1.6454 \times 10^{-5} & 5.887 \times 10^{-7} \\ 5.887 \times 10^{-7} & 1.7579 \times 10^{-6} \end{bmatrix} \begin{bmatrix} -0.9941328 \\ 0.1081728 \end{bmatrix}$$

i.e. the standard error of the length of BG is 0.00405.

In the same way we could estimate the standard error of a bearing such as AG using the formula for the coefficients in the **A** matrix

$$\frac{(E_G - E_A)}{AG} \quad \text{and} \quad \frac{(N_G - N_A)}{AG}$$

The rule is to derive the coefficients of the equation as if for an observed direction. In fact any function of the coordinates, such as an area (or in the case of three dimensions, a calculated volume), can be treated in this way. Many of the procedures described here can be completed without any observations at all. A pre-analysis can be made when designing a measurement programme, before even going into the field. Such an analysis can usually save money and wasted effort.

Note: A graphical way to display statistical results is by error ellipses, as described in Appendix 6.

Chapter 7
Satellite Surveying

7.1 Introduction

So far we have been dealing with surveying systems that are local and ground based over which the surveyor generally has complete control. Among other issues, he or she has to make decisions about the following:

1. what measurements to make;
2. what quality to accept;
3. how to process results.

Even with the use of commercial software packages, the surveyor usually has to make choices.

Alternatively, when a global surveying system such as Global Positioning System (GPS) is used, the situation is different. The surveyor operates instruments and processing software supplied by manufacturers which exploit the commercial opportunities of satellite systems provided by the governments of the USA, Russia (and in future by the EEC) following internationally agreed protocols. GPS is generally referred to as the most typical. At various stages, there is some element of choice based on cost of equipment, operation time in the field, availability of post-measurement analysis and a choice of coordinate system output, to achieve a quality of output to suit the particular needs of the survey. Although in making these choices the surveyor is guided by the equipment supplier, he or she is also responsible to the client.

It is therefore good practice to adopt a careful, standard survey approach and incorporate some measure of checks into the work, such as also measuring a few lines by conventional Electronic Distance Meter (EDM) as part of a GPS survey and checking on marked points at different times of the survey. It is also good practice to carry out a sample field check for quality of results before leaving the site, and to accept the coordinates output to the basic International Terrestrial Reference Frame (ITRF) system as an archival backup for possible future reference, even if the current output is in some derived form such as plane projection coordinates.

Although in the latter case, positions are determined from measurements of length made in three dimensions, there is a very significant difference between them and static work, in that the distance measurements are time related. This aspect of relating time from site to site is continually being addressed by the provision of radio links between reference ground stations and the roving stations used by surveyors to carry out their work. It must be realized that each time a receiver begins to determine position, an iterative process takes place to establish a good starting point for the subsequent relative calculations.

Global Positioning System—GPS

To guide the surveyor in making sensible choices, a basic understanding of GPS is required. GPS is available on a worldwide basis to provide coordinated positions for many purposes. It consists of about 24 satellites and some ground control stations from which a receiver can measure its position to varying degrees of accuracy, from a few metres to better than a centimetre.

Each satellite has its own signature and carries its position for use by the ground receiver. This information can be updated later for more precise calculations.

The satellite to ground distance S is obtained from the equation

$$S = VT$$

where V is the speed of radio signals and T the time taken to travel the distance (only about 0.1 sec). T is measured in two ways, one approximate and the other very accurate as described in Chapter 13 where it is shown that the above equation can be written in the form

$$S = n\lambda + \Delta\lambda$$

where n is the integer number of wavelengths λ contained in the line and $\Delta\lambda$ represents the fractional part. Using a coarse acquisition code (C/A), an attempt is made to find n by directly timing the interval between the satellite and the receiver. This gives an approximate value for S which is accurate to about 50 m. All that is then needed to fix a point is at least three distances from three satellites intersecting at an angle between 30° and 150°.

Since the receiver clock does not function according to the satellite atomic clock, its error has to be determined. This is carried out by measuring distance to a fourth satellite at the same time. (Receivers can operate on several channels simultaneously.) The redundant length enables the receiver clock error to be eliminated by subtraction (see below). To achieve greater accuracy in timing, the two carrier waves of the radio signals are employed, as described in Chapter 13.

The receiver position in terms of the ITRF is found by variation of coordinates as described below.

Before proceeding further, it is important to stress that several things can go wrong with the measurements which do not normally apply to traditional EDM. The technology depends on matching the wave patterns of two signals by tuning circuits to each other just as when tuning a radio receiver to a station. When they are matched they are said to be locked. The radio signals can 'lose lock' so that the initiation process (C/A code) fails and has to be repeated. The satellites may be positioned in such a way to give a bad intersection at the receiver. The signals may be reflected from surrounding structures or be subject to interference from other radio sources. It is therefore vital that the set of field procedures and processing techniques recommended by the instrument manufacturers are strictly adhered to in all GPS work.

Traditionally, point positions on the surface of the Earth were established from observations made of the stars (especially the Sun), planets and the Earth's moon by optical instruments at known times. With the advent of artificial Earth satellites different systems came into use, including stellar photography, electromagnetic distance measurement (Sequential Collation of Range, SECOR and GPS), laser ranging, range difference measurements (Transit Doppler) or interferometric systems (Minitrack or Very Long Baseline Interferometry). Direct heights are also obtained by satellite radar and from the stereoscopic analysis of remote sensing imagery and aerial photography.

These have been superseded by GPS which provides three-dimensional coordinates of points worldwide on the ITRF (see Iliffe 2000 for more details).

Most instrument providers will also output results in a wide variety of formats to suit users, such as a local datum or a national map projection. The ITRF values should always be archived for security.

The Earth and its artificial satellites are in constant motion, so to give any meaning to distance measurements between them we must give serious consideration to timing and its possible errors. To set the scene, a point on the Earth's equator is moving through space at a speed of 464 m s^{-1} and a GPS surveying satellite is moving in orbit at 7 km s^{-1}. Combining the two with an accuracy of one millimetre therefore requires a timing accuracy of 10^{-10} sec.

Unlike ground based EDM, which uses the time for a signal to travel to and from each end of a line, GPS uses the single time of flight measured by two separate clocks. It is clear that a receiver clock cannot achieve the accuracy attainable by an atomic clock or by selective ground stations controlling the operation. Since it is assumed that each receiver clock is wrong, this error has to be eliminated by some observing technique, usually differencing.

Other phenomena affect the whole operation, such as time delays in passing through the Earth's atmosphere, also dealt with by mathematical modelling. Some means of timing links have to be established between ground stations, and roving surveyors fixing the points of a survey in their work. With the most sophisticated modelling it is possible to yield the coordinates of points on the ground to a relative precision of a millimetre within a particular reference system.

The satellites are monitored by a control segment consisting of ground stations which transmit positional information (ephemerides) back to the satellites. These in turn relay this information to the receivers at the roving stations. Depending on the mode of operation (one or two frequencies), datum consideration and observational routines, the system is capable of a maximum accuracy of 2–3 mm. Simpler arrangements with single receivers provide position with an accuracy of about 50 m.

However, there remain two problems for the geospatial surveyor. The first is how to relate the satellite coordinate reference system to historically established systems used for mapping and other civil purposes such as land certification. The second is how to relate heights to the Earth's level surface and sea level.

The former problem is resolved by coordinate transformation (see Chapter 4) and the latter by knowledge of the geoid-spheroid separation (see Chapter 9).

7.2 How GPS satellites are used by the surveyor

The Global Positioning System is adopted by many commercial system providers serving a wide variety of users, ranging from professional geospatial surveyors and map makers, environmentalists, public services such as police, ambulance and fire services, to motorists and ramblers. These providers are generally commercial companies producing a range of products of varying sophistication. The geospatial surveyor usually operates at the upper end of this technical spectrum. The aim of this chapter is to explain the basic principles of the GPS and other related phenomena.

Point fixation

The position of a single point is of very little use to a surveyor. At least two are normally required, giving direction (azimuth/bearing) as well as position and inter-point distance i.e. a vector.

A GPS survey, based on a control network covering the area within which minor points are interpolated (see Chapter 2), normally consists of pairs of vectors forming the basic design unit. These vector pairs are derived from three receivers fixing from the same satellites at the same time. For consistency, these vectors link together into a connected network, with most stations visited at least twice.

Once the control has been established, the breakdown proceeds with single visits to points, referenced to a site based receiver, part of a reference network of receivers established by large organizations, usually governments, via a real time radio link. Such practices are continually being improved. See the **Summary of Operational Practice** at the end of this chapter.

Basic geometrical principles of point positioning

Figure 7.1 illustrates part of the GPS system in which the known positions of satellites S_1 to S_4 are shown in orbit round the Earth. The 24 satellites, at heights of about 20 000 km, are spread in six evenly spaced almost circular orbits with orbital periods of 12 hours. Their Ascending nodes have Right Ascensions evenly spaced at 60°

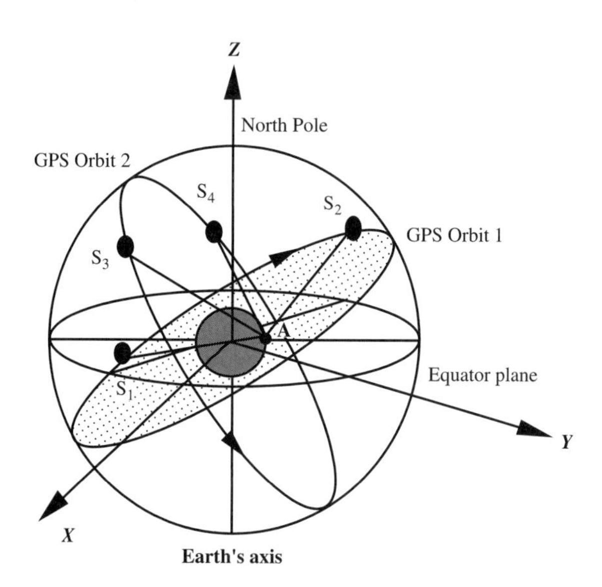

Figure 7.1

intervals round the equator. (Note that one sidereal day is about 4 minutes shorter than a solar day.)

To establish the coordinates of a point A on the surface of the Earth, distances are measured simultaneously from at least four satellites to A. Four distances are required, as the receiver clock is not sufficiently accurate for the timing process. Its error on GPS time has to be resolved for each determination of position. There are four equations linking the five points in space, described in a three dimensional Cartesian system *XYZ*. These are of the form

$$S^2 = \Delta X^2 + \Delta Y^2 + \Delta Z^2 \quad \text{or} \quad S^2 - \Delta X^2 - \Delta Y^2 - \Delta Z^2 = 0 \qquad (7.1)$$

Specifically, the equations for all four distances are

$$\begin{aligned}
({}_AS_1 + E + K)^2 - ({}_A\Delta X_1)^2 - ({}_A\Delta Y_1)^2 - ({}_A\Delta Z_1)^2 &= 0 \\
({}_AS_2 + E + K)^2 - ({}_A\Delta X_2)^2 - ({}_A\Delta Y_2)^2 - ({}_A\Delta Z_2)^2 &= 0 \\
({}_AS_3 + E + K)^2 - ({}_A\Delta X_3)^2 - ({}_A\Delta Y_3)^2 - ({}_A\Delta Z_3)^2 &= 0 \\
({}_AS_4 + E + K)^2 - ({}_A\Delta X_4)^2 - ({}_A\Delta Y_4)^2 - ({}_A\Delta Z_4)^2 &= 0
\end{aligned} \qquad (7.2)$$

where ${}_AS_1$ is the measured distance between ground station *A* and satellite number 1, ${}_A\Delta X_1 = X_1 - X_A$ etc., *E* is the unknown receiver clock error and *K* is a refraction error. (Note: A distance such as ${}_AS_1 + E + K$ is known as a *pseudo range*.)

Since the coordinates of the four satellites can be calculated from the time of observation (see below) and the four pseudo-ranges are known from the measurement system, to obtain the coordinates of A we solve the simultaneous Equations (7.2). To simplify the arithmetic, the examples given here deal with a two-dimensional treatment.

7.3 Differential ranging

Figure 7.2 illustrates two new points A and B. Ranges *S* to satellites are measured and reduced to a common epoch. Clock index and residual tropospheric errors at points A and B are denoted by K_A and K_B respectively. Clock and ionospheric errors at the satellites are represented by K^{18} etc. Since the ground distance AB (60 km) is very small compared to the Earth–satellite distances (26 000 km), the error model is representative. For example, the range from A to satellite 18 can be expressed in the form

$$S_A^{18} = r_A^{18} + K_A + K^{18} \qquad (7.3)$$

Subscripts are used for ground stations and superscripts for high satellites. The range r_A^{18} is the crude distance measurement, often called the pseudo-range. This would be the same as measuring an EDM distance without applying index and refraction corrections.

The measurement and computation is usually carried out in three dimensions (*X*, *Y*, *Z*). Here we reduce the problem to two dimensions (*E*, *N*) as listed in Table 7.1 below, for simplicity. The principles involved are valid in either two or three dimensions.

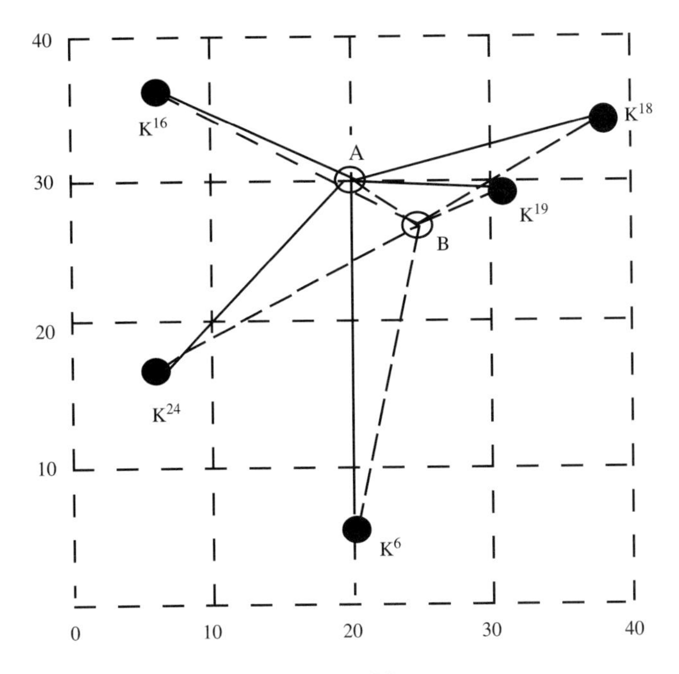

Figure 7.2

Adopting the usual procedure for Least Squares estimation based on provisional coordinates according to Equation (6.1)

$$\mathbf{Ax} = \mathbf{L} + \mathbf{v}$$

where L_i represent the 'observed minus provisional' terms and v_i the residuals about the mean. Usually, the satellite coordinates are considered fixed, so we have the typical observation equation. We obtain linear equations of the form:

$$(\mathrm{d}E^{18} - \mathrm{d}E_{\mathrm{A}})\sin U_{\mathrm{A}}^{18} + (\mathrm{d}N^{18} - \mathrm{d}N_{\mathrm{A}})\cos U_{\mathrm{A}}^{18} - K_{\mathrm{A}} - K^{18} = L_{\mathrm{A}}^{18} + v_{\mathrm{A}}^{18} \quad (7.4)$$

Table 7.1

Data points			*Horizontal distances from*	
N	*E*	*Satellite/Point*	*A*	*B*
3431.47	3730.53	18	1783.5079	1781.4292
2946.97	3011.82	19	1013.2087	1009.3850
1824.27	739.19	24	1723.9440	1722.7334
3521.06	829.67	16	1281.0838	1286.6094
885.13	2000.00	6	2114.8700	2109.2740
3000.00	2000.00	*A*	0	6.6304
2994.40	2003.55	*B*	6.6304	0

where U is the bearing of the line. Usually, the satellite coordinates are considered fixed, so we have

$$-\mathrm{d}E_{\mathrm{A}} \sin U_{\mathrm{A}}^{18} - \mathrm{d}N_{\mathrm{A}} \cos U_{\mathrm{A}}^{18} - K_{\mathrm{A}} - K^{18} = L_{\mathrm{A}}^{18} + v_{\mathrm{A}}^{18} \tag{7.5}$$

In the example below, the following error values in metres have been assigned:

$K_A = 0.2$; $K_B = -0.3$; $K^{18} = 10$; $K^{19} = 11$; $K^{24} = 12$; $K^{16} = 13$; $K^{6} = 14$

Table 7.1 gives the standard coordinates from which the other data such as bearings are calculated.

Table 7.2 demonstrates how the errors are applied to the horizontal distances. For example, the range from A to satellite 18 is changed by

$$K_{\mathrm{A}} + K^{18} = 0.2 + 10$$

We will demonstrate that the effects of such errors can be eliminated from the solution by the method of differencing.

Provisional coordinates adopted for A and B are: (3000.1, 2000.1) and (2994.4, 2003.55) respectively, in the (E, N) reference frame.

The observation Equations (7.5) provide the direct solution for $(\mathrm{d}E_{\mathrm{A}}, \mathrm{d}N_{\mathrm{A}})$ and $(\mathrm{d}E_{\mathrm{B}}, \mathrm{d}N_{\mathrm{B}})$ from the pseudo ranges given in Tables 7.3 and 7.4.

Table 7.2

	Distance error	*Obs A*	*Obs B*	*Obs S_A = Obs A + K_A + K^{sat}*	*Obs S_B = Obs B + K_B + K^{sat}*
K_A	0.2				
K_B	–0.3				
K^{18}	+10	1783.5079	1781.4292	1793.7079	1791.1292
K^{19}	+11	1013.2087	1009.3850	1024.4087	1020.0850
K^{24}	+12	1723.9440	1722.7334	1736.1440	1734.4334
K^{16}	+13	1281.0838	1286.6094	1294.2839	1299.3094
K^{6}	+14	2114.8700	2109.2740	2129.0700	2122.9740

Table 7.3 *Values of observation equations (Equations 7.5)*

Satellite	*–cos U_A*	*–sin U_A*	*Obs S_A*	*L_A*
18	–0.241 9881	–0.970 2792	1793.707 9	10.272 8332
19	0.052 2454	–0.998 6343	1024.408 7	11.305 0924
24	0.681 9402	0.731 4079	1736.144 0	12.195 0590
16	–0.406 7699	0.913 5307	1294.283 9	13.067 9709
6	1	4.7286×10^{-5}	2129.070 0	14.299 9976

Table 7.4 *Values of observation equations (Equations 7.5)*

Satellite	*–cos* U_B	*–sin* U_B	*Obs* S_B	L_B
18	–0.24534790	–0.9694351	1791.1292	9.7
19	0.04698901	–0.9988954	1020.0850	10.7
24	0.67922871	0.73392667	1734.4334	11.7
16	–0.40933930	0.91238223	1299.3099	12.7
6	0.99999858	0.00168304	2122.9730	13.7

The solutions are of the form as given in Equation (6.7):

$$\mathbf{Nx} = \mathbf{b}$$

which can be written as a series of n linear equation in n unknowns (see Chapter 6). For point A,

$$\mathbf{x}_\text{A} = \begin{bmatrix} dE_\text{A} \\ dN_\text{A} \end{bmatrix} = \begin{bmatrix} 1.69179189 & 0.30984896 \\ 0.30984896 & 3.30820811 \end{bmatrix}^{-1} \begin{bmatrix} 15.4053774 \\ -0.3989376 \end{bmatrix} = \begin{bmatrix} 9.2874 \\ -0.9904 \end{bmatrix}$$

and for point B

$$\mathbf{x}_\text{B} = \begin{bmatrix} dE_\text{B} \\ dN_\text{B} \end{bmatrix} = \begin{bmatrix} 1.691311 & 0.317625 \\ 0.317625 & 3.308690 \end{bmatrix}^{-1} \begin{bmatrix} 14.5712547 \\ 0.1055529 \end{bmatrix} = \begin{bmatrix} 8.7674 \\ -0.8097 \end{bmatrix}$$

Summarizing these results,

A = (3000.1 + 9.29 = 3009.39, 2000.1 – 0.99 = 2000.01)

B = (2994.4 + 8.77 = 3003.17, 2003.55–0.81 = 2002.74)

Unsurprisingly, there are considerable errors in coordinates due to the length errors and the failure to model them.

7.4 Single differences

To eliminate the effect of the large errors at each satellite, the equations for A and B are differenced before solving, giving a typical range-difference equation:

$$\begin{aligned} &-dE_\text{A} \sin U_\text{A}^{18} - dN_\text{A} \cos U_\text{A}^{18} + dE_\text{B} \sin U_\text{B}^{18} + dN_\text{B} \cos U_\text{B}^{18} - K_\text{A} + K_\text{B} \\ &\quad = L_\text{A}^{18} - L_\text{B}^{18} + v_\text{A}^{18} - v_\text{B}^{18} \\ &\quad = L_\text{AB}^{18} + v_\text{AB}^{18} \end{aligned} \tag{7.6}$$

The observation equations of this type are given in Table 7.5.

Table 7.5 *Single difference*

Satellite	*–cos U_A*	*–sin U_A*	*–cos U_B*	*–sin U_B*	L_{AB}
18	–0.241 988	–0.970 279	0.245 347	0.969 435	0.5728
19	0.052 245	–0.998 634	–0.046 989	0.998 895	0.6051
24	0.681 940	0.731 407	–0.679 228	–0.733 926	0.4951
16	–0.406 769	0.913 530	0.409 339	–0.912 382	0.3679
6	1.000 000	4.72×10^{-5}	–0.999 998	–0.001 683	0.5999

Solutions are of the form $\mathbf{Nx} = \mathbf{b}$, where

$$\mathbf{N} = \begin{bmatrix} 1.691791 & 0.309848 & -1.691525 & -0.313451 \\ 0.309848 & 3.308208 & -0.314027 & -3.308443 \\ -1.691525 & -0.314027 & 1.691311 & 0.317625 \\ -0.313451 & -3.308443 & 0.317625 & 3.308688 \end{bmatrix}$$

therefore

$$\mathbf{x}_{\mathrm{AB}} = \begin{bmatrix} dE_{\mathrm{A}} \\ dN_{\mathrm{A}} \\ dE_{\mathrm{B}} \\ dN_{\mathrm{B}} \end{bmatrix} = \begin{bmatrix} 23878.098 & 30322.594 & 23821.688 & 30295.636 \\ 30322.594 & 331639.785 & 29620.216 & 331644.309 \\ 23821.688 & 29620.216 & 23767.371 & 29593.180 \\ 30295.636 & 331644.309 & 29593.180 & 331649.176 \end{bmatrix} \begin{bmatrix} 0.680913 \\ -0.461803 \\ -0.673519 \\ 0.459671 \end{bmatrix} = \begin{bmatrix} 137.526 \\ -7.478 \\ 137.161 \\ -7.477 \end{bmatrix}$$

Summarizing these results,

$$\mathrm{A} = (3000.1 + 137.53 = 3137.63,\ 2000.1 - 7.48 = 1992.63)$$

$$\mathrm{B} = (2994.4 + 137.16 = 3131.56,\ 2003.55 - 7.48 = 1996.07)$$

The very large changes are due to the assigned errors at A and B being large in relation to the short line AB and no attempt being made to model them. Such large errors occur in practice on account of the uncertainties in ionospheric and tropospheric refraction in satellite systems, or instability of reference marks in monitoring surveys.

This indicates the magnitude of the positional errors that can be present in single simple receiver results.

However, the differences in coordinates between A and B are preserved and are little affected by the large errors introduced.

7.5 Double differences

If each equation for the single difference is now subtracted from one reference equation at a station, in this case the first, the remaining un-modelled station errors are eliminated. These double difference equations are given in Table 7.6.

Table 7.6 *Each row is obtained by subtracting the first row of Table 7.7 from each individual row of Table 7.7*

18	*0*	*0*	*0*	*0*	*0*
19	0.294 23	–0.028 35	–0.2923	0.029 46	0.032 25
24	0.923 92	1.701 68	–0.9245	–1.703 36	0.07777
16	–0.164 78	1.883 80	0.1639	–1.881 81	0.20486
6	1.241 98	0.970 32	–1.2453	–0.971 11	0.027 16

Solutions are of the form **Nx** = **b** as before, where

$$\mathbf{N} = \begin{bmatrix} 2.509904 & 2.458610 & -2.513986 & -2.461143 \\ 2.458610 & 7.386816 & -2.464515 & -7.386712 \\ -2.513986 & -2.464515 & 2.518084 & 2.467052 \\ -2.461143 & -7.386712 & 2.467052 & 7.386616 \end{bmatrix}$$

therefore

$$\mathbf{x}_{AB} = \begin{bmatrix} dE_A \\ dN_A \\ dE_B \\ dN_B \end{bmatrix} = \begin{bmatrix} 120999.3 & -28692.7 & 120891.1 & -28753.7 \\ -28692.7 & 218515.9 & -28988.1 & 218640.4 \\ 120891.1 & -28988.1 & 120784.1 & -29049.5 \\ -28753.7 & 218640.4 & -29049.5 & 218765.1 \end{bmatrix} \begin{bmatrix} 0.005129 \\ -0.492825 \\ -0.004947 \\ 0.492561 \end{bmatrix} = \begin{bmatrix} 0.100865 \\ -0.101241 \\ 0.000865 \\ -0.001241 \end{bmatrix}$$

In this case, the solution is almost perfect, exactly as predicted, showing that the effects of the very large imposed errors have been eradicated.

In practice, it is important to make simultaneous measurements with two or more receivers to allow the differencing process to proceed, and to ensure that a sufficient number of satellites have been observed to secure the necessary redundancy of measurement.

Guidance as to potential accuracy is obtained from the pseudo-range pre-analysis using the orbital prediction methods outlined below. Practical systems include computer software for this prediction.

Other differencing strategies could be adopted, such as swapping the receivers over and repeating measurements at a later epoch, or differencing the double differences between two epochs i.e. triple differencing. The latter helps to isolate an individual error in a line due to an integer miscount (also referred to as cycle slips; see Chapter 13).

Summary

In addition to the differencing strategies, the various schemes to reduce timing errors by comparison with base stations linked by radio, and to the field by Blue Tooth technology, has improved the accuracy of GPS and other similar systems to the 5 mm mark.

GPS receivers are now employed in a wide variety of hosts, such as aircraft to determine tilts, ships to relate to sea-bed transponder calibration, and in many direct ways for simple navigation.

New systems (such as Galileo) are being developed which, having stronger signals, will penetrate within buildings and into shallow tunnels, thus opening up new applications and reducing even further the need for total stations.

7.6 Basic orbital mechanisms

Background

We now use a simple spherical model to explain those mechanisms which underlie past, present and future satellite systems. A brief study of the Earth–Sun relationship assists with an understanding of basic concepts, if only for the reason that the Sun and stars are readily visible, whereas most artificial satellites are generally not. The Sun's position is also of interest for general civil use, such as finding the direction of Mecca and studying the regime of sunlight on buildings.

The parameters (the ephemerides) needed to locate the positions of various satellites are available on the internet or from system providers, and stellar information from the Star Almanac for Land Surveyors (SALS). Satellite information is also transmitted by satellite systems in real-time (broadcast ephemeris), or can be obtained later as a precise ephemeris.

Rotating systems

A spinning wheel, such as a child's top or gyroscope, maintains its axis of spin at a constant direction in space unless subjected to any other forces which disturb this direction. This can be verified by moving a toy gyroscope around, always keeping the axis of spin parallel to its original direction. Thus the Earth spins around its axis while moving round the Sun in an orbital plane (the ecliptic), always keeping its north–south axis in the same direction (within about one second of arc).

Thus we see the stars apparently rotating round on a 24-hour cycle (the Sidereal Day) maintaining relatively identical positions throughout the year defined by one orbital trip, in approximately 366 sidereal days.

The Earth 'spinning' round the Sun is also a gyroscopic system maintaining the orbital plane at a fixed direction in space. Any attempt to change these directions of spin brings into action forces which resist this change. This fact gives rise to very small changes in the Earth's behaviour (daily wobble) and a rotation of the orbital plane (annular precession).

In a similar way, natural satellites such as the Earth's Moon, or artificial Earth satellites such as GPS vehicles, orbit around their central attracting mass.

The most important elementary fact about all these spinning systems is that they maintain axes with relatively fixed directions in space. This fact is used to establish the primary mathematical models used for various calculations. In most professional applications, these simple models have to be modified to yield the necessary accuracy for geospatial surveying.

Orbits

It was established by Kepler and others that an object orbiting around a central mass behaves as follows:

1. The orbit lies in a plane approximately fixed *in direction* in space. It is usually defined by the inclination i of its orbital plane to the plane of the equator.
2. The shape of the orbit is an ellipse with the central mass at one focus.
3. The period T of the rotating object is directly linked to its distance r from the central attracting mass and a constant C by the expression:

$$T^2 = Cr^3$$

4. The radius vector r, joining the focus of the ellipse to the satellite, sweeps out an area at a constant rate. The *areal* velocity is constant.

Practical departures from these Keplerian rules arise from the action of other forces on satellites and the fact that the Earth's gravity field is irregular. Since most satellites used for geodetic purposes lie in nearly circular orbits, whose treatment is simple and illuminating, we shall adopt a very simple model for our discussions here.

7.7 Elliptical orbits

According to the Keplerian theory of idealized orbits, a satellite moves in an ellipse (i.e. a plane curve) in such a way that the rate of change of the area, swept out by the vector joining the satellite to a focus, is constant. That is, the *areal* velocity is constant.

Consider a satellite moving with constant speed through space unaffected by gravity. In Figure 7.3, the satellite S moves from A to B in unit time and from B to C in unit time. Relative to a reference point P the vector PS sweeps out an area at a constant rate, as the areas of triangles PAB and PBC are equal.

Consider now the satellite at B moving towards P, under the influence of gravity alone. Suppose in unit time it moves through a distance BD. The net effect of both motions on the satellite is to move it along BE. The area of triangle PBE = area of triangle PCB = area of triangle PAB. Since each of these triangles is swept out in unit time, this means that the areal velocity is constant.

Referring to Figure 7.4, the satellite S moves around the orbital ellipse with the Earth located at the focus F. The point of closest approach (perigee) is at P. The

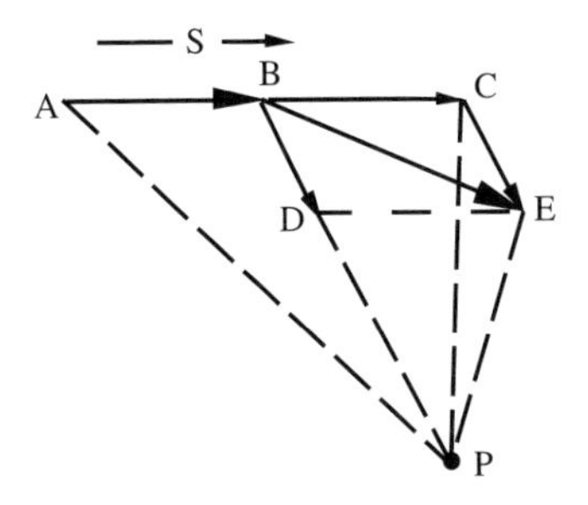

Figure 7.3

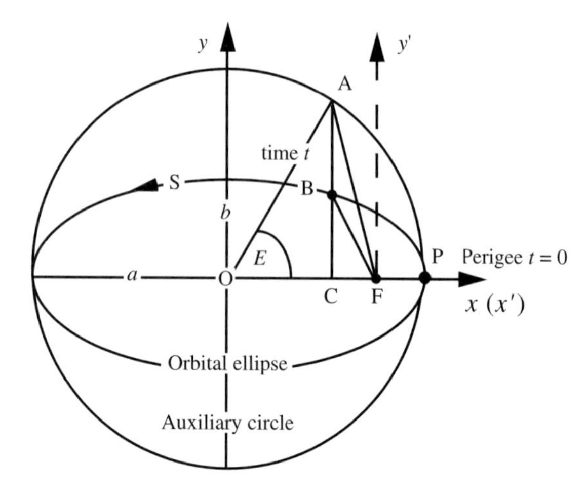

Figure 7.4

satellite's period T and the time of observation t (when the satellite is at position B) can be measured.

To derive the Cartesian coordinates of the satellite referred to the Earth at F, we require to calculate the angle E corresponding to the time of observation t. Notice that E is the angle AOF where A is the projection of CB on to the auxiliary circle. Once we have obtained E we may obtain the coordinates of B (x', y') relative to the origin at F from

$$x' = a(\cos E - e) \tag{7.7}$$

$$y' = b\sin E \tag{7.8}$$

The connection between the *eccentric anomaly* E and the time of observation t is given by Kepler's equation

$$M = E - e\sin E \tag{7.9}$$

where the *mean anomaly*

$$M = \frac{2\pi t}{T}$$

Equation (7.9) is transcendental since we wish to find E from M using

$$E = M + e\sin E \tag{7.10}$$

Equation (7.10) is solved by iteration, starting from

$$E' = M + e\sin M \tag{7.11}$$

Note: This is the most significant equation in the whole orbital calculation, and its solution has been given much attention in the past. Today, the computer has made it commonplace.

Kepler's equation

To derive Kepler's key equation, consider Figure 7.4. Let the area of triangle BFP = Δ. From the areal velocity law, we have

$$\frac{M}{\Delta} = \frac{2\pi}{\pi\, ab} \quad \text{or} \quad M = \frac{2\Delta}{ab} \tag{7.12}$$

Since the ellipse is a circle uniformly flattened in the ratio b/a and OF = ae,

$$\Delta = \frac{b}{a}\text{area AFP} = \frac{b}{a}(\text{area OAP} - \text{area OAF}) = \frac{b}{a}\left(\frac{1}{2}a^2E - \frac{1}{2}a^2 e \sin E\right)$$

Substituting for Δ in (7.12) gives Kepler's Equations (7.9) and (7.10).

7.8 Cartesian coordinates

Readers unfamiliar with positional astronomy should refer to Appendix 7 before reading further.

Once the position of the satellite (x', y') in the plane of the orbit has been found, its Earth-centred Cartesian coordinates (X, Y, Z) oriented on the equator and Greenwich (see Figure 7.5) are given in terms of the *perigee argument* ω, the *inclination* of the orbit i and the *right ascension of the ascending node* Ω by the transformation:

$$\begin{bmatrix} X \\ Y \\ Z \end{bmatrix} = \begin{bmatrix} \cos\Omega & \sin\Omega & 0 \\ -\sin\Omega & \cos\Omega & 0 \\ 0 & 0 & 1 \end{bmatrix} \begin{bmatrix} 1 & 0 & 0 \\ 0 & \cos i & \sin i \\ 0 & -\sin i & \cos i \end{bmatrix} \begin{bmatrix} \cos\omega & \sin\omega & 0 \\ -\sin\omega & \cos\omega & 0 \\ 0 & 0 & 1 \end{bmatrix} \begin{bmatrix} x' \\ y' \\ 0 \end{bmatrix}$$

The orbital parameters a, e, Ω, ω, i and T form the *ephemeris* (plural *ephemerides*) of the satellite. These are usually transmitted to the receiver at the time of observation, or are obtained later in a more precise form.

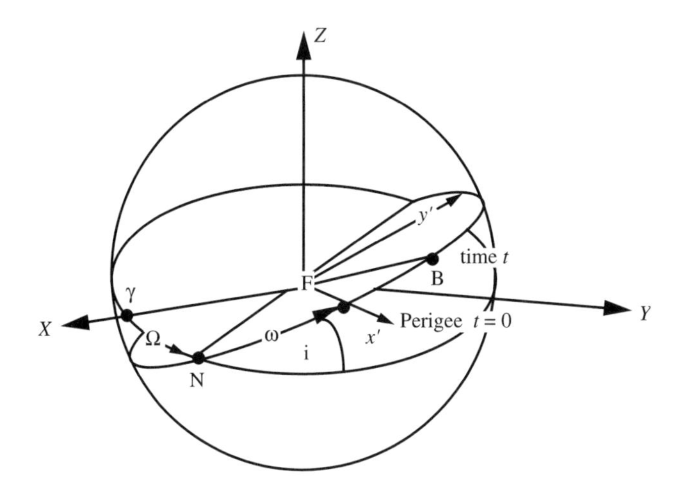

Figure 7.5

7.9 Motion in a circular orbit

Following Kepler's law of areal velocity, if the orbit is a circle with constant radius r, then the satellite of mass m moves around the central mass M, the Earth, with a constant angular speed of ω radians per second. Thus the force F away from M due to the motion is

$$F = m\omega^2 r$$

But $\omega = 2\pi/T$, where T is the period of rotation, so

$$F = \frac{4\pi^2 mr}{T^2} \tag{7.13}$$

Newton's law of gravitational attraction states that F is also given by

$$F = \frac{\mathrm{kM}m}{r^2} \tag{7.14}$$

where k is the gravitational constant and M is the mass of the Earth. When r = R (the radius of the Earth), $F = m$g and hence kM = gR^2, where g is the gravitational constant. Thus we have the useful relationship

$$T^2 = \frac{4\pi^2 r^3}{gR^2} \tag{7.15}$$

Substituting g = 9.81 m s^{-2} and π and converting time to minutes we obtain the useful working formulae:

$$T = 84.6\left(\frac{r}{\mathrm{R}}\right)^{3/2}$$
$$r = \mathrm{R}\exp\left(\frac{2}{3}\ln\frac{T}{84.6}\right) \tag{7.16}$$

Setting R = 6.4 × 10^3 km, we calculate the distance r for a GPS satellite from Equations (7.16) with a 12-hour orbit as 26 678 km, giving the height $h = r - \mathrm{R}$ = 2.02 × 10^4 km. The Earth's moon with a 28-day period is about 384 000 km distant.

Likewise we find from Equations (7.16) that an earlier system Doppler satellite with 108-min period was at a height of about 1000 km.

7.10 Example of satellite prediction

Although Earth satellites are no longer observed by angular methods, their positions relative to the measurement station are required to calculate good geometry for range or range-difference measurements.

The declinations and right ascensions (see Appendix 7) of satellites vary more quickly than those of the stars (which only vary slowly) or the Sun. The period of a GPS satellite is 12 hours, and its angle of inclination is 55°.

The key information about a satellite is the right ascension of its ascending node (AN). This relates the orbit to the meridian of Greenwich via the right ascension of Greenwich, better known as Greenwich Sidereal Time (GST).

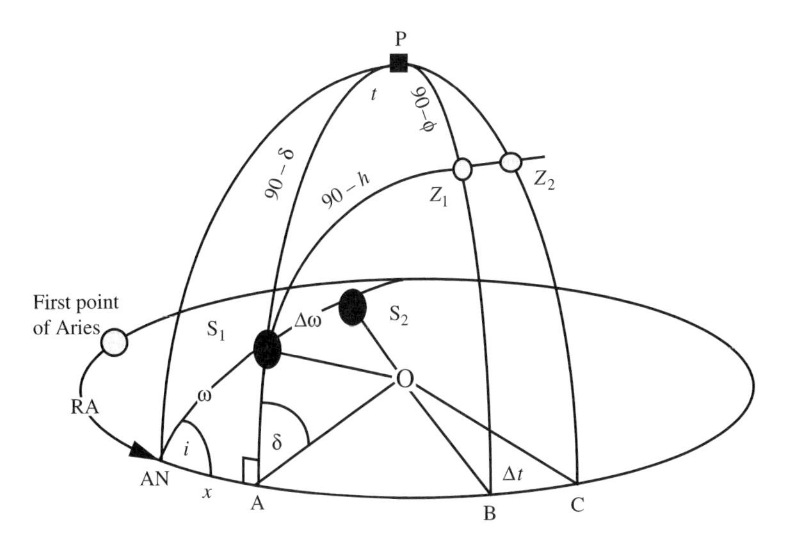

Figure 7.6

Figure 7.6 shows the basic elements of a simple model to predict GPS satellite positions for use in an error prediction model. If a satellite's geocentric (corresponding to an observer situated at the centre of the Earth) altitude h and its azimuth z are known at a position whose zenith is at Z_1 (latitude ϕ), its declination δ and hour angle t (see Appendix 7 for definitions) can be calculated from

$$\sin\delta = \sin\phi\sin h + \cos\phi\cos h\cos z$$

$$\cos t = \sin h\sec\phi\sec\delta - \tan\phi\tan\delta$$

(see Appendix 2) and the GST is found from the Universal Time (UT): GST = UT + R where R (Right Ascension of the Mean Sun, not to be confused with the Earth's radius) is obtained from the Sun's ephemeris tabulated by SALS (see Appendix 7.2).

The final missing link is the position of the ascending node, AN. This is found from the inclination of the orbit i in the right-angled spherical triangle formed by AN, S_1 and A (see Appendix 2). The distances labelled x and ω are derived from the following equations using Napier's rule for circular parts (see Appendixes 2 and 7):

$$\sin x = \tan\delta\cot i \tag{7.17}$$

$$\sin\omega = \sin\delta\,\text{cosec}\, i \tag{7.18}$$

Thus the right ascension of the ascending node is the RA of the satellite minus x. The angle ω is the distance along the orbit which the satellite has travelled since the node. After a known time Δt, the satellite will travel a further angular distance $\Delta\omega$ given by

$$\Delta\omega = \frac{\Delta t}{T}$$

where T is the period of the satellite (approx 12 hours ST for a GPS satellite). Thus the new value of its declination and RA can be calculated from a second right-angled triangle using formulae (7.17) and (7.18).

Table 7.7

UT	*H*(°)	*z*(°)	*h*(°)	δ(°)	*t*(°)	*x*(°)	ω(°)
07.30	60	116	67	36	26	31	46
08.00	61	93	68	44	31	42	58
09.00	48	60	57	55	55	84	87
10.00	18	58	31	43	91	138	123

The zenith of the observing position has moved to Z_2 during time Δt. Finally the new altitude and azimuth of the satellite are calculated from

$$\cot z' = \frac{\cos\phi \tan\delta - \cos(t+\Delta t)\sin\phi}{\sin(t+\Delta t)}$$

$$\sin h' = \sin\phi \sin\delta + \cos\phi \cos\delta \cos(t+\Delta t)$$

The topocentric altitude is computed via

$$\tan H = \tan h - \frac{\mathrm{R}}{r\cos h}$$

(see Appendix 7.5). The range to the satellite is computed from the radius of the Earth R and the known orbital distance r. This range is then compared with the measured range to give the absolute term of a variation of coordinates estimation process. An example in two dimensions is given in Table 7.7, illustrating the basic principles of position determination by range difference used by GPS. The data in Table 7.7 are for successive positions of the GPS satellite vehicle number 19 on 30th March 1992.

Sky charts

The azimuth and altitude data for a satellite or star are often used to compile a sky chart depicting the geometry of the observation period. Figure 7.7 depicts the data in Table 7.8 for satellite 19 together with four others used in a GPS fixation.

The cuts of the position lines meet at various azimuths at an angle between 30° and 150°, giving planimetric strength, and the different altitudes indicate good height fixes.

Positional dilution of precision (PDOP)

The quality of a GPS position is often described in terms of the square root of the trace of the dispersion matrix of the final coordinates, or by

$$\frac{\sigma_x^2 + \sigma_y^2 + \sigma_z^2}{\sigma_0^2},$$

referred to as the positional dilution of precision. A factor of 3 is often taken as the upper bound of acceptability, or, in other words, if the positional variance is three

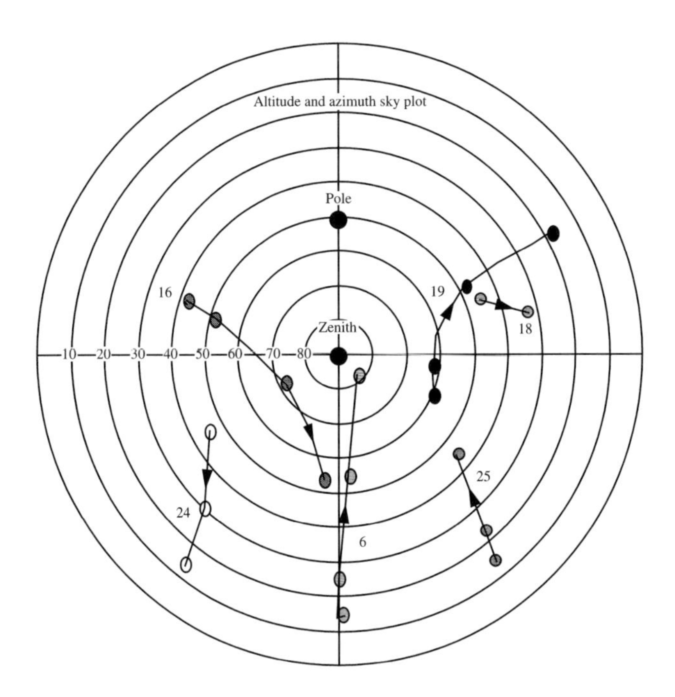

Figure 7.7

Table 7.8

Satellite	*UT*	*H(°)*	*Z(°)*
18	07.30	47	70
	08.00	37	76
25	08.45	15	145
	09.00	21	143
	10.00	44	123
24	07.30	47	239
	8.00	39	227
	09.00	15	210
16	07.30	43	296
	08.00	53	294
	09.00	74	248
	10.00	56	185
4	07.45	18	179
	08.00	25	180
	09.00	55	182
	10.00	84	121

times worse than the observational variance, the work is unacceptable. Clearly the geometry of the intersecting lines is critical. The PDOP can be pre-computed in planning a network (see Appendix 6).

Summary of operational practice

It is clear that to obtain reliable coordinates from the GPS or other systems, great care is needed by the surveyor. All suppliers of equipment as well as professional institutions issue user guidance on best practice.

Advice on how best to proceed with GPS surveys is bound to change over time (see RICS Guidance Notes), however the basic principles are:

1. Independent checking should always be carried out, even if it is merely an overview of the consistency of results, such as establishing points on an obvious straight line.
2. Attention to point initialisation is always worthwhile even if it delays the survey and results in the moving of reference points until an acceptable quality is obtained e.g. to avoid multi-path degradation of results.
3. Always archive the ITRF values upon which the system operates.
4. When building up a network of control vectors between pairs of points, in groups of three receivers for example, avoid using correlated vectors as if they are independent when incorporating results into a network. If a dispersion matrix is available, then it should be incorporated in the network (see Chapter 8 for an example).
5. Keep clear records of the events of the field observations e.g. times at which lock of signal occur, and meteorological data.
6. Control points should always be fixed from at least five satellites.
7. Note whether or not the integer ambiguity has been resolved i.e. if a floating point solution is being used to get a better value to an integer. Points should be marked for revisiting in case of any problems with post processing and unacceptable quality of results.

Chapter 8
Survey Computations

8.1 Introduction

Although the computation of coordinates from measured data is almost always carried out with the help of software packages and turnkey systems, Excel spreadsheets can be very useful for immediate computations of small tasks. There is the occasional need to resort to the hand calculator, if only to check key results. Also, it is important to be able to verify that a software package is correct.

Therefore, knowledge of the procedures involved is paramount, especially if the geospatial surveyor is called upon to act as an expert witness in a legal case. We outline here some basic concepts. Most of the examples have been calculated using spreadsheets.

GPS results, which are output from a complex processing package, include a quality assessment in the form of a dispersion matrix. Refer to Chapter 6 for how to incorporate such results into a survey.

Map projections are much less used than previously. Computations involve the distortion of directions by the arc to chord (t – T) corrections, and distances by line scale factors (see Appendix 8). Once these corrections have been applied, the formulae are the same as for the plane computations treated here.

Computations on the surface of the spheroid (Allan 1997) have also largely been replaced by three-dimensional plane geometry.

In this chapter, we cover the following topics:

1. Traverses (simple two-dimensional traverse; Least Squares treatment; blunder detection);
2. Intersection;
3. Lateration;
4. Resection.

8.2 Traverse computation

Simple Traverse

The traverse is a series of points joined together by a chain of distances and directions. Figure 8.1 depicts the layout of the plan view of a typical traverse. It is usual to calculate the plan positions (Eastings and Northings) separately from the heights. For the separate calculation of heights see Chapter 9. A strictly three-dimensional treatment is usually reserved for high precision industrial work (see Chapter 12).

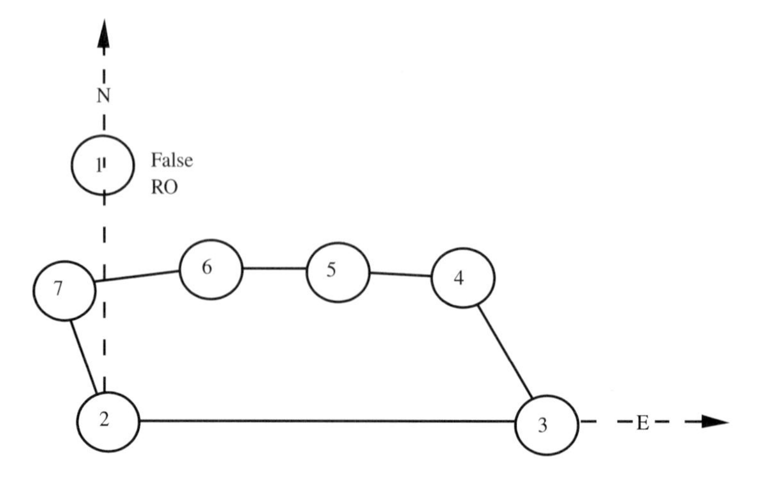

Figure 8.1

Coordinate system and datum

Figure 8.1 shows a typical loop traverse, commencing at point (2), progressing through points (3)–(7) before closing back on itself. If the traverse is an isolated one, for example to control the mapping of a building site, an arbitrary system of coordinates is used.

The longest line (2)–(3) is adopted as the reference direction, here called Easting assumed at a bearing of 90°, and one terminal at point (2) is adopted as the origin of coordinates, here (2000 m, 1000 m).

Note: 1. It is good practice to select such widely differing values for Eastings and Northings in order to avoid confusion when copying down figures. 2. There is a case for not using 'Eastings' and 'Northings' unless the grid is actually oriented on these directions, but the use of X and Y can be even more confusing because there is no universal convention for the direction of the X-axis. Whatever system is used, it is wise to state on any documentation the convention that has been adopted.

We also adopt an arbitrary point (1) in the direction of 'North', which is clearly at 90° from the direction of line (2)–(3). The length of (1)–(2) is of no interest but for convenience we adopt 1000 m.

This device has been introduced to suit software that may allow for a traverse to be linked into an existing network, where points (2) and (1) will have given coordinates on the existing system.

We now treat the following topics:

1. The simple traverse computation;
2. A modification to deal with the misclosures by Least Squares;
3. An example to show how a mistake in a bearing can be detected;
4. An example to show how a mistake in a length can be detected.

8.3 Example 1: Simple traverse

The first stage is to prepare the crude measurements for calculation. Thus we reduce all observed angles to the final values to be used in the computation. By this we mean that any corrections for tilted axes, or projection distortions (t – T) have been applied, and also perhaps the rejection of bad observations, before the final means are taken. Slope distances are also reduced to the horizontal, amended for instrument calibration, and possibly for projection scale factors. The final reduced data are then ready for use either in some customized software system or, as here, in a spreadsheet.

Note on precision of calculation

Computer and spreadsheet systems operate to far more decimals of precision than the original measurements warrant. The angles involved in this traverse are recorded to one second of arc or 0.0003 degree and the distances to 0.001 m. Hence it has some meaning to record results to this precision. The spreadsheets work to very many more figures than this. Hence the intermediate results tabulated here, which have been truncated, may be a little inconsistent if used separately by a reader as the basis for further calculation.

Tables 8.1 and 8.2 show the reduced internal clockwise angles A, e.g. at station (2) between (1) and (3) as 123, and horizontal distances S tabulated in a spreadsheet, with calculation of the coordinate differences $\Delta E = S \cos U$ and $\Delta N = S \sin U$ providing accumulated coordinates E and N, discussed in detail below.

Table 8.1 *Simple traverse computation*

Internal angle	*A (sexag.deg.)*	*Bearings*	*U*
		2 → 1	0
		1 → 2	180
1 2 3	90.0000	2 → 3	90
2 3 4	59.7200	3 → 4	329.7200
3 4 5	127.2489	4 → 5	276.9689
4 5 6	166.3600	5 → 6	263.3289
5 6 7	183.3817	6 → 7	266.7106
6 7 2	42.8619	7 → 2	129.5725
7 2 3	140.4008	2 → 7	309.5992
Sum	719.9733		
Theory	720.0000	Comp	309.5725
Misclosure	0.0267	C–O	–0.0267
Seconds	96.12		–96.12

Table 8.2

Distances S	*Easting E*	*Northing N*	*Station*
	2000	2000	1
	2000	1000	2
263.209	2263.2090	1000	3
41.593	2242.2367	1035.9185	4
90.828	2152.0797	1046.9387	5
118.03	2034.8489	1033.2272	6
70.294	1964.6707	1029.1938	7
45.829	1999.9966	999.9982	2*
	0.0034	0.0018	closure
	0.00385648	vector	

Analysis of angle closure ΣA

For N stations, the sum of the internal angles ΣA should be $2N - 4$ right angles or 720°. (In this case, $N = 6$ because station (1) is not part of the loop traverse.) The misclosure is 0.0267° short of this ideal, or 96.12 seconds of arc. Does this indicate a mistake or is it to be expected on statistical grounds? See 'Checks for blunders' below for how to answer such a question. Suffice it to say that we are satisfied with this result and can proceed without hesitation.

Note For approximate calculation in the field, the angular misclosure can be spread equally to the angles to make them consistent with theory. In the proper treatment by Least Squares Estimation no such 'corrections' are made.

Calculation of bearings *U*

The next stage is to obtain the bearings of all lines (see Table 8.1) commencing with the assumed bearing of the line (2–3), $U_{2-3} = 90°$. Then the reverse bearing,

$$U_{3-2} = U_{2-3} + 180° = 270°$$

The next bearing

$$U_{3-4} = U_{3-2} + A_{234} = 270 + 59.7200 = 329.7200°$$

The reverse bearing

$$U_{4-3} = U_{3-4} + 180° = 329.7200 + 180 = 509.7200°$$

Since this exceeds one cycle we can subtract 360° to give $U_{4-3} = 149.7200°$. (Notice this is the same as $329.7200 - 180 = 149.7200$.)

However, computer software is capable of calculating the trigonometrical functions of many cycled functions such as 509.7200° so there is often no need to subtract 360° except for visual inspection.

This process is repeated until all bearings are obtained, including the last bearing (2–7) carried through the traverse.

$$U_{2-7} = 309.5725°.$$

Since this bearing will be used later to compute the provisional coordinates of point (2) carried round the traverse, it is called the 'computed value' of the bearing U_{2-7} denoted by U^*_{2-7}. Since the angle (723) was also observed we can get a second value for the bearing from

$$U_{2-7} = U_{2-3} - A_{723} + 360° = 309.5992°$$

This observed bearing will not be used to compute provisional coordinates so is called the observed value denoted by U^o_{2-7}. The two versions of this bearing clearly differ by the angular misclosure (Table 8.1):

$$U^*_{2-7} - U^o_{2-7} = C - O = 309.5725° - 309.5992° = -0.0267$$

Calculation of coordinates

We are now in a position to compute the coordinates of all points (Table 8.2), including the starting point (2) taken round the traverse. The coordinate differences are obtained from equations

$$E_3 = E_2 + S_{2-3} \cos U_{2-3} \tag{8.1}$$

$$N_3 = N_2 + S_{2-3} \sin U_{2-3} \tag{8.2}$$

For example,

$$0 = 1000 + 263.209 \times \cos 90$$

$$2263.2090 = 2000 + 263.206 \times \sin 90$$

Note: If using Excel, angles must first be converted to radians via SIN(U*PI()/180).

This concludes the simple traverse computation; no attempt is made to 'distribute' the misclosure of 0.0034 in Eastings and 0.0018 in Northings.

Analysis of positional closure

The two values of point (2) indicate the quality of the work. In this case these values of 3 and 2 mm are very good indeed. In ordinary work, closures of cm are more common. In this particular case, the fieldwork was carried out to very stringent standards for a deformation study so the high quality result is not unexpected.

Checks for blunders

There are two checks which can be carried out:

1. That the bearings taken round in a loop agree to within an expected limit.
2. That the extra distance connecting the last two points agrees within expected limits.

Simple traverses are therefore not very reliable in a statistical sense (see Chapter 6).

As a guide in the field, bearings usually have to be within $\pm 3\sigma\sqrt{(2n)}$ where there are n observation stations, and a *direction* standard error is σ. This value of sigma is selected from previous experience, or from a computer simulation of the task in hand. For example if $n = 6$ and $\sigma = 10''$, bearings have to agree to 104". If they do not, a blunder is suspected, and search made for the likely source. In this case the misclosure of 96" gives no cause for investigation.

In fact the traverse ran along a sea front from (2–3), and along a cliff top from (3–7). Both of the short legs (3–4) and (7–2) were very steep (about 41°) and could be a source of cross tilt error ($i \tan H$; see Section 13.5) even although this was guarded against by careful levelling. If a mistake is made in recording a direction at one station, it can be detected by computing the traverse clockwise and anticlockwise. (See example 3 below). The station containing the mistake will have identical coordinates in both calculations.

The distance check should meet the criterion of $3\sigma\sqrt{n}$, where σ is the standard error of a typical length. For example if $\sigma = 2.5$ mm and $n = 6$, the limit of tolerance is 18 mm.

If a blunder has been made in one length, and no other mistakes are present, the final compared distances will differ by a vector which has the same direction as the line in which the blunder lies (see example 4).

8.4 Computation by Least Squares

The principle of Least Squares estimation is used to resolve what to do with redundant information, such as the bearing and position misclosures. If a Least Squares treatment is to be given, no alteration of the observation data should be made, because the hand computation is only to provide provisional coordinates.

There is some justification for reconciling small discrepancies within the closure bounds, such as distributing the bearing error, recalculating and then distributing the final vector misclosure (Bowditch method). If this so-called adjustment makes a material difference to the result, it should not be done, and the work should be repeated to a proper standard.

Some software systems require that the theodolite circle be set to an approximate bearing at each station as the work proceeds, as it helps to identify quadrants.

Example 2

Consider the traverse of Figure 8.1. Two fixed points are assumed to be (2), the initial traverse point, and (1), a fictitious point not actually observed. The reason for adopting this procedure is that the software was written on the assumption that the traverse would always be tied to a fixed control point to give it orientation, as in cadastral work. This computational device does not affect the relative results in any way. In this example the traverse is oriented by an arbitrary direction.

All directions have been assigned an optimistic 10" standard error, being the means of three rounds taken on a *total staton*. All distances have been reduced for slope and index and have been given standard errors of 5 mm.

When approximate coordinates have been calculated via the simple computation described above, direction and distance equations are formed, and the solution obtained.

The direction equation for AG developed in Chapter 6 (Equation 6.20) is

$$\delta E_A - \delta E_G - \delta N_A \tan \overset{o}{U}_{AG} + \delta N_G \tan \overset{o}{U}_{AG} + (N_G - N_A) \sec^2 \overset{o}{U}_{AG} \, \delta U_{AG} + K_{AG} = 0 \tag{6.20}$$

where

$$K_{AG} = F(E_A^*, N_A^*, E_G^*, N_G^*, \overset{o}{U})$$

and K_{AG} is obtained by substituting the provisional coordinates and the observed bearing into the expression

$$\Delta E = \Delta N \tan U$$

A simpler alternative form of coefficients of Equation (6.20)

Defining L = computed – observed parameter as usual, we have in this case

$$L = U^* - \overset{o}{U} \quad \text{or} \quad \overset{o}{U} = U^* - L$$

By definition,

$$K = \Delta E^* - \Delta N^* \tan U$$
$$= \Delta E^* - \Delta N^* \tan(U^* - L)$$

Now since L is small, expanding by Taylor's theorem we have

$$K = \Delta E^* - \Delta N^* \tan(U^* - L) = \Delta E^* - \Delta N^* \tan U^* + \Delta N \sec^2 \overset{o}{U} L$$

But

$$K = \Delta N \sec^2 \overset{o}{U} L$$

for

$$\Delta E^* - \Delta N^* \tan U^* = 0$$

Thus Equation (6.20) can be written

$$\delta E_A - \delta E_G - \delta N_A \tan \overset{o}{U}_{AG} + \delta N_G \tan \overset{o}{U}_{AG} + (N_G - N_A) \sec^2 \overset{o}{U}_{AG} \, \delta U_{AG}$$
$$+ (N_G - N_A) \sec^2 \overset{o}{U}_{AG} L_{AG} = 0 \tag{8.3}$$

Putting the length of the line AG = S_{AG} Equation (8.3) is easily simplified to

$$\frac{\cos U}{S_{AG}} \delta E_A - \frac{\cos U}{S_{AG}} \delta E_G + \frac{\sin U}{S_{AG}} \delta N_A - \frac{\sin U}{S_{AG}} \delta N_G + L_{AG} = \nu_{AG} \tag{8.4}$$

because

$$-\delta U_{AG} = \breve{U} - \hat{U} = v_{AG}$$

is the residual.

Equation (8.4) is the more commonly used version of a direction equation.

Since the direction of the example in the above Equations (8.3) and (8.4) was derived from angles measured relative to a direction between fixed points, there is no unknown orientation parameter Z at the observation station.

This situation applies only to station (2) of the traverse where there is a fixed direction to station (1). At the five stations (3–7), allowance has to be made for an error in orientation by adding a parameter Z at each observed station. For example, the direction equation from (3–4) with the addition parameter Z_3 is

$$\frac{\cos U}{S_{34}}\delta E_3 - \frac{\cos U}{S_{34}}\delta E_4 + \frac{\sin U}{S_{34}}\delta N_3 - \frac{\sin U}{S_{34}}\delta N_4 + Z_3 + L_{34} = v_{34} \tag{8.5}$$

Since the point (2) is held fixed there are no variables for its coordinates. Also, since the bearing (2–1) is fixed, there is no unknown orientation parameter Z at point (2). The equation for the line (2–3) is therefore

$$-\frac{\cos U}{S_{23}}\delta E_3 - \frac{\sin U}{S_{23}}\delta N_3 + L_{23} = v_{23} \tag{8.6}$$

Finally, since we have decided to calculate the bearings in the sexagesimal system, the values of L are in seconds of arc. Thus all coefficients of the coordinate variables have to be scaled by 206265 (the number of seconds in a radian). For example, the coefficient of δN_3 in Table 8.3 below is

$$206265 \times \frac{\sin 90}{263.209} = 783.654814$$

Distance Equations

The distance equations are of the form of Equations (6.28), no allowance being made for an instrument index (see Chapter 13). The equation for the line (3–4) is therefore

$$\frac{(E_4 - E_3)}{S_{34}}\delta E_3 + \frac{(N_4 - N_3)}{S_{34}}\delta N_3 - \frac{(E_4 - E_3)}{S_{34}}\delta E_4 - \frac{(N_4 - N_3)}{S_{34}}\delta N_4 \tag{8.7}$$
$$v_{34} + L_{34} = 0$$

The coordinates of the other five points are free to change giving ten variables, and each station has an unknown orientation parameter Z i.e. 15 variables in total (N).

Table 8.3

$i\ j$	δE_i	δN_i	δE_j	δN_j	L
Direction coefficients					
2 3	-1.74×10^{-13}	783.654814	1.7401×10^{-13}	–783.65481	0
2 7	–2868.8394	–3467.9297	2868.83938	3467.92969	0
3 2	-1.74×10^{-13}	783.654814	1.7401×10^{-13}	–783.65481	0
3 4	–4282.5619	–2500.5222	4282.56193	2500.52219	0
4 3	–4282.5619	–2500.5222	4282.56193	2500.52219	0
4 5	–275.53454	–2254.1633	275.534536	2254.16332	0
5 4	–275.53454	–2254.1633	275.534536	2254.16332	0
5 6	203.014123	–1735.7321	–203.01412	1735.73207	0
6 5	203.014123	–1735.7321	–203.01412	1735.73207	0
6 7	168.369236	–2929.4843	–168.36924	2929.48428	0
7 6	168.369236	–2929.4843	–168.36924	2929.48428	0
7 2	2867.22301	3469.2662	–2867.223	3469.2662	–96.12
Distance coefficients					
2 3	–1	2.2204×10^{-16}	1	-2.22×10^{-16}	0
3 4	0.50422621	–0.8635716	–0.5042262	0.86357161	0
4 5	0.99261216	–0.1213306	–0.9926122	0.12133057	0
5 6	0.99322937	0.11616977	–0.9932294	–0.1161698	0
6 7	0.99835245	0.05737933	–0.9983524	–0.0573793	0
7 2	–0.7708191	0.6370541	0.7708191	–0.6370541	0.0015

There are 12 observed directions and 6 observed distances, giving 18 observation equations (M). Thus the redundancy is $M - N = 3$.

The coefficients of all the variables are listed in Table 8.3.

Since we used the observed measurements to obtain the provisional coordinates, only the last direction and last distance exhibit absolute terms L other than zero (see last column). As it is not possible to display the coefficients of this (18 × 15) matrix in full form, we simply show its pattern in Table 8.4. The coefficients of Z's are all 1, the asterisks * show where the numbers from Table 8.3 are located and the values of L are listed in Table 8.3. All other terms are zero.

Table 8.5 shows the actual values of the first and last three columns of Table 8.4.

Table 8.4

	E_3	N_3	Z_3	E_4	N_4	Z_4	E_5	N_5	Z_5	E_6	N_6	Z_6	E_7	N_7	Z_7	L
2 7													*	*		*
2 3	*	*														*
3 2	*	*	1													*
3 4	*	*	1	*	*											*
4 3	*	*		*	*	1										*
4 5				*	*	1	*	*								*
5 4				*	*		*	*	1							*
5 6							*	*	1	*	*					*
6 5							*	*		*	*	1				*
6 7										*	*	1	*	*		*
7 6										*	*		*	*	1	*
7 2													*	*	1	*
2 3	*	*														*
3 4	*	*		*	*											*
4 5				*	*		*	*								*
5 6							*	*		*	*					*
6 7										*	*		*	*		*
7 2													*	*		*

Weighted equations

The observation equations are of the form **Ax** = **L** + **v** and the normal equations are

$$A^{\mathrm{T}}WAx = A^{\mathrm{T}}WL$$

where **W** is a weight matrix. In this case **W** has no covariance terms so this weight matrix is a diagonal matrix. Also the weight of a direction is $1/10^2$ and of a distance is $1/0.005^2$.

The standard error of unit weight is computed via

$$\text{Std Error} = \frac{1}{2}\left(\frac{\mathbf{v}^{\mathrm{T}}\,\mathbf{W}\,\mathbf{v}}{M-N}\right) = 1.64$$

The station corrections Z are of little interest. The initial coordinates E', N', their corrections and final estimates are listed in Table 8.6.

Finally the set of coordinates submitted to a client are rounded to a realistic precision of 1 mm. The estimate of the standard error or unit weight of 1.64 indicates a reasonable estimate of original weights.

Table 8.5 *Observation Equations*

	E_3	N_3	Z_3	E_7	N_7	Z_7	L
2	0	0	0	2868.83938	3467.92969	0	0
2	1.7401×10^{-13}	–783.65481	0	0	0	0	0
3	1.7401×10^{-13}	–783.65481	1	0	0	0	0
3	–4282.5619	–2500.5222	1	0	0	0	0
4	–4282.5619	–2500.5222	0	0	0	0	0
4	0	0	0	0	0	0	0
5	0	0	0	0	0	0	0
5	0	0	0	0	0	0	0
6	0	0	0	0	0	0	0
6	0	0	0	168.369236	–2929.4843	0	0
7	0	0	0	168.369236	–2929.4843	1	0
7	0	0	0	2868.83938	3467.92969	1	96.12
2 3	1	-2.22×10^{-16}	0	0	0	0	0
3 4	0.50422621	–0.8635716	0	0	0	0	0
4 5	0	0	0	0	0	0	0
5 6	0	0	0		0	0	0
6 7	0	0	0	–0.9983524	–0.0573793	0	0
7 2	0	0	0	0.7708191	–0.6370541	0	0.0015

Table 8.6

Stat	E'	dE	E	N'	dN	N
2	2000	0	2000	1000	0	1000
3	2263.209	–0.0013	2263.2076	1000	0.0071	1000.00711
4	2242.23672	–0.0015	2242.23518	1035.91853	0.0071	1035.92564
5	2152.07974	-4.85×10^{-05}	2152.07969	1046.93875	0.0102	1046.94899
6	2034.84888	–0.0006	2034.84821	1033.22723	0.0227	1033.25000
7	1964.67069	0.0013	1964.67202	1029.19381	0.0014	1029.19522

Since there is little redundancy, the *reliability* of a simple traverse is never high. If, however, additional intersections are made to a visible 'up station' such as a church tower, the reliability can be vastly improved even if these extra observations add nothing to the accuracy of the actual traverse. This practice is never possible in many cases, however, such as inside a tunnel or oil pipe (see Chapter 6).

Note on simplifying the calculations

It is possible to simplify the calculation of the normal matrix by first scaling the observation equations by the reciprocals of their standard errors, and forming the normals directly from these scaled equations. This stems from the fact that the weight matrices are diagonal. We can write

$$\mathbf{A}^{\mathrm{T}}\mathbf{W}\mathbf{A} = \mathbf{A}^{\mathrm{T}}\sqrt{\mathbf{W}}\sqrt{\mathbf{W}}\mathbf{A}$$

Thus the diagonal terms of weight matrix $\sqrt{\mathbf{W}}$ will be either $\sqrt{(1/10^2)}$ or $\sqrt{(1/0.005^2)}$ i.e. 0.1 or 200. If we multiply each direction equation by 0.1 and each distance equation by 200 we obtain scaled observation equations

$$\sqrt{\mathbf{W}}\mathbf{A}\mathbf{x} = \sqrt{\mathbf{W}}\mathbf{L} + \sqrt{\mathbf{W}}\mathbf{v}$$

Operating on these new equations in the same way as the original observation equations gives the weighted normal equations, for

$$\mathbf{A}^{\mathrm{T}}\sqrt{\mathbf{W}}^{\mathrm{T}}\sqrt{\mathbf{W}}\mathbf{A}\mathbf{x} = \mathbf{A}^{\mathrm{T}}\sqrt{\mathbf{W}}^{\mathrm{T}}\sqrt{\mathbf{W}}\mathbf{L} + \mathbf{A}^{\mathrm{T}}\sqrt{\mathbf{W}}^{\mathrm{T}}\sqrt{\mathbf{W}}\mathbf{v}$$

As $\mathbf{W}$ is a diagonal matrix, so also is $\sqrt{\mathbf{W}}$, therefore

$$\sqrt{W}^{\mathbf{T}} = \sqrt{W}$$

Therefore we have once more

$$A^{\mathbf{T}}WAx = A^{\mathbf{T}}WL$$

as $\mathbf{A}^{\mathrm{T}}\mathbf{W}\,\mathbf{v} = 0$.

It is also possible to reduce the number of equations to be solved by applying Schreiber's method described in Section 8.10 below.

8.5 Traverse blunders

By downloading data directly to a laptop or other data logger, the chances of a blunder in the work are greatly reduced. However, if they do occur, the following computational strategies are useful to detect single mistakes.

Traverse blunder in distance

If a mistake is made in measuring one leg, the effect can usually be detected and the line checked. Table 8.7 shows the effect of introducing a metre error into the line (3–4). The misclosure vector is one metre plus random error, and at the same bearing as the line (3–4).

The process can therefore be used in reverse to locate the likely leg in which a mistake is made. The bearing of the misclosure is arctan(–0.862, 0.508) = 329°.

Table 8.7

	U	*U*	*S*	*E*	*N*
1				2000.000	2000.000
1–2		180.0000		2000.000	1000.000
2–3	90.000	90.0000	263.209	2263.209	1000.000
3–4	59.7200	329.7200	42.593	2241.732	1036.782
4–5	127.2489	276.9689	90.828	2151.576	1047.802
5–6	166.3600	263.3289	118.03	2034.345	1034.091
6–7	183.3817	266.7106	70.294	1964.166	1030.057
7–2	42.8619	129.5725	45.829	1999.492	1000.862
2–7	309.5992	309.5725	closure	0.508	–0.862
	Bg closure	0.0267	vector	1.002	
		96.12"			

Traverse blunder in bearing

In a similar manner, the effect of a mistake of 1 degree at station (4) swings the rest of the traverse to create a 4 m misclosure perpendicular to the line (4–2); see Table 8.8. Thus the perpendicular bisector of the vector misclosure passes through the point at which the mistake was made, and so the blunder station can be found.

Table 8.8

	U	*t*	*S*	*E*	*N*
1				2000.000	2000.000
1–2		180.0000		2000.000	1000.000
2–3	90.0000	90.0000	263.209	2263.209	1000.000
3–4	59.7200	329.7200	41.593	2242.236	1035.919
4–5	128.2489	277.9689	90.828	2152.286	1048.511
5–6	166.3600	264.3289	118.03	2034.833	1036.847
6–7	183.3817	267.7106	70.294	1964.596	1034.039
7–2	42.8619	130.5725	45.829	1999.407	1004.231
	309.5992	310.5725	vector	0.593	–4.231
	Bg closure	–0.9733		4.273	
		–3503.88			

Another way to locate this mistake is to compute the traverse in the reverse direction, from (2–7) and so on, clockwise back to (2). Only point (4) will have the same coordinates in both calculations, thus proving it is the guilty point. Some surveyors compute every traverse in both directions to give some idea of its intermediate consistency.

Traversing is often the only way to fix control, particularly inside buildings, tunnels or large pipes and in urban streets and forests. It therefore has to be done with care and a system of quality checking the basic data. If it is inevitable that the traverse cannot be closed, another return route should be taken to close back on the starting point. Inside a tunnel this should be a mirror image of the inward route, to balance the effects of refraction on lines of sight, which can be as much as 10" per station in a systematic manner.

8.6 Error propagation in traverses

The precise way to determine the errors propagated through a traverse or any network is by considering the error ellipses at points and between points. The former are dependent upon the choice of datum, usually the starting point of the traverse, while the latter give information between pairs of points.

For example, the positional error ellipses, for a straight unclosed traverse of n equal legs starting at a fixed point A with a fixed bearing, grow larger from zero at the fixed starting point to a maximum at its end. All directions and lengths are assumed of the same weight (Figure 8.2).

As a rough guide in the field, the simple analysis of a straight traverse can be of assistance in planning work. Consider a straight traverse of n legs each of length S along the x-axis with angular errors at each station of σ_q. The linear standard error in x between the ends of the traverse is

$$\sigma_x = \sigma_s \sqrt{n}$$

and in y is

$$\sigma_y = S\sigma_q \sin 1 \sqrt{\frac{1}{6} n(n+1)(2n+1)}$$

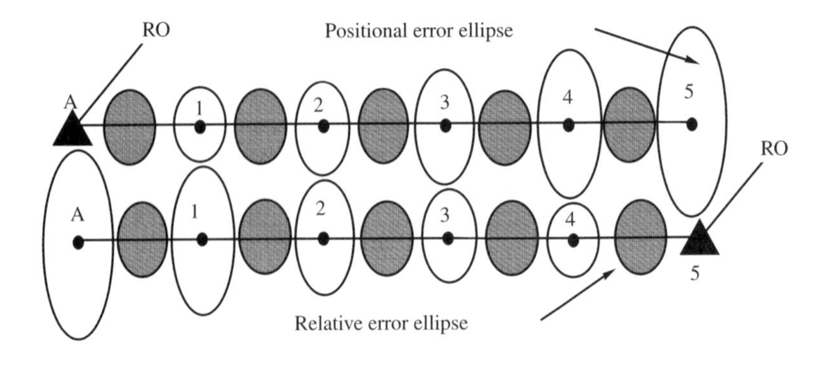

Figure 8.2

Suppose the standard error of a leg is 2.5 mm, the 16 legs are each 50 m long and the standard error in an angle is 10", then we have $\sigma_x = 10$ mm and $\sigma_y = 9.4$ mm.

8.7 Intersection

Intersection is when a point is fixed by two intersecting angles from known points. In most ordinary work, the problem is treated in two parts: plan (E, N) followed by height H. In very precise industrial work the problem has to be treated properly in three dimensions (see Chapter 12).

We consider plan only. If only two rays are present there is no redundancy. If several rays are used, a Least Squares approach is employed. In this case, the observation equation to be used is Equation (8.3) above. Generally, the treatment is identical, but the orientation error Z should be very small as a known bearing is used as the reference direction. Many surveyors set the theodolite circle to a bearing for this purpose and to ease the arithmetic. An example of this type of equation was given for the traverse computation above.

The provisional coordinates of an intersected point can be obtained graphically or by direct solution using two fixed points.

Consider Figure 8.3 in which the new point P is to be fixed by intersection from the known points A and B. The points are listed in clockwise order ABP. The known quantities are angles α, β, E_A, N_A, E_B and N_B.

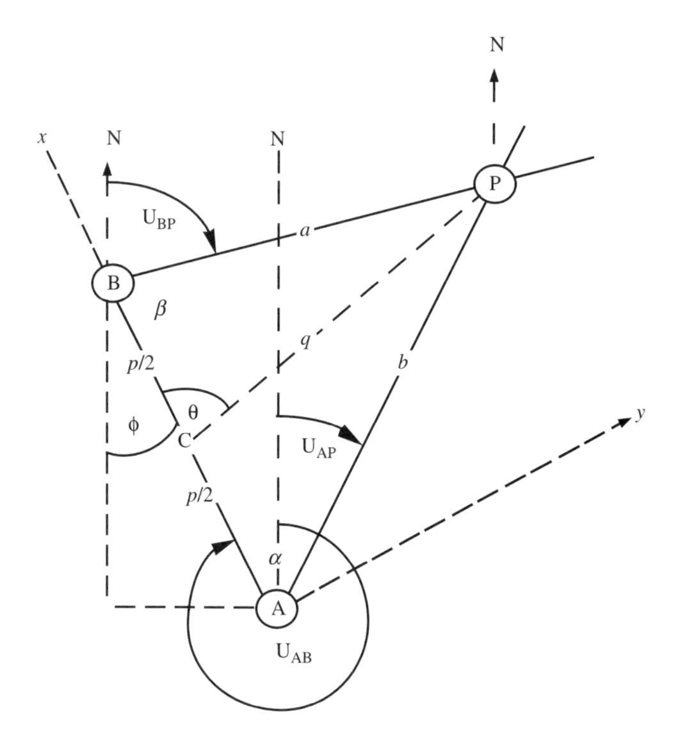

Figure 8.3

The computation is carried out in the following stages.

(a) Obtain the bearing of AB from

$$U_{AB} = \arctan(N_B - N_A, E_B - E_A)$$

(b) Obtain the bearings of AP and BP from

$$U_{AP} = U_{AB} + \alpha$$

$$U_{BP} = U_{AB} + 180 - \beta$$

(c) Calculate the coordinates of P from the formulae

$$N_P = \frac{N_A \tan U_{AP} - N_B \tan U_{BP} - E_A + E_B}{\tan U_{AP} \tan U_{BP}} \tag{8.4}$$

$$E_P = \frac{E_A \cot U_{AP} - E_B \cot U_{BP} - N_A + N_B}{\cot U_{AP} \cot U_{BP}} \tag{8.5}$$

These formulae are derived from rearranging the expressions:

$$\tan U_{AP} = \frac{E_P - E_A}{N_P - N_A} \quad \text{and} \quad \tan U_{BP} = \frac{E_P - E_B}{N_P - N_B}$$

Example 3

We use the data of Table 6.3 for points B and C to fix G by intersection using bearings U. In the notation of Equations (8.4) and (8.5), point C of Table 6.3 becomes A, B remains B and G is the new intersected point P. Hence in the notation of Figure 8.3, the data are

	E	*N*	*U*	*U (deg)*	*tan U*	*cot U*
A	564.439	487.776	AP	231.1238	1.2403675	1.24036751
B	567.895	443.275	BP	276.3248	–9.0220816	–9.0220816

Hence the coordinates of P(G) are $(E_P, N_P) = (516.330, 448.990)$.

Alternative formulae for angles

Alternative formulae for intersection in terms of the angles α and β are:

$$N_P = \frac{N_A \cot\beta + N_B \cot\alpha + E_A - E_B}{\cot\alpha + \cot\beta} \tag{8.6}$$

$$E_P = \frac{E_A \cot\beta + E_B \cot\alpha - N_A + N_B}{\cot\alpha + \cot\beta} \tag{8.7}$$

Again, we adopt a clockwise convention when listing the points. Note also that the various products involve data from both points, unlike in the bearings formulae e.g. compare $N_A \cot \beta$ with $N_A \tan U_{AP}$.

Example 4

The data for the above example are

	E	*N*	*Angle*	*Angle (deg)*	*cot*
A	564.439	487.776	α	55.5645	0.68562362
B	567.895	443.275	β	79.2344	0.19013702

giving as before the coordinates of P(G) to be $(E_P, N_P) = (516.330, 448.990)$.

Error analysis

As mentioned already, guidance in the analysis of error can be derived from a simple graphic treatment, considering the angle of cut between the rays, and their likely lateral shifts. A precise calculation can then be done, by the Least Squares method, to examine the error ellipses in the most interesting areas, such as close to the base line AB.

Figure 8.4 shows the usable space within which a satisfactory result may be obtained from the intersection of points from the two fixed stations A and B. The plotted criterion is usually the sum of the variances in E and N, called the *Positional Dilution of Precision*, or PDOP.

$$\text{PDOP} = \sigma_E^2 + \sigma_N^2 = a^2 + b^2$$

where a and b are the semi-axes of the error ellipse. Sometimes the maximum error, a, is used as an alternative criterion to illustrate the bounds of the area within which the required precision will be achievable.

For a complete example of the error ellipse see Section 8.10 below.

8.8 Fixation by distances only (lateration)

Fixing position by lengths alone is called *lateration*. It is more usually known as *trilateration* when three sides are measured to fix a triangle. The computational procedure is similar to other methods once the measured distances have been reduced to a common reference system, usually the ellipsoid or on a projection. The treatment of distance equations by Least Squares can be read in Chapter 7.

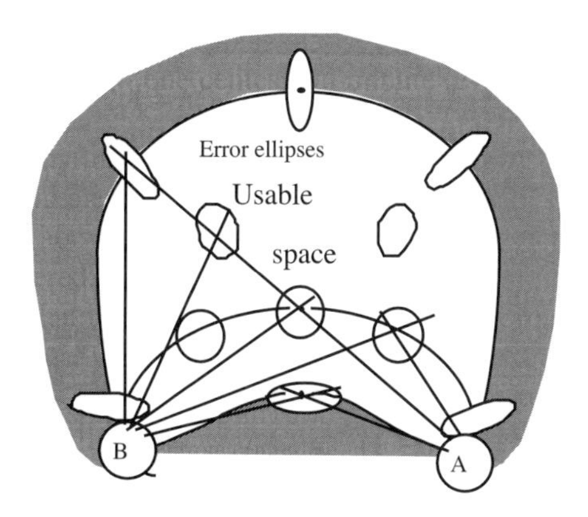

Figure 8.4

Direct solution of lateration coordinates

Although provisional coordinates may be obtained from a solution of the triangle ABP, the following direct formulae are convenient.

$$E_P = \frac{1}{2}(E_A + E_B) + \frac{1}{2p^2}(a^2 - b^2)(E_A - E_B) - \frac{2\Delta}{p^2}(N_A - N_B) \qquad (8.8)$$

$$N_P = \frac{1}{2}(N_A + N_B) + \frac{1}{2p^2}(a^2 - b^2)(N_A - N_B) - \frac{2\Delta}{p^2}(E_A - E_B) \qquad (8.9)$$

where

$$\Delta = \sqrt{S(S-a)(S-b)(S-p)} \quad \text{and} \quad 2S = a + b + p$$

For a proof of these formulae refer to Figure 8.3 in which C is the mid-point of AB, AB = p and PC = q. The coordinates of P are given by

$$\begin{aligned} E_P &= E_C + q\sin(\theta - \phi) \\ N_P &= N_C + q\cos(\theta - \phi) \end{aligned} \qquad (8.10)$$

and the area of the triangle ABP = Δ is calculated as

$$\Delta = \frac{1}{2} pa \sin\beta$$

Then we have

$$\sin\theta = \frac{2D}{pq}, \quad \cos\theta = -\frac{a^2 - b^2}{2pq}, \quad \sin\phi = \frac{E_A - E_B}{p}, \quad \cos\phi = -\frac{N_A - N_B}{p}$$

Substituting these expressions into Equations (8.10), remembering that C is the midpoint of AB, we obtain Equations (8.8) and (8.9).

Example 5

Using the same basic data i.e. $(E_A, N_A) = (564.439, 487.776)$ and $(E_B, N_B) = (567.895, 443.275)$ the point P is fixed from A and B by the distances AP and BP. The calculations are displayed in the following extract from a spreadsheet.

AB = p	AP = b	BP = a	$2S$	S
44.6349968	61.7960237	51.8802233	158.311244	79.1556219
$S-p$	$S-b$	$S-a$	Δ	
34.5206251	17.3595982	27.2753986	1137.45859	
		Term 1	Term 2	Term 3
		566.167	0.97766536	–50.814113
	E	516.330		
		465.5255	–12.588856	–3.9462838
	N	448.990		

8.9 Resection in two-dimensions

If directions in the horizontal plane of a theodolite circle are observed at an unknown station P to three known points A, B, and C, the coordinates of P in the horizontal plane can be calculated, provided all four points do not lie on a vertical cylinder. Considering only the horizontal plane through the theodolite, this condition means that all four points, P and the projections of A, B and C on this plane, must not lie on a circle.

Before the widespread availability of total stations, this surveying technique, known as *resection,* was very convenient in the field when fixing photo control points, positions at sea or on engineering sites where sudden decisions have to be made to fix a point with reference to visible controls without visiting them. Coordinated points such as church spires are excellent for this control. It has less importance with the advent of reflectorless EDM, but is still worth considering.

The computation is usually effected by Least Squares and coordinate variation, if software is available, and more than three control points are observed, as a check. We shall give a full example of this technique as a further illustration of the Least Squares method, and for the purposes of

1. demonstrating the use of a device, *Schreiber's method,* to remove unwanted parameters from a solution, and
2. computing a positional error ellipse in full.

Table 8.9 lists the coordinate data to be used and shown in Figure 8.5. The problem is to find the coordinates of A by angles only.

The observation equation for a direction is (see Equation (8.3))

$$\mathbf{P}\,\mathrm{d}E + \mathbf{Q}\,\mathrm{d}N + \mathbf{Z} = \mathbf{L} + \mathbf{v} \tag{8.11}$$

For convenience, the value of **L** for the reference direction, A to 18, is set as zero, by assuming the initial observed bearing to be the same as the computed bearing. Other observed bearings are obtained by adding the observed angles to this initial bearing. Using the provisional coordinates of A we compute the bearings to the controls as shown in column 2 of Table 8.10. These bearings are reduced to a zero setting at

Table 8.9 *Plan coordinates of data points*

Point	*x (N)*	*y (E)*
18	3431.47	3730.53
19	2946.97	3011.82
24	1824.27	739.19
16	3521.06	829.67
6	885.13	2000.00
A	3001.00	2001.00

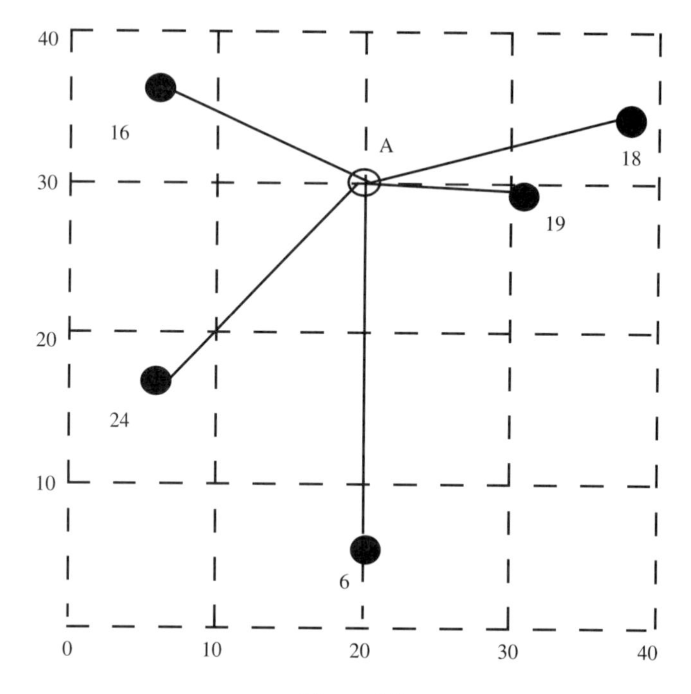

Figure 8.5

station 18 in column 3, and converted to sexagesimal values in column 4. The observed directions referred to a zero setting on station 18 are shown in column 5. The absolute vector **L** (computed – observed bearings) is shown in column 6.

The full observation equation coefficients and the absolute vector are given below, computed from Equation (8.11). Note that the precision of the figures for the coefficients can be misleading: the equations are not linear to this precision.

It will be noticed there is a sixth equation, marked with asterisks. This Schreiber equation is explained in Section 8.10. The normal equations i.e. **Nx** = **b** (see Chapter 6) are given on the assumption that the directions are all of an equal weight still to be found.

Table 8.10

Station	*Bearing*	*With ref. to station 18*	*Sexagesimal values*	*Observed directions*	*Absolute vector* $L = C - O$
18	76.0234	0	0	0° 00' 00"	00.00
19	93.0597	17.0363	17° 02' 10.46"	17° 00' 32"	–98.46
6	180.0271	104.0037	104° 00' 13.23"	104° 00' 05"	–8.23
24	226.9982	150.9748	150° 58' 29.36"	151° 00' 04"	94.64
16	293.9408	217.9174	217° 55' 02.60"	217° 59' 59"	296.40

Table 8.11

1 (P)	*2 (Q)*	*3 (Z)*	*4 (L)*
–27.951	112.304	1	0
10.876	203.476	1	–98.46
97.485	–0.046	1	–8.24
81.535	–87.430	1	94.64
–65.310	–147.098	1	296.40
*43.216	*36.316	*2.236	*127.16

Table 8.12

1	2	3	4	5
21 316.22	1547.78	96.63	–13 514.98	–0.979
1547.78	83 296.26	81.21	–71 909.17	–0.934
96.63	81.21	5	284.34	90.95

Table 8.12 shows the normal equations in columns 1 to 4, and the solution in column 5.

The final coordinates are:

$$(E_A, N_A) = (2001.00 - 0.979 = 2000.021, 3001.00 - 0.934 = 3000.066).$$

The residuals, in seconds of arc, are

$$\mathbf{v} = \begin{bmatrix} 13.43 \\ -11.12 \\ 3.73 \\ -1.88 \\ -4.77 \end{bmatrix}$$

giving the standard error 2.603" (see Chapter 6).

8.10 Schreiber reduction

The orientation parameter Z is of very little interest to the surveyor. At every station where horizontal angles are observed, one such parameter has to be included, thus increasing the size of the matrix of normal equations unnecessarily. The parameter can be eliminated by Schreiber's method, as follows.

Compile an additional observation equation whose coefficients are the respective sums of all coefficients in each column, including the absolute term, divided by the square root of the number of directions. For example, the first coefficient of this Schreiber equation marked in the previous section is

$$43.216 = \frac{-27.951 + 10.876 + 97.485 + 81.535 - 65.310}{\sqrt{5}}$$

The Schreiber normal equations are formed from the Schreiber observation equation in the usual way, giving

$$\begin{bmatrix} 1867.63 & 1569.45 & 96.63 \\ 1569.45 & 1318.87 & 81.21 \\ 96.63 & 81.21 & 5 \end{bmatrix} \mathbf{x} = \begin{bmatrix} 5495.44 \\ 4618.04 \\ 284.34 \end{bmatrix}$$

If the Schreiber normal matrix is subtracted from the original normal matrix, the equation with the orientation correction Z is eliminated, as shown:

$$\begin{bmatrix} 19448.59 & -21.67 & 0 \\ -21.67 & 81977.39 & 0 \\ 0 & 0 & 0 \end{bmatrix} \mathbf{x} = \begin{bmatrix} -19010.42 \\ -76527.21 \\ 0 \end{bmatrix}$$

Solving the smaller 2×2 matrix (neglecting the zero terms) provides the same answer as before, as illustrated below. The inverse of the 2×2 matrix is given to show that the error ellipse parameters are unchanged.

$$\mathbf{x} = \begin{bmatrix} 5.1418 \times 10^{-5} & 1.359 \times 10^{-8} \\ 1.359 \times 10^{-8} & 1.2198 \times 10^{-5} \end{bmatrix} \begin{bmatrix} -0.979 \\ -0.934 \end{bmatrix}$$

Schreiber's method can be used in all cases where such a parameter is present, such as in distance measurement. It is useful where the capacity of the computer is being stretched to the limit.

8.11 Error ellipse and pedal curve

The quality of the fixation may be represented graphically by an error ellipse and its pedal curve. First we find the direction T of the maximum variance from

$$\begin{aligned} 2T &= \arctan(2\sigma_{EN}, \sigma_N^2 - \sigma_E^2) \\ &= \arctan(2 \times 1.359 \times 10^{-8}, 1.2198 \times 10^{-5} - 5.1418 \times 10^{-5}) \\ &= 180.0^\circ \end{aligned}$$

Therefore $T = 90°$.
Now a and b, the semi-major and semi-minor axes of the error ellipse, are given by

$$2a^2 = \sigma_N^2 + \sigma_E^2 + \Delta \quad \text{and} \quad 2b^2 = \sigma_N^2 + \sigma_E^2 - \Delta$$

where Δ is given by either of the equations

$$\Delta = 2\sigma_{EN} \operatorname{cosec} 2T \quad \text{or} \quad \Delta = (\sigma_N^2 - \sigma_E^2)\sec 2T$$

From the second equation we find $\Delta = 3.922 \times 10^{-5}$. Therefore, a and b can be derived from the above equations. We find $a = 0.072$ and $b = 0.035$.

The equation of the standard error σ_U at any direction U is given by

$$\sigma_U = \frac{\sqrt{6.3616 \times 10^{-5} + 5.1418 \times 10^{-5} \cos 2(U - 90)}}{1.414}$$

8.12 Direct resection solution

The Snellius method has been found to give the most universal solution to the resection problem. Because the once-popular Tienstra solution fails in the common important case when the three control points are collinear, it is not advocated.

The angles α, β and γ observed at P to three control points A, B and C and the side lengths a and b are known. The problem is first to find an unknown orienting angle such as x or y, then to solve for the coordinates of P. From Figure 8.6 we have

$$x + y = 360^\circ - (\alpha + \beta + C) = S$$

say, which is known.

$$x = S - y \quad \text{and} \quad \frac{b \sin x}{\sin b} = \text{PC} = \frac{a \sin y}{\sin a}$$

therefore

$$\frac{\sin x}{\sin y} = \frac{a \sin b}{b \sin a} = K, \text{ say}$$

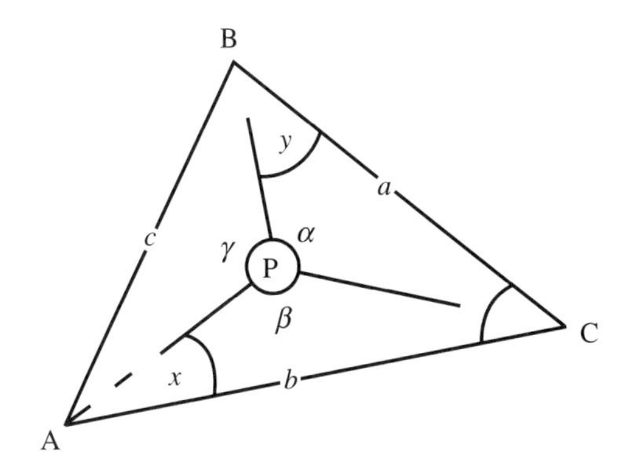

Figure 8.6

which is also known, therefore

$$K = \frac{\sin(S-y)}{\sin x} = \sin S \cot y - \cos S$$

and finally

$$\cot y = \frac{K + \cos S}{\sin S} \tag{8.12}$$

Note that although it is possible to set

$$\frac{\sin A}{\sin B} = \frac{a}{b},$$

this ratio, calculated from angles, is indeterminate when C lies on AB but not if the sides are used. Once y, and therefore angle PCB, has been found, we calculate the coordinates of P by intersection from B and C.

Example 6

The control points A, B and C to fix point P are:

Point	*E*	*N*
A	3730.53	3431.47
B	2000.00	885.13
C	829.67	3521.06

The extract from a spreadsheet calculation of (*E*, *N*) is displayed below.

Angles	*a*	*b*	*C*	$x + y = S$
Deg	113.998333	142.000278	64.2902377	39.7111512
Rad	1.98964626	2.47837239	1.12207632	0.69309034
K	a sin b / b sin a			
K	0.66969017			
cot *y*	(*K* + cos *S*)/sin *S*			
cot *y*	2.25219262			
tan *y*	0.44401176	deg		
y	0.41786295	23.9417837	check	
x	0.27522738	15.7693675	15.7693675	
PC	PC = a sin y / sin a			
PC	1281.11838			
angle PCA	0.38799288	22.2303547		

Bearing CP	113.999312			
Rad	1.98966334			
dE	1170.36613	2000.036	E	Final Solution
dN	−521.06374	2999.996	N	

It is most important to be systematic about the signs of the various angles, particularly if a computer algorithm is written. When the new point P lies outside the control triangle ABC, special care is needed. The following convention is acceptable:

1. Letter the triangle ABC in a clockwise manner.
2. Assign the angles α, β and γ also clockwise. For example α is the clockwise angle between the directions PB and PC (i.e. those not involving A).

It is also prudent to anticipate what will happen to the arithmetic process when the point P lies on the *danger* circle through ABC. In this case, $x = y$, $K = 1$ and $S = 0$, therefore $\sin S = 0$ and $\cot y$ is indeterminate. No unique solution is possible.

8.13 Error analysis

Combining a graphic figural treatment, to obtain general information, with a Least Squares analysis of the interesting cases, is the best method of performing an analysis of resection errors.

Resection consists in general of the intersection of two circles defined by the angles subtended at two control points. In Figure 8.7 we have the two angles α and β subtended by the pairs of points A and B and C and D.

The angle α defines the circle, centre O_1 radius r, through the controls A and B. The other circle is defined by β. The point P lies at the intersection of these two circles. There are two solutions, one of which is correct. Usually this can be chosen from other information such as estimated distances. Of course, if the angle BPC is also observed, there is no ambiguity. At sea, two angles are usually observed by two different sextants. The two position-circles aid the error analysis.

At P, the solution consists of the intersection of the two tangents, which are perpendicular to their radii. Thus in the detailed diagram on the right of Figure 8.7 we see that P is fixed by two lines intersecting at an angle θ. This angle is subtended by P at the two centres, O_1 and O_2. Thus the resection problem has been reduced to an equivalent distance intersection problem. If we decrease the angle α by $\delta\alpha$ the tangent at P moves away from the centre by dr, given by

$$\mathrm{d}r = \frac{ab}{p}\mathrm{d}\alpha \sin 1''$$

where a, b and p are the lengths of the sides of the triangle ABP. In this way we can calculate the shifts to the radii, and estimate the strength of the positioning of P. If some close comparison is required of two values, the numerical Least Squares method is needed.

The figural approach is useful as a management tool in the field. It is also a useful way to locate a buried mark for which previous angles to the controls are known.

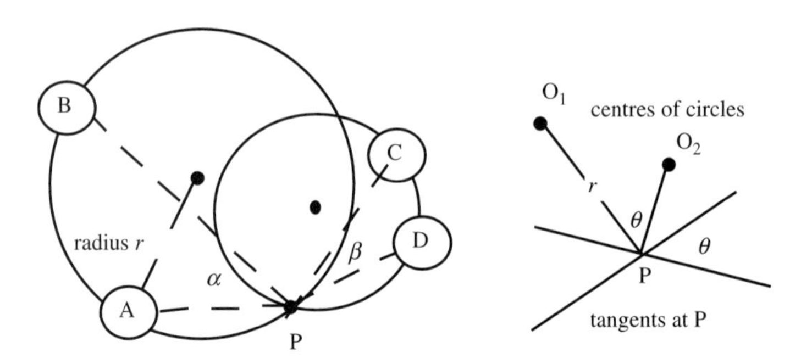

Figure 8.7

Angles are taken at a likely point P close to the mark. The tangents can be marked out by strings because the angle BPO_1 = BAP – 90°. The shift is computed from the change in the angle between the old and new, and thus the position line for the mark laid out by an offset string. The same approach to the other angle β gives a second string line and the location of the mark. This procedure is much easier and more accurate than coordinating the new point and calculating the bearing and distance to the old mark. It should also be noted that the coordinates of the control points are not required, only the old angles to them.

Chapter 9
Heights and Levels

9.1 Introduction

Knowledge of the heights and levels of points on the surface of the Earth is required to:

1. enable survey measurements, aerial photographs and satellite imagery to be reduced to a datum surface such as sea level, or a reference ellipsoid;
2. create digital terrain models (DTMs) to be used to solve such problems as inter-visibility between points on the ground;
3. enable drainage and other water works to be surveyed so that water may flow in desired directions;
4. enable relief to be depicted on topographical and air maps;
5. enable roads and railways to be constructed in such a manner that steep hills are avoided;
6. provide information for scientists concerning the shape and structure of the Earth, and tectonic movements on its surface.

The vertical and levels

The vertical is the direction which a plumb line takes when it hangs freely under the effect of the Earth's gravitational pull. A horizontal surface is a *level surface* at right angles to the vertical. The height of a point may be defined as the linear distance from the point up or down the vertical through the point, to a reference horizontal surface or datum. Later in this chapter height is defined purely in terms of gravitational force. If the horizontal datum surface covers a small portion of the Earth it may be sufficient to consider it as a plane surface throughout its entirety, but when treating the surface of the Earth as a whole, the horizontal surface will be curved and be everywhere at right angles to the verticals, as in Figure 9.1.

The horizontal datum surface for heights is one which closely approximates to mean sea level (MSL), determined from observations on tide gauges averaged over a period of 19 years. This period allows the various tide-raising forces to repeat. Allowance also has to be made for long-term trends.

Since most vertical angles between ground stations are of the order of 1° or less, computational approximations are usually permissible. Most heighting processes deal with differences in height, observed over comparatively short lines, carried forward over long distances by a chain of many measurements. For this reason, small systematic errors may often accumulate in serious proportions and must be avoided where possible. Accuracies of 2 mm km^{-1} are possible by levelling.

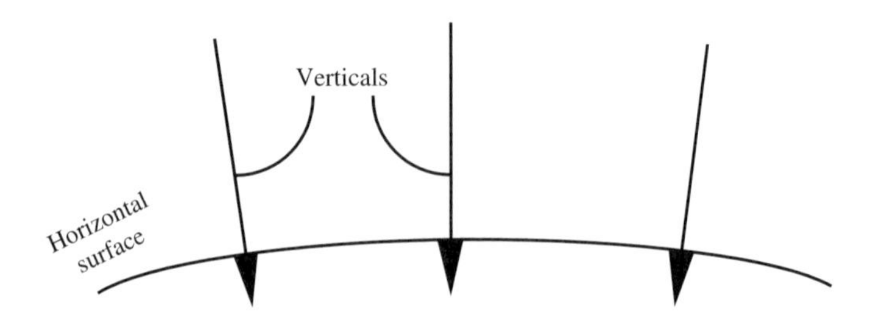

Figure 9.1

Artificial Earth satellites can provide geometrical height differences to about 3 mm accuracy but need to be connected to a level datum, either by transformations to fit levelled points, or to lower accuracy, by a knowledge of the Earth's gravity field.

Before the advent of the GPS system of position fixing, lines of levels were run very carefully over whole countries and continents between tide gauges where sea level was determined. Permanent bench marks were established to which surveyors fixed their heights by the means described below. Today, however, the GPS system, together with knowledge of the Earth's gravity field, enables absolute heights to be determined replacing these reference bench marks. See Chapter 7 and Section 9.10 below.

Methods of determining relative heights

The following four ground surveying methods are used to determine the relative heights of survey points:

1. hydrostatic levelling;
2. spirit levelling;
3. trigonometrical heighting;
4. barometric heighting.

Depending on the flying height of the aircraft and other factors, topographic stereo-photogrammetry, the principal method of deriving heights covering large tracts of land, can regularly achieve practical accuracies of about 0.1 m. Ground survey heights are usually needed to control this technique (see Appendix 5).

9.2 Hydrostatic levelling

This is probably the oldest form of levelling. A tube containing water will readily define two points A and B at the same height due to the balance set up by the water under the force of gravity (see Figure 9.2). The heights found by this method are referred to as dynamic heights.

This method of levelling was much used in Holland where differences in height are very small and great accuracy is essential. Pipes of up to 10 km long have been used. For convenience, flexible pipes are laid out along the bed of a canal by a survey ship. The main problems associated with this method are to do with eliminating air

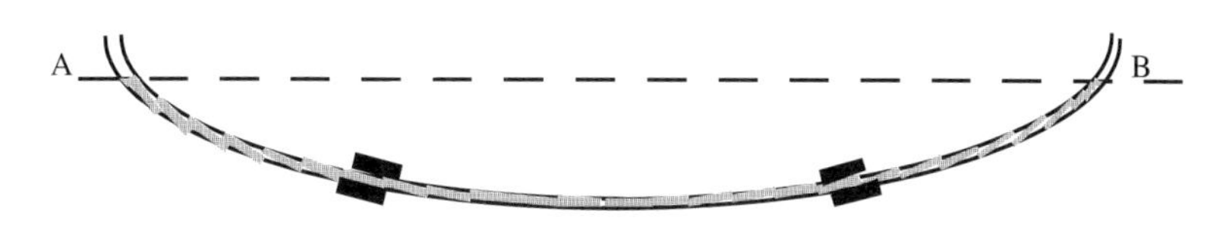

Figure 9.2

bubbles from the pipe when filling it with water or dual liquids, and controlling the temperature of the fluids.

Miniature versions of high accuracy hydrostatic levelling are employed in industrial surveying and an ordinary water hose is useful for work in forested areas, for example to mark the future shoreline of an artificial lake.

9.3 Spirit levelling

In this method, a horizontal line of sight is established by the observer with the aid of a spirit bubble, plumb line or a freely suspended compensator system, which enables the line to be sighted through a telescope in a horizontal direction. Although compensators have replaced bubbles in all but the most precise work, the term *spirit levelling* is retained here. The height of the instrument station is generally not determined in the procedure.

For ease of operation, spirit levelling is used in conjunction with one or two graduated rods or staves to determine the difference in height between two points. In Figure 9.3 the height of B with respect to that of A is found by taking readings R_A and R_B on the vertical staves at A and B. For

$$R_A + H_A = R_B + H_B$$

therefore

$$H_B = H_A + R_A - R_B$$

Since A is the backward point if we are moving from A to B, the reading on A i.e. R_A is referred to as the 'back sight reading' or simply the *backsight*. The reading on B i.e. R_B is called the *foresight*. It will be noted that the height difference between A and B

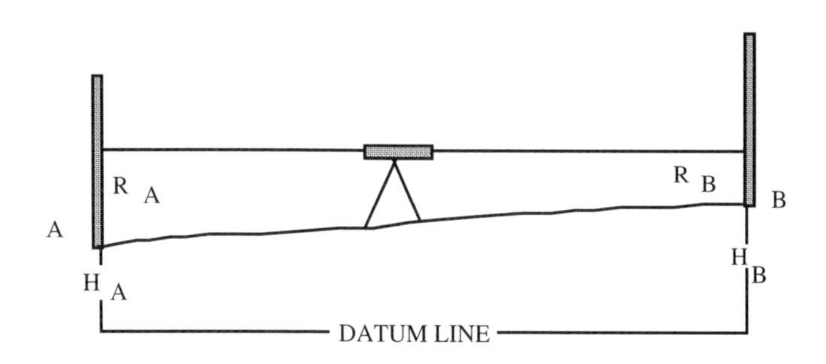

Figure 9.3

is given both in magnitude and sign. We always consider the height difference in the sense

$$\text{Difference in height} = \text{Backsight} - \text{Foresight}$$

After making a reading on B, the staff at B is turned round to face the new instrument position and the staff at A is moved forward to the new instrument position C. Thus the new backsight and foresight readings are r_B and r_C. The foot of the staff at B should not change height in turning the staff round, hence it is necessary that some point whose height is well defined should be chosen for the *change point* B. Such a point would be e.g. the head of a rivet on a drain cover or a kerbstone. In open fields a large screwdriver driven into the ground makes a good change point. Special plates are also available. If a point is heighted but not used as a change point, it is called an intermediate point, and the sight to it is an intermediate sight (IS).

When the height of a ceiling or roof is required the staff is held upside down and pressed hard against the surface. In this case the backsight reading is recorded as negative.

9.4 Trigonometrical heighting

In Figure 9.4 the difference in height between the points A and B is given by

$$\Delta H = S \tan \theta$$

where S is the horizontal distance AC, and θ is the vertical angle recorded by a theodolite circle against a horizontal datum, determined by a bubble or compensator. Since tan θ is negative when θ is negative and positive when it is positive, for $\theta < 90°$ the equation gives both the magnitude and sign of ΔH. If the height of either A or B is known, the height of the other may be calculated. In precise work, vertical angles are observed at both ends of the line i.e. at both A and B. The distances sighted in this method may be anything up to 50 km, though accurate results are limited to lines of 10 km or less. If lines are restricted to 200 m accuracies of 2 mm can be maintained.

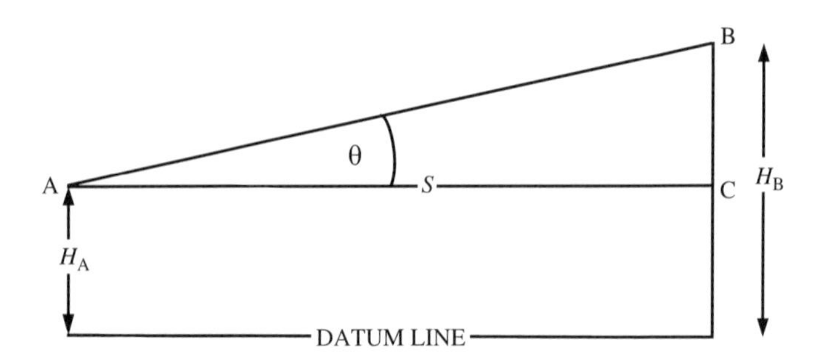

Figure 9.4

9.5 Barometric heighting

This method of measuring height differences depends on the variation of atmospheric pressure with height above sea level. Pressure P is related to the acceleration due to gravity g, the density of the air ρ and the height H by the functional expression

$$P = \text{fn}\,(g, \rho, H)$$

If P is measured at two places, where g and ρ are sensibly constant, the pressure height relationship for two points is

$$\Delta H = H_2 - H_1 = \text{K} \ln \frac{P_1}{P_2}$$

where K is a constant. If one height is known we may find the other.

The barometric method of heighting is especially convenient where points are not intervisible, such as in forested country, or where rapid, but relatively inaccurate results are required. Accuracies of about 1 m are attainable with care (Allan 1997).

9.6 Accuracies of heighting methods

It is impossible to give figures for the accuracy attainable by a survey technique which will apply to all circumstances. The accuracy of the method depends, to some extent, on the care with which the survey is carried out, the time spent on the operation and various factors such as the weather conditions at the time of the measurements. However, as a guide, the following accuracies may be expected if a normal technique is adopted.

1. Hydrostatic and geodetic levelling: a standard error of 1.5 $\sqrt{K}$ mm, where K is the length of the line in kilometres.
2. Ordinary spirit levelling: a standard error of 6 $\sqrt{K}$ mm.
3. Trigonometrical heighting: a standard error of 60 $\sqrt{K}$ mm, although accuracies as high as for geodetic spirit levelling can be obtained with special care.
4. Barometric heighting: about 1.5 m, provided operations are limited to an area of about 10 km^2 and elevation difference to not more than 500 m.

Benchmarks

A benchmark is a permanent survey mark whose height is known, and from which levelling and height are controlled. Wherever possible a line of levels or series of trig heights should be tied into benchmarks; but it is essential that barometric heighting is controlled by benchmarks. Great care is needed in the siting of these marks on ground that is thought to be stable and relatively free from vandalism. With GPS it is possible to establish these benchmarks provided the geoid-spheroid separation is known accurately, say to 5 mm.

9.7 Spirit levelling: principles and procedures

The principle of levelling with the aid of a staff or staves has been explained above. The same simple technique is common to all types of spirit levelling irrespective of the accuracy required. The following section explains the basic fieldwork involved, the manner in which the observations are booked in the field, and the various simple checks that are applied to the arithmetical operations required to produce the final heights of the stations. In many books, the subject of geodetic levelling is described after that of simple or engineering levelling. Since these have so much in common, we shall treat the most accurate work first and highlight those operations that are unnecessary if a lower standard of result is sufficient.

Types of level instrument

The various types in common use are described in Section 13.6. It should be remembered that a theodolite and a total station may be used as a level on occasion.

Types of staff

A large number of different types of staff are available to the surveyor. Folding or telescopic staves are easier to carry, however rigid staves are more accurate since they maintain their length better over a period of time. Geodetic levelling staves are constructed of wood with the graduations on a strip of invar held securely at the bottom of the staff but which is free to expand elsewhere along its length. The coefficient of expansion of invar is so small that temperature corrections need not be applied as a rule. The figuring of staves is not consistent. Modern staves with bar-coded scales have revolutionized levelling, by making it consistently more accurate and cost effective.

Some geodetic staves have two different scales side by side so that a gross error in reading the staff may be immediately detected and the reading repeated. To ensure that the staff height at a change point is not altered when it is turned around, a special steel foot with a dome-shaped top is employed.

For high precision industrial work, a short rule is often used.

Booking of readings

The process of levelling requires a very orderly approach to recording data because so many readings are taken. Data recorders are widely used with robust software to detect omissions and assess quality as the work proceeds. However, the principles are the same for hand booking which is now outlined with respect to the example shown in Figure 9.5 and Table 9.1.

Figure 9.5 shows a cross-section and plan along the route taken by a surveyor levelling from point 1 to point 11. The instrument positions are at A, B, C, and D, and the change points are 5, 9 and 10. Table 9.2 shows a typical set of readings for the levelling illustrated in Figure 9.5. The algebraic version given in Table 9.1 is to assist in the explanation.

The instrument is set up at A and carefully levelled. A staff is held at point 1 and the reading R_1 is taken. The staffman then moves in turn to points 2, 3, 4 and 5 where readings R_2, R_3, R_4 and R_5 are made.

The intermediate sights are booked in the second column of Table 9.1, with the change points in columns 1 and 3 for backsights and foresights respectively. The

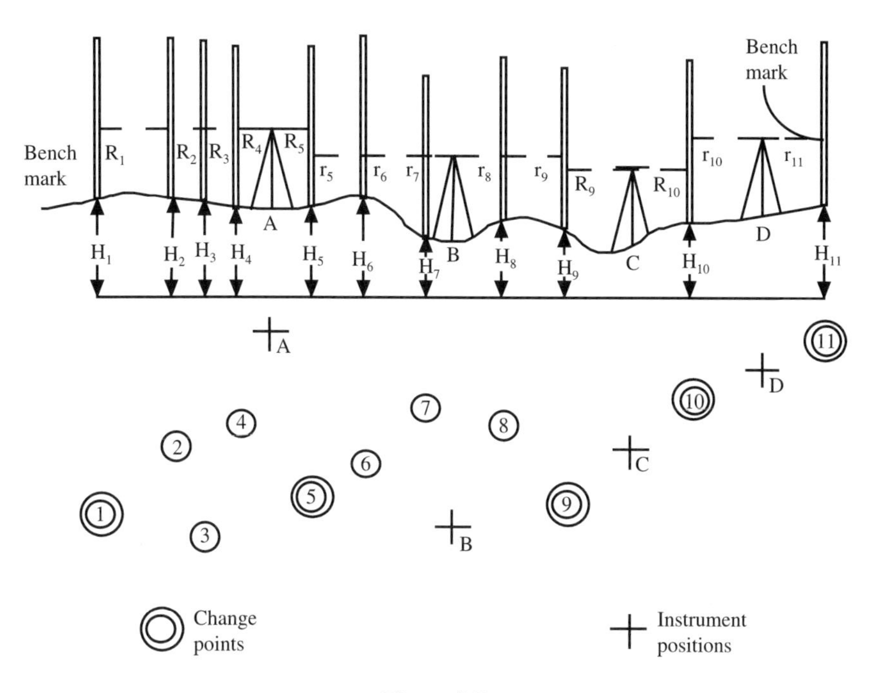

Figure 9.5

Table 9.1 *Rise and Fall Method: algebraic version*

Backsight	*Intermediate sight*	*Foresight*	*Rise*	*Fall*	*RL*
R_1					H_1
	R_2		$R_1 - R_2$		H_2
	R_3			$R_2 - R_3$	H_3
	R_4		$R_3 - R_4$		H_4
r_5		R_5		$R_4 - R_5$	H_5
	r_6		$r_5 - r_6$		H_6
	r_7			$r_6 - r_7$	H_7
	r_8		$r_7 - r_8$		H_8
R_9		r_9		$r_8 - r_9$	H_9
r_{10}		R_{10}	$R_9 - R_{10}$		H_{10}
		r_{11}	$R_{10} - R_{11}$		H_{11}

Table 9.2 *Rise and Fall Method: numerical version*

Backsight	*Intermediate sight*	*Foresight*	*Rise*	*Fall*	*RL*	*Notes*
1.198					36.469	BM1
	0.811		0.387		36.856	
	1.305			0.494	36.362	
	1.283		0.022		36.384	
0.920		1.588		0.305	36.079	CP
	0.884		0.036		36.115	
	1.109			0.225	35.890	
	0.655		0.454		36.344	
1.393		2.097		1.442	34.902	CP
1.603		1.015	0.378		35.280	CP
		0.887	0.716		35.996	BM_2
Checks						
5.114		5.587	2.193	2.466	0.473	
		0.473		0.473		

differences in height between each point are recorded in the 'rise' and 'fall' columns where appropriate.

The instrument is then changed to position B while the staff is kept at point 5. Readings from position B are denoted by r_5, r_6, r_7, r_8 and r_9 with the staff at points 5, 6, 7, 8, and 9 respectively. The backsight reading is r_5 and the foresight reading is r_9.

Note that only the change points (CP) have two readings. The reduced level of each point in turn is obtained by adding the rise or subtracting the fall from the height of the point to which the difference of height was related i.e. the point above it in the field book. At the foot of each field book page an arithmetical check is made.

The following four columns are summed separately and the totals entered at the foot of each column:

1. the sum of backsight readings;
2. the sum of foresight readings;
3. the sum of rises;
4. the sum of falls.

The difference in height between points 1 and 11 in this example is

$$H_{11} - H_1 = 0.473$$

The various checks then are:

$H_{11} - H_1$ = sum of the backsights – sum of the foresights

$H_{11} - H_1$ = sum of the rises – sum of the falls

This can easily be proved from the algebraic example.

The totals at the foot of one levelling page are carried forward to the top of the next page, taking great care not to make a mistake in the transfer. It must be stressed that these checks in no way verify the readings but merely their abstraction into heights.

Normally the line of levels will start and close on a benchmark, although a loop may be run which begins and ends on the same point. The adjustment of any misclosure is considered below. The method of booking given above is the 'rise and fall' method.

Another method worthy of mention is the 'height of collimation' method. In this method we book the height of the line of sight and from it subtract the individual readings to obtain each ground height. For example, in Figure 9.5 the line of sight of the instrument in position A is $H_1 + R_1$.

A typical booking by height of collimation of the readings of Tables 9.1 and 9.2 is given in Table 9.3. The sole arithmetical check on the booking is

$$H_{11} - H_1 = \text{sum of the backsights} - \text{sum of the foresights}$$

There is no arithmetical check on calculating the heights of the intermediate points. For this reason, this method is less favoured than the rise and fall method. It

Table 9.3 *Height of Collimation Method*

Backsight	*Intermediate sight*	*Foresight*	*Height of collimation*	*RL*	*Notes*
1.198			37.667	36.469	BM1
	0.811			36.856	
	1.305			36.362	
	1.283			36.384	
0.920		1.588	36.999	36.079	CP
	0.884			36.115	
	1.109			35.890	
	0.655			36.344	
1.393		2.097	36.295	34.902	CP
1.603		1.015	36.883	35.280	CP
		0.887		35.996	BM_2
Checks					
5.114		5.587		0.473	
		0.473			

is however more convenient in the setting out of height points on the ground and in recording heights below roofs or arches in engineering work.

9.8 Errors in levelling

Because levelling is a chain process in which small errors accumulate, attention has to be paid to errors which, although small in themselves, are significant in total. The following are the various sources of error in levelling.

Gross reading error

In precise work each staff has two scales engraved upon it so that a mistake will be detected by the booker at once. Alternatively, the readings to the stadia hairs are used for the same purposes. Bar-coded staff reading levels result in few mistakes, if any.

In all levelling it is standard practice to close the work, either out and back or round in loops, to detect mistakes. The prudent surveyor identifies semi-permanent change points to avoid having to re-level too much work.

Staff datum error

A staff may have a small datum or zero error. This means that the complete scale on the staff is moved through a small amount with respect to the base of the staff. If a staff becomes badly worn at the base, such an error is introduced. If only one staff is used to height two points, the staff error from this source will not affect the difference in height since it is common to both readings. If two staves are used for speed, as in geodetic levelling, the error in the difference in height of one set-up is the difference between the two datum errors of the staves. After every pair of set-ups the error is eliminated if the back staff is leap-frogged to the forward position. Hence if two staves are used, there should be an *even number* of set-ups between each benchmark.

Graduation error

In precise work, the errors in the graduations of each staff are determined and applied to the staff readings where applicable. Each staff is placed upon a specially designed bench along which is a very accurately divided tape or scale and against whose graduations those of the staff are compared with the assistance of a travelling microscope. For best results a polynomial is fitted to the calibration data and corrections are applied to every reading: not an arduous task if the bookings are originally by data logger.

Non-verticality of the staff

If a staff is held at an angle θ to the vertical the reading is Secθ instead of S. Hence the reading error is approximately

$$S\sec\theta - S = \frac{1}{2}S\theta^2$$

This is a serious source of error in all types of levelling. In ordinary work, the staff is swung backwards and forwards about its base so that it moves through the vertical position at some stage. The observer then obtains a correct minimum reading at this point. In most staves a bubble is fitted to the back of the staff to enable it to be held vertically. The adjustment of a staff bubble should be checked every week against a plumb line. A geodetic levelling stave is provided with two rods which allow it to be held vertically for some time. Some surveyors prefer a staff clamped to an adapted survey tripod for precise work.

Warpage of the staff

If the staff becomes warped the effect is similar to non-verticality; a reading becomes too high by an amount proportional to the reading itself.

Temperature errors

In refined work, the graduations are engraved or painted on a strip of invar attached to the bottom of the staff. Thus the scale is free to expand and contract without stresses being set up. Since the coefficient of linear expansion of invar is very small (of the order of 14×10^{-7} °C) temperature corrections need not be applied as a rule, although in the tropics they may be significant.

Staff illumination

It has been found that engraved staves may exhibit a small systematic error under certain conditions. If one staff is constantly illuminated by the sun, while the other is constantly shaded, a small error of about 60 μm per shot can arise. The reading on the illuminated staff is too low because the bottom portion of the graduation is not seen. If a line of levelling is run from south to north, the south-facing staves (the foresight staves) are consistently read too low and the difference in height (back – fore) is therefore too large if positive and too small if negative. Thus the northern benchmarks give too great a height. The same effect will be obtained when running a line of levels east to west, although the effect will tend to be balanced if the line is levelled in one direction throughout the day. Painted staves do not exhibit this effect to any marked extent.

The illumination of bar coded staves may also have a systematic effect, for example in a tunnel illuminated by artificial light.

Bubble and compensator sensitivity

In a level, the bubble or compensator should be sufficiently sensitive to permit staff readings of the required accuracy to be read, while at the same time they should not be so sensitive that much time is wasted in levelling the bubble. In a geodetic level the bubble is capable of being levelled with a standard error of 0.25", equivalent to 0.1 mm at about 100 m. Vibration can be a nuisance in a workshop floor.

Temperature effects

Most good instruments are designed so that temperature does not affect their performance to any extent. However, in precise work, the instrument is always shaded from the sun by a survey umbrella to prevent errors arising from differential heating and twisting of the tripod.

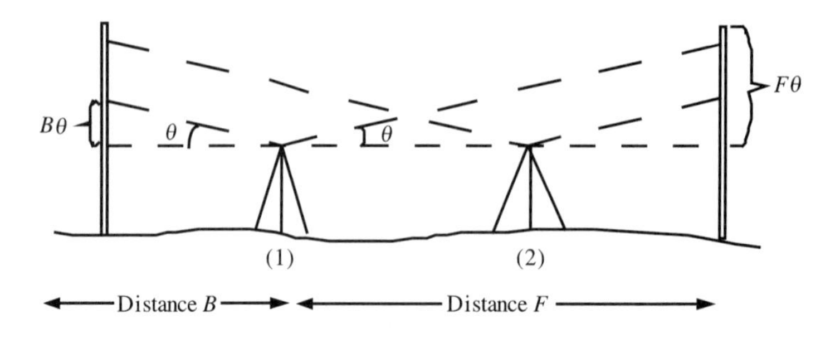

Figure 9.6

Collimation errors

If a level has a collimation error of θ the reading error on a staff will depend on the length of the line of sight. In Figure 9.6 the backsight distance B is less than the foresight distance F with the result that the staff errors in reading are $B\theta$ and $F\theta$.

The error dH in the height difference is given by

$$dH = \theta(B - F)$$

If the lengths of the back and foresights are equal there is no error in the height difference. Hence in all types of levelling these sights should be equated where possible, and the collimation error should be made as small as possible by adjusting the level by the two-peg test described in Section 13.6.

Sinking or rising of the staff

If the staff is set up on soft ground it will gradually sink while the observer is working. In Figure 9.7 the levelling staff was at a position F when the reading R_B was taken from position I, but it had sunk to position F' by the time that the reading was taken from position i to give a reading r_B which is in error by e.

There is therefore a strong argument for reading onto a staff in quick succession to minimize the time during which the staff may sink. As will be seen below, this conflicts with the effect of the sinkage of the instrument. Both effects can be avoided if all precise work is carried out along a hard surface such as the kerbstones of a road. In hot weather, a tarred surface will permit the staff to sink, while pegs

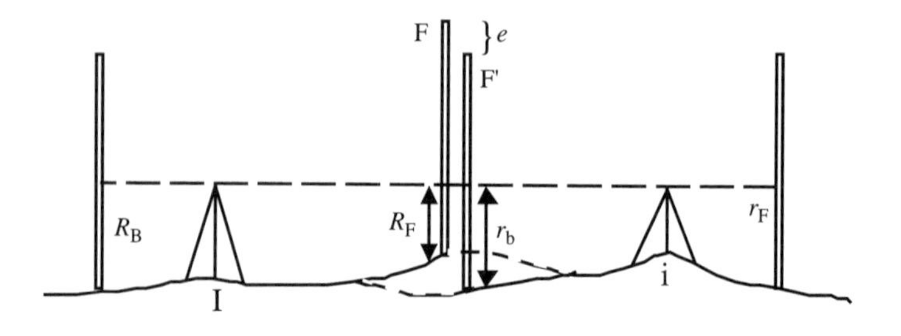

Figure 9.7

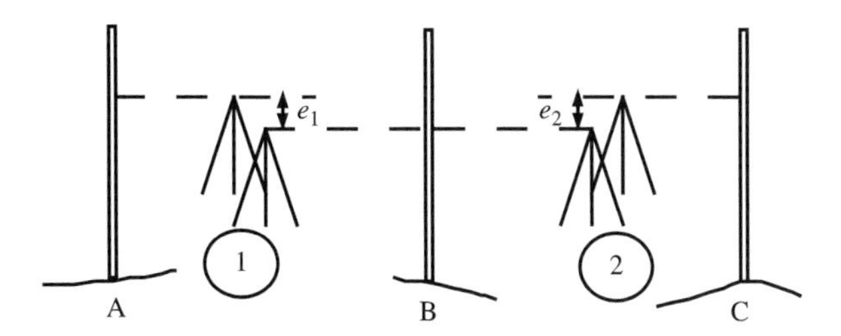

Figure 9.8

driven into the ground will rise for a time after being driven, producing the opposite effect.

Sinking of the instrument

In a similar manner to staff sinkage, the instrument may sink into the ground while the observations are being made. In Figure 9.8 the instrument sinks a distance e_1 in position 1 and e_2 in position 2. If the same degree of sinking occurs at both positions and the observer takes the same time to take the reading, there is a strong argument for reading first to the backstaff at A then to the forestaff at B, during which time the error e_1 has occurred. After changing to position 2, the first reading should be to the forestaff at C and finally to the backstaff at B, during which the error e_2 has occurred. The chances are that these errors e_1 and e_2 will be nearly equal and therefore the height difference between A and C will be correct.

The consequence of this effect means that there should be an even number of set-ups between the benchmarks. Since it is argued that the instrument is more likely to sink than the staff on account of its greater weight, the procedure is to read alternately to back and fore staves in precise work, or (which is the same thing) to always read to the same staff first, the staves being leap-frogged forward.

Effect of the Earth's curvature

In Figure 9.9, AB' is a horizontal line with line of sight at A. Due to the curvature of the surface of the Earth, a staff reading at B will be too great by BB'. This is a very small amount compared with R, the radius of the Earth. To explain the geometry of the figure, refer to Figure 9.9, but remember that the true dimensions are more similar to those shown on the right. Let AB = AB' = S and BB' = x. Then in triangle AB'O

$$\mathrm{B'O}^2 = \mathrm{AO}^2 + S^2$$

$$(\mathrm{R} + x)^2 = \mathrm{R}^2 + S^2$$

$$\mathrm{R}^2 + 2\mathrm{R}x + x^2 = \mathrm{R}^2 + S^2$$

$$\Rightarrow x = \frac{S^2}{2\mathrm{R}}$$

since x^2 is negligible.

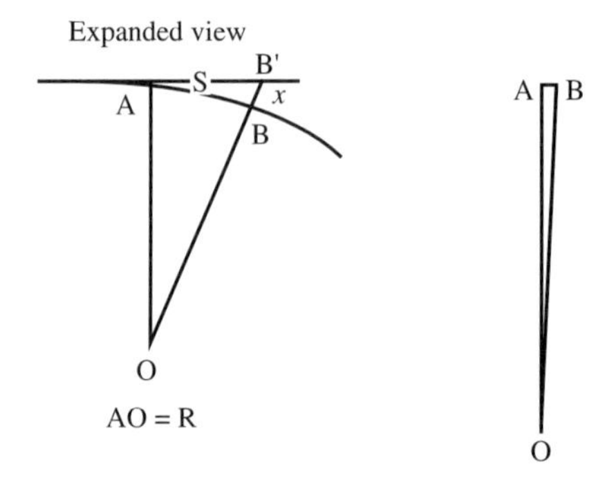

Figure 9.9

It will be obvious that if the instrument is set up halfway between the two staves both the backsight and foresight readings will be in error by the same amount, and therefore the height difference is free from error. However, the instrument need not be set up each time at the mid-point of the staves to eliminate this error; if the total sum of the backsights equals the total sum of the foresights, the total curvature error over the whole line of levelling will be nil.

Effect of refraction

Two points B and F in Figure 9.10 are at the same height for simplicity, and R_B, S_B etc. are the staff readings and distances. We neglect the curvature of the Earth. Since the light ray from the instrument to the staff is refracted as it passes through the atmosphere, a recorded staff reading is too low. The errors on the backsight and foresight due to refraction are e_B and e_F respectively.

If α is the angle subtended by AB = S at the centre of the Earth, whose radius is R (Figure 9.9), and the coefficient of refraction is K, the angle of refraction θ_B is given by

$$\theta_B = Ka = K\frac{S^2}{R}$$

The coefficient of refraction K, about 0.07, varies considerably and in some cases can even be negative. The main factor affecting its value is the rate of change of

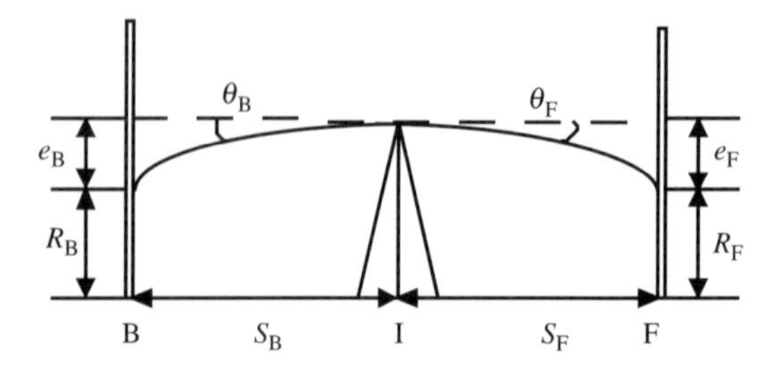

Figure 9.10

temperature with height above the ground. The greater this rate of change, the smaller the refraction. If K is constant during the time that both sights are observed, the effect on the difference in height will be nil if the instrument is positioned mid-way between the staves. In many cases, however, K is not constant over a given period, hence a small systematic error will be introduced into the line of levels. If a line is levelled from early morning to noon, the refraction becomes progressively less. In this case, always reading to the same staff first will help to eliminate the refraction error since the change in refraction will alternately affect each height difference with an opposite sign, each even number of height differences being relatively free of error.

Combined curvature and refraction

It is usual to combine the curvature and refraction corrections into one formula

$$C + r = \frac{S^2}{2R}(1 - 2K) \qquad (9.1)$$

Taking average values and converting S to km, this correction is 67 S^2 mm. Normally it does not need to be applied because backsights and foresights are equated. In some special cases and in trigonometrical heighting it will be applied.

9.9 Practical levelling

As a result of considering the above sources of errors in levelling, it is apparent that for all types of work, most errors will be avoided if the following three rules are adopted.

1. To detect gross errors, a line of levels should be run between two points of known height, twice over the same line or using a staff with two scales.
2. The staff should be vertical when a reading is taken; this is achieved either by swinging it through the vertical position, or with the aid of a bubble, which should be tested.
3. The length of the backsight should equal the length of the foresight at every set-up.

In ordinary work on engineering sites it is not feasible to have backsights and foresights exactly equal, and in intermediate sights no compensation for curvature, refraction and collimation takes place. Hence it is imperative that the level has very little collimation error, otherwise tedious corrections will have to be applied.

Errors due to curvature and refraction are negligible for most work with sights not exceeding 100 m. Again it is important to avoid soft muddy ground, but for obvious reasons this is not possible on an engineering site. Pegs knocked into the ground will continue to rise up for several minutes afterwards.

The future of geodetic levelling

It is difficult to predict how much geodetic levelling will be carried out in the future. On one hand the GPS system can be used to interpolate between benchmarks and trigonometrical heighting can achieve equal standards. On the other hand, the new bar-coded staff reading levels are so quick and accurate that they must be given serious consideration. The likely outcome is a hybrid of all three methods.

Booking and recording readings

With conventional levels and staves, modern practice is to use data loggers to book readings. These loggers process the data for statistical rejection in real-time to alert the observer. Generally several readings are taken, following an orderly scheme, to produce sufficient redundancy for these tests.

9.10 Spheroidal, orthometric and dynamic heights

It should be noted that levelling takes place on the surface of the Earth affected by gravity and has nothing directly to do with the reference spheroid adopted by GPS. However, it is now regular practice to link satellite-determined spheroidal heights to levelling via the geoid-spheroid separation N.

Spheroidal to orthometric heights

As was stated in Chapter 7, heights based on a reference spheroid can now be determined to a relative accuracy of a few millimetres by satellite systems. To link them to a level surface datum (geoid) requires knowledge of the separation N between the two surfaces. This information is available with varying accuracies from national and international sources.

Figure 9.11 shows the contours of N with respect to a reference spheroid for the United Kingdom. Here the differences are of the order of 50 m.

In the UK the Ordnance survey can provide values of N to 0.02 m as a general rule. This data often comes in the form of values for every intersection on a 5-minute graticule from which individual values can be interpolated by the user. The interpolation may be carried out by the method described in detail in Section 10.9.

Example

Values of the geoid-spheroid separation N and their tabular differences are listed below for points at the intersections of a 5-minute graticule within which N for a point A has to be interpolated. The arguments are $n = 0.2$ in the north–south direction and $m = 0.6$ east–west. The scheme shows the values of N in bold type and their first and second differences (see Section 10.9). The origin of coordinates is at the bottom left ($N = 47.153$).

	45.001				
	-0.885				
0.382	**45.886**	**1.234**	**47.120**		
	-1.267	0.023	-1.244		
	47.153	**1.211**	**48.364**	**0.415**	**48.779**
			-0.796		

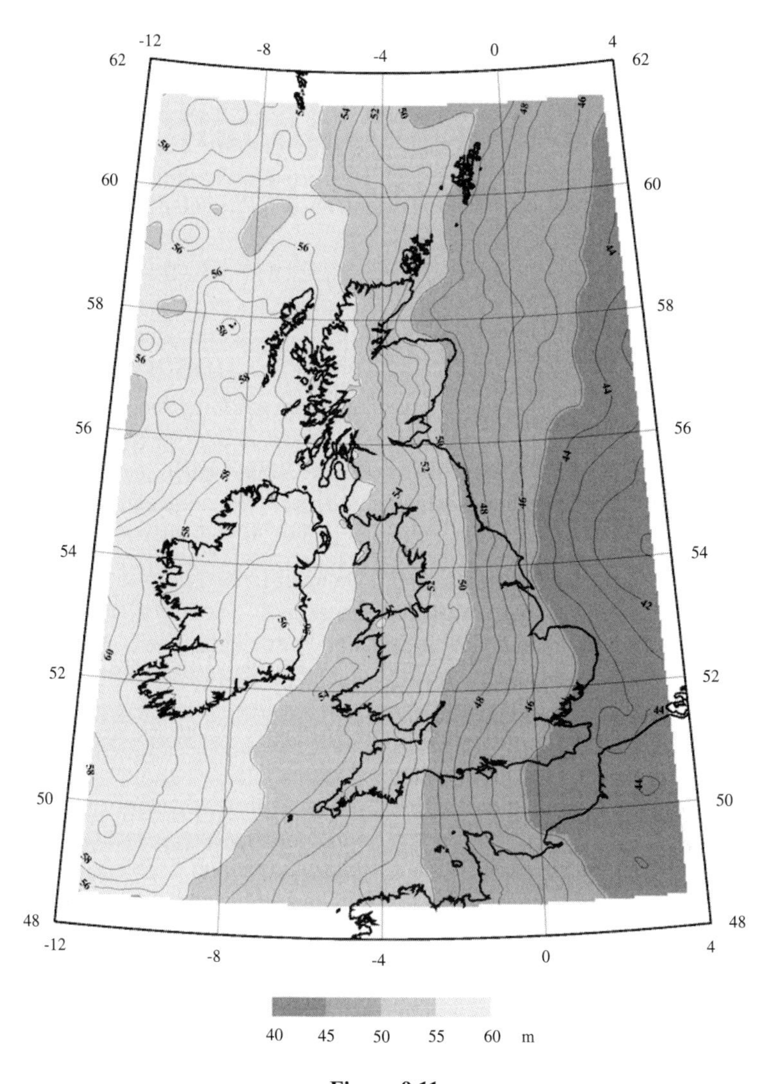

Figure 9.11

Applying the standard interpolation formula for two dimensions,

$$N_A = N_T + n\,\Delta N' + m\,\Delta E' + n\,m\,\Delta EN'' + \frac{1}{2}n(n-1)\Delta N'' + \frac{1}{2}m(m-1)\Delta E''$$

$$= 47.153 + 0.2\times(-1.267) + 0.6\times1.211 + 0.2\times0.6\times0.023 + \frac{1}{2}\times0.2\times(-0.8)\times0.382$$

$$+\frac{1}{2}\times0.6\times(-0.4)\times(-0.796)$$

$$= 47.69392$$

If the spheroidal height of A is 500.000 m, its orthometric height is 547.693 m.

The following sections do not relate at all to spheroidal heights.

Orthometric and dynamic heights

Earlier we defined the height of a point to be the linear distance from a datum surface at mean sea level to the point. However, the concept of 'height' may be considered from yet another different point of view in terms of the force of gravity acting on a mass situated near the surface of the Earth. Before we had detailed knowledge of the Earth's gravity field, allowance was made for its shape by the use of theoretical models. The following is a brief treatment of this classical approach.

The gravitational potential at a point at a distance D from the centre of the Earth is given by $d\mathrm{P} = g\ \mathrm{d}D$, where g is the acceleration due to gravity. An equipotential surface is one on which all the points have the same gravitational potential i.e. the potential P is constant. Since g varies with latitude, being greater at the poles than at the equator, an equipotential surface is closer to the centre of the Earth at the poles than at the equator.

Figure 9.12 shows the shape of two equipotential surfaces defined by their potentials P_1 and P_2. If H_0, H_{45} and H_{90} are the linear separations between these equipotential surfaces at the latitudes of 0°, 45°, and 90° respectively; g_0, g_{45} and g_{90} their respective gravities, we can define a constant 'height' difference by

$$\mathrm{d}P = \text{constant}$$

$$g_0 H_0 = g_{45} H_{45} = g_{90} H_{90}$$

Since no work is done against gravity, if a point moves over an equipotential surface, such a surface is dynamically flat. For example if a pipeline followed P_1 in Figure 9.12, no water would flow along it under the action of gravity. But if the pipeline took the path of the broken line, equidistant from P_1, water would flow from A to B under the force of gravity due to the difference of potential at these points. Thus, although A and B are at the same *linear* height above P_1 in a *dynamic sense* there is a slope from A to B. Points defined by linear distances along the vertical are called

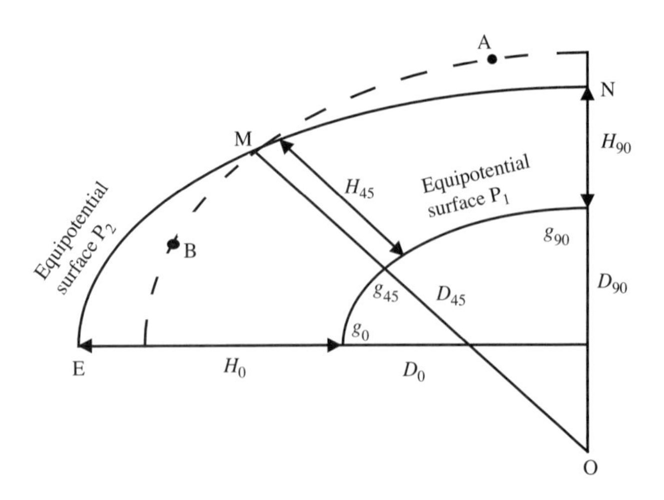

Figure 9.12

orthometric heights, and those defined in terms of potential are called *dynamic heights*.

To relate the concept of potential to a linear system of height, a standard latitude is chosen (45° for a world system, 53° for England and Wales) at which both types of height are defined to be equal. For example in Figure 9.12 the linear separation h between P_1 and P_2 at latitude 45° is used to define the potential surface P_1 with respect to P_2 as datum in linear units. Hence the *dynamic heights* of points E, M, and N all equal h, the orthometric height of M above P_1 at the standard latitude 45°. The *orthometric height* of E is H_0 and of N is H_{90}. The orthometric height at the equator is greater than the dynamic height, and vice versa at the poles, the two being equal at 45°.

Relation between orthometric and dynamic heights

To relate the two concepts we use the fact that $g_0 H_0 = g_{45} H_{45} = g_{90} H_{90}$. Then

$$h = H_{45} = g_0 \frac{H_0}{g_{45}}$$

It can be shown that theoretical gravity g_f at latitude ϕ is related to the gravity g_0 at the equator by the simplified expression

$$g_f \approx g_0(1 + 0.0053 \sin^2 \phi)$$

therefore

$$h \approx \frac{H_0}{1.00265}$$

and for any point at an intermediate latitude

$$h \approx \frac{H_\phi(1 + 0.0053 \sin^2 \phi)}{1.00265} \tag{9.2}$$

If the orthometric height H_f is known, its equivalent dynamic height h may be derived from this expression.

Example

If the orthometric height of a point on the equator is 1000 m its dynamic height is 997.4 m i.e. there is a difference of 2.6 m.

Orthometric heights from levelling

In the actual process of levelling, the bubble sets itself tangential to the equipotential surface through the instrument, while the readings on the staves are orthometric operations.

In Figure 9.13 the divergence between the orthometric and dynamic surfaces through the instrument gives rise to an error dH in the backsight reading r_B such that the correct reading is

$$R_B = r_B + \mathrm{d}H$$

when levelling from north to south in the northern hemisphere.

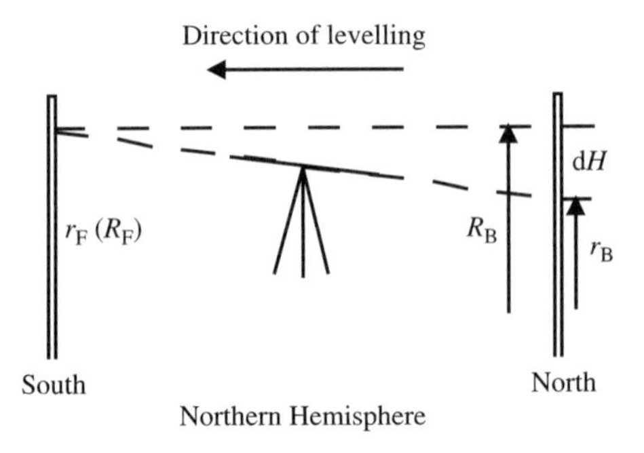

Figure 9.13

Defining the difference in latitude over the line $d\phi = \phi_B - \phi_F$ i.e. in the same sense as dH, dϕ is positive; from south to north it is negative. Over the equipotential surface, gH is constant, therefore

$$H\,dg + g\,dH = 0$$

$$dH = -\frac{H\,dg}{g}$$

$$= 0.0053H\sin^2\phi\,d\phi$$

Thus to obtain the orthometric height difference from the levelled height difference, a correction of +dH is applied. Because this correction is very small for one set-up, it is normally applied to lines of levels at about every 10 km or at convenient section points along the line. The total effect for the line is the summation of all the individual corrections i.e. the corrections are applied by numerical integration over the whole line.

Example

If the average height of a section of levels, 10 km long running from south to north at latitude 10°N, is 1000 m, and the indicated height of the northern benchmark was 1026.9384 m, calculate its orthometric and dynamic heights, assuming that the height of the southern bench mark was orthometric. In this case dϕ is negative and

$$dH = \frac{-1000 \times 0.0053 \times 0.3420 \times 10}{6400} = -0.0028 \text{ m}$$

Hence the orthometric height of the required benchmark is 1026.9356 m, and the dynamic height is given by

$$h = \frac{1026.9356(1 + 0.0053 \times 0.03014)}{1.00265} = 1024.3850$$

Geopotential numbers

The acceleration due to gravity varies not only with latitude ϕ but with height above sea level h and with the composition of the Earth. Allowance is made for these factors in the form of a system of heights referred to as *geopotential numbers*. The system, devised by the International Association of Geodesy, defines the geopotential number of a point A to be

$$C_A = \int_0^A g\Delta h$$

The units to be used are the kilogal/metre i.e. g is in kilogals and Δh in metres above sea level (one gal is an acceleration of 1 cm s^{-2}). It is desirable that observed values of g are used, although theoretical values based on some hypothesis concerning the structure of the Earth may be used for want of observational data (Torge 2001).

9.11 Trigonometrical heighting

A common method of determining height difference ΔH is to observe the vertical angle θ to a target or object and compute the difference in height between the theodolite and the target by the relationship

$$\Delta H = S \tan\theta$$

where S is the horizontal distance between the stations, or from

$$\Delta H = S' \sin\theta$$

where S' is the slant range as measured by EDM.

In Figure 9.14, H_A and H_B are the heights of the stations A and B above a horizontal datum line DE which is considered straight, i_A is the height of the trunnion axis of a theodolite at A with respect to some reference mark on the station at A, g_B the height of the target at station B, the vertical angle observed at A is θ_A and the horizontal distance DE is S. These quantities are related by the expression

$$H_A + i_A + S\tan\theta_A = g_B + H_B$$

or, setting $\Delta H = H_B - H_A$,

$$\Delta H = i_A + S\tan\theta_A - g_B \tag{9.3}$$

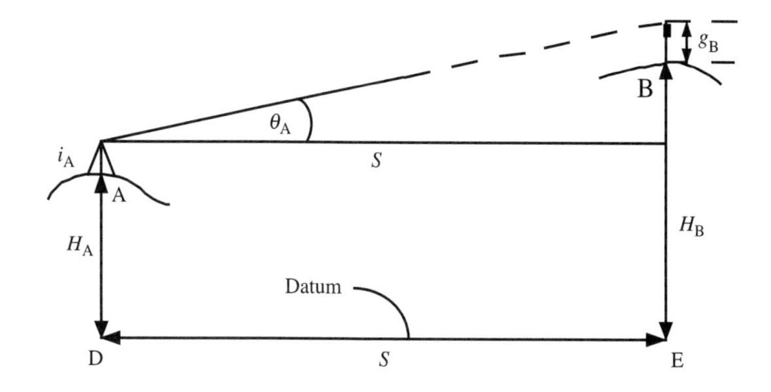

Figure 9.14

Note that this equation holds for all values provided the following conventions are adopted:

1. heights above a station are considered *positive*, those below are *negative*;
2. the height difference ΔH from A to B is defined as $H_B - H_A$;
3. an angle of elevation is positive, and a depression is negative.

These conventions are commonly used in everyday life and should not cause any difficulty.

Example

If $H_A = 100$ m, $S = 1000$ m, $\theta_A = 1° 25' 57''$ i.e. an elevation, $i_A = 1.5$ m and $g_B = 10.00$ m, the height H_B is derived as follows:

$$H_B = 100 + 1.5 + (1000 \times 0.02501) - 10.00 = 116.51$$

It should be noted that although i_A and g_B are normally positive they can have negative values, for example, if an instrument or beacon is erected below the reference mark on a ceiling or tunnel roof.

It is usual to observe a vertical angle on face left and on face right so that the effect of the vertical index error of the theodolite is eliminated on taking the mean. Whenever possible, observation should be taken from A to B, and from B to A. It is always better to compute the height difference by two separate computations using Equation (9.3), rather than a formula which combines the observations from both ends. The reason for this is that a gross error may be detected from any discrepancy between the two values. Great care should be taken to avoid mistakes in recording beacon and instrument heights.

Earth curvature

The reference datum line DF is not straight but is curved as in Figure 9.15. The linear amount of the Earth's curvature FE $= x$ is given by

$$x \approx \frac{S^2}{2R}$$

where R is the radius of the Earth. This curvature effect may also be considered as an angular correction α to the observed angle of

$$\alpha \approx \frac{x}{S} \approx \frac{S}{2R}$$

$$\alpha'' \approx 206265 \frac{S}{2R}$$

We then use the effective angle $\theta + \alpha$ in calculations to allow for the Earth's curvature. Note that α is always positive, whereas θ may be either positive or negative.

Refraction effect

In Figure 9.16 the light path from theodolite to target J is shown as a curved line instead of a straight line. This is so in practice, due to the refractive effect of the

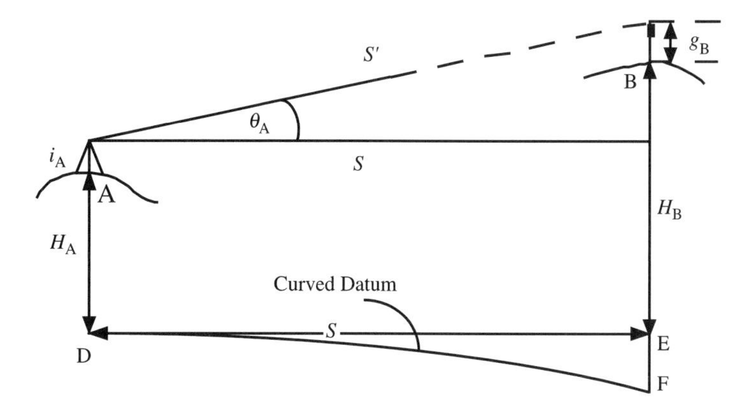

Figure 9.15

atmosphere. The telescope points to G in the direction of the tangent to the curved refracted ray instead of along the direction of the chord, and therefore records an angle which is greater than that used in the computation. To obtain the equivalent angle free from refraction, the angle β has to be *subtracted* from the observed angle θ. The refraction angle β is given by

$$\beta \approx \frac{KS}{\mathrm{R}}$$

The coefficient of refraction K is approximately 0.07, although it varies according to the density of the air through which the light passes and should be determined for the average conditions of the survey area.

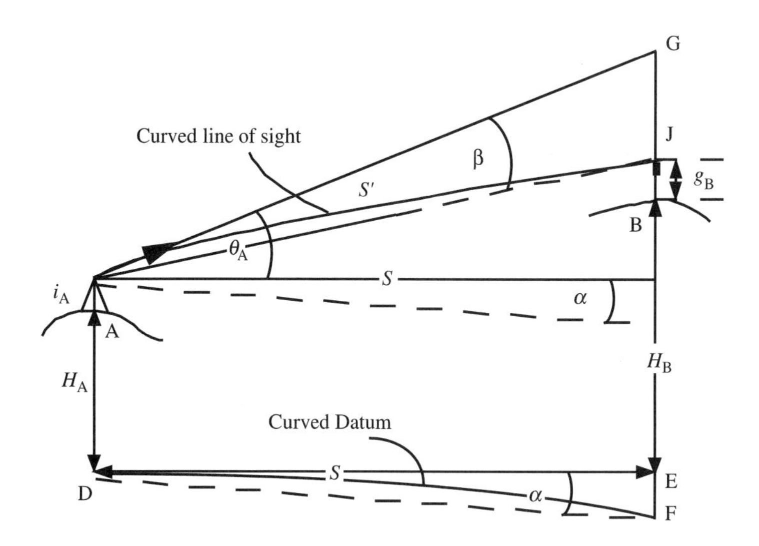

Figure 9.16

Combined curvature and refraction

To reduce an observed angle θ to its equivalent plane angle ϕ we have:
Corrected plane angle = observed angle + curvature – refraction

$$\phi \approx \theta + \alpha - \beta$$

or

$$\phi \approx \theta + 206265(1-2K)\frac{S}{2\mathrm{R}} \tag{9.4}$$

In topographical surveying, the combined correction is approximately 14" per km, setting $K = 0.07$ and R = 6400 km. This combined curvature and refraction correction is normally positive because the curvature correction exceeds the refraction correction. Since the latter varies with the prevalent air conditions, the combined correction will also vary with location and with time of day. Since refraction is lowest at the mid-day period, usually taken from 12:00 to about 15:00, and since its value during this period varies little from one day to the next, vertical angles are best observed during these hours. However, if two observers are situated at each end of the line, truly simultaneous vertical angles may be observed, which enables the mean value to be almost free of refraction error. Such observations may be taken at almost any time of the day.

In practical computation of trigonometrical heights a few reliable observations, taken over several long lines, are used to compute the combined curvature and refraction correction in seconds of arc per kilometre, and thereafter this value is used to compute all heights by single rays.

In some industrial locations, such as a tunnel, the coefficient of refraction can be negative due to unusual heat gradients. An empirical analysis should be carried out to locate such anomalies.

Example

A line EG observed between two pillars is chosen from which to compute the combined curvature and refraction correction. The height difference from E to G is computed without applying a correction for curvature and refraction. This is repeated for the direction G to E; see Table 9.4 below.

The variation in these two height difference values is due to twice the effect of curvature and refraction, which is calculated from

$$\frac{8.644 \times 206265}{2 \times 8024.90} = 111''$$

Table 9.4

Line	*Vertical angle θ*	*Distance S (m)*	*i*	*g*	*ΔH*
EG	+00° 07' 39"	8024.90	0.207	0	+18.065
GE	–00° 11' 33"		0.253	0	–26.709

Table 9.5

Line	*Vertical angle θ*	*Distance S (m)*	*i*	*g*	*ΔH*
EG	+00° 07' 39"		0.207	0	
	+00° 01' 51"				
Cn	+00° 09' 30"	8024.90			22.383
GE	–00° 11' 33"		0.253	0	
	+00° 01' 51"				
Cn	–00° 09' 42"	8024.90			22.390

This is the correction to be applied to a single ray observation over the line. The value of the correction to be used for the area of the survey is therefore 13.843" per 1000 m. As a check on the arithmetic and to illustrate the method of single ray calculation, these two observations are recalculated separately using this value for the combined correction (see Table 9.5). The combined correction is

$$13.843 \times 8024.90 = 111'' = 01'51''$$

The accepted mean difference in height is 22.387 m. It will be noticed that this same value is obtained from the mean of the two values 18.065 and 26.709 computed without recourse to the correction at all. To adopt this latter method is dangerous because there is no clear indication that the results are free from gross error. The method of single ray computation, with an applied curvature and refraction correction, allows the quality to be assured.

Linear value of the curvature and refraction correction

It is useful to evaluate the linear effect of the combined curvature and refraction over the distance *S*. The linear correction, in metres, is

$$S(\alpha - \beta) = S^2 \frac{(1-2K)}{2R} = 0.0672\,S^2 \tag{9.5}$$

where *S* is in kilometres.

This formula is useful in working out intervisibility problems. For example, if the distance *S* is 8 km, the combined correction is approximately 4.3 m.

Precision of measurements

We now consider briefly the precision required in the various measured quantities to achieve a given result for the difference in height. Denote tanθ by t, the distance by S and the height difference ΔH by h, then

$$h = S\,t$$

Since the height of instrument and beacon directly affect the value obtained for h they must be measured to at least the same precision desired in the final result. In normal work, these quantities are measured to 0.005 m by taking readings to a tape held vertically as with a level staff, and by direct measurement of the height of theodolite from the secondary axis, applying a small alignment offset if necessary.

Consider small changes dt in t, dS in S and dh in h, related by

$$\mathrm{d}h = S\,\mathrm{d}t + t\,\mathrm{d}S$$

If dt = 0, θ = 1° and dh is to be 0.01 m,

$$0.01 = \frac{\mathrm{d}S}{60} \quad \text{and} \quad \mathrm{d}S = 0.6 \text{ m}$$

Thus distances are not required to any great accuracy.
Again, if dS = 0 and if S = 1000 m,

$$\mathrm{d}\theta'' = \frac{0.01 \times 206265}{1000} = 2''$$

To guard against mistakes in reading the vertical angle, one reading should be taken on each of the stadia hairs, thus obtaining an independent check from the mean.

9.12 Least Squares estimation of heights

The estimation of trigonometrical heights and of level networks by the method of Least Squares is similar. The difference in height brought through a line of levels is treated in the same manner as a height difference computed for a long line from vertical angle observations. The ends of such a line of levels are referred to as junction points. The method is to obtain best estimates for these junction points or trig stations, and in the case of a line of levels, to adjust the individual points along the line by proportion if required.

Dispersion matrix or weights

Some thought has to be given to an estimation of the quality of the observations, or to the *a priori* dispersion matrix of the height differences. For each line of levels or trig height difference, expectations are taken of the error equation dh = Sdt + tdS giving variance estimates

$$\sigma_h^2 = S^2\sigma_t^2 + t^2\sigma_S^2 \tag{9.6}$$

Assigning reasonable values for the variances in the angles and distances yields the initial dispersion matrix elements. In levelling, the distance error does not matter because $t = 0$, so we have

$$\sigma_h^2 = S^2\sigma_t^2$$

The degree of sophistication chosen for the error model depends on circumstances, but normal practice is to work with a reasonable estimation per set up between staves of, say, 2 mm in ordinary work. It is further assumed that the length of sight is kept more or less constant. The line has n set-ups yielding a line estimate of variance $2\sqrt{n}$

mm yielding, for 25 set-ups per km, about 10 mm km^{-1} for ordinary work. Geodetic levelling can achieve 1.5 mm km^{-1}. In trigonometrical work, the side length S is crucial. It is usual to calculate variances for every line according to the variance formula Equation (9.6).

In Chapter 6, Section 6.5, the Least Squares treatment of a small network of levels by the observation equations method is given. Although these problems can be handled very simply by the method of condition equations (see Appendix 3.4), observation equations have been preferred because of their generality and the ease of computer programming.

After the heights of junction points have been estimated, the heights of individual set-up points, if required, are obtained by adjusting the change to the observed line height difference in proportion to the number of set-ups in the line.

9.13 Intervisibility problems

Initially we adopt a simple model which neglects curvature and refraction. Figure 9.17 shows two stations A and C whose heights above a horizontal straight line datum are H_A and H_C respectively. Considering for the moment that the line of sight from A to C is straight, we wish to establish the height H_B of an intermediate point B lying on this line.

This problem arises when a survey reconnaissance is being carried out or when a tower has to be erected to a computed height above datum so that observations may be made along a line, such as a traverse. Equating values of $\tan\theta$,

$$\frac{H_A - H_C}{A'C} = \frac{H_A - H_B}{A'B'} = \frac{H_B - H_C}{B'C}$$

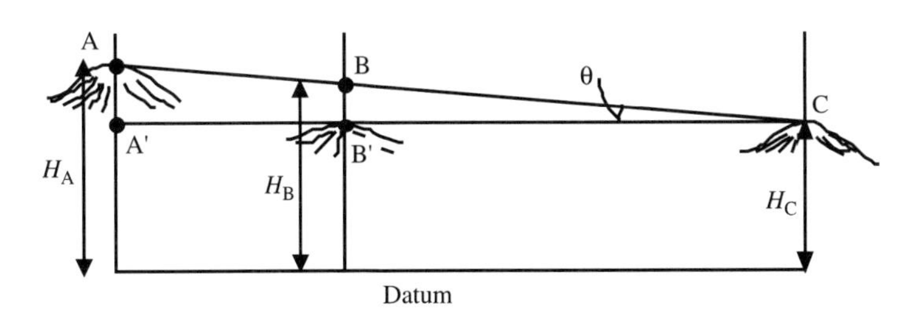

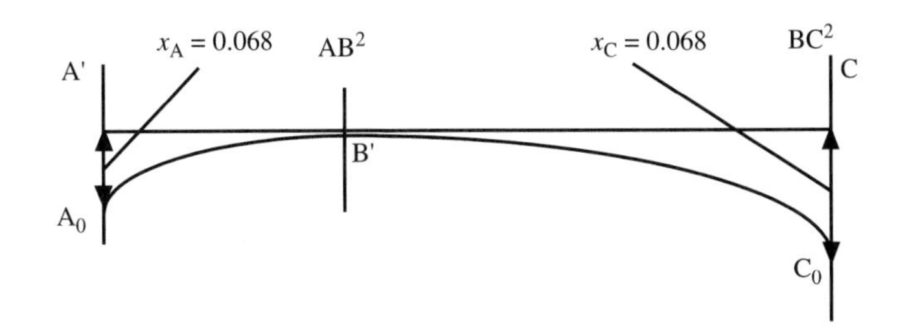

Figure 9.17

Example

If $H_A = 100$ m, $H_C = 50$ m, AC = 10 000 m and BC = 8000 m, the height of B is given by

$$H_B - H_C = \frac{H_A - H_C}{A'C'} B'C = \frac{50 \times 8000}{10000} = 40$$

Therefore H_B is 90 m above datum. If the hill at B' lies below this height, say at 85 m, line of sight will pass from A to C and vice versa. To erect a tower on line AC, its height has to be BB' = 5 m.

In practice, because of refraction of the line of sight and the curvature of the datum surface, the simple case above has to be modified.

Figure 9.17 illustrates how the net effect of these factors – curvature and refraction – is to lower the datum at A and C with respect to B by x_A and x_C. Hence, if the heights of A and C are decreased by their respective corrections, the problem is reduced to the simple model treated above.

Example

If the heights of A, B and C are 417.27 m, 273.10 m, and 214.58 m respectively and AB = 24.30 km, BC = 12.23 km, what height of tower should be erected at C so that the line of sight from a theodolite at A, 1.2 m above ground, will clear the ground at B by 6.1 m?

The effective height at B is 279.2 m. The curvature and refraction corrections to be applied to H_A and H_C are respectively 40.15 m and 10.17 m.

The effective height at A is then 417.27 + 1.2 – 40.15 = 378.32. The effective height C is then given by

$$H_B - H_C = \frac{H_A - H_B}{A'B'} B'C = \frac{99.12 \times 12.23}{24.30} = 49.89$$

Since the ground height of C is 214.58 m above datum and x_C is equal to 10.17 m, the tower should be 18.8 m tall.

Practical applications

In practice, a section has to be drawn from map contours and a likely position for a station is chosen. Any doubtful intervisibility problems are solved as above. Intervisibility problems, now tackled by computer software, are of some interest in the production of maps of visibility and obscurity required for optical communications, flight paths into airports, planning applications and military purposes.

Comment on heights and levels

The importance of very precise heights and levels to engineering and hydrological work has to be stressed. Geometrical methods such as satellite, photogrammetric and theodolite systems cannot directly compete for accuracy in the previous analysis because of uncertain knowledge of the Earth's gravity field at the millimetre level. Combinations of systems can be effective in which the geometrical systems are used to interpolate values controlled by levelling.

The two-plus-one coordinate system will continue to be of importance, while some purely geometrical applications in engineering and construction may be treated in a three-dimensional Cartesian coordinate system which ignores the vertical altogether.

Chapter 10
Maps and Map Data Processing

10.1 Reconnaissance

The first, and perhaps the most important stage of survey is the reconnaissance, or 'recce'. Experience of the full technical and administrative factors of surveying is required to carry out an effective recce. This experience includes the whole production chain subsequent to fieldwork so that potential problems and difficulties can be avoided.

A pre-analysis of any information related to the task or its specification is valuable, such as knowledge of any previous surveys, availability of controls and datum information and copies of old maps and aerial photography. The site is visited to make decisions about how to carry out the work, where to locate points and marks, the availability of accommodation and possible labour and services.

A computer simulation of the technical proposal may also be feasible if software is available. A final decision has to be taken on how the task will be executed, by whom, and with what equipment. Suitable documentation has to be prepared for the surveyors who will actually carry out the task.

Documentation

Clear documentation of surveys is required at all times. This includes station descriptions, field books or records of data loggers used and any changes to original objectives. The whole data set should be clearly cross-referenced and kept in a clear form. This documentation may be the only evidence that the work was properly carried out and may have to be submitted to a court of law as evidence. The copying of field notes without retention of the originals is tantamount to fraud.

Station marking

There is much wisdom in marking stations, even though it may not be part of the specification to do so. Certainly, while the task is in progress, marks should be located if only with wooden pegs. It is also wise to reference any station marks by witness marks. This will enable the station point to be recovered should it be damaged or removed, as happens frequently on engineering sites as construction work progresses. It has become common practice to place reference marks on walls rather than on the ground since they are far less likely to be disturbed. Although two marks are sufficient to re-establish a point, a third acts as a check.

10.2 Introduction to detail surveying

The methods described here are those used to make plans at large scales from given control, to revise existing plans and to complete plans made in part by air survey methods. Every survey is based on a rigid framework which is regarded as correct until proved to be in error. Generally speaking, all details such as structures and natural features can be shown to scale on a *plan* but not on a *map*.

This rule is often broken in special cases, for example when a forester wishes named trees which are too small to plot to scale to be located in position. They have to be depicted by conventional signs. Map detail is classified under the following headings:

1. *Soft detail*: has an outline incapable of exact definition, or is likely to change, such as vegetation and crops.
2. *Hard detail*: is clearly defined, such as a building line where the building enters the ground. A useful rule is that anything that would obstruct the normal passage of a bicycle is shown as a full line, while that would not, such as a road kerb, is shown by a pecked line.
3. *Overhead detail*: constitutes no obstruction at ground level, such as power lines, or tree canopies shown by conventional signs.
4. *Interior detail*: is surveyed if it is of special interest, such as a party boundary delimiting ownership. Building interiors have separate plans for each floor linked together by vertical correlation.

Depiction of detail

The manner of depicting features varies a great deal according to map scale and purpose, for example from topographic maps, building interiors and archaeological maps to thematic plans of a rail network. The reader should inspect a range of these plans to see the symbols used, many of which are obvious. Special plans showing street furniture (lamps, gullies, man-holes, etc.) are drawn to the specification of the client. Every user has particular requirements which must be discussed and agreed upon, particularly whether the item is to be shown to scale or not.

10.3 Detail surveying

In surveying detail point positions we need to record sufficient information to enable a map to be plotted from them. As the dot plot of Figure 10.1 is the only information from which to plot the map it needs to be completely comprehensive and intelligible. When operating in the field, the plotting process must be kept in mind, so that points can be plotted from a minimum of two position lines. The most common method of data capture is by radiation.

Points need to be classified as well as positioned, so that their common identity can be used to delineate special features such as house corners. It is important to know how the map is to be plotted, before embarking on the capture of field data.

If the plotting is to be manual, the collection of data can be flexible and inconsistent. If the data collection serves a computer software system, a rigorously ordered approach is essential.

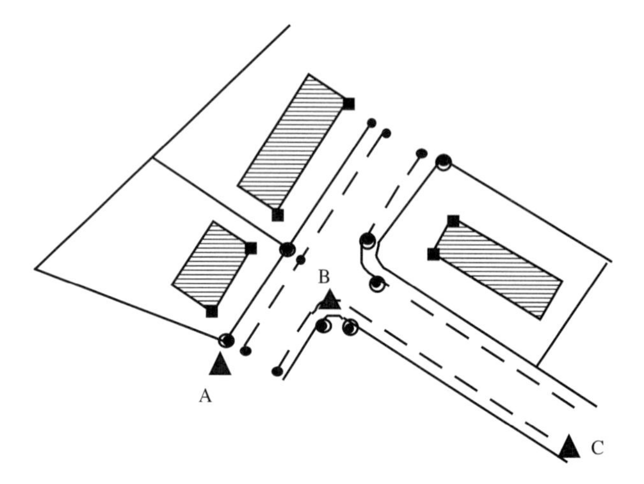

Figure 10.1

Figure 10.1 shows a typical part of an urban area to be mapped by radiation from a traverse station B. It will be seen that *nodes*, where two or more lines meet, are recorded by a different key from single point features which lie on a *string*, such as marks the road kerbs. Some points will be both nodes and string points. The curved corners of the road kerb, to be drawn by a circular arc, are marked by the curve end points, which are shown as special nodes. Even when a computer plotting system is used, most surveyors prefer to draw field sketches to label the points and identify their types.

Not all details need be picked up by radiation. The enlarged diagram depicted in Figure 10.2b shows detail such as lamp posts, which can be tied in by taking rectangular off-sets to a straight line, such as the kerb. This *line and off-set* method uses a local x, y coordinate system based on the kerb.

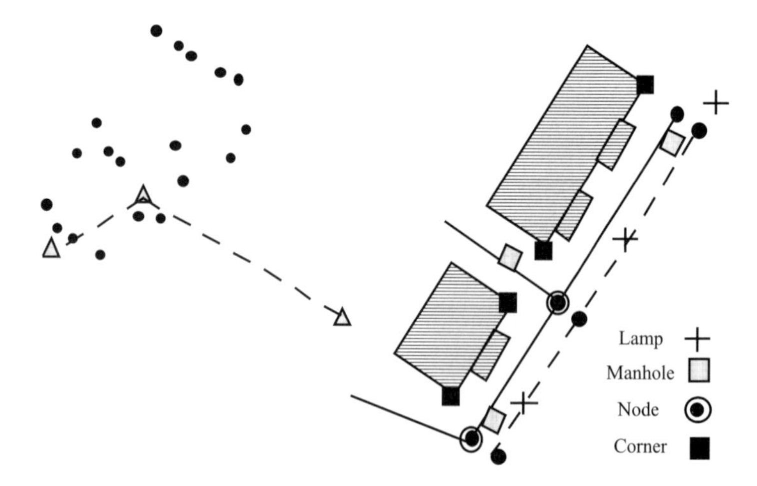

Figure 10.2

Ancillary methods of detail fixing

In addition to the radiation and line and off-set methods of fixing details, other geometrical principles can be used to advantage, particularly as check measurements. These methods include:

1. Theodolite intersection of inaccessible points such as trees, and corners of buildings.
2. Sighting tangents to irregular or rounded shapes such as tanks, ponds and clumps of trees.
3. Off-sets from radiation to the left or to the right, or plus on-line measurements to reach inaccessible points such as hedge centres.
4. Intersecting rays to already established lines such as building frontages, useful when mapping building elevations.
5. Direct taped measurements of key dimensions such as buildings, spaces between buildings and gates.
6. Dimensions to points where face-lines cut out on others.

Field completion

No matter how much care is taken to avoid mistakes and omissions in map-making, some do escape the checking systems used by the surveyor and cartographer. It is important that a map should be given a final field check, mounted either on a plane table or small board such as that used for map revision. The checker works systematically round the horizon at selected key stations, noting all plotted features and looking for those that are omitted. Graphic methods usually suffice to make corrections.

Plots compiled from aerial photographs show the roofline instead of the building line. A tedious but necessary process of measuring the roof overhangs has to be carried out to complete the map by off-setting from the plot. Parts of ground not visible from the air, such as forest paths, also have to be added by ground methods.

10.4 Graphic map revision

Laptop computers are now commonly used for map completion and revision in the field, where the traditional graphic has been replaced by a flat screen computer display. However, the methods employed are still fairly traditional, although the addition of a hand-held reflectorless laser ranger has proved to be an effective asset.

The principle of lining in and cutting back from established points is basic to the simple graphic method of completing or revising maps. Instruments used are an optical square, a plastic tape or a laser ranger, a lightweight field drawing board or laptop computer and a set square and ruler. The reviser works alone on the original map, or part of it, fixed to a small board or displayed on the computer screen. Although the following explanation may seem complicated, the actual process is rather easy and interesting to carry out once the idea has been grasped. The technique is best learned by experiment, rather than from a book.

Suppose the new building A of Figure 10.3 has to be drawn on the map which already shows buildings B, C, D and E and fence lines BF and FG. The reviser checks

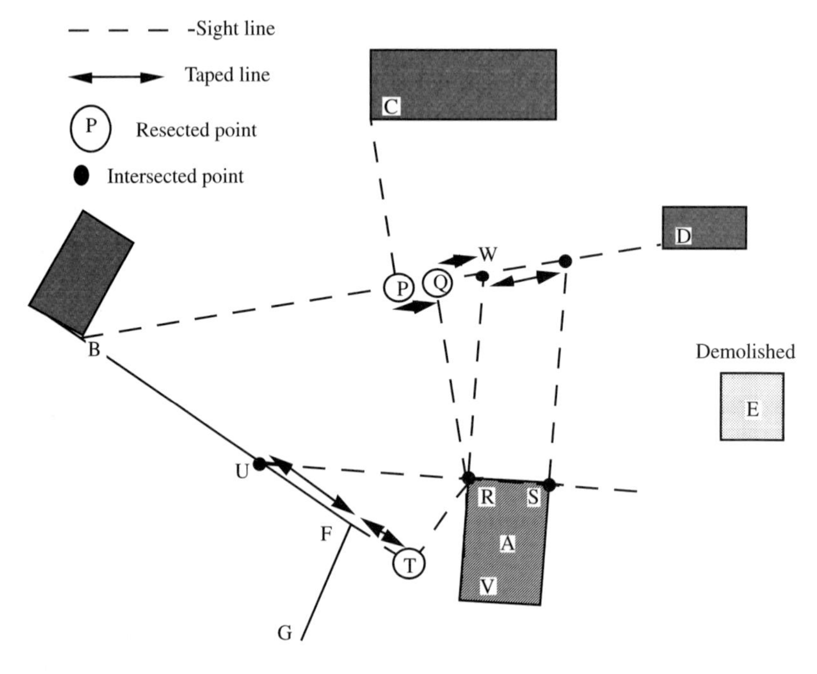

Figure 10.3

the original map to ensure that these buildings are as plotted. Suppose it is found that building E has been demolished. It will be recorded as missing. Otherwise the map is correct. To fix building A, lines of sight to others will be used with the minimum distance measurement. Points which can be occupied on the ground are established by the intersection of identifiable lines. For example, the surveyor can occupy point P by cutting in from C to the line BD using the optical square. By moving to Q, the perpendicular off-set to corner R can be drawn after measuring the short distance PQ shown with a heavy line. A similar process allows a line to be drawn to S. If the distance QR is more than a tape length, or outside the range of the EDM it will not be measured to fix R. However, before leaving this place it was noticed that the building face line VR cuts out on BD at W, so distance QW is measured.

A perpendicular off-set from T to R will be used to fix R, after locating T on the line BF produced and FT measured. To fix the direction of RS, the point at which this line cuts the fence BF in U is established and FU taped. Then S is fixed by UR produced. Usually the building dimension RS will be taped to check. The direction of RV was already noted while at Q.

10.5 Plotting details

If the detail points are plotted by computer, their coordinates are converted to rectangular Cartesian form for output to a coordinate plotter. The most accurate plotters employ a flat-bed system in which the *x–y* slides move over a table at great speed responding to the commands of the software.

Other plotters use a rotating drum carrying the paper to create the abscissa x and a fixed slide for the ordinate y. Although satisfactory for most work, these drum plotters are not good enough for multicoloured map-making for which accurate colour registration is essential.

Map digitising

Much input to a geographical information system (GIS) is from maps digitised on special tables which operate like plotters in reverse. A cursor is worked over the map to digitise selected discrete points in much the same way as a surveyor would capture the original points and features. A coding system is also adopted with nodes and points, as well as any attributes such as tree types.

A key procedure in digitising is the control of scale and distortion by digitising four grid corner points or more if necessary. Within the software, transformation formulae (see Chapter 4) are set up. In this way, the digitised shapes can be corrected geometrically.

Mapping at small scales

Because many important features such as roads, buildings, etc. cannot be plotted at the scale of the survey, conventional signs have to be used to depict them. Wherever possible, the centre of the sign is used to indicate the exact map position of the feature. For example, if a road is represented on a map at a scale of 1:25 000 by two lines 1.5 mm apart, the corresponding ground width is 37.5 m. The centre-line is in the correct position. However, there is a knock-on effect. A house at the side of the road has to be misplaced laterally with respect to the road centre-line. In urban areas, this process is impossible, so the whole set of buildings has to be conventionalized into a block.

Plan information

All maps and plans should bear the following information:

1. title;
2. scale;
3. grid;
4. key to map symbols;
5. datum information (plan and height);
6. sheet history (date sources etc.);
7. orientation (possibly with magnetic north);
8. graticule, if relevant.

Although it has been customary in the past to write the map scale in words or as a representative fraction, there is a trend away from this practice because of the dangers arising from the widespread use of inferior photocopying devices. Grids show the correct scale irrespective of the distortion and scale change, while line scales give some indication of the effect in one direction only.

10.6 Heights and contours

A *contour line* is a line drawn on a map through points at equal heights above some datum or reference point. If we imagine the land surface to be flooded to some height above sea level, say 100 m, the shore line of the flooded area will trace out the 100 m contour line. The task of the surveyor is to fix enough points on this line to enable it to be drawn on a map. The mathematical modelling process consists of selecting as few discrete points as possible to carry out this mapping. The density of points will depend on how complex the ground is, and how accurate a result is required.

The specification for contours has to be discussed with the client. Matters of importance are:

1. the datum to be adopted (national levelling or arbitrary);
2. the vertical interval between contours;
3. the accuracy of contours, often expressed as a fraction of the vertical interval.

Contouring methods

There are five methods of producing contours:

1. tracing out the contour by direct levelling;
2. interpolating contours between *spot heights* at the corners of a grid;
3. interpolation from spot heights at carefully chosen positions at changes of slope and direction;
4. plotting by stereo-photogrammetry;
5. plotting from airborne laser rangers.

The first method is used in engineering works where great accuracy is required, such as in irrigation and other water-related problems. A point on the contour is established by levelling from a *benchmark* (reference height point) and thereafter the staff is located by trial and error on the contour. The contour is pegged out on the ground. In forested areas sometimes a hosepipe is used to augment levelling and to avoid excessive tree cutting.

In the second method, a grid of *spot heights* is set out on the ground at intervals according to the complexity of the surface and accuracy required, and the grid intersections are levelled or heighted. Figure 10.4 shows such points at the numbered grid intersections. Since both plan and heights are known, the contours can be interpolated within the grid.

In the third method, the surveyor anticipates how the interpolation from a dot plot will be carried out, and supplies spot heights by total station at carefully selected locations. This is a more efficient process than the grid method, which requires more skill from the surveyor.

The photogrammetric and laser ranging techniques usually require some height control points fixed by ground methods to relate the stereo-models to the ground. Photogrammetry, as with the first method, has the great advantage of tracing out contours directly, thus giving much better shape. The practical limit of accuracy by this method is about 0.1 m. The laser rangers produce clouds of points within which contours are selected.

Contour drawing

Except in photogrammetry or direct surveying, contours have to be interpolated from discrete points. This can be done manually or by computer software. To draw the

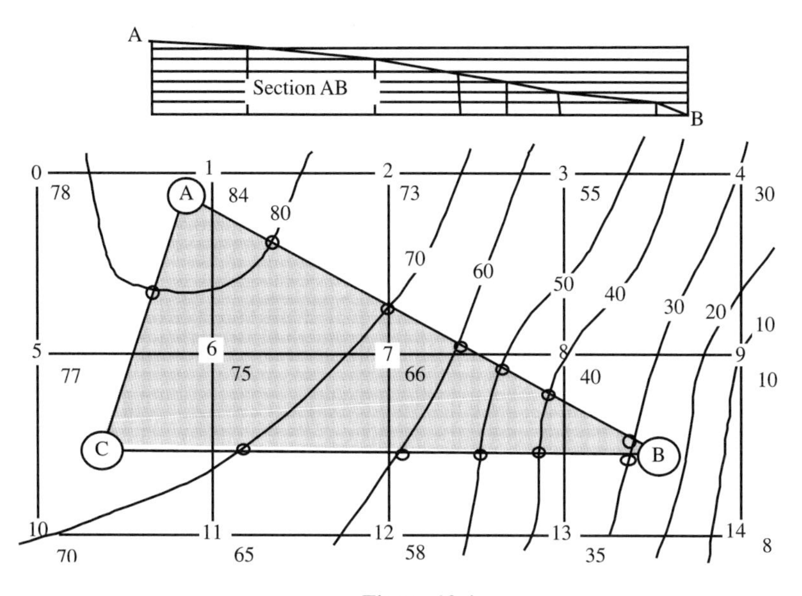

Figure 10.4

contours, the approach is to join up the spot heights, such as A, B and C of Figure 10.4, by a series of non-overlapping triangles and interpolate linearly along all sides to give a series of points on the required contours, marked by dots in the figure. When plotting by hand, the position of a contour on a sectional line, between two points not on contours (shown by the open circles in Figure 10.4), can be interpolated with the assistance of an elastic band marked at even intervals to act as a convenient analogue computer. Otherwise the process is carried out by linear or non-linear numerical interpolation.

These discrete points on contours are joined up by curves, paying due regard to the actual shape of the ground, which will have been recorded in a field sketch.

Because it is much easier for computer software to operate on a grid, heights may be captured in grid form; alternatively, a grid of points may be interpolated from the points of the selective method. The selection of most nearly equilateral triangles to fit the surveyed points can be made automatically by the Delaunay triangulation algorithm. Thereafter, the contours are interpolated according to various schemes of varying complexity. An important test of computer software is that the contours must preserve the integrity of the original heights from which they have been derived. In other words, a spot height must be on the correct side of a contour line.

10.7 Calculation of areas

The area contained within a closed figure is of interest in engineering and cadastral surveying. Basic to most methods is the calculation of Δ, the area of a triangle ABC, from one of the following formulae:

$$\Delta = \frac{1}{2} bc \sin A \tag{10.1}$$

$$\Delta = \sqrt{s(s-a)(s-b)(s-c)} \tag{10.2}$$

$$2\Delta = \begin{vmatrix} E_A & N_A & 1 \\ E_B & N_B & 1 \\ E_C & N_C & 1 \end{vmatrix} \tag{10.3}$$

The area of a polygon is evaluated as the summation of a series of adjacent triangles, as indicated in Figure 10.5.

Instead of summing up the determinants (10.3) of each triangle individually, we use the *Rule of Sarrus* to cross-multiply the coordinates taken in clockwise order around the figure according to the scheme

$$2\Delta = -\sum_{i=1}^{n} E_i \sum_{i=2}^{1} N_i + \sum_{i=2}^{1} E_i \sum_{i=1}^{n} N_i \tag{10.4}$$

To minimize the numbers in the calculation it may be necessary to transform all coordinates to the centroid before evaluating the area.

The areas of plots bounded by curved boundaries are calculated by combining the areas of polygons and irregular shapes. Off-sets from the polygon sides are taken at regular intervals to compute the excess areas by Simpson's rule for approximate integration. Figure 10.6 illustrates a method of calculating the area of a typical field. To the area of polygon ABCDE are added the extra small areas measured by regular off-sets y from the perimeter lines such as AB.

Successive application of Simpson's rule to the n off-sets y separated by even intervals x gives the area A as

$$A = \frac{1}{6}x(y_0 + 4y_1 + 2y_3 + \ldots)$$

until the end of the line. If n is odd, the last term is y_n, but if n is even the area of the last figure, a trapesium, is

$$\frac{1}{2}x(y_{n-1} + y_n)$$

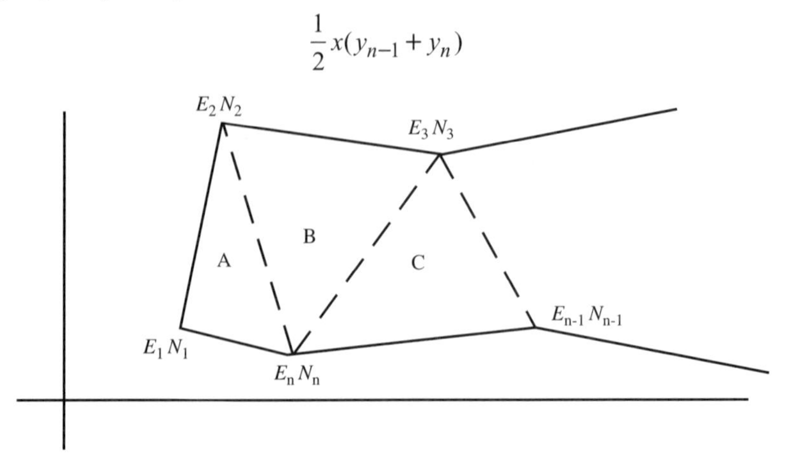

Figure 10.5

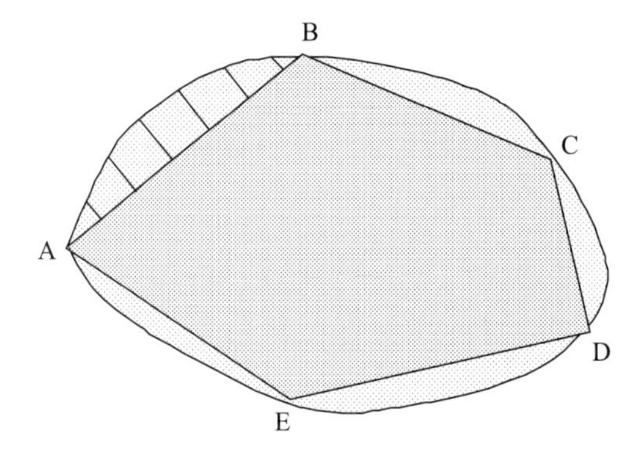

Figure 10.6

Planimeters

Irregular areas on maps are also measured by digital planimeters. The operator traces the boundary of the parcel whose area is recorded. Planimeters have to be calibrated from time to time.

10.8 Subdivision of land parcels

In cadastral and planning surveys, problems arise with the subdivision of plots. These problems vary from simple to very complex. A general approach is to obtain a solution by variation of coordinates. At its simplest, the problem might be to divide the plot ABCDEF depicted in Figure 10.7 into two agreed portions and the road frontage AF equally at G. Suppose the portions have to be equal. Let the total area be A_T. Since point G is given, the variables are the coordinates of H. The first stage is to find which line of the plot is cut by GH. This is carried out by calculating the areas of the triangles with sides radiating from G. Suppose we find that area GABC is less than 0.5 A_T and that area GABCD is greater than 0.5 A_T. In that case, we know that H lies on CD. This process of selection can be programmed.

Although this simple problem has a particular direct solution obtained by remembering that H lies on the line CD and that the area GABCH = 0.5 A_T, a more general approach to the solution of such problems is needed. If we assign approximate coordinates to H, selected perhaps as the mid-point of CD, the general method is as follows.

Suppose the equation of the line CD, in which both m and c are known, is given by

$$E - mN - \mathrm{c} = 0$$

Then we have

$$\mathrm{d}E - m\,\mathrm{d}N = 0 \tag{10.5}$$

The provisional computed area of figure GABCH subtracted from the known required value gives the absolute term k of the differential area equation

$$\mathrm{P}\,\mathrm{d}E + \mathrm{Q}\,\mathrm{d}N = \mathrm{k} \tag{10.6}$$

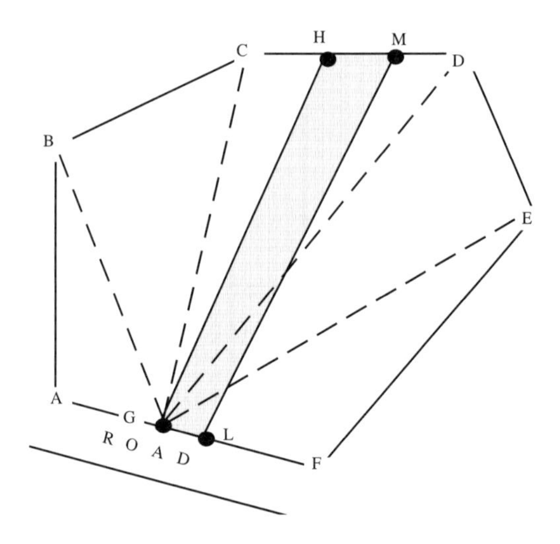

Figure 10.7

where

$$P = -(N_G - N_C) \text{ and } Q = (E_G - E_C)$$

Equations (10.5) and (10.6) provide the solution dE and dN. If the changes to the provisional coordinates are very large, iteration may be needed.

Complex example

In a more complex example where the sub-division may include a new road (2 units wide) occupying an unknown area A_R, the solution is more difficult. Consider the coordinate data of Table 10.1. We assume that the subdivision has to be made from GH and LM preserving these as parallel roadsides (at a provisional bearing of zero), and that the areas of the triangles GCH and LMD have to be equal. From the initial coordinates of G and H these areas are 40 and 26.4 square units respectively, giving a difference of 13.6.

Table 10.1 provides the solution for G and H, together with the checks that the roadsides have the same gradient, that M and H lie on CD (whose gradient is 1.25) and that the required areas are equal.

The method is to form the observation equation for area difference, and to link it with the constraints of the problem by Lagrange's method, as explained in Appendix 3. The observation equation is formed by considering the changes to be made to the areas of the provisional triangles GCH and LMD to equate them, that is

$$(N_G - N_C)\mathrm{d}E_H - (E_G - E_C)\mathrm{d}N_H + (N_L - N_D)\mathrm{d}E_M - (E_L - E_D)\mathrm{d}N_M$$

$$= 2(A_{GCH} - A_{LMD}) \quad (10.7)$$

Two constraint equations ensure that the points H and M lie on CD, that is

Table 10.1

Point	*E*	*N*
G	5.00	0.00
C	0.00	12.00
D	10.00	20.00
L	7.00	0.00
H*	5.00	16.00
M*	7.00	17.60
Solution		
H	4.23	15.38
M	6.15	16.92
Check bearings		
HG	ML	MH
–0.05	–0.05	1.25
Check areas		
GCH	33.85	
LMD	33.85	

$$dE_H - 1.25\ dN_H = 0$$

$$dE_M - 1.25\ dN_M = 0 \tag{10.8}$$

Two other constraints preserve equal changes dm to the gradients of GH and LM, originally assumed zero. That is

$$dE_H - \mathrm{m}\ dN_H + N_H\ \mathrm{dm} = 0$$

$$dE_M - \mathrm{m}\ dN_M + N_M\ \mathrm{dm} = 0 \tag{10.9}$$

The numerical values of the coefficients and absolute term are given in Table 10.2.

To obtain a solution we introduce Lagrange's method (see Appendix 3, Equations (3.15) and (3.16)). The normal equations plus the Lagrangian additive with vector **k** are

$$\mathbf{Nx} + \mathbf{E}^{\mathrm{T}}\mathbf{k} = \mathbf{b}$$

where $\mathbf{A}^{\mathrm{T}}\mathbf{A} = \mathbf{N}$. These are constrained by the equations

$$\mathbf{Ex} - \mathbf{d} = \mathbf{0}$$

Table 10.2

Observation equation ABS					
–12	–5	–20	3	0	27.2
Constraint equations					
1	0	0	0	16	0
0	0	1	0	17.6	0
1	–1.25	0	0	0	0
0	0	1	–1.25	0	0

which can be put together as the hypermatrix

$$\begin{bmatrix} \mathbf{N} & \mathbf{E}^{\mathrm{T}} \\ \mathbf{E} & \mathbf{0} \end{bmatrix} \begin{bmatrix} \mathbf{x} \\ \mathbf{k} \end{bmatrix} = \begin{bmatrix} \mathbf{b} \\ \mathbf{d} \end{bmatrix}$$

or

$$\begin{bmatrix} 144 & 60 & 240 & -36 & 0 & 1 & 0 & 1 \\ 60 & 25 & 100 & -15 & 0 & 0 & 0 & -1.25 \\ 240 & 100 & 400 & -60 & 0 & 0 & 1 & 0 \\ -36 & -15 & -60 & 9 & 0 & 0 & 0 & 0 \\ 0 & 0 & 0 & 0 & 0 & 16 & 17.6 & 0 \\ 1 & 0 & 0 & 0 & 16 & 0 & 0 & 0 \\ 0 & 0 & 1 & 0 & 17.6 & 0 & 0 & 0 \\ 1 & -1.25 & 0 & 0 & 0 & 0 & 0 & 0 \\ 0 & 0 & 1 & -1.25 & 0 & 0 & 0 & 0 \end{bmatrix} \begin{bmatrix} \mathbf{x} \\ \mathbf{k} \end{bmatrix} = \begin{bmatrix} -326.4 \\ -136 \\ -544 \\ 81.6 \\ 0 \\ 0 \\ 0 \\ 0 \\ 0 \end{bmatrix}$$

The solution is:

$$dE_{\mathrm{H}} = -0.77;\ dN_{\mathrm{H}} = -0.61$$

$$dE_{\mathrm{M}} = -0.84;\ dN_{\mathrm{M}} = -0.67$$

Giving the final coordinates of Table 10.1 as

$$E_{\mathrm{H}} = 4.24;\ N_{\mathrm{H}} = 15.39$$

$$E_{\mathrm{M}} = 6.16;\ N_{\mathrm{M}} = 16.93$$

The areas of the triangles GCH and LMD both equal 33.85 square units as required, and the bearings preserved.

10.9 Interpolation within a grid

If a function F is dependent on two variables tabulated at regular grid intersections, usually in cartography E and N, or latitude and longitude i.e. F = fn (E, N), intermediate values of F may be interpolated from a double entry table or spreadsheet from its regular arguments. As an example we find the Universal Transverse Mercator projection Eastings interpolated from the intersections of a five minute graticule in the vicinity of a point A with coordinates: longitude λ = 35°18'16.7559" and latitude ϕ = 15° 52' 29.9096". The nearest tabular entry below A is at (35° 15', 15° 50').

The various differences between the tabulated values for the ‘square’ in which A lies follow the scheme

	F_3				
	$\Delta N'$				
$\Delta N''$	F_2	$\Delta E'$	F_4		
	$\Delta N'$	$\Delta EN''$	$\Delta N'$		
	F_1	$\Delta E'$	F_5	$\Delta E'$	F_6
				$\Delta E''$	

where

$\Delta E'$ is the first difference in the table in the Eastings direction,

$\Delta N'$ is the first difference in the table in the Northings direction,

$\Delta E''$ is the second difference in the Eastings direction,

$\Delta N''$ is the second difference in the Eastings direction,

$\Delta EN''$ is the second difference in the Eastings direction between the first differences to the left and right. (Note: it is also the second difference in the northings direction between the first differences above and below it.)

The corresponding numerical values for UTM Eastings at the five-minute graticule containing point A is:

740789.1				
$\Delta N'$ = –99.6				
$\Delta N''$ = –0.4	**740888.7**	$\Delta E'$		**749814.7**
$\Delta N'$ = –99.2	$\Delta EN''$ = –3.6	$\Delta N'$ = –102.8		
740987.9	$\Delta E'$ = 8929.6	**749917.5**	$\Delta E'$ = 8930.1	**758847.6**
				$\Delta E''$ = 0.5

The interpolating fractions are

$$n = \frac{2'29.9096''}{5} = \frac{2.498\,493}{5} = 0.499\,699$$

$$m = \frac{3'16.7559''}{5} = \frac{3.279\,265}{5} = 0.655\,853$$

Setting the nearest tabular entry to A (35°15', 15°50') by E_{T} the interpolation formula is

$$E_{\mathrm{A}} = E_{\mathrm{T}} + n\Delta N' + m\Delta E' + nm\Delta EN'' + \frac{1}{2}n(n-1)\Delta N'' + \frac{1}{2}m(m-1)\Delta E'' = 746\,793.7 \qquad (10.10)$$

$$E_A = E_T + T_1 + T_2 + T_3 + T_4 + T_5 = 740987.9 - 49.570108 + 5856.50495 - 1.1798239 + 0.10006027 - 0.0860368 = 746793.7$$

10.10 Calculation of volumes

The calculation of volumes of earthworks is of major importance in civil engineering work. The topic is dealt with by various commercial computer packages which model the problem with as little intervention from the user as possible. However, the irregularities encountered in civil engineering operations and the complex nature of the ground surface, mean that the imposition of some *a priori* constraints is necessary, such as the break lines marking the edge of an embankment or cutting.

Although the accuracy with which ground surfaces can be modelled is a matter of current investigation, there is no escaping the fact that it depends on the density of points used to describe the surface. A very varied surface needs many points, but only a few are needed for smooth surfaces. Accuracy is also related to costs and need. The surfacing of an airfield runway, or motorway paving and their related volumetric-cost calculations demand much tighter geometrical accuracy than in ordinary roadwork, or in quarrying.

In this chapter, we concentrate upon traditional roadwork calculations which illustrate many of the basic problems involved, both in the calculation of volumes and in the setting out of formations.

Prismoidal and end-area formulae

Because land surfaces are invariably irregular, all earthwork calculations must of necessity be approximate, the precision obtained being entirely dependent on the density of the measured ground heights in relation to the nature of the surface.

Volumes are usually obtained by calculating the volume of the regular geometrical solid whose surfaces most nearly represent the actual ground surfaces. The geometrical solid most frequently used for this purpose is the prismoid, which is defined as a solid figure having plane parallel ends and plane sides (see Figure 10.8).

It will be clear that the area A of a cross-section, parallel to the ends, will be a function of the perpendicular distance x of the cross-section from one of the ends, and

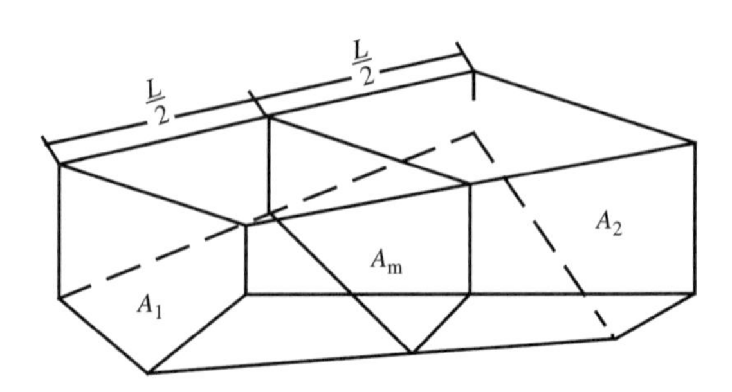

Figure 10.8

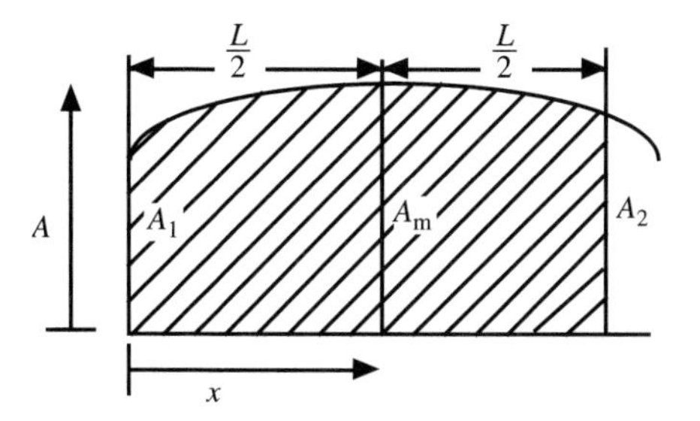

Figure 10.9

that if the cross-section area A is plotted against the distance x (see Figure 10.9), the volume will be equal to the area under the curve i.e.

$$V = \int A\,\mathrm{d}x$$

An approximation to the area under the graph can be found using Simpson's rule, by taking three ordinates, A_1, A_m and A_2 equally spaced at a distance $L/2$ apart i.e.

$$V = \frac{L}{6}(A_1 + 4A_\mathrm{m} + A_2)$$

where A_m is the cross-section area midway between A_1 and A_2 (not the mean of A_1 and A_2).

For a true prismoid, the cross-sectional area is a quadratic function of x, the distance from one end, so that the above relationship is exact and the equation is known as the *prismoidal rule*. When this rule is applied to the portion of a normal road or railway cutting or embankment, between two parallel cross-sections as shown in Figure 10.10 where three of the side faces are planes, if the ground surface is also a plane then the figure is a true prismoid. Since the rule is also exact when the cross-sectional area is a cubic function of x, the rule will also give the true volume, if either:

1. the slope of the ground at right angles to the centre-line is uniform (i.e. PQ is a straight line) and the central height h is a quadratic function of x (i.e. the longitudinal profile on the centre-line is parabolic), or
2. the transverse profile PQ is parabolic, and h is directly proportional to x (i.e. the longitudinal profile is a straight line).

Thus it will be seen that, provided the ground surface curves smoothly, the rule will give a very good approximation to the true volume for this type of solid.

A less accurate estimate of the volume of a prismoid can be obtained by using the trapezoidal rule to find the approximate value of the area under the curve in Figure 10.10 i.e.

$$V \approx \frac{1}{2}L(A_1 + A_2)$$

which, when applied to volumes, is known as the *end-area rule*. The best accuracy will be obtained from this rule when A_1 is approximately equal to A_2 and the accuracy

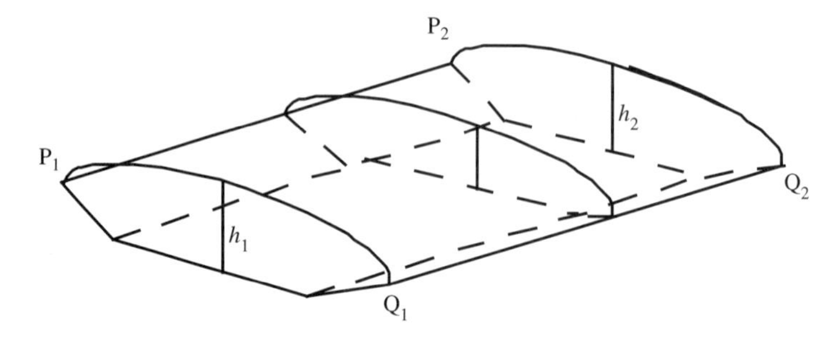

Figure 10.10

will decrease as the difference between A_1 and A_2 increases. As an extreme example, for a pyramid or cone where one end-area is zero, and the true or prismoidal volume is base × height/3, the rule gives base × height/2 i.e. an error of 50 %.

10.11 Determination of volumes from contours

This accuracy of this method depends largely on the contour vertical interval and on the accuracy with which the contours have been determined and plotted. It is extremely valuable where very large volumes are involved e.g. reservoirs, land reclamation schemes, open-cast mining etc., and is also very useful in the preliminary stages of route projects (roads, railways and canals) for making initial estimates of cost, comparison of alternative routes and selection of best profile.

The method consists of splitting the solid along the contour planes into a series of horizontal slabs, each slab then being regarded as a prismoid whose length is the contour vertical interval, and whose end-areas are the areas enclosed by the contour lines at the height in question, the areas being taken off the map or plan by planimeter or by calculation from coordinates.

If the prismoidal rule is to be applied to each slab individually, it will be necessary to interpolate the contour lines midway between those already plotted, in order to obtain the mid-area of each slab. It is usually adequate, however, to take the slabs in pairs and find the volume by Simpson's rule. Any portions of the solid which are not embraced by two contour planes will have to be treated separately and the volume of the nearest appropriate geometrical solid (usually a pyramid or wedge) found.
Referring to Figure 10.11 (vertical interval = 10 m) by end-area rule,

$$V = \frac{10}{2}(A_{120} + 2(A_{130} + A_{140} + A_{150}) + A_{160}) + \frac{7}{3}A_{160}$$

By prismoidal (or Simpson's) rule,

$$V = \frac{10}{2}(A_{120} + 4A_{130} + 2A_{140} + 4A_{150} + A_{160}) + \frac{7}{3}A_{160}$$

If the new profile (i.e. the ground profile after the works have been carried out) is anything other than a horizontal plane, the new contour lines for the finished work

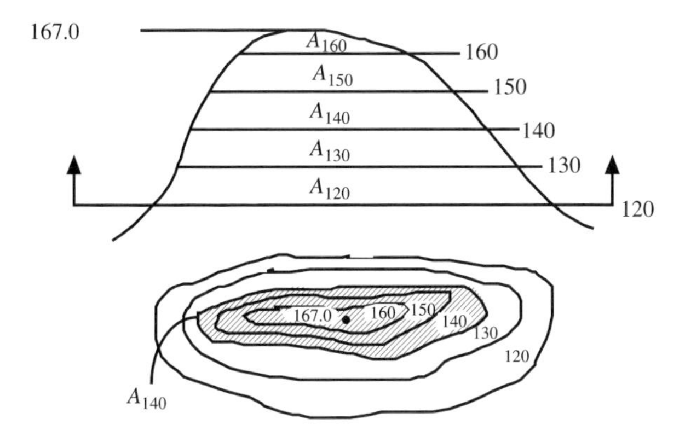

Figure 10.11

will have to be superimposed over the existing ground contours, and the end-areas of the prismoids will then be the areas on the plan enclosed between the new and the old contour lines for the height in question.

In Figure 10.12 (not to scale) the new contours for the benching shown in the cross-section have been superimposed onto the contours for a natural hillside, and the prismoidal end-areas at heights 20 m and 25 m above datum indicated by hatching.

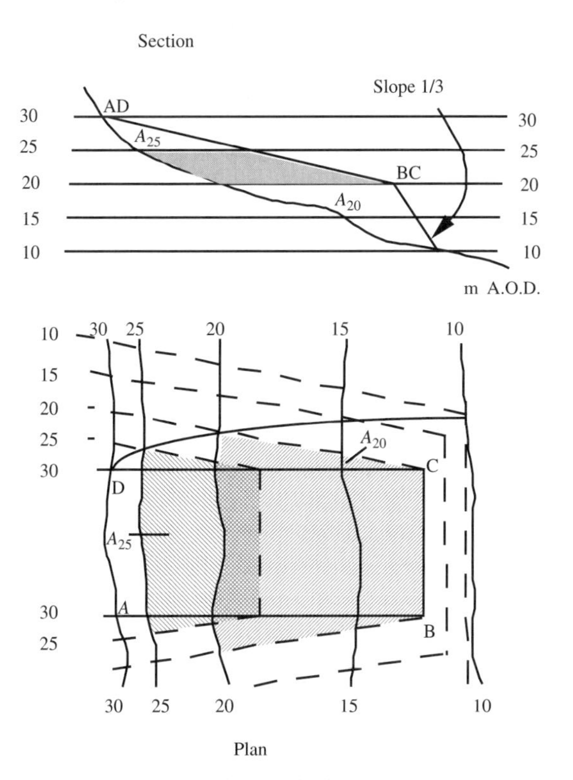

Figure 10.12

In all but the simplest earthwork projects, the plotting of the contours for the finished work may require some thought, and Figure 10.12 also illustrates some of the principles involved.

It will be assumed that the embankment surrounding the benching on sides AB, BC and CD has a slope of 1/3 (i.e. 1 vertical in 3 horizontal) and that the vertical interval is 5 m as shown. If at any point of known height on the plan we move away along the line of greatest slope for a distance of $3 \times 5 = 15$ m, the level will have dropped 5 m and a point on the next contour line will have been found. For example, if from points B and C (both at 20 m above datum) we move out 15 m in a direction normal to BC, two points on the 15 m contour will have been found; if we move out 30 m, two points on the 10 m contour will have been found, and so on. As for a plane, the contours will be a series of equidistant parallel lines, and the contours for the embankment falling away from the line BC can now be plotted.

Similarly, if we move away 15 m from C in a direction normal to CD, and 45 m away from D in the same direction, two further points on the 15 m contour for the embankment will have been found. Since the normal to CD at D, in this case, lies along the 30 m contour, the point we have just located on the 15 m contour for the embankment lies below existing ground level, and the 15 m contour for the finished work will only extend part of the way along the line between the two points by which its position has been established. Clearly at the point where the new and the old 15 m contours intersect, the new and existing ground levels are the same and the actual new contour line will not extend beyond this point.

By linking up all such points where new and existing contours of the same value intersect, the position of the toe of the embankment can be established. This is also shown in the figure.

Figure 10.13 shows the same principle applied to a section of road running in cutting and on embankment, with again some of the prismoid end-areas marked to indicate the areas that will have to be calculated or taken off with a planimeter. It should be noted that, in both these cases, the cross-sectional view is not required for the purposes of the volume calculations, and is only included to illustrate the meaning of the construction lines on the plan. A very small part of a project, drawn to a large scale, has been used for the sake of clarity, but it must be realized that in the ordinary way, due to the difficulty of obtaining sufficiently accurate contours, the accuracy of the volumes obtained under these conditions would be considerably inferior to those obtained by the cross-section method described below.

In the road example illustrated in Figure 10.13, the road gradient is uniform, and the plan centre-line straight. If the vertical profile of the road is laid to a vertical curve, the new contours crossing the road will no longer be equidistant, and the side-slope contours will cease to be equidistant parallel lines and become instead equidistant parallel curves. Unless the change in gradient is very acute, the curvature of the side-slope contours is very slight.

The case where the road centre-line is curved on plan is different, as will be seen from Figure 10.14. The line of greatest slope of the side-slope is usually normal to the edge of the road, so that the same technique of stepping off intervals of n times the vertical interval (where the side-slope is $1/n$) from points of known height on the edge of the road still applies, as shown in Figure 10.14.

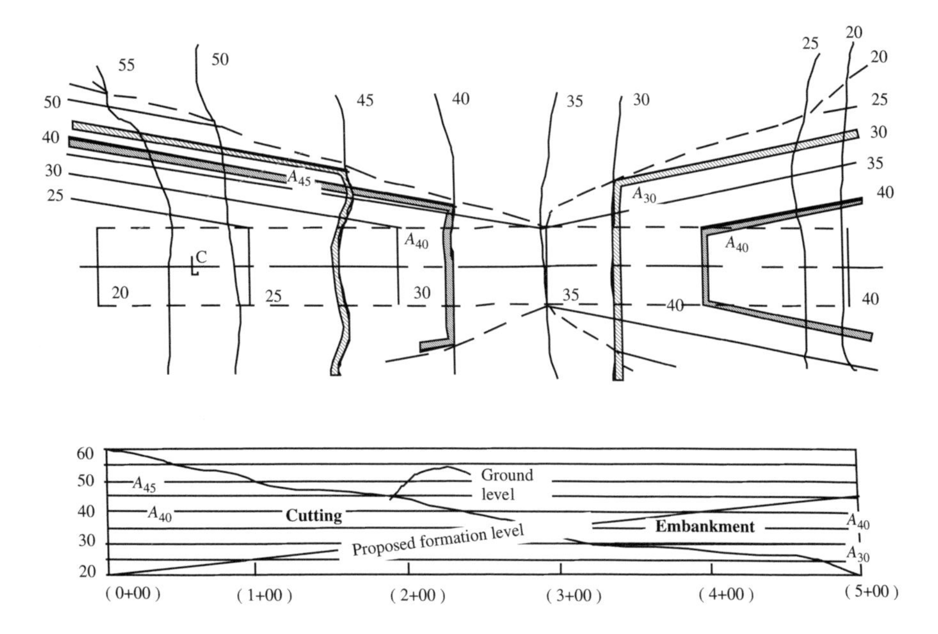

Figure 10.13

Figure 10.15 demonstrates one further application of contour geometry to the solution of earthwork volume problems. Here a tunnel 16 m in diameter, and with its axis horizontal, is to be driven into a hillside with contours as shown.

In order to be able to plot the position of the interpenetration curve between the tunnel and the hillside and the shape of the end-areas of the horizontal slabs on the plan view, it is necessary in this case to draw the front elevation of the mouth of the tunnel. From this view, the true-length (such as *ab*) of the horizontal distances from the centre-lines to the lines of intersection of the contour planes with the walls of the tunnel can be obtained. In the figure, this has been achieved by drawing the auxiliary view of the tunnel mouth, so that these intersection lines can be obtained by simple

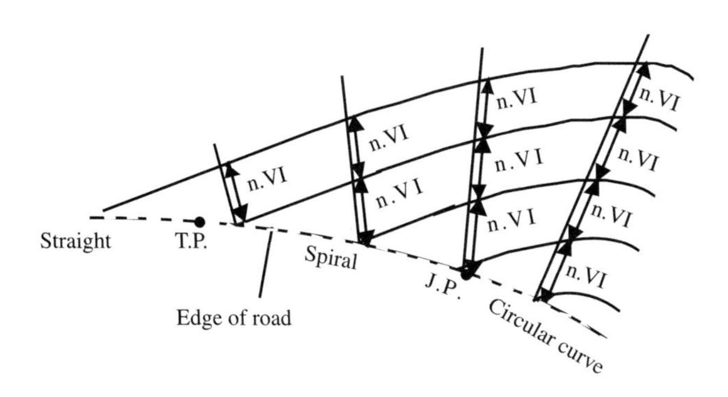

Figure 10.14

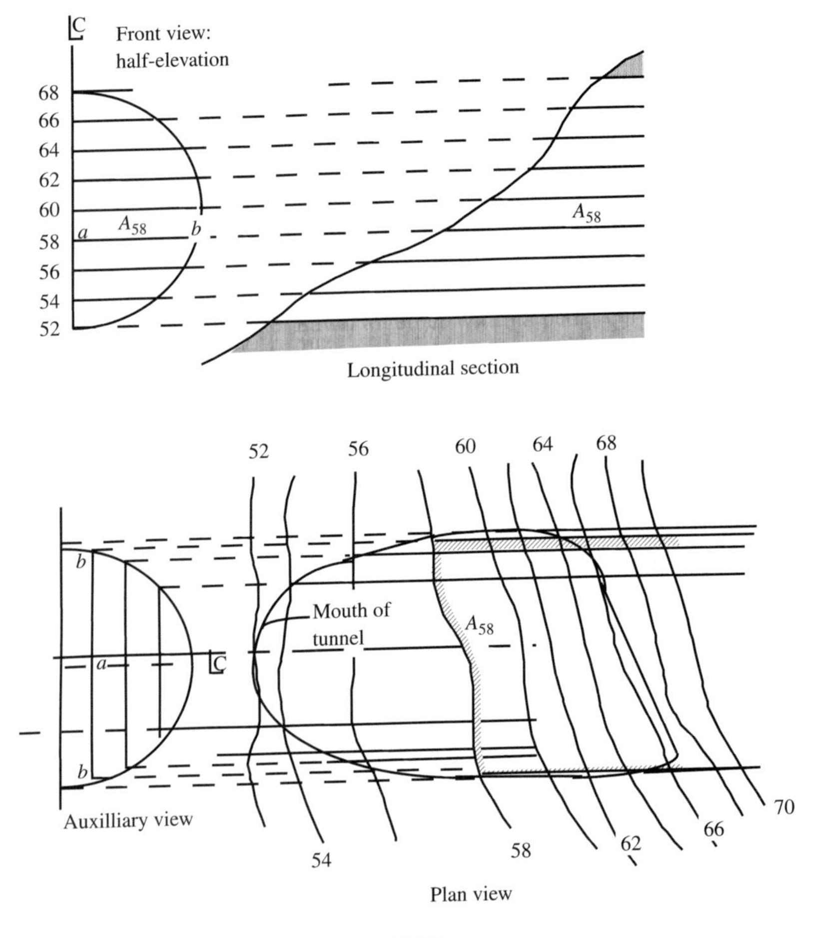

Figure 10.15

projection. The positions of points on the plan view of the tunnel mouth denoted by the intersection of the old and new contour lines at 58 m above datum, and the shape of the slab end-area at the same height, are marked on the diagram. The embankment carrying the road up to the tunnel mouth has been omitted, as the construction for this would be similar to that shown in Figure 10.15.

10.12 Determination of volume from spot heights

The accuracy of this method depends only on the density of the levels taken and is usually used for large open excavations such as reservoirs, for ground levelling operations such as parks or playing fields or for building sites. The use of photogrammetric methods of heighting from air photographs and advanced data processing techniques has opened up a field in which, by enormously increasing the number of spot heights that can be used, the method can be extended to almost any earthwork project for which photogrammetrical heights are sufficiently accurate.

The coordinates and heights of points may also be rapidly obtained by total station, usually from carefully selected points at changes of slope or bends in plan. Efficient gathering of data by field methods is still a skilled task. Because such points appear at irregular intervals over the ground, it has become common to describe them as *random points*. This terminology is most unfortunate because they are *systematically* selected to model the ground and its break lines and are certainly not random. The technique of gathering height data from a regular grid is *statistically random*.

In normal ground methods, the site is gridded by a series of lines forming squares (or occasionally rectangles) and the ground levels are determined at the intersections of the grid lines, together with such additional points (referenced to the grid by offsets or tie lines) as may be necessary to pick up break lines: special boundaries or exceptional ground irregularities or discontinuities. The spacing of the grid lines will depend on the nature of the ground, and should be sufficiently close for the ground surfaces between the lines to be reasonably regarded as planes.

There are two reasons for using grids:

1. they are simple to use in the field, and
2. subsequent computer calculations are regularized.

The proposed formation levels at the corners of the grid squares are obtained from the designer's drawings, and the volume within each square is taken as being the plan area of the square, multiplied by the average of the depths of excavation (or fill) at the four corners of the square. The volumes of the portions lying between the outermost grid lines and the random boundaries of the site are taken as the plan areas of the nearest equivalent trapezia or triangles, each multiplied by the average of their corner depths (see Figure 10.16).

Because all the depths at the internal intersections of the grid lines are used in the calculation of the volume of more than one square, a formula can be developed of the form

$$V = \frac{l^2}{4}(h_1 + 2h_2 + 3h_3 + 4h_4 + \dots) + R$$

where l is the length of side of square, h_1 the depths such as at a and e which are used once, h_2 the depths such as at b, c and d which are used twice, h_3 the depth such as at f which is used three times, h_4 the depth such as g, h etc. which are used four times,

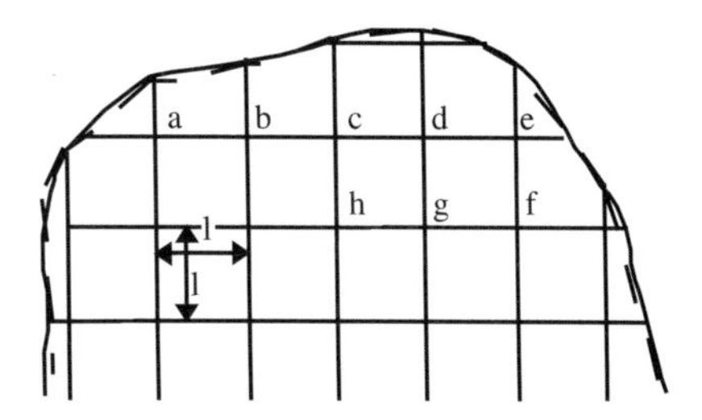

Figure 10.16

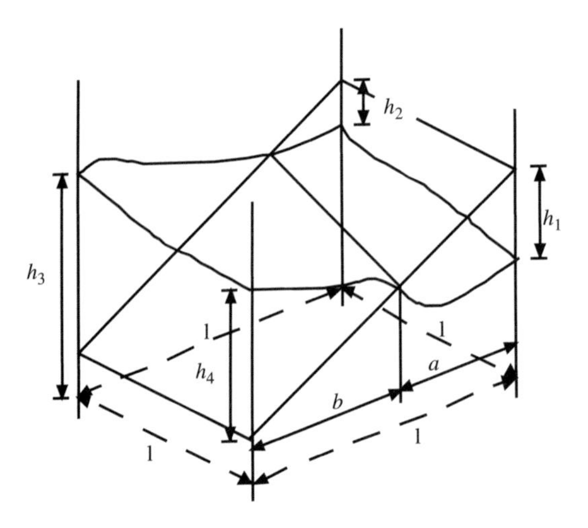

Figure 10.17

and R the volume of extra-peripheral trapezia and triangles, which must, of course, be calculated individually.

Certain difficulties will arise if there is a changeover from cut to fill, or vice versa, within one grid square (see Figure 10.17).

If $h_1 = -h_4$ and $h_2 = -h_3$, the calculation

$$V = \frac{l^2(h_1 + h_2 + h_3 + h_4)}{4}$$

will produce zero volumes for the square which is patently incorrect, and whatever the values of the depths the calculation will only produce the net volume cut or fill: whichever is the larger.

If the grid lines are fairly close together, the formulae

$$V_{\text{cut}} = \frac{l^2(h_1 + h_2 + 0 + 0)}{4}$$

$$V_{\text{fill}} = \frac{l^2(0 + 0 + h_3 + h_4)}{4}$$

will not produce excessive inaccuracy.

To avoid mistakes and to assist in the tabulation of the figures, it is often convenient to extract the ground levels from the field book and the formation levels from the drawings, and to record them in colour on a gridded plan of the site in the manner indicated in Figure 10.18.

The first figure is the ground level, the second below it is the proposed formation level, the difference where positive (red) indicates cut, and where negative (blue) indicates fill.

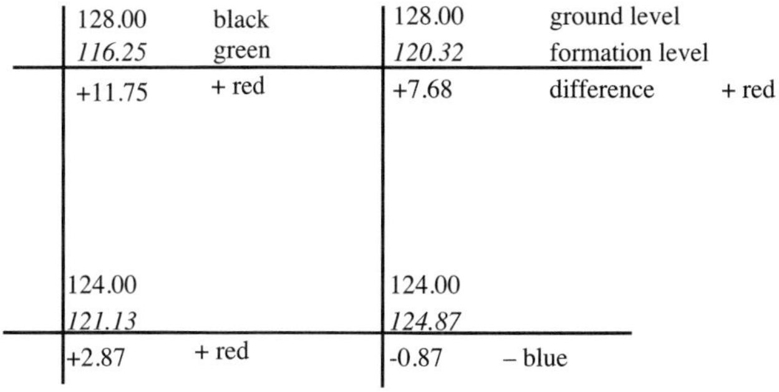

128.00	black	128.00	ground level	
116.25	green	*120.32*	formation level	
+11.75	+ red	+7.68	difference	+ red
124.00		124.00		
121.13		*124.87*		
+2.87	+ red	-0.87	– blue	

Figure 10.18

10.13 Determination of volume from cross-sections

This method is widely used for engineering projects of all types, and lends itself particularly well to route projects such as roads, railways and waterways. It illustrates general principles and has applications in small tasks. Much work formerly undertaken in the field, such as cross-sectioning after the centre-line has been set out, can now be achieved from the computer model provided it is sufficiently accurately formed. Work practices vary according to the technology available.

Cross-sections are taken on lines at right angles to the main job centre-line and the volume between adjacent cross-sections is found for a prismoid with the areas of the cross-sections as its end and, if necessary, its mid-areas. For route surveys it is usually convenient to take the cross-sections at the round number chainage points, with additional sections at points of unusual irregularity or discontinuity.

When the ground is very irregular it is best to plot the profiles of the existing ground and the proposed formation to a convenient scale and obtain the areas of the cross-sections by planimeter or counting squares. In this connection it is important to note that it is customary, for clarity, to plot cross-sections to an exaggerated vertical scale, and suitable allowance must be made for this if the areas of the cross-sections are to be found by any of the above methods.

When the ground surface is reasonably regular and the ground profiles at the cross-sections can be represented by uniform slopes, it is more expeditious to make use of suitable formulae for calculating the cross-sectional areas.

As referred to in the previous section, photogrammetrical heighting, coupled with computer processing, can greatly assist with this method. In such cases, the area of the cross-sections can be obtained from a series of equidistant spot heights. As will be seen from Figure 10.19, if the depths are determined at sufficiently close intervals dl, the areas will be given by

$A_{cut} = \mathrm{d}l$ (for +ve h); $\qquad\qquad$ $A_{fill} = \mathrm{d}l$ (for –ve h)

and because depths taken beyond the end of the section will be zero, there is no need to determine the overall width of the cross-section independently.

When normal ground methods of survey are employed and the ground slopes can be regarded as uniform, the following calculation methods will be found useful.

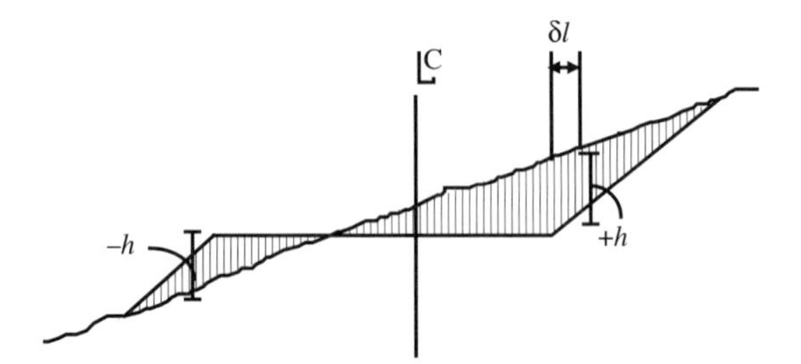

Figure 10.19

It will be noted that the general case considered here is the *three-level* cross-section, in which the ground slopes are assumed to be uniform from A to E and from E to D, as in Figure 10.20.

Although a sudden change of slope exactly on the route centre-line will rarely occur in practice, this assumption gives a much closer approximation to the true ground profile when this is a smooth curve, as shown in Figure 10.20, than would be given by the assumption of a straight line from A to D. Where the ground surface approximates to a plane, however, the *two-level* cross-section is quite justifiable.

In Figure 10.20 slopes are expressed as tangents i.e. 1 vertical in *n* or *s* horizontal, the rest of the symbols being self-explanatory.

Calculation of half breadths

Unless the field method of *slope staking* has been used, the half breadths w_1 and w_2 and the corresponding h_1 and h_2 will not be known, and they will have to be determined from the other data. The ground slopes can be obtained by direct measurement with a clinometer or gradienter, or by calculation from two spot heights at known positions on the slope. The formation breadth and level, and the values of the side slopes will have been supplied by the designer.

$$h_1 = h + \frac{w_1}{s_1}$$

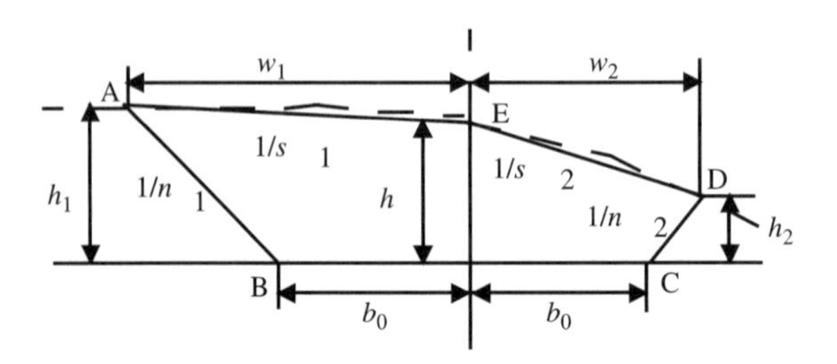

Figure 10.20

therefore

$$w_1\left(\frac{1}{n_1}-\frac{1}{s_1}\right)=h+\frac{b_0}{n_1}$$

giving

$$w_1=\frac{\left(h+\frac{b_0}{n_1}\right)s_1 n_1}{s_1-n_1}$$

applying when the ground rises from the centre-line, and similarly

$$w_2=\frac{\left(h+\frac{b_0}{n_2}\right)s_2 n_2}{s_2+n_2}$$

applying when the ground falls away from the centre-line.

It will be realized that, if a slope falling away from the centre-line is regarded as being negative, one equation will suffice; e.g. for slopes as shown in the figure, if the formation breadth is 40 m, the central depth is 7 m, both the side slopes are 1/5, and the ground slopes are both 1/20, we have

$$w_1=\frac{\left(7+\frac{20}{5}\right)\times 20\times 5}{20-5}=73.3$$

$$w_2=\frac{\left(7+\frac{20}{5}\right)\times 20\times 5}{20+5}=44.0$$

10.14 Calculation of cross-sectional areas

Once the half breadths w_1, w_2 and the side heights h_1, h_2 have been found, the area A of the cross-section can be found by inspection, or from coordinates taking the origin at any point such as F, giving

$$2A=h(w_1+w_2)+b_0(h_1+h_2)$$

It should be clear that by turning the diagrams upside down, all the above equations apply equally to embankments having the same characteristics.

In addition to the above, an additional type of cross-section (shown in Figure 10.21) must be considered.

It should be noted that in this case it is more common for n_1 and n_2 to be unequal, the slope for the fill usually being flatter than for the cut. Referring to Figure 10.21, for the left-hand side we have the same equation as before i.e.

$$w_1=\frac{\left(h+\frac{b_0}{n_1}\right)s_1 n_1}{s_1-n_1}$$

and for the right-hand side

$$w_2 = \frac{\left(-h + \frac{b_0}{n_2}\right) s_2 n_2}{s_2 - n_2}$$

Clearly, if P lies to the left of the centre-line, the above equations will be interchanged. Since FP = $h\ s_2$, the coordinates of all points with respect to an origin at F are known, and therefore the areas of the cut and of the fill can be evaluated to give

$$2A_{\text{cut}} = -b_0 h_1 - w_1 h - h^2 s_2$$
$$2A_{\text{fill}} = -b_0 h_2 + h s_2 h_2$$

It should be clear, by inverting Figure 10.20, that if P lies to the left of the centre-line, all these equations for cut and fill must be interchanged.

10.15 Curvature correction

When the centre-line of a road or railway is curved in plan, the portion between two adjacent cross-sections is no longer a true prismoid, and a correction is required. Consider first a portion of cutting or embankment of constant but unsymmetrical cross-section area A having an arcual length L, and whose centre-line has a constant horizontal radius R, as illustrated in Figure 10.22.

If the cross-section is unsymmetrical, its centroid G will be situated at a horizontal distance e from the centre-line, where e (referred to as the eccentricity) is regarded as being positive if G lies on the opposite side of the centre-line from the centre of curvature.

If θ is the angle subtended at the vertical axis at the centre of curvature by the two vertical planes containing the identical cross-sections at the ends of the solid, then the volume of the solid will be that generated by rotating the cross-section A through an angle θ. By the theorem of Pappus, the volume of this solid will be the product of the area A and the length of the path of the centroid i.e.

$$V = A(R+e)\theta = AL + \frac{AeL}{R}$$

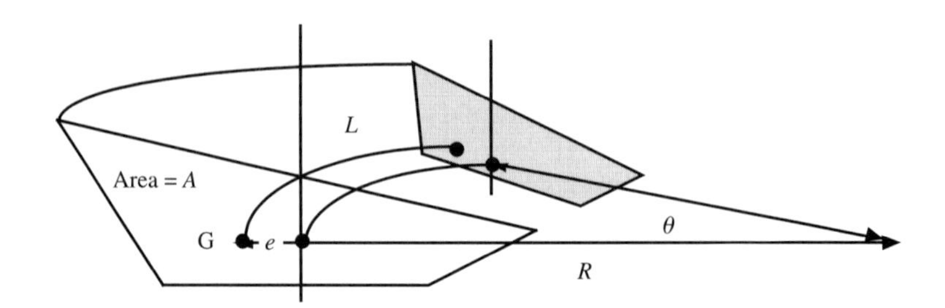

Figure 10.22

In this expression, AL is the volume of a prismoid of length L, and the curvature correction is AeL/R, being positive or negative according to whether e is positive or negative. If the cross-section is symmetrical, G lies in the centre-line and the correction is zero. In the general case, the shape of the cross-section will not be constant, so that neither A nor e will be constant, and if the horizontal curve is a spiral, the radius R will not be constant.

As in practice the ratio e/R will always be comparatively small, it is usually sufficient to calculate the correction for each cross-section and then to use either Simpson's rule, the prismoidal rule or the end-area rule as appropriate, to determine the volume in the normal manner.

In calculating the value of the correction, it is useful to remember that for the clothoid spiral the radius of curvature is inversely proportional to the distance round the spiral from the origin, and that the product of these two quantities is equal to the LR value for the curve.

Calculation of eccentricity

Because the correction term to be applied to the cross-sectional area A is Ae/R, equations for Ae are derived in this section. If the actual eccentricity itself is required for any purpose, it can always be obtained numerically by dividing Ae by A, which will already have been found.

Three cases only will be considered: the three-level cross-section with unequal side-slopes, the same with equal side-slopes, and the part cut-part fill cross-section. All other cross-sections can be very simply derived from these. In each of these cases, the expressions will be found in terms of the half-breadths, which themselves can be found as before.

It will be remembered that the perpendicular distance of the centroid of a triangle from its base is 1/3 of its height, and that the area is half the base times the height; see Figure 10.23.

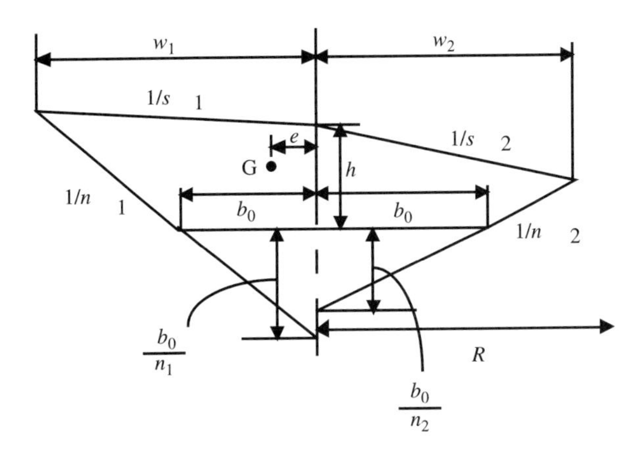

Figure 10.23

Assuming e to be positive when G lies to the right of the centre-line, taking moments about the centre-line we have

$$Ae = \left(h - \frac{b_0}{n_1}\right)\frac{w_1^2}{6} - \frac{b_0^3}{6n_1} + \left(h - \frac{b_0}{n_2}\right)\frac{w_2^2}{6} - \frac{b_0^3}{6n_2}$$

If $n_1 = n_2 = n$ this simplifies to

$$Ae = \frac{1}{6}\left(h - \frac{b_0}{n}\right)(w_1 + w_2)(w_1 - w_2)$$

For the part cut-part fill section in Figure 10.24, by the same process, we have

$$Ae_{\text{cut}} = \left(h - \frac{b_0}{n_1}\right)\frac{w_1^2}{6} - \frac{b_0^3}{6n_1} + s_2^2\frac{h^3}{6}$$

$$Ae_{\text{fill}} = \left(h - \frac{b_0}{n_2}\right)\frac{w_2^2}{6} - \frac{b_0^3}{6n_2} - s_2^2\frac{h^3}{6}$$

10.16 Earthworks from computer models

Where points have been carefully selected at breaks of slope or changes of direction in plan, more efficient estimates of volume can be obtained by forming a system of triangles connecting these points together, and calculating the volume for each triangular column from the formula:

$$V = \frac{A}{3}(h_1 + h_2 + h_3)$$

where A is the area of the triangle. The volumes of all triangular columns are added to give required cuts and fills.

The selection of triangles has been automated by an algorithm devised by Delauney, which selects a network of the most nearly equilateral triangles from any given set of points, according to the rule that the circum-circle of any triangle cannot

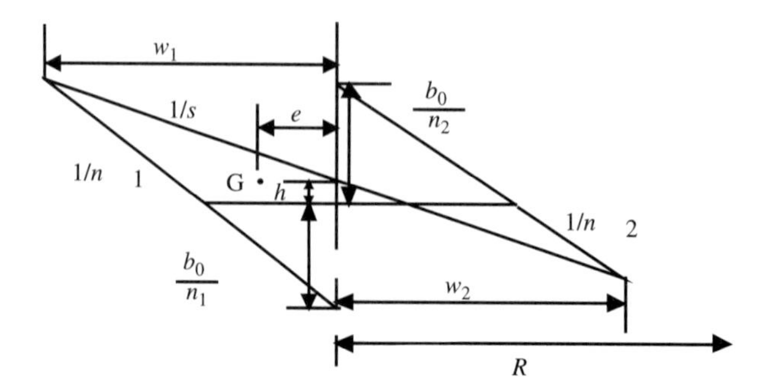

Figure 10.24

contain another point. The system produces a unique set of triangles except where there are four con-cyclic points. In this case, an arbitrary rule makes the automatic choice from two possibilities. The algorithm starts at any outside point, proceeds to find the nearest two neighbours by examining distances, then moves to one of them to proceed to finding nearest neighbours. The selection of triangles is easy to do manually, but takes a little thought and much calculation to program for a computer (see Section 12.7).

If the surface of the ground has been modelled by computer, the volumes of earthworks are calculated in a different manner. The surface model, described by a series of triangles, is intersected by the geometrically primitive surfaces of the formation, usually consisting of planes and lines. Figure 10.25 illustrates a typical problem. It shows ABC, a typical triangular facet of the ground surface, intersected, in the line ST by the side slope plane of the embankment. This side slope is part of the plane PQR, which is defined by the road design. The plane ABC is given by its coordinated points. There are three objectives:

1. to calculate the coordinates of S and T;
2. to find the equation of the line ST, for plotting by computer, and setting-out in the field from coordinates;
3. to calculate the volume of the solid RABTS, a portion of the embankment.

The complete task consists of many similar calculations. All data recorded by coordinates in the computer files can then be used to create cross-sections and three-dimensional visualizations as required. The whole process is converted to a computer algorithm based on the transformation of the coordinates of the solid. Equations are written for the projecting lines, and the coordinated points on the plane of the drawing are determined.

Alternatively, isometric views may be drawn using skew coordinate axes as in Figure 10.26.

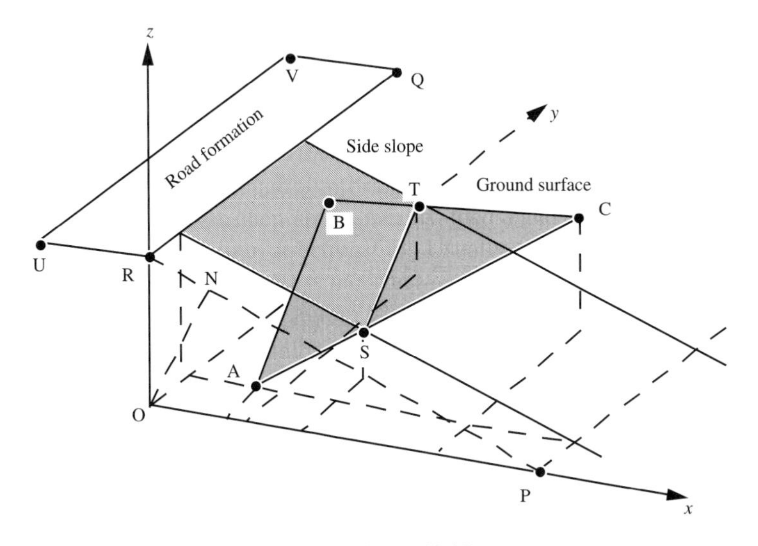

Figure 10.25

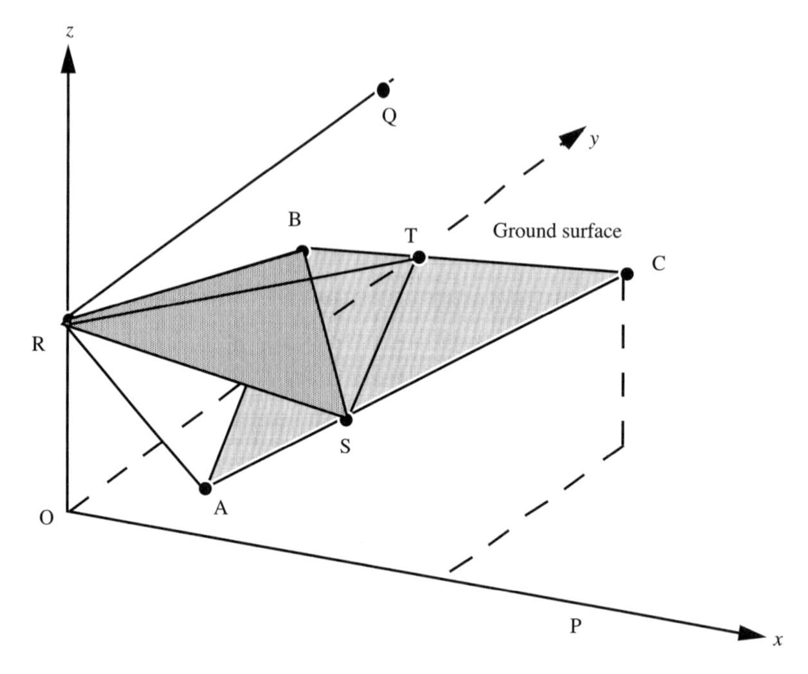

Figure 10.26

Example

The procedure is best explained by an example using the data of Table 10.3 in which the coordinates of the ground surface triangle ABC and of the formation plane PQR are listed. See Figures 10.25 and 10.26.

Equation of the plane PQR

The first stage is to find the equation of the plane PQR from the coordinates of these points. This equation is written

$$Ax + By + Cz = D$$

Table 10.3

Point	x	y	z
P	50	0	0
Q	0	100	10
R	0	0	10
A	10	10	0
B	0	50	5
C	70	30	10

where the coefficients A, B, C and D are given by the determinants

$$A=\begin{vmatrix} y_P & z_P & 1 \\ y_Q & z_Q & 1 \\ y_R & z_R & 1 \end{vmatrix};\quad B=\begin{vmatrix} z_P & x_P & 1 \\ z_Q & x_Q & 1 \\ z_R & x_R & 1 \end{vmatrix};C=\begin{vmatrix} x_P & y_P & 1 \\ x_Q & y_Q & 1 \\ x_R & y_R & 1 \end{vmatrix};\quad D=\begin{vmatrix} x_P & y_P & z_P \\ x_Q & y_Q & z_Q \\ x_R & y_R & z_R \end{vmatrix}$$

Substituting the above values,

$$A=\begin{vmatrix} 0 & 0 & 1 \\ 100 & 10 & 1 \\ 0 & 10 & 1 \end{vmatrix};\quad B=\begin{vmatrix} 0 & 50 & 1 \\ 10 & 0 & 1 \\ 10 & 0 & 1 \end{vmatrix};C=\begin{vmatrix} 50 & 0 & 1 \\ 0 & 100 & 1 \\ 0 & 0 & 1 \end{vmatrix};\quad D=\begin{vmatrix} 50 & 0 & 0 \\ 0 & 100 & 10 \\ 0 & 0 & 10 \end{vmatrix}$$

from which we obtain the equation of the plane PQR as

$$1000x+0y+5000z=50000$$

Note that, although this equation can be simplified to

$$x+5z=50$$

and normalized to

$$0.1961x+0.9806z=9.8058$$

it has been retained in the above form taken within the computer software.

The reader may like to verify, from the section ORP, that the perpendicular distance ON from the origin of coordinates O (Figure 10.25) to the plane is 9.8058, and that its direction cosines are 0.1961 and 0.9806.

Coordinates of T and S

The equations of the line BC are

$$x+(x_B-x_C)t=x_B$$
$$y+(y_B-y_C)t=y_B$$
$$z+(z_B-z_C)t=z_B$$

where t is the ratio BT/BC. Since this line also intersects the plane PQR at T, the coordinates of T also satisfy the equation

$$1000x+0y+5000z=50000$$

These four equations are solved to give the coordinates of T and the scalar t. The numerical values of the coefficients and their solution are given in Table 10.4 for the two points T and S.

Example The coefficients of these equations, listed in Table 10.4, are used to find the coordinates of point T on BC and S on AC. Each set of four equations is solved separately. The points are

T(12.07, 46.55, 5.86) and S(25.00, 15.00, 2.50)

Table 10.4

Point T	*Equations*	*Soln T*				
x	1	0	0	–70	0	12.07
y	0	1	0	20	50	46.55
z	0	0	1	–5	5	5.86
Plane PQR	1000	0	5000	0	50000	$t = 0.17$
Point S	*Equations*	*Soln S*				
x	1	0	0	–60	10	25
y	0	1	0	–20	10	15
z	0	0	1	–10	0	2.5
Plane PQR	1000	0	5000	0	50000	$t = 0.25$

Calculation of volumes

The volume of fill is calculated from the addition of a series of tetrahedra such as RABS, RBTS etc. Taking the example for RABS, the volume V is evaluated from the determinant

$$6V = \begin{vmatrix} x_R & y_R & z_R & 1 \\ x_A & y_A & z_A & 1 \\ x_B & y_B & z_B & 1 \\ x_S & y_S & z_S & 1 \end{vmatrix}$$

The numerical values are given in Table 10.5, giving $V = 1375$ cubic units.

Equations of line ST

To plot points on the toe of the new embankment, or set them out in the field, we use the equations of the line ST.

$$x + (x_S - x_T)t = x_S$$
$$y + (y_S - y_T)t = y_S$$
$$z + (z_S - z_T)t = z_S$$

$$x + 12.93t = 25.0$$
$$y - 31.55t = 15.0$$
$$z - 3.36t = 2.5$$

Thus the coordinates of a point on a cross-section perpendicular to the centre-line at $y = 20$ can be found from

$$t = \frac{-5}{-31.55} = 0.1585$$

Table 10.5

	x	y	z
R	0	0	10
A	10	10	0
B	0	50	5
S	25	15	2.5

giving $x = 22.95$ and $z = 3.03$. From these coordinates, the bearing and distances from an instrument set-up can be calculated for setting out. Normally all these computations are carried out by menu-driven software mounted on portable computers for use on site.

10.17 Mass-haul diagrams

Mass-haul diagrams are used for route earthwork projects to assist in designing the best profile and in organizing the actual work in the most economical manner. Such diagrams relate only to the longitudinal movement of earth along the route, and take no account of transverse movement of material. The mass-haul diagram is a graph showing the cumulative volume of excavation (as ordinate) plotted against the centre-line chainage (as abscissa) and is usually (although not necessarily) plotted to the same horizontal scale as, and projected up from, the longitudinal section (see Figure 10.27).

The ordinate (aggregate volume) can be regarded as a measure of the volume of earth contained in the bowl of a hypothetical scraper of infinite capacity, as it moves in the direction of increasing chainage, along the route centre-line.

When this imaginary machine is cutting, the volume of earth in the bowl increases, and the greater the depth and/or width of the cut, the greater will be the rate of

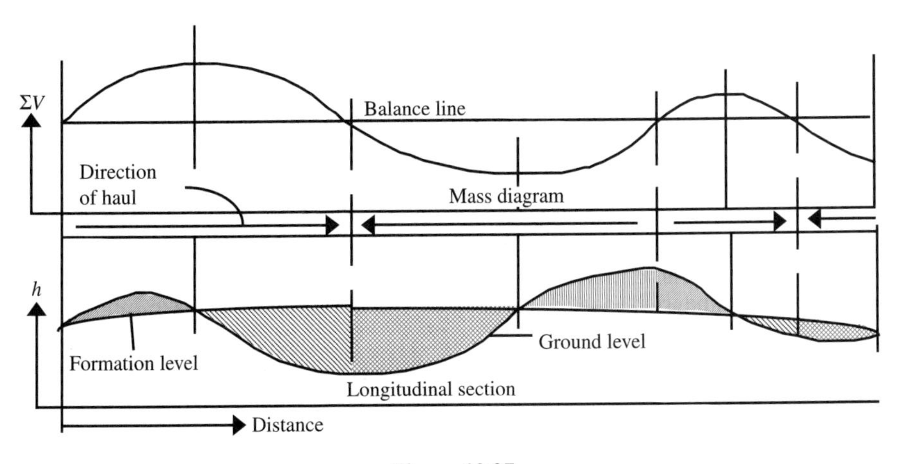

Figure 10.27

increase of volume of material in the bowl, and the steeper will be the gradient of the plotted curve. When the machine stops cutting, and starts spreading, the ordinate will cease to increase and commence to decrease, and a local maximum (zero gradient) will occur on the curve. Similarly, when the machine changes over from filling to cutting, a local minimum will occur on the curve.

From the above, it will be apparent that a positive gradient to the curve indicates cut, and a negative gradient indicates fill, and in fact, since

$$V = \int A \, dx$$

the gradient of the mass-haul curve at any point will be equal in sign and magnitude (subject to a possible scale factor) to the cross-sectional area of the cutting or embankment.

The shape of the curve for a particular project i.e. a horizontal line when there is no cut or fill, an inclined straight line when the cross-sectional area is constant and a curve with a changing gradient when the cross-sectional area is varying, will soon become apparent from the table of cross-sectional areas from which the volumes and hence the cumulative volume, are calculated.

Although certain indications i.e. whether it is cut or fill (hence the sign of the mass-haul curve gradient), the position of changes from cut to fill and hence the position of local maxima on the curve, may be obtained from the longitudinal section, care must be taken in trying to relate gradients on the curve with the depths of cut and fill on the section. If cuttings and embankments with side slopes are used, the areas of the cross-sections will not be directly proportional to the depths of cut and fill, and if the formation width on the value of the side slopes varies, there will be even less correspondence between the gradients of the mass-haul diagram and the ordinates of the longitudinal section.

Since the ordinates on the mass-haul diagram represent cumulative volumes, it will be further apparent that for any two points on the curve having the same ordinate, the volume in the bowl of the hypothetical machine will be the same, and therefore the volumes that have been dug and spread between these two chainages will be equal.

Two points on a continuous curve that have the same ordinate must, of course, embrace at least one local maximum or minimum, and the vertical distance between the points and this highest or lowest point will represent the volume which has been dug and deposited again i.e. is equal to the volume of the embankment between the two chainages. Two points on the curve having the same ordinate will, of course, be indicated by the points at which a horizontal line drawn on the diagram cuts the curve. Any such horizontal line is referred to as a balance line since there is a balance of cut and fill between the chainages of all the points where it cuts the curve (see Figure 10.27).

Since a positive gradient indicates cut, and a negative gradient indicates fill, it should be clear that when the curve lies above the balance line the direction of the movement of the earth is forwards (i.e. increasing chainage) and when the curve lies below the balance line, the direction of movement is backwards. Thus for a continuous balance line, as in Figure 10.27, the points where it cuts the curve also indicate the chainages at which the direction of haul changes.

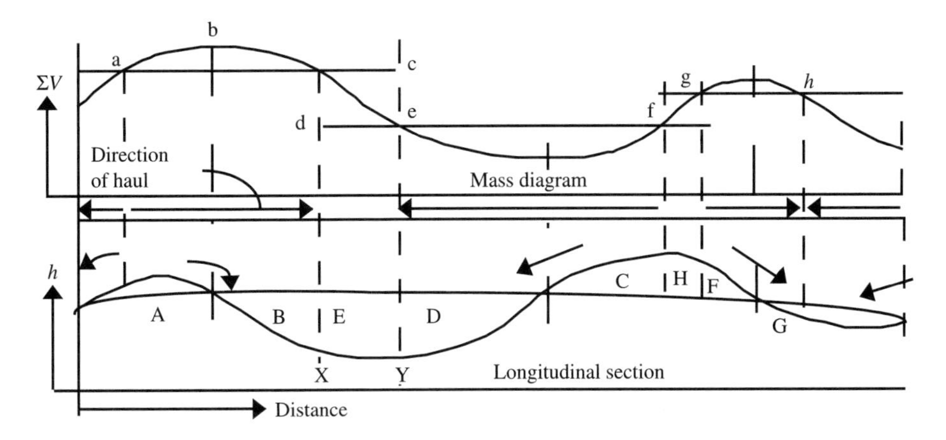

Figure 10.28

If two, or more, balance lines are drawn on the diagram, as shown in Figure 10.28, the earth from the cutting A is moved forwards and just fills the embankment B. Similarly the spoil from cutting C is moved backwards and just fills embankment D. If this scheme is adopted, there is no material available to fill the embankment E between chainages X and Y, and filling will have to be obtained from a borrow pit alongside the route in the vicinity of E, or otherwise imported on to the job. Similarly, as the spoil from cutting F when run forward will be sufficient to fill G, the earth that is excavated from cutting H will have to be dumped alongside the route adjacent to H, or otherwise removed from the job.

Borrowing and wasting

When material is either brought on to or removed from the job, it is customary to refer to this as *borrowing* or *wasting*, respectively. It will be seen that when the balance line jumps down (as from ab to ef) borrowing must take place, and when it jumps up (as from ef to gh) wasting must occur.

If the borrowed material is brought into the embankment E, transversely, all along the route between the chainages X and Y, no balance lines need be drawn for this portion. If, however, the material is imported at one point, the appropriate lines can be drawn. The continuation of the balance line ab to c, and then jumping down to e, would indicate that a volume equal to the ordinate ec was being imported at Y, and since the curve between X and Y lies below the balance line ac the direction of haul would be backwards.

If, however, the line ab jumps down to d, and then continues to e and f, the material is being imported at X and since the curve now lies above the balance line, is being moved forwards. The material could, of course, be brought in at any single point between X and Y, which would be represented by a balance line somewhere between ac and df.

It has been seen that drawing more than one balance line on the diagram introduces the necessity to waste or borrow, which apart from that required at one end of the job to accommodate any overall imbalance in the cut and fill, will involve an additional amount of excavation and filling, and before this can be justified some further factors will have to be investigated.

The area under the mass-haul curve, the product of volume and distance hauled, is known as the haul H, where

$$H = \int V \, dx$$

The units of V are cubic metre-metres (m^4).

Examples

Two simple examples will demonstrate how the mass-haul curve can be used to find out whether intermediate borrowing or wasting is desirable, and if so, the best positions for it to be carried out.

Referring to the material that has to be imported to fill embankment E in Figure 10.29, the total volume required is given by the ordinates db or ec. If it is brought in at x and run forwards, the haul will be given by the area bde, enclosed between the curve and the balance line de. If it is brought in at y and run backwards, the haul will be given by the area bce, enclosed between the curve and the balance line bc. As the second of these two areas (due to the shape of the curve) is smaller, importing at y and running backwards is the more economical procedure. It should be noted that the average haul distance for this work can be found (if required) by dividing the haul by the volume i.e. dividing the area bce by the ordinate ec.

The whole job is represented by the curve in Figure 10.30, which has an overall excess volume of cut over fill equal to the ordinate qs, which will have to be wasted somewhere. If the single balance line pq is used, the direction of haul will be forwards, the whole of the excess will be wasted at the end of the job, and the haul will be given by the area enclosed between the curve and the balance line pq. If, however, the balance line rs is used, the directions of haul will be both backwards and forwards from x, the whole of the excess volume will be wasted at the beginning of the job, and the haul will be given by the sum of the two areas enclosed between the curve and the balance line rs, which is considerably less expensive than if the balance line pq were used. The area of the rectangle prsq represents the haul involved if the whole of the excess volume were moved from one end of the job to the other.

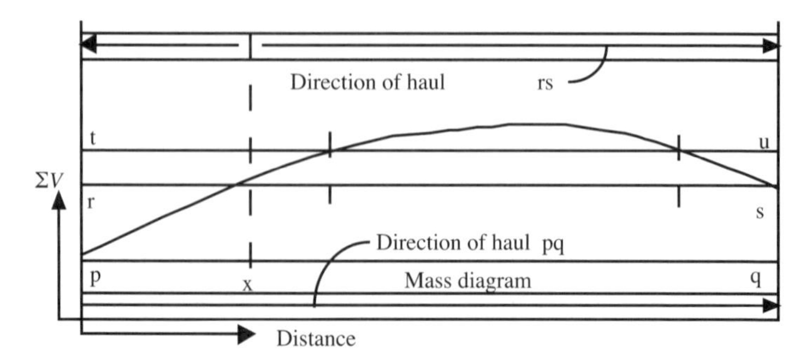

Figure 10.29

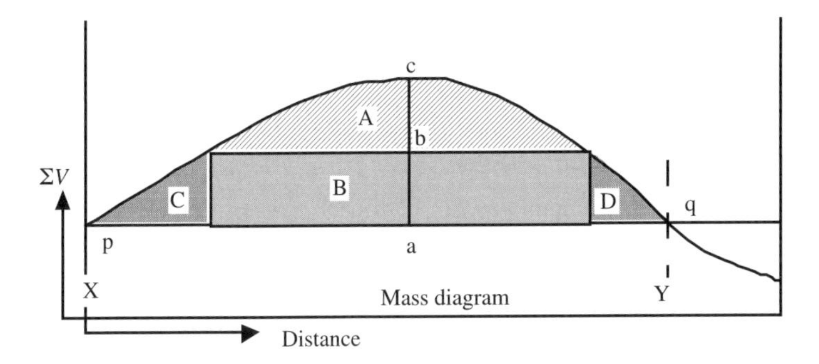

Figure 10.30

Costing

In this project the most economical scheme, from the point of view of haul, is one employing the balance line tu, in which the sum of the areas enclosed between the curve and the balance line is the minimum possible. This, however, involves increasing the volume to be wasted by the ordinate rt, and introducing an equal volume (su) to be borrowed at the end of the job. To decide whether this will in fact produce a cheaper scheme than that of employing the balance line rs will require a knowledge of the actual costs of borrowing, wasting, and hauling so that the cost of an increase in borrow and waste can be compared with the saving arising from a decrease in haul.

In looking for the most economical scheme a further factor must be taken into account.

Free haul and overhaul

When the excavated material is loaded into lorries and dumpers for transport, the cost is not directly proportional to the haul (i.e. product volume × distance) since the cost of loosening, getting out, loading, dumping, spreading and consolidating will be constant per cubic metre regardless of the distance run by the vehicles. When the earth has to be moved short distances, it may not be loaded into vehicles at all, or only into short-run vehicles.

For these reasons, it is common practice to divide the material into two categories.

1. Material that is to be moved through distances less than an agreed amount known as the free-haul distance (fixed by the type of plant envisaged for the job), such material being charged at a unit price per cubic metre, regardless of the distance moved (provided of course it is less than the free-haul distance). This volume of material is known as the *free-haul volume*.
2. Material that has to be moved through distances greater than the free-haul distance, and which is charged at the same rate per cubic metre as the free-haul volume, plus an extra charge at a unit rate per m^4 of haul for the distance it is moved in excess of the free-haul distance. This volume of material is known as the *overhaul volume* and the product of the volume and the excess distance through which it is moved is referred to as the *overhaul*.

Example

Refer to Figure 10.30 on which an additional balance line equal in length to the free-haul distance (FHD) (say 100 m) has been drawn. Between chainages X and Y the total volume of material to be moved is given by the ordinate ac (1000 m^3) at P pounds sterling per m^3. Of this volume bc (300 m^3) is the free-haul volume and ab (700 m^3) is the overhaul volume.

The total haul between X and Y is given by the whole area between the curve and the balance line pq. Of this, the area A (21 000 m^4) represents the free-haul for the free-haul volume bc, all of which is moved a distance less than 100 m, the average haul distance for this material being 21 000 / 300 = 70 m.

The area B represents the free-haul for the overhaul volume ab, being the haul for moving it the first 100 m of the journey, and included in the rate of P pounds per m^3. The areas C and D (43 000 m^4 in total) represent the haul for the overhaul volume ab, being moved through a distance beyond the first hundred metres, and so the overhaul, which is charged at £Q per m^3. The total cost of this part of the job therefore comprises:

Free-haul volume + overhaul volume = 1000 at £P per m^3 = 1000 × £P

Overhaul = 43 000 at £Q per m^4 = £43 000 Q

Total cost = £(1000 P + 43 000 Q)

If required, the average distance moved by the overhaul volume is

$$\frac{(43\,000 + 70\,000)}{700} = 161\text{ m}$$

The average overhaul distance is

$$\frac{43\,000}{700} = 61\text{ m}$$

Therefore, 161 m minus FHD = 161 – 100 = 61 m and the average distance moved by all the material is

$$\frac{(43\,000 + 70\,000 + 21\,000)}{1000} = 134\text{ m.}$$

Bulking and consolidation

In practical earthwork projects, one further factor has to be taken into account, which however does not in any way affect the foregoing arguments. If 1 m^3 of solid rock is excavated and subsequently used as fill, however carefully the consolidation is carried out it will occupy a larger volume, perhaps 1.3 m^3 in its new position. This is known as bulking, and the material is said to have a *bulking factor* of 1.3. If, on the other hand, 1 m^3 of clay is excavated and is consolidated carefully at its optimum moisture content when used as filling, it may only occupy, say, 0.9 m^3 in which case it is said to have a *consolidation factor* of 0.9.

All material, of course, when loosely loaded into a vehicle for transport, will occupy a greater volume than it did on the ground, and this is also referred to as bulking, the bulking factor in this case being somewhat larger than that referred to above. This bulking of material in transit, however, although it affects the constructor in deciding his haulage costs, does not affect the mass-haul curve, for which only volumes in the ground are considered.

The bulking and consolidation factors first referred to affect the mass-haul diagram, which aims at balancing cut and fill. All that is necessary to allow for this phenomenon is, prior to calculating the cumulative volumes from which the curve is plotted, to convert all volumes of fill into *equivalent cut*. This is most conveniently done by dividing all the cross-sectional areas for fill by the bulking or consolidation factor, since to fill a void of 1 m^3 with rock fill having a bulking factor of 1.3 will only necessitate $1/1.3 = 0.77$ m^3 of virgin rock being excavated. Similarly, to fill a void of 1 m^3 with earth having a consolidating factor of 0.9, it will be necessary to excavate $1/0.9 = 1.11$ m^3 of virgin soil.

Conclusion

From the foregoing discussions, it can be seen that the mass-haul diagram can be used in two ways.

1. To assist the designer in determining the most suitable profile and horizontal layout for the job. In this case, a diagram is drawn for a trial profile and layout, the curve is analysed, trial balance lines are drawn and modifications made to the profile and/or lay-out (as far as controlling heights and ruling gradients allow) in order to obtain balances of cut and fill, keep the total volume of earth moved to a minimum and, in particular, keep the overhaul as small as possible. It must be realized that any modification of the profile or lay-out will necessitate the recalculation of the individual and cumulative volumes and the re-plotting of the curve.
2. To assist the constructor in planning and organizing the actual work. In this case the profile has been settled and the volumes calculated and plotted. Trial balance lines and the resultant calculation of haul volumes, hauls, haul distances and directions of haul are then carried out in order to find the most economical sequences of construction, determine the requirements for plant and transport and locate the best positions for borrow pits and spoil heaps.

The treatment of all complex issues of design, setting-out and costing is available to the designer and surveyor as suites of computer software to carry out the tasks described in this chapter. Computer software is also available to create three-dimensional models which may be viewed from any vantage point for the illustration of projects to prospective clients or to support arguments in public enquiries. Animated videos of sequential graphics are also used to advantage. However, if the ground has been described in terms of triangular facets of a sufficiently small size to describe the ground adequately and volumes are calculated from three-dimensional coordinates, accurate results can be obtained.

Chapter 11
Construction and Curves

This chapter is concerned with the use of ground methods to locate points in pre-designed positions on engineering structures and works. The geometrical part of the task usually occurs in three stages: (1) site survey; (2) planning and design, and (3) setting out or building. At all stages the tolerances of measurement have to be assessed and controlled. This is now possible with on-line data processing from electronic total stations linked by radio to a site office computer.

It is common practice to base all dimensional problems on a numerical coordinate system developed and processed by computer software, and to present results in digital form for future use. Methods of working have become very versatile as a result. Essential principles, enabling the surveyor to apply common sense to a computer-dominated working medium with a special relationship with the construction industry, will be described here.

11.1 Setting out works

Setting out engineering works involves the siting on the ground of the various elements of the works in accordance with the dimensioned plans and drawings supplied by the designer. Most significant dimensions are readily available, as are coordinates from a computer model, although on-the-spot calculations performed by field computers are not unheard of.

Since it is often not possible to set out the whole of the works before construction commences, the accurate positioning of each element independently is highly important and errors or mistakes can be very expensive. It is important to remember that inaccurate preliminary surveys are not unknown, nor are dimensions scaled from distorted drawings or prints, so it is advisable to check all leading dimensions on the site before commencing any setting out. Calibrated EDM and tapes are essential for these checks.

Normally the dimensions of individual elements (buildings, roads, bridges etc.) will be fully figured on drawings, and their relative positions given in computer output. However, it is advisable to look for any controlling factors that will influence the actual positioning of the element. For example, if space is to be left between two houses for two prefabricated garages, each 2.5 m wide, then these houses must be set out 5.0 m apart against a possible scaled dimension of 4.9 m.

11.2 Marks

The usual practice is to mark key points (corners of buildings, centre-lines or kerbs of roads) with semi-permanent marks such as wooden pegs. A small nail in the top of the peg can be used for greater precision. A pipe nail can be driven into tarmac or asphalt, while it may be necessary to cut a cross with a cold chisel on stone or concrete or drive in a nail. Attaching bright fluorescent tape to buried marks greatly assists in finding them again.

Much wasted labour can be avoided by establishing permanent reference points (by a peg surrounded with concrete or by driving a short length of small diameter tube) adjacent to the works but secure from damage by excavating machinery or construction traffic, from which the marks can be re-established quickly and easily if they are lost. Generally GPS fixation plus radial survey by total station enables marks to be placed away from works.

11.3 General procedures

The general procedure in setting out is to establish a main control framework (a single base line will often suffice), usually by GPS and traversing, from which the detail can be set out by means of off-setting, tie lines, radiation or intersection. In the case of 'route' works (railways, roads, waterways or pipe-lines) the control framework is usually a traverse and it is convenient to utilize the main intersecting straights for this purpose, leaving only the curves to be established by other means.

For small sites, or when a high precision is not required, line and offset methods will often be adequate. The 3:4:5 triangle for setting out right angles and the principles of equality of diagonals for checking the squareness of rectangles can be employed. Where greater precision is required, instrumental methods are preferred.

Basic techniques

No matter what the structure may be, whether railway line, sports stadium, complex building or harbour wall, the task consists of:

1. the setting up of the instrument at a known position from which the radial data are known, and
2. the actual positioning of the formation points which the builder uses to complete the work.

If the instrument station has to be in a predetermined position whose coordinates are given, the process is iterative. The instrument is set up in approximately the correct location and surveyed in from the fixed starting points. The shifts to the correct position are calculated and a new point established. This can then be marked using a large peg with a nail in approximately the final position, say. Fresh measurements are then taken for a final calculation to verify if the nail is correctly located. If not, it has to be moved until the coordinates are within specified limits. For precise work, the setting out can be assisted by a special tripod mounted tribrach capable of small X-Y movements. The height can then be established above the correct point.

Care should be taken not to mix up slope and horizontal distances, both of which are readily available from a total station.

Several rounds of angles and values of the distances will probably be required in precise work. The centring should be carried out with great care, with full attention paid to the levelling process (simply looking through the plummet is no check). The effects of tilts on the line of sight should also be remembered, and key points should be set out on both faces.

If the software is available, it is much quicker to establish an initial point to specification then compute the radials from this new point.

Setting out the detail is again iterative, depending on the accuracy required. A provisional position is established, resurveyed and the point moved by small offsets to its correct location. If a large component is to be positioned, which has to jacked into position, the surveyor prefers to get the base at the right height first, then move the component on the level base to its correct plan position. This procedure may not always be possible.

A far as possible, points should be set out individually, and not from one another as in a chain, otherwise errors will accumulate.

To guard against blunders, whatever the feature to be set out, common sense should be used to see if the expected shapes are developing.

Structures to be set out vary enormously, from simple rectangles to complex three-dimensional surfaces such as the Sydney Opera House. Typically the most common works deal with communication lines such as pipelines, roads and railways. We shall now consider some aspects of curve surveying to illustrate some general principles.

11.4 Curve surveying

In route surveying, it is usual to treat the horizontal shape separately from the vertical. In all setting out, a system of coordinates is obviously called for e.g. rectangular coordinates for off-setting methods and polar coordinates for radiation methods. Setting out elements consisting of straight lines and curves are based on rectangular coordinates converted to radials from any suitable station. Many instruments now have a speech modulation system or a digital display to enable the operator to communicate with the prism holder.

In setting out road and railway works it is usual to establish pegs at 10 or 20 m intervals along the actual centre-line of the route, these pegs being consecutively numbered from the commencement of the route. Thus any point on the centre-line may be identified by its *chainage* i.e. its distance along the centre-line from the commencement. In setting out curves it is convenient that some of the pegs *preserve running chainage*. In all examples presented here, the chainage is assumed to be increasing from left to right.

Proceeding in the direction of increasing chainage, the first tangent to a curve is known as the running-on tangent, and the second as the running-off tangent. The junction of a straight line and a curve is known as a tangent point (TP), the junction of two portions of a compound or combined curve as a junction point (JP), the point at which the running-on and running-off tangents meet as the intersection point (IP) and the external angle between the tangents as the intersection angle. (The internal angle, known as the apex angle, is rarely used in curve calculations.) These terms are illustrated in Figures 11.1.

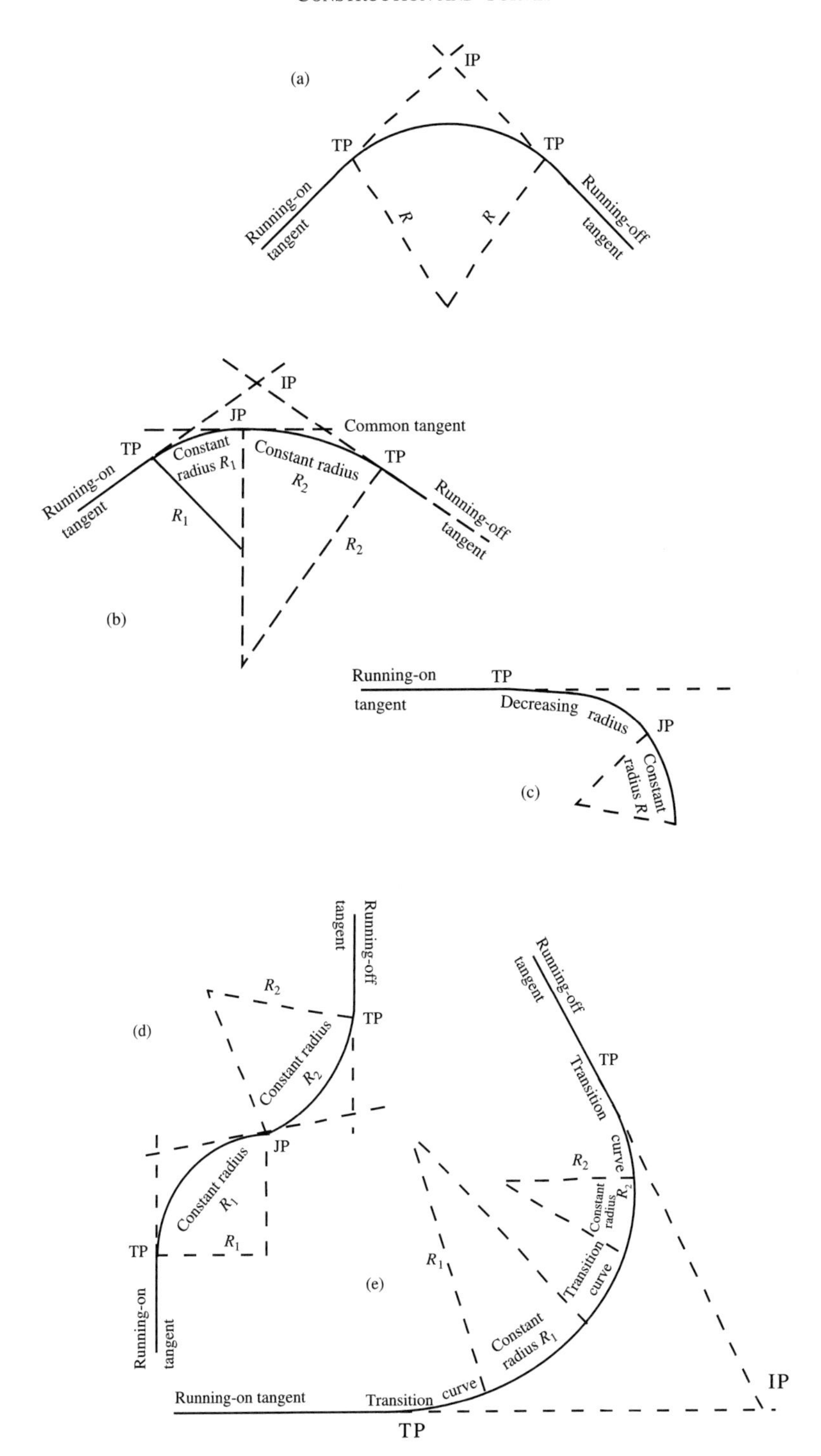
(a)
IP
TP
TP
Running-on tangent
Running-off tangent
R
R
(b)
IP
JP
Common tangent
TP
TP
Constant radius R_1
Constant radius R_2
Running-on tangent
Running-off tangent
R_1
R_2
(c)
Running-on tangent
TP
Decreasing radius
JP
Constant radius R
(d)
Running-off tangent
R_2
TP
Constant radius R_2
JP
Constant radius R_1
R_1
TP
Running-on tangent
(e)
Running-off tangent
TP
Transition curve
R_2
Constant radius R_2
Transition curve
R_1
Constant radius R_1
Transition curve
Running-on tangent
TP
IP

Figure 11.1

11.5 Types of horizontal curve

The types of horizontal curve which the surveyor normally has to deal with in connection with building and civil engineering works are (see Figure 11.1):

(a) *Simple curves*: circular curves of constant radius.
(b) *Compound curves*: two or more consecutive simple curves of different radii.
(c) *Transition curves*: curves with a gradually varying radius (often referred to as 'spirals').
(d) *Reverse curves*: two or more consecutive simple curves of the same or different radii with their centres on opposite sides of the common tangent.
(e) *Combined curves*: consisting of consecutive transition and simple circular curves. This is the usual manner in which transition curves are used in road and railway practice, to link a straight and a circular curve or two branches of a compound or reverse curve.

The problems connected with horizontal curves are: firstly the design and fitting of curves of suitable radius or rate of change of radius to suit site conditions, vehicle speeds, etc., and secondly the setting out of the designed curves in the field. The first of these problems is generally the concern of the design engineer. The computation of setting out data and the field setting out falls to the surveyor. For this reason, these last two areas of the work will be dealt with first, but design and fitting will also be considered briefly.

11.6 Circular curve geometry

Circular curves can be defined by radius which is self-explanatory, or by degree. The degree D of a circular curve is defined as being the angle in degrees, subtended at the centre by a *chord* 100 m long, so that in Figure 11.2 we have

$$D = 2\arcsin\frac{50}{R} \quad \text{or} \quad R = \frac{50}{\sin(D/2)}$$

If R (in m) is large, $D = 180 / \pi = 5729.6 / R$.

From Figure 11.3 we have:

1. Since A = C = 90°, ABCO is a cyclic quadrilateral CBD = AOC = 2θ

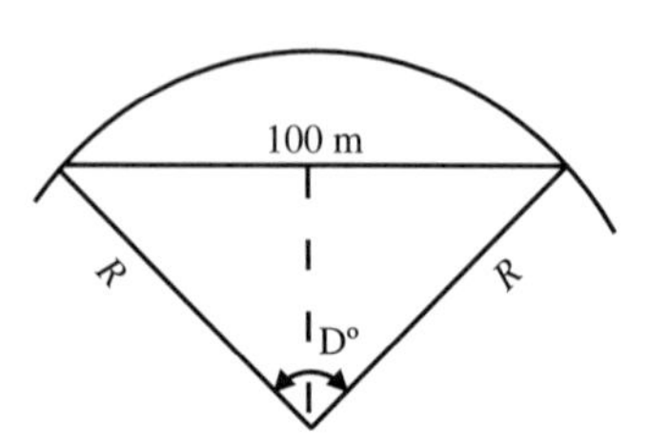

Figure 11.2

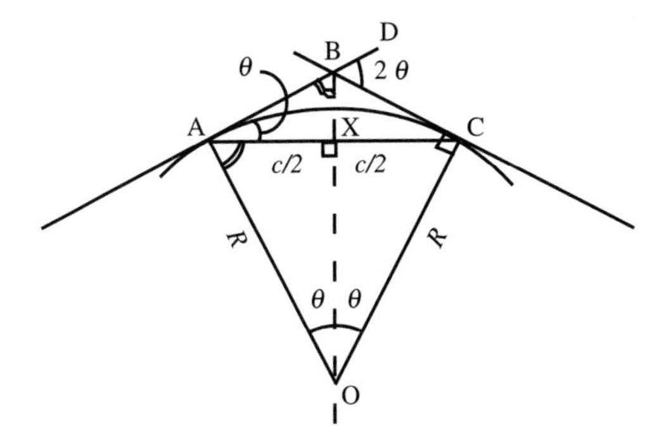

Figure 11.3

2. In the right-angled triangles ABX and OBA, angle ABO is common, therefore the triangles are similar and hence, AOB = BAX = θ. (NB for 100 m chord, θ = D / 2.)
3. From triangle AOX,

$$\sin\theta = \frac{c}{2R}$$

From Figure 11.4 we have ABO = CBO = (90 − θ), therefore

$$\phi = 180 - (180 - 2\theta) = 2\theta$$

The above relationships are used often in curve surveying.

11.7 Transition curves

A transition curve is one in which the curvature varies uniformly with respect to arc, in order to allow a gradual change from one radius to another (a straight being merely a circular curve of infinite radius) and to permit a gradual change in the super-

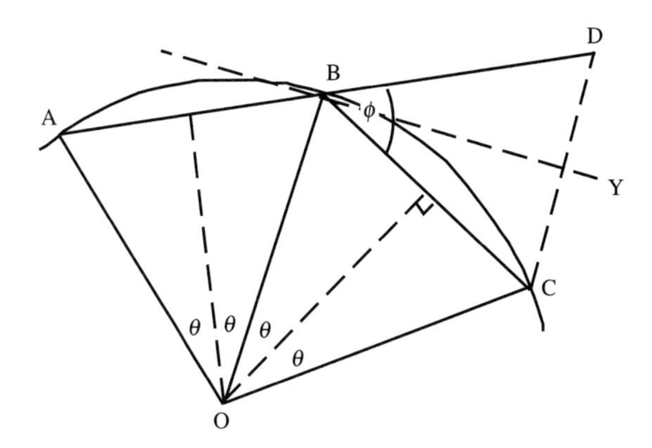

Figure 11.4

elevation (see 11.18). It must, of course, have the same radius of curvature at its ends as the circular curves that it links.

By definition, the transition curve must have a constant rate of change of curvature with respect to arc i.e. if ϕ is the tangential angle and s is the arc,

$$\frac{\mathrm{d}^2\phi}{\mathrm{d}s^2} = \mathrm{K}$$

where K is a constant. Consider the case of a transition curve linking a straight and a circular curve of (constant) radius R (see Figure 11.5) where

$$\frac{\mathrm{d}\phi}{\mathrm{d}s} = \int \mathrm{K}\,ds = \mathrm{K}s + \mathrm{K}_1$$

Since the curvature at the origin O is zero,

$$\frac{\mathrm{d}\phi}{\mathrm{d}s} = 0$$

when $s = 0$ and therefore $\mathrm{K}_1 = 0$ and

$$\frac{\mathrm{d}\phi}{\mathrm{d}s} = \mathrm{K}s$$

Integrating again,

$$\phi = \int \mathrm{K}s\,\mathrm{d}s = \frac{1}{2}\mathrm{K}s^2 + \mathrm{K}_2$$

but, since $\phi = 0$ when $s = 0$, $\mathrm{K}_2 = 0$. The fundamental equation of the transition curve therefore becomes that of the mathematical curve, the *clothoid*

$$\phi = \frac{1}{2}\mathrm{K}s^2 \tag{11.1}$$

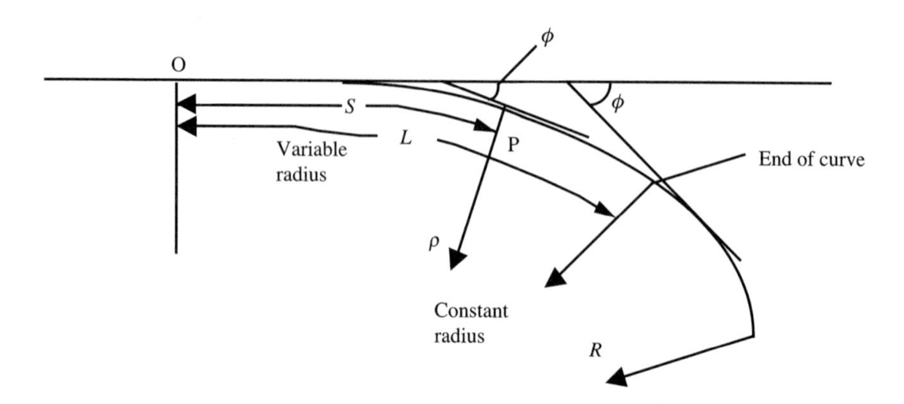

Figure 11.5

This may also be written

$$s = \mathrm{C}\sqrt{\phi}$$

where

$$\mathrm{C} = \sqrt{\frac{2}{\mathrm{K}}}$$

If L is the total length of the curve, then when $s = L$,

$$\frac{\mathrm{d}\phi}{\mathrm{d}s} = \frac{1}{R} = \mathrm{K}L \quad \therefore \quad \mathrm{K} = \frac{1}{LR}$$

and the equation of the curve becomes

$$\phi = \frac{s^2}{2LR}$$

or

$$s = \sqrt{2LR\phi} \tag{11.2}$$

Note that since

$$\frac{\mathrm{d}\phi}{\mathrm{d}s} = \mathrm{K}s = \frac{s}{LR} = \frac{1}{\rho}$$

where ρ is the radius of curvature corresponding to the arc s, then

$$s\,\rho = L\,R = \text{ constant} \tag{11.3}$$

This is an important and very useful property of the clothoid.

When $s = L$ (i.e. at the end of the curve) the total tangential, or 'spiral' angle for the curve Φ is given by

$$\Phi = \frac{L}{2R}$$

This is half the tangential angle turned through by a circular curve of the same length, and of radius R. The above equations do not lend themselves readily to the computation of curve components or to the field setting out of the curves and it is necessary to arrange them in a more convenient form. Three methods of representing the curve are commonly used for these purposes, namely systems:

1. of rectangular coordinates;
2. of polar coordinates;
3. utilizing deflection angles and chords.

All of these systems involve the use of approximations, the extent of the approximation being dependent upon the degree of accuracy required.

In this connection it must be remembered that even the maximum curvature of a practical transition curve is comparatively small and that the curvature shown in the diagrams is grossly exaggerated for the sake of clarity.

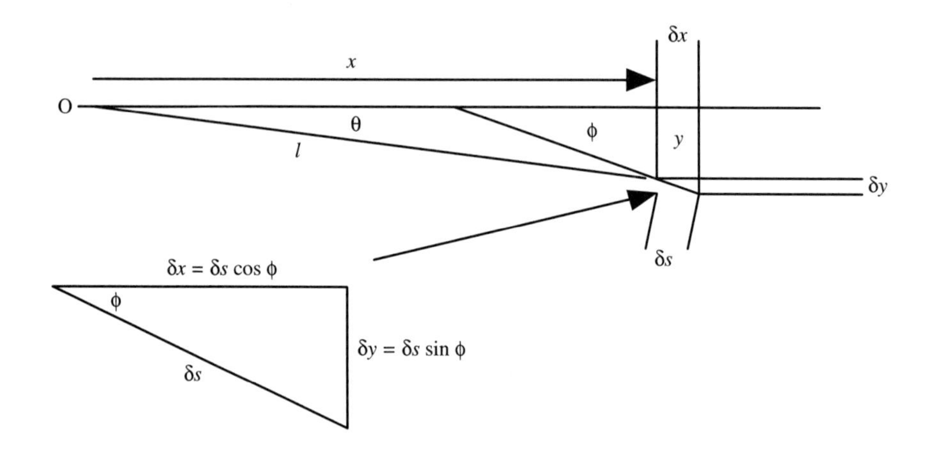

Figure 11.6

11.8 Rectangular coordinates

It is customary to use the tangent point as the origin, the continuation of the tangent as the X-axis, and offsets from this continuation as the Y-axis. Considering a small element of arc ds at a distance s from the origin, it will be clear from Figure 11.6 that

$$x = \int \cos\phi \, \mathrm{d}s \quad \text{and} \quad y = \int \sin\phi \, \mathrm{d}s$$

In order to perform these integrals it will be necessary to substitute for ϕ in terms of s or vice versa. The choice of the most convenient substitution depends on the relationship we require to establish. One illustration should suffice.

$$\phi = \frac{s^2}{2LR}$$

therefore

$$x = \int \cos\left(\frac{s^2}{2LR}\right) \mathrm{d}s \quad \text{and} \quad y = \int \sin\left(\frac{s^2}{2LR}\right) \mathrm{d}s$$

Since $x = 0$ when $s = 0$,

$$x = s - \frac{s^5}{40(LR)^2} + \frac{s^9}{3456(LR)^4} + \ldots \tag{11.4}$$

Since $y = 0$ when $s = 0$

$$y = \frac{s^3}{6LR} - \frac{s^7}{336(LR)^3} + \frac{s^{11}}{42240(LR)^5} + \ldots \tag{11.5}$$

By substituting

$$s = \sqrt{2LR\phi}$$

expressions can be found for x and y in terms of ϕ. The power series for x and y, since s is less than L, approximate to

$$x = s \quad \text{and} \quad y = \frac{s^3}{6LR}$$

or

$$\tan\theta = \frac{y}{x} \quad \text{and} \quad L = x\sec\theta \tag{11.6}$$

This is the *cubic parabola*, which is commonly used, without correction, as a transition curve where the final curvature is small. It should be noted, however, that it has a very important limitation. When the ratio y / x exceeds about 0.15 (equivalent to a deflection angle of 8° 34') the curvature ceases to increase as the distance round the curve increases and begins to decrease again, so that the cubic parabola is useless as a transition curve outside this range.

The coordinate system actually used for the setting out may be based on any suitable point to which the curve based coordinates (x, y) are transformed by the methods of Chapter 3. This procedure gives great flexibility if field software is available.

11.9 Polar coordinates

It is convenient to take the tangent point as the origin and the continuation of the tangent as the initial line. The amplitude θ is then referred to as the deflection angle, and the radius vector L as the long chord. It can be seen from Figure 11.6 that

$$\tan\theta = \frac{y}{x} \quad \text{and} \quad L = x\sec\theta$$

In practical problems, θ and L can be calculated from x and y.

11.10 Computation of curve components

The initial conditions for setting out any curve involve:

1. the location of the straights and their intersection points; and
2. the determination of the intersection angles.

This information may be supplied by the planning engineer, determined in the field by direct measurement or determined indirectly from field measurements (e.g. a traverse). It will be necessary for some of the curve components (e.g. radius or tangent length) to be fixed, and these again will normally be supplied by the engineer or must be determined from traffic considerations or by site controls (e.g. property boundaries).

The present section assumes that the above data has been obtained, and will deal with the determination of the remaining components and the location of the tangent points, junction points and, where necessary, subsidiary intersection points.

11.11 Circular curves

If the intersection angle β is fixed, the only 'free' components are the radius R, the length of the curve L and the two equal tangent lengths T_1I and IT_2. Figure 11.7 shows that $L = R\beta$ and that

$$T_1I = IT_2 = R\tan\frac{\beta}{2}$$

The following examples illustrate the necessary computations.

Example 1

Suppose $\beta = 75°$, $R = 1000$ m and chainage of intersection point I = 28 + 63.2 (Figure 11.7). The chainage is the total distance in metres from the starting point of the works. It is the total running length of the centre-line, and is used for further calculations, especially volumes. In this case the chainage of 2863.2 m is purely arbitrary. We have to determine tangent length, curve length and chainage of tangent points. The first tangent point will be at a lower chainage than that given for the intersection. The convention is to work in units of 'chains' of 100 m e.g. 28 + 63.2 ≡ 2863.2 m. The practice stems from the early use of a measuring chain (66 ft) in the 19th century.

$$T_1I = IT_2 = 767.3 \text{ and } L = R\beta = 1309.0$$

Chainage I = 28 + 63.2; T_1I = 7 + 67.3
Chainage T_1 = 20 + 95.9; L = 13 + 9.0
Chainage T_2 = 34 + 04.9

Example 2

Suppose that $\beta = 75°$ (see Figure 11.7), chainage I = 28 + 63.2 m and chainage T_1 = 24 + 36.6 m.

Determine R, the length of curve, and the chainage of T_2.

(Chainage I = 28 + 63.2) – (Chainage T_1 = 24 + 36.6) = 2863.2 – 2436.6 = 426.6

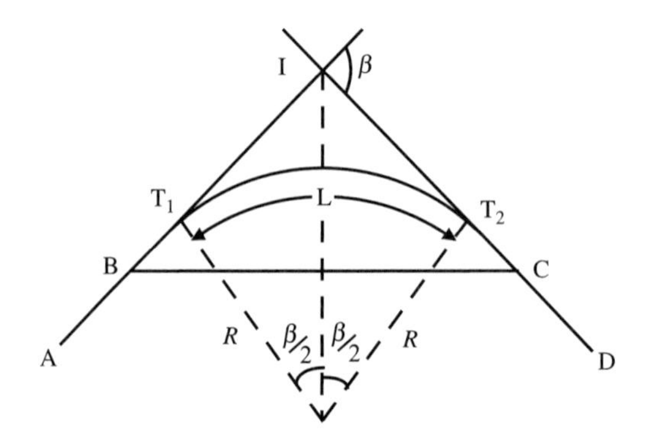

Figure 11.7

$$R = \frac{T_1I}{\tan(\beta/2)} = \frac{426.6}{\tan 37.5} = 556.0 \text{ m}$$

$$L = R\beta = 727.8 \text{ m}$$

$$\text{Chainage } T_2 = \text{chainage } T_1 + L = 31 + 64.4$$

Example 3

Suppose that ABC = 149° 44' 40", BCD = 137° 43' 40", BC = 1023.3 m (Figure 11.7). A circular curve of radius 850 m is to be inserted, tangential to AB and CD. If the chainage of B is 30 + 15.6 m, find the chainages of the initial and final tangent points.

$$\text{IBC} = 180° - 149° \, 44' \, 40'' = 30° \, 15' \, 20''$$

$$\text{ICB} = 180° - 137° \, 43' \, 40'' = 42° \, 16' \, 20''$$

$$\beta = 72° \, 31' \, 40''$$

$$\text{BI} = \text{BC}\frac{\sin \text{ICB}}{\sin\beta} = 721.6 \text{ m}$$

$$\text{IC} = \text{BC}\frac{\sin \text{IBC}}{\sin\beta} = 540.5 \text{ m}$$

$$L = R\beta = 1076.0 \text{ m}$$

$$\text{IT} = 623.6 \text{ m}$$

$$\text{Chainage } T_1 = 31 + 13.6$$

$$\text{Chainage } T_2 = 41 + 89.6$$

11.12 Compound circular curves

As can be seen from Figure 11.8, the centres O_1 and O_2 lie on a straight line O_2J which is perpendicular to the common tangent PJQ. Note that in general, O_2J does not pass through I.

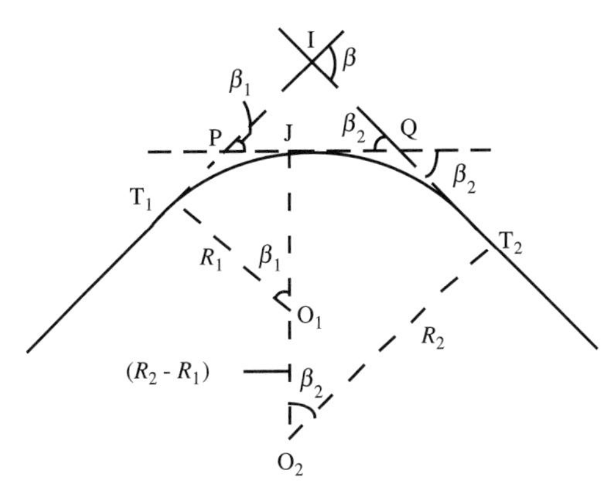

Figure 11.8

Assuming that the main intersection angle β is fixed, there are eight remaining components, R_1, R_2, T_1I, T_2I, β_1, β_2, L_1 and L_2. Any three of these must be fixed for the curve to be defined uniquely. Five equations must therefore be established in order to obtain the remaining components. Three are obtained simply from

$$\beta = \beta_1 + \beta_2$$

$$L_1 = R_1\beta_1$$

$$L_2 = R_2\beta_2$$

Consider the polygon $O_1O_2T_2IT_1$, in which the algebraic sum of the projections of the sides onto any one side must be zero.

Consider first the projection onto O_2T_2. We have

$$R_2 + 0 - IT_2 \sin\beta - R_1 \cos\beta - (R_2 - R_1) \cos\beta_2 = 0$$

and

$$IT_1 \sin\beta = R_2 - R_1 \cos\beta - (R_2 - R_1) \cos\beta_2$$

Similarly, projecting onto O_1T_1,

$$IT_2 \sin\beta = R_1 - R_2 \cos\beta + (R_2 - R_1) \cos\beta_1$$

Example 4

Given $\beta = 75°$, $\beta_1 = 30°$, $R_1 = 800$ m and $R_2 = 1000$ m, determine β_2, L_1, L_2, IT_1 and IT_2.

$$\beta_2 = 45°, L_1 = 418.9, L_2 = 785.4, IT_1 = 674.5 \text{ and } IT_2 = 739.6$$

11.13 Reverse circular curves

From traffic considerations it is desirable to introduce a section of straight and transition curves between the branches of a reverse curve, so the pure reverse curve is rarely used in practice. It must be considered, however, as the treatment is the same in principle as applied to compound curves.

It will be seen from Figure 11.9 that the centres O_1 and O_2 lie on the straight line O_1JO_2 which is perpendicular to the common tangent PJQ. O_1JO_2 does not in general, pass through the main intersection point I of the running-on and running-off tangents T_1P and QT_2. Note that I may be at some considerable distance (infinite if T_1P and QT_2 are parallel) from the curves, and will commonly be inaccessible. The value of the main intersection angle will be known, being the difference in bearing between the lines PT_1 and QT_2.

There are eight remaining components of the compound curves (R_1, R_2, T_1I, T_2I, β_1, β_2, L_1 and L_2) of which three must be fixed to define a unique curve system. Five equations must be established to obtain the remaining components. From triangle IPQ (Figure 11.9),

$$\beta_2 = \beta_1 + \beta$$

In all cases β will be the difference (+ve) between β_1 and β_2.

As before,

$$L_1 = R_1\beta_1$$

$$L_2 = R_2\beta_2$$

As with compound curves, the sum of the projections of the sides of the polygon $O_1O_2T_2IT_1$ onto any one side will be zero, leading, by an exactly similar process as that used for the compound curve, to the equations

$$IT_1 \sin\beta = R_2 + R_1 \cos\beta - (R_2 + R_1) \cos\beta_2$$

$$IT_2 \sin\beta = R_1 + R_2 \cos\beta - (R_2 + R_1) \cos\beta_1$$

One particular case of reverse curves, when the running-on and running-off tangents are parallel, requires special treatment because the standard equations are not applicable.

From Figure 11.10, the following relationships can be derived.

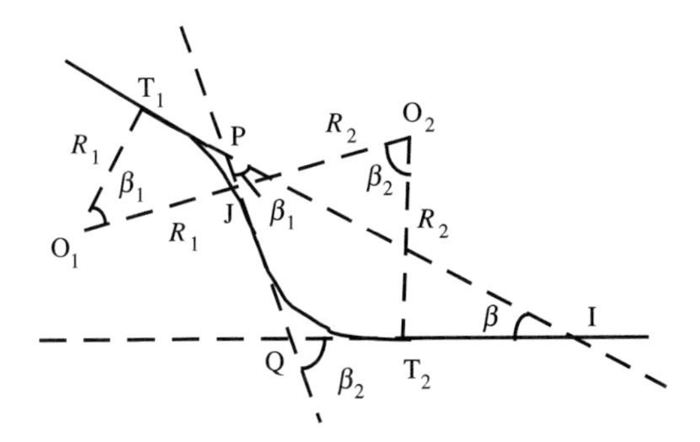

Figure 11.9

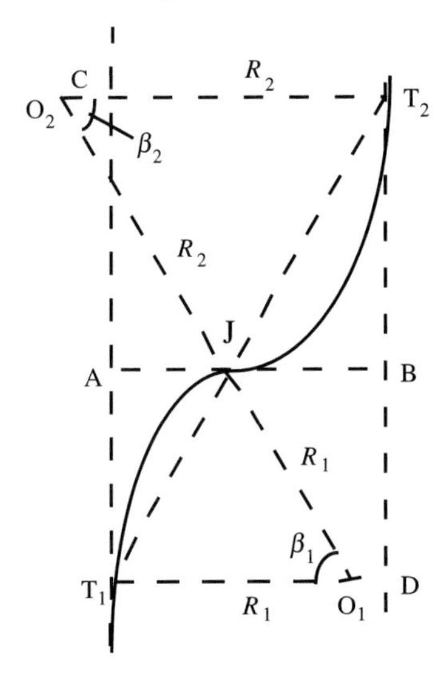

Figure 11.10

$$\beta_1 = \beta_2 = \beta \text{ ; AJ} = R_1(1 - \cos\beta)$$

$$JB = R_2(1 - \cos\beta)$$

$$AB = AJ + JB; T_1A = R_1 \sin\beta$$

$$BT_2 = R_2 \sin\beta$$

$$T_1C = T_2D = (R_1 + R_2) \sin\beta$$

$$T_1J = 2R_1 \sin(\beta/2); JT_2 = 2R_2 \sin(\beta/2)$$

$$T_1T_2 = T_1J + JT_2$$

11.14 Transition and combined curves

Once the rate of change of curvature of a transition curve (hereafter referred to as a spiral) has been fixed (by the methods discussed later), the only component remaining to be found to define the curve uniquely is its length. Consequently, the components of combined curves will be investigated in this section. The following notation will be used in the following discussion, referring to Figure 11.11.

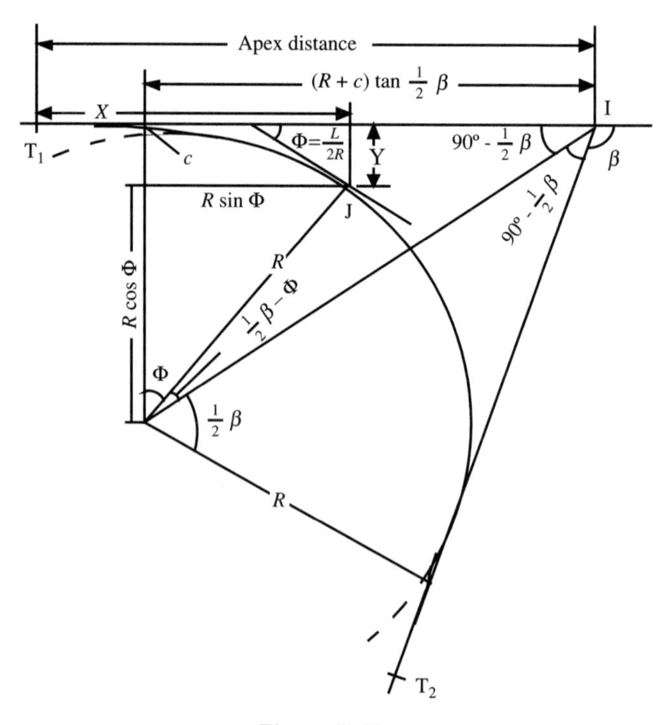

Figure 11.11

TJ = L is the total length of each spiral;

R is the radius of the central circular arc (and of course the radius of curvature of the spiral at its junction with the circular arc);

c is the shift;
T denotes springing point of spiral from straight;

J denotes the junction point between spiral and circular arc;

ϕ is the angle between two tangents to the curve separated by an arc of length s;

x, y (θ , l) are the Cartesian (polar) coordinates of a point on the curve relative to the initial springing point as origin and the original straight as x-axis;

Φ, X, Y, and Θ refer to the particular values of ϕ, x, y, θ at the point J (i.e. when $s = L$).

To avoid confusion with the tangent lengths for purely circular curves, the distances IT_1, IT_2 etc. will be referred to as 'apex distances'.

In this section, the spiral used will be the true transition curve (the clothoid), which, if the correction terms are ignored, becomes the cubic parabola. Some of the relationships developed below are true whatever curve is used. Where this is so it will be indicated in the text.

Shift

For it to be possible to insert a spiral between a straight and a circular curve it is necessary to move the circular curve away from the straight by an amount known as the shift. Similarly, in order to insert a spiral between two branches of a compound curve, it is necessary to move the circular curve with the smaller radius inwards or the circular curve with the largest radius outwards i.e. to separate the two curves. It should be noted that in this latter case, the effect of inserting a spiral is to change the position of the common tangent. In the unusual circumstances that this is not practicable, it will be necessary to substitute a completely new curve system.

For the single circular curve it will be seen at once from Figure 11.11 that

$$R + c = R\cos\Phi + Y$$

therefore

$$c = Y - R(1 - \cos\Phi) \tag{11.7}$$

This is true whatever spiral is used. For the clothoid,

$$y = \frac{s^3}{6LR} + \text{higher order terms}$$

so that when $s = L$,

$$Y = \frac{L^2}{6R} + \ldots$$

But

$$\phi = \frac{s^2}{2LR}$$

therefore

$$\Phi \approx \frac{L}{2R}$$

But

$$1 - \cos\Phi \approx \frac{\Phi^2}{2} \approx \frac{L^2}{8R^2}$$

Thus we get

$$c \approx \frac{L^2}{24R} \tag{11.8}$$

Notice when $s = L/2$,

$$y = \frac{L^2}{48R} = \frac{c}{2}$$

Thus the shift and curve approximately bisect each other: a useful check when setting out.

Apex distance

For the circular curve it can be seen from Figure 11.11 that

$$\mathrm{T_1I} = X + (R + c)\tan(\beta/2) - R\sin\Phi \tag{11.9}$$

This is true whatever spiral is used. For the clothoid,

$$x = s + \text{higher order terms}$$

therefore

$$X = L + \text{higher order terms}$$

and

$$\Phi = \frac{L}{2R}$$

giving

$$X - R\sin\Phi = \frac{L}{2} + \text{higher order terms}$$

$$\mathrm{T_1I} = \mathrm{T_2I} = L/2 + (\mathrm{R} + \mathrm{c})\tan(\beta/2) + \text{higher order terms} \tag{11.10}$$

It can be shown (Allan 1997) that the first small term in Equation (11.10)

$$\frac{L^3}{240R^2}$$

is generally negligible.

Length of combined curve

It has been shown that the angle consumed by one (clothoid) spiral $\Phi = L / 2R$ so that the angle consumed by two identical spirals $2\Phi = L / R$, leaving the angle available for the central circular curve as $(\beta - L / R)$. The length of the central circular arc is therefore

$$R(\beta - L/R) = R\beta - L$$

and the total length of the combined curve is

$$R\beta + L$$

Example 5

Referring to Figure 11.11 and employing the notation of the previous sections, assume $\beta = 75°$, $R = 500$ m, $L = 400$ m (LR = 200 000) and chainage I = (28 + 63.20) m.

It is required to find the position and chainage of the tangent points, and the chainage of the junction points.

Shift:

$$c = \frac{L^2}{24R} \approx 13.333$$

The omitted correction is –0.076.
Apex distance:

$$T_1I = T_2I = \frac{L}{2} + (R + c)\tan\frac{\beta}{2} - \frac{L^3}{240R^2} = 200 + 393.84 - 1.07 = 592.77$$

Length of circular arc:

$$R\beta + L = 654.50 - 400 = 254.50$$

Chainages:
T_1 = 28 + 63.20 – (5 + 92.77) = 22 + 70. 43
J_1 = 22 + 70. 43 + (4 + 0) = 26 + 70. 43
J_2 = 26 + 70. 43 + (2 + 54.50) = 29 + 24.93
T_2 = 29 + 24.93 + (4 + 0) = 33 + 24.93

11.15 Wide roads and dual carriageways

For roads having a large formation width, in particular for dual carriageways, it will be necessary to set out the individual kerbs and embankment edges independently as the method of off-setting from the centre-line is only applicable to narrow roads. This makes no difference to the setting out of the curves themselves, but does affect the calculations of the curve components. Examples include:

1. where the carriageway retains constant width throughout;
2. where the carriageway is widened gradually along the spiral but maintains constant (increased) width round the circular portion. This is to meet traffic engineering considerations.

In both cases, the circular arcs remain truly concentric. (It should be noted that these have the effect that spirals used for each kerb utilize different speed (*LR*) values; methods which use the same speed values are possible but do not satisfy the other conditions set out above.) These matters are mentioned to illustrate the highly complex design of motorway interchanges. The provision of the setting out data is something for which computer processing is necessary to calculate coordinates for use by the setting out surveyor.

11.16 Field setting out

As mentioned at the beginning of the chapter, a system of coordinates is required for all setting out. The three most convenient are rectangular coordinates, polar coordinates and chords and deflection angles. Each of these will now be considered briefly.

Although in practice route curves are nearly always combined curves, in the following sections the setting out procedure for circular curves and spirals will be kept separate as they are set out separately. The procedure at the junctions between them will be dealt with in the section relating to spirals.

Circular curves by rectangular coordinates

These methods are convenient for simple equipment using tapes and optical squares for short curves, and are particularly useful for such situations as urban road intersections where the centres are inaccessible. The disadvantage of using tapes, however, is that the chords are not equal in length.

With total stations, rectangular coordinates are converted to local polar coordinates referred to any suitable reference target.

Taped offsets from the tangent

From Figure 11.12 using the tangent point as the origin, the distance along the tangent as the x coordinate and the rectangular off-set as the y coordinate,

$$R^2 - x^2 = (R - y)^2 = R^2 + y^2 - 2Ry$$

$$y \approx \frac{x^2}{2R}$$

When R is large compared with chord length C, $x \approx C$ and

$$y \approx \frac{C^2}{2R}$$

and the correction for x is

$$x \approx C - \frac{y^2}{2C}$$

If $R = 100$ m and $C = 20$ m, then $y = 2$ and $x = 19.90$. The approximation gives 1.98, an error of 0.02 m. This is a suitable method for small tasks or infilling detail already set out.

Taped off-sets from the chord

This method is particularly useful for infilling additional chord points on any circular curve once the main chord points have been established. Refer to Figure 11.13; the midpoint of the chord is taken as origin, distances along the chord as the X coordinates, and rectangular off-sets as the Y coordinates.

From the previous method,

$$y' = \frac{x^2}{2R}$$

Since $y' = Y'$ when $x = C / 2$,

$$Y' = \frac{C^2}{8R}$$

and any value of y

$$y = Y' - y' = \frac{C^2 - 4x^2}{8R}$$

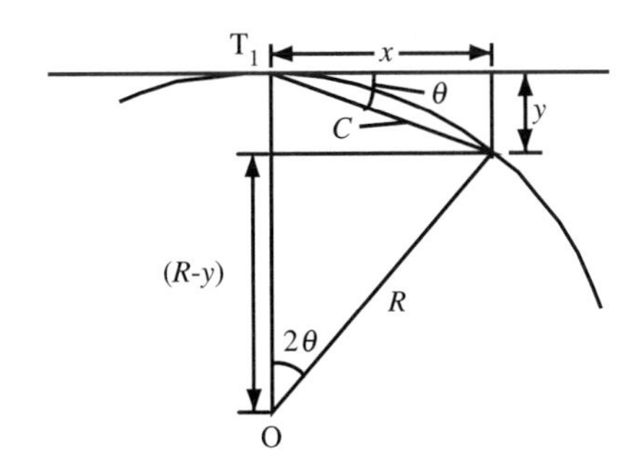

Figure 11.12

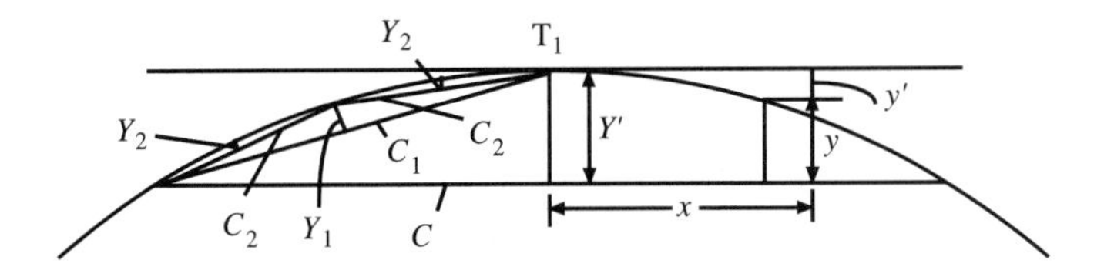

Figure 11.13

Then since chord $C_1 \approx C / 2$,

$$Y_1 \approx \frac{Y'}{4}$$

and so on for

$$Y_n = \frac{Y_{n-1}}{4}$$

giving the general rule of 'halving the chord and quartering the offsets'.

11.17 Circular curve ranging by theodolite

The first step in setting out the curve must be to establish the tangent points, which will involve measuring or deriving the intersection angle, calculating the tangent distances and pegging the tangent points and determining their chainage, as has already been described.

Ranging the curve

The basic geometry shown in Figure 11.14 applies to three methods of setting out a circular curve:

1. From one theodolite situated at a tangent point such as T_1 and taping chords C along the curve;
2. From two theodolites situated at each tangent point;
3. By radial survey from a total station at either tangent point.

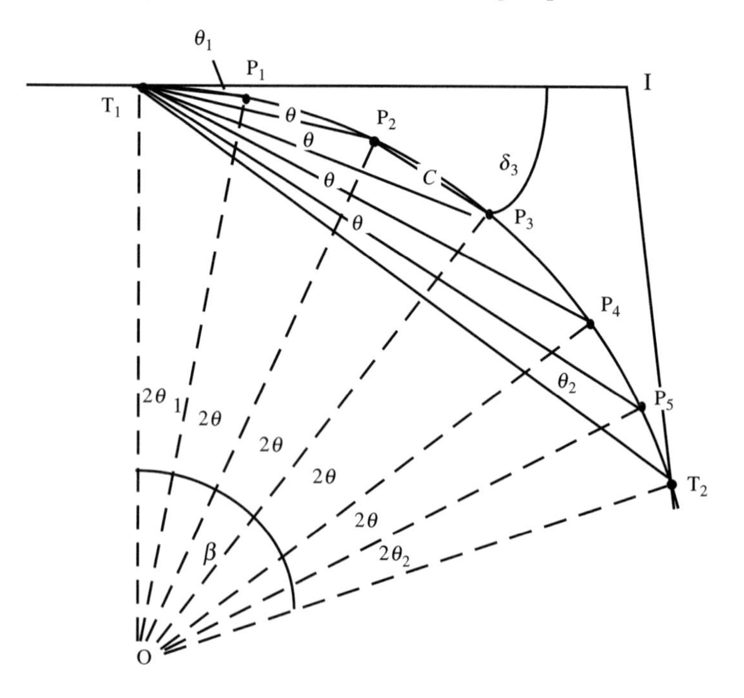

Figure 11.14

The last two methods can be used from any points whatsoever, as stated earlier. The traditional methods of working close to the curve will be described, as they are still useful and also serve to describe the general procedure.

We will focus the description on the first method, by one theodolite and taped chords. It will be assumed that running chainage is to be maintained, and that the theodolite is to be set up at one of the two tangent points. Not shown in Figure 11.14 are $C_1 = T_1P_1$ and $C_2 = T_2P_2$, the initial and final subchords. All the remaining chords are equal to C. The deflection angles δ are referred to the running-on tangent T_1I, and P_1, P_2 etc. are the chord points. Only δ_3 is shown in the figure.

The deflection angles have to be calculated so that the theodolite circle may be set on I and each angle turned as the chords are taped from a previously fixed point. This chain process is bound to accumulate some slight error, but if a mistake is made, a clear indication of it will be given. Thus the work is quality assured and controlled. (If intersection point I is not visible, a reference point at a calculated bearing from I is used instead.)

The fact that the sum of deflection angles is equal to half the intersection angle is useful as a check on the arithmetic of the calculations.

If the circular curve is defined by its ‘degree’ instead of its radius, and $C = 100$ m (which is common) then as previously shown $\theta = 0.5$ D and the calculation of the deflection angles is considerably simplified. This is the main advantage of the degree method. From Figure 11.14,

$$T_1OP_1 = \theta_1 = \arcsin(C_1 \ / \ 2R)$$

$$P_1OP_2 = \theta = \arcsin(C \ / \ 2R)$$

$$P_5OT_2 = \theta_2 = \arcsin(C_2 \ / \ 2R)$$

Then the deflection angles δ are

$$\delta_1 = \theta_1$$

$$\delta_2 = \theta_1 + \theta$$

$$\delta_3 = \theta_1 + 2\theta$$

and so on until all sum to $\beta/2$. From the above equations, a table of deflection angles for setting out the curve can be prepared as in the following example.

Example 6

Two straights making a deflection angle of 75° intersect at I (chainage 28 + 6.32 m). A circular curve of 100 m radius, deflecting left, is to be set out in 10 m chords. Tabulate the setting out data assuming that the instrument is set up at T_1 and that the initial reading of the horizontal circle, when the instrument is bisecting I, is zero.

Tangent length = 60.88

Arc length = 130.90

Chainages: $T_1 = 22 + 5.44$; $T_2 = 35 + 6.34$

$$\theta = 2° \ 51' \ 58''$$

$$C1 = 10 - 5.44 = 4.56; \ C_2 = 6.34$$

$$\theta_1 = \theta \frac{C_1}{C} = 1°18'23''$$

$$\theta_2 = \theta \frac{C_2}{C} = 1°48'59''$$

The curve can be set out from the data displayed in Table 11.1.

Once the instrument has been set up over the tangent point and the intersection point bisected, it is advisable to bisect the other tangent point, and measure the total deflection angle to check that it is in fact equal to half the intersection angle before proceeding any further.

This having been done, lay off the first deflection angle, stretch the tape a distance C_1 from T_1, line in the end of the tape and drive in a peg at P_1. Lay off the second deflection angle, stretch the tape a full chord's length C from P_1, line in the end of the tape and drive in a peg at P_2. Proceed in this manner until the last peg has been driven, and then measure the final subchord C_2. The amount by which this differs from the calculated value will be a measure of the accuracy with which the curve has been set out.

Table 11.1

Chainage 10 m chords	*Deflection angles*
22 + 5.44	
23	1° 18' 23"
24	4° 10' 16"
25	7° 02' 09"
26	9° 54' 02"
27	12° 45' 55"
28	15° 37' 48"
29	18° 29' 41"
30	21° 21' 34"
31	24° 13' 27"
32	27° 05' 20"
33	29° 57' 13"
34	32° 49' 06"
35	35° 40' 59"
35 + 6.34	37° 39' 00"

The accuracy of the method is obviously dependent on the care which is taken in taping the chords and in lining in the pegs, and by the angle at which the two locating position lines meet. During the actual setting out the surveyor has an immediate idea of accuracy. If unhappy about the cut, a remedy has to be used such as moving the theodolite to another station.

Setting out by theodolite intersection

If the ground is unsuitable for taping it may be preferable to use two theodolites (one at each tangent point), lay off the deflection angles from both tangent points simultaneously and locate the chord points by intersection. This of course presupposes that the chord points are all visible from both tangent points. Where the ground is undulating and accurate lining is difficult (whether using one theodolite or two), it will be preferable to change the position of the instrument as in the procedure for obstructed curves. This procedure is also necessary when the intersection angle between the line of sight and the tape becomes small. When the theodolite is set up at T_2 a fresh tabulation of deflection angles must be made, using, of course, the same values of θ_1, θ_2 and θ.

Setting out by polar coordinates

Circular curves can of course be set out using polar coordinates with the tangent point as the origin and the tangent as the reference direction. This is obviously the most convenient method when using a total station. It must be remembered that the design *horizontal* chords l have to be obtained from the *slant* EDM distances S i.e. from $S' = l \sec\alpha$ where α is the slope angle. Many instruments incorporate suitable computer algorithms for this calculation directly in the field, and some permit the prism holder to read the distance at their end of the line. Increasingly, GPS differential fixes are being used to assist conventional methods with total stations in the setting out of planimetric data such as curves. It is most important to tie into a base station surveyed on the same datum as the general design, and transform the GPS coordinates into the same reference system.

Setting out transition and other curves

Although it is feasible to set out complex curves by deflection angles in a similar manner to that described for the circular curve, modern practice is to use a total station and a system of polar coordinates from almost any convenient position close to the site but free from the actual works themselves. These polar coordinates are computed from the Cartesian values of the curves. It should be remembered that mistakes may occur so checks on key points from a second position are desirable and completing a visual check of the curve should be common practice.

Where numerous obstructions exist, or where earthworks or construction plant prevent the use of any permanent ground marks on the curve itself, it is more convenient to set out the curve from stations remote from the curve. In such cases, once the calculations have been made and tabulated, the curve can be quickly re-established whenever required.

The rectangular coordinates of all the chord points on the curve must be calculated, and the coordinates of the permanent stations on the same axes obtained, usually by traversing or by resection. The curve can then be set out either by intersection, or by radiation.

11.18 Design and fitting of horizontal curves

It has already been mentioned that the design of road and railway layouts is the province of the specialist engineer so only general principles are discussed here. It has also been shown that both circular curves and clothoid spirals are defined uniquely if any two of their properties are fixed. For circular curves these two will usually be selected from radius (or degree), length of arc, intersection angle, or tangent length. For spirals they will be selected from minimum radius, length of curve, maximum spiral angle and shift. The combined curve, consisting of two identical spirals and a central circular arc, is the case most commonly met in practice, and as purely circular curves are very much simpler to deal with, such a combined curve only will be assumed in general in this section.

On severely restricted sites, curves may have to be designed from purely geometric considerations, but design is based on traffic requirements for safety and comfort, possibly modified by aesthetic considerations or site restrictions. In most practical cases, the overall intersection angle is predetermined by the general layout, and the problem is finally resolved to determining a suitable radius for the circular arc, and length for the spiral, from the traffic viewpoint.

It will be obvious that from traffic considerations the largest possible radius and longest possible spiral are desirable, but that some restriction will always arise from site conditions or cost. The first problems to tackle, therefore, are the determination of suitable minima for the central radius and the length of the spiral for given traffic speeds.

As the speed of vehicles using a particular road or railway is a variable quantity, a 'design speed' must be defined, which is normally taken as the 85 percentile speed, i.e. the speed that will not be exceeded by 85 % of the vehicles using, or are expected to use, the road.

Minimum safe radius for circular curves

A vehicle travelling round a curve will be subject to a centrifugal acceleration, which must be combated by the super-elevation of the track combined with the reaction of the rails on the wheel flanges in the case of a railway, or by the frictional force between the tyres and the road surface.

In general, the track will have to accommodate vehicles travelling at a wide range of speeds, so the super-elevation ($\tan\theta$) cannot be so large that it will be uncomfortable for slow-travelling or stationary vehicles. For railways, where the vehicles have a low centre of gravity, a maximum value of 0.1 is often adopted, and for roads 1/14.5 (= 0.069).

In Figure 11.15, we resolve the normal force N and the tangential force F horizontally and vertically to give

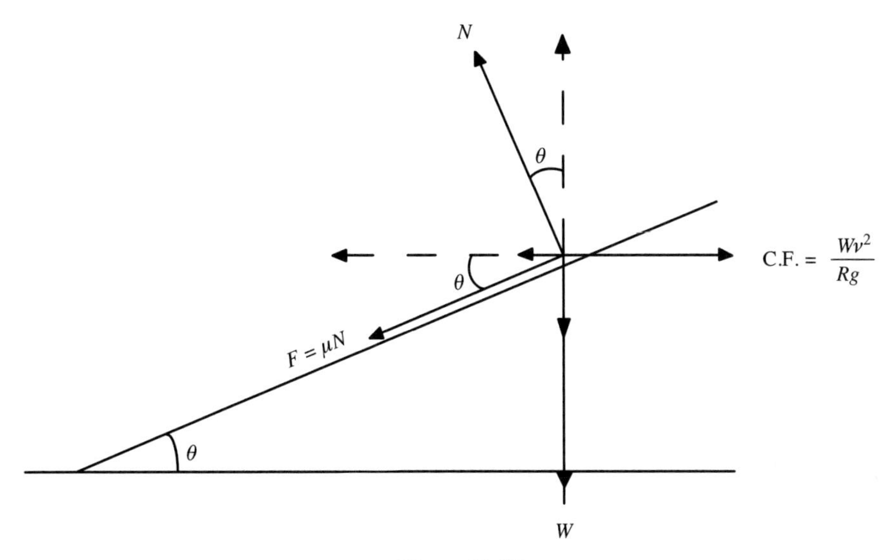

Figure 11.15

$$\frac{Wv2}{gR} = N\sin\theta + F\cos\theta \quad \text{and} \quad W = N\cos\theta - F\sin\theta \tag{11.11}$$

Replacing F by $\mu\ N\ =\ N\tan\lambda$ (where μ is the coefficient of friction and λ the angle of friction) and dividing gives

$$\frac{v^2}{gR} = \frac{\sin\theta + \cos\theta\tan\lambda}{\cos\theta - \sin\theta\tan\lambda} \tag{11.12}$$

Dividing top and bottom of the right hand side by $\cos\theta$ leads to

$$\frac{v^2}{gR} = \tan(\theta + \lambda) \tag{11.13}$$

An ideal situation arises if, for a given speed, there is no tangential force i.e. $F\ =\ 0$ i.e. $\mu\ =\ 0$ and $\lambda\ =\ 0$, where

$$\tan\theta = \frac{v^2}{gR}$$

In practice, some assistance from the frictional force can be expected, and use is made of the empirical formula

$$\tan\theta = \frac{v^2}{gR} - 0.15$$

For design speed 64 km hr^{-1}, and maximum super-elevation 1/14.5, a minimum safe radius is $R = 147$ m.

Minimum length of spiral

Various theories have been put forward for finding a suitable length of spiral, but the most widely accepted is one that limits the rate of change (with respect to time) of centrifugal acceleration to a value between 0.15 m s^{-3} and 0.6 m s^{-3}, depending on the degree of comfort required. A value of 0.3 m s^{-3} is most commonly used.

For the clothoid spiral the rate of change of curvature ($1/\rho$) is constant, so that for constant speed the rate of change of centrifugal acceleration a, where a is given by

$$a = \frac{v^2}{r}$$

will be constant and equal to the change in acceleration divided by the time in which the change takes place i.e.

$$\frac{\mathrm{d}a}{\mathrm{d}t} = \frac{\left(\frac{v^2}{R} - 0\right)}{t}$$

where R is the radius at the end of the curve and t is the time taken by a vehicle travelling round the curve at constant speed v. But t = L/v, so that

$$\frac{\mathrm{d}a}{\mathrm{d}t} = \frac{v^3}{LR}$$

$$\frac{\mathrm{d}a}{\mathrm{d}t} = \frac{(V/3.6)^3}{LR} \qquad (11.14)$$

where V is in kilometres per hour. It has already been shown that the product of arc and radius LR is constant for the spiral, and fixes the shape of the curve. It is known as the 'speed value' of the curve.

Once the central radius R has been fixed, the minimum length of the curve can be found from the minimum LR value (or speed value). Assuming $\mathrm{d}a/\mathrm{d}t = 0.3$ m s^{-3} for $v = 64$ km hr^{-1}, $LR = 18\,728$ m = 18 800 approximately.

If the radius of central circular curve is 300 m, the minimum length of spiral = 62.7 m. If the radius of central circular curve is the minimum for 64 km hr^{-1} = 147 m, the minimum length of spiral is

$$\frac{18800}{147} = 127.6 \text{ m}$$

Provided site conditions allow, it will always be preferable for both L and R to exceed the minima.

11.19 Other design considerations

The intersection angle being assumed fixed, and the minimum values of R and LR having been found, the problem remaining is to find a combination of L and R to suit the particular case. Two factors will influence the final choice of L and R.

If there are no site restrictions, the ideal combined curve is one in which the central circular arc is approximately equal to the length of each of the spirals. In this case, we have the angle consumed by one spiral is $L \;/\; 2R$ and by the circular arc is $L \;/\; R$ so that the intersection angle β is given by

$$\beta = \frac{L}{2R} + \frac{L}{R} + \frac{L}{2R} = \frac{2L}{R}$$

If there are site restrictions, there may not be space for a central circular arc, in which case the angle consumed by the spirals will equal the intersection angle i.e.

$$\beta = \frac{L}{R}$$

11.20 Vertical curves

Whenever two gradients intersect on a road or railway it is obviously necessary from the traffic point of view that they should be connected by a vertical curve, so that the vehicles can pass smoothly from one gradient to the next. Although in practice road and rail gradients are comparatively flat and it is therefore relatively unimportant what type of curve (circular, parabolic or sinusoidal) is used, it is usually assumed that a curve having a constant rate of change of gradient (i.e. a parabola) is the most satisfactory, and it so happens that this is the easiest to calculate. For modern high-speed roads, some engineers prefer a curve in which the rate of change of gradient is uniform along the curve, but as roads designed to these standards have also very flat gradients, this is usually regarded as an unnecessary refinement. The simple parabola only will be considered.

Gradients are expressed as percentages, for example a +4 % gradient is one in which the level rises 4 units vertically in 100 units horizontally. A positive gradient indicates a rising slope and a negative gradient a falling slope. This method of representing gradients not only has the advantage of simplifying the calculation of rises and falls in level (e.g. a +4 % gradient 260 m long rises $0.04 \times 260 = 10.4$ m), but also, apart from a factor of 1/100, such gradients are identical with those used in coordinate geometry or calculus. For example, if the X axis denotes the horizontal and the Y axis the vertical, then a 4 % gradient is a mathematical gradient of 0.04 or $dy/dx = 0.04$.

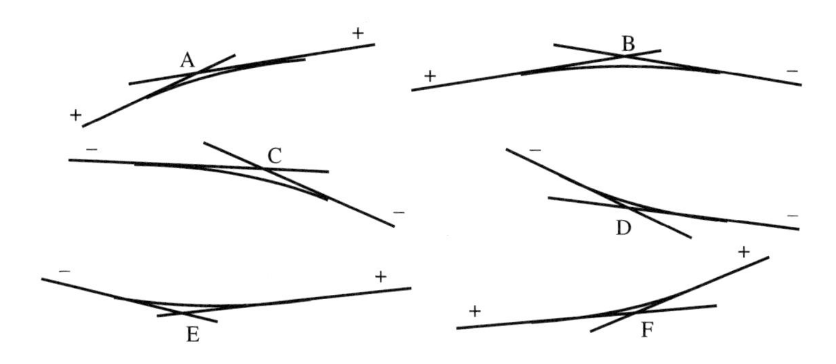

Figure 11.16

Because of the well-known simple geometry of the parabola, many different methods are put forward for calculating the properties of vertical curves, but the approach here is selected because it is felt that it is the simplest to understand and because it will readily yield a solution to any problem that arises. In some cases a quicker solution can be found to a particular problem by utilizing simple geometry, but the same methods applied to another problem may become very involved.

11.21 Types of vertical curve

Vertical curves can take various different forms depending on the sign and magnitude of the intersecting gradients, passing from left to right over the curves (Figure 11.16).

It will be observed that when there is a decrease in gradient, a curve convex downwards is formed, and when there is an increase in gradient a curve concave upwards is formed. The former are known as 'summit' curves, and the latter as 'valley' curves. It should be noted, however, that a true summit (highest point), or a true valley (lowest point), can only occur when there is a change of sign between the two gradients. It will be obvious that the gradient at the highest or lowest point is zero.

11.22 Simple parabola

Assume rectangular coordinates, with X-axis horizontal and Y-axis vertical as in Figure 11.17.

The basic requirement for the curve is that the rate of change of gradient (with respect to distance) is constant. This can be expressed in two different ways:

1. $$\frac{d^2y}{dx^2} = K$$

2. $$\frac{q-p}{L} = K$$

To explain (1), *since the rate of change of gradient is constant*, it is given by

$$\frac{\text{the change in gradient}}{\text{the distance over which it takes place}}$$

where the initial and final gradients dy/dx are denoted by p and q respectively, and the horizontal length of the curve by L.

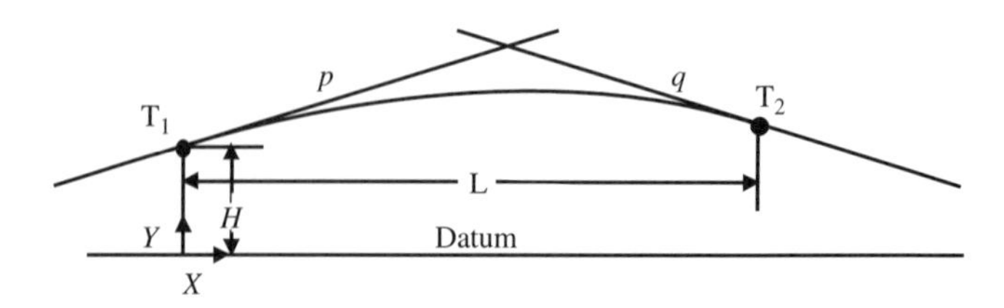

Figure 11.17

The gradient at any point on the curve can be found from

$$\frac{\mathrm{d}y}{\mathrm{d}x} = \int \frac{\mathrm{d}^2 y}{\mathrm{d}x^2}\,\mathrm{d}x = \int \mathrm{K}\,\mathrm{d}x = \mathrm{K}x + A$$

If point T_1 is taken as the origin for the X axis, then at T_1, $x = 0$ and $\mathrm{d}y/\mathrm{d}x = p$, where $A = p$ and the equation of the curve becomes

$$\frac{\mathrm{d}y}{\mathrm{d}x} = \mathrm{K}x + p$$

or

$$\frac{\mathrm{d}y}{\mathrm{d}x} = \frac{(q-p)x}{L} + p$$

Note that when $x = L$, $q = \mathrm{K}L + p$, i.e.

$$\mathrm{K} = \frac{q-p}{L}$$

The level of any point on the curve can be found from

$$y = \int \frac{\mathrm{d}y}{\mathrm{d}x}\,\mathrm{d}x = \int (\mathrm{K}x + p)\,\mathrm{d}x = \frac{1}{2}\mathrm{K}x^2 + px + \mathrm{B} \qquad (11.15)$$

If T_1 is taken as the origin for the Y-axis, then at T_1, $x = 0$, $y = 0$, and so $\mathrm{B} = 0$. It is usually more convenient to take the level datum as the origin for the Y-axis, so that when $x = 0$, $y = \mathrm{B}$ = reduced level of T_1, H say.

In this case, the equation becomes

$$y = \frac{(q-p)x^2}{2L} + px + H \qquad (11.16)$$

This is an equation of the form $y = \mathrm{a}x^2 + \mathrm{b}x + \mathrm{c}$ (a parabola with its axis vertical) where a is 0.5 × rate of change of gradient, b the initial gradient at T_1 and c the reduced level of T_1. In evaluating these coefficients, care must be taken with their signs. The sign of b and c will be perfectly obvious in a practical problem, but the following will be of use in ensuring the correct sign for a.

The equation $y = \mathrm{b}x + c$ ($= \mathrm{p}x + H$) is the equation of the straight line passing through the point T_1 with a gradient p i.e. the equation of the initial gradient tangent. If the curve is a summit curve it will fall away from this tangent and so the sign of a will be negative. If the curve is a valley curve, it will rise above the initial tangent and so the sign of a will be positive. This will be demonstrated by numerical examples for each case, illustrated in Figures 11.16.

(a) If $p = +4$ %, $q = +3$ %, then

$$\mathrm{K} = \frac{3-4}{100L} \quad \text{and} \quad y = -\frac{x^2}{200L} + \frac{4x}{100} + H$$

(b) If $p = +4$ %, $q = -3$ %, then

$$K = \frac{-3-4}{100L} \quad \text{and} \quad y = -\frac{7x^2}{200L} + \frac{4x}{100} + H$$

(c) If $p = -3$ %, $q = -4$ %, then

$$K = \frac{-4-3}{100L} \quad \text{and} \quad y = -\frac{x^2}{200L} + \frac{3x}{100} + H$$

(d) If $p = -4$ %, $q = -3$ %, then

$$K = \frac{-3+4}{100L} \quad \text{and} \quad y = \frac{x^2}{200L} - \frac{4x}{100} + H$$

(e) If $p = -4$ %, $q = +3$ %, then

$$K = \frac{3+4}{100L} \quad \text{and} \quad y = \frac{7x^2}{200L} - \frac{4x}{100} + H$$

(f) If $p = +3$ %, $q = +4$ %, then

$$K = \frac{4-3}{100L} \quad \text{and} \quad y = \frac{x^2}{200L} + \frac{3x}{100} + H$$

11.23 Setting out data for vertical curves

It will be clear from the above derivation of equations that when the gradients have been fixed, only one other factor (either the horizontal length of the curve or the rate of change of gradient) is required to define the curve uniquely. The choice of these factors is dealt with later and for the remainder of the present section it will be assumed that the curve has been defined, and that all that remains is to determine the setting out data.

Two properties of the parabolic vertical curve must be established to assist in calculating the data. In Figure 11.18, let the horizontal distance from T_1 to the intersection point I be l and the equation of the curve be as given in Equation (11.16) i.e.

$$y = \frac{(q-p)x^2}{2L} + px + H$$

Substituting L for x, the reduced level of point T_2 becomes

$$\frac{1}{2}(p+q)L + H$$

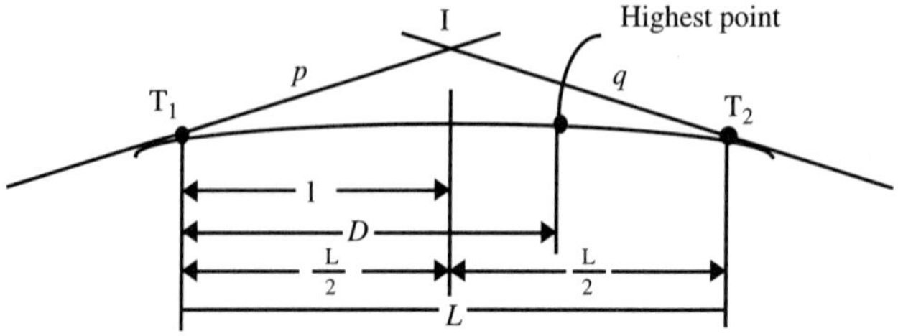

I
Highest point
p
q
T_1
T_2
l
D
$\frac{L}{2}$
$\frac{L}{2}$
L

Figure 11.18

But the reduced level of T_2 is also given by

$$H + pl + q(L - l)$$

Equating these two expressions for the reduced level of T_2 gives

$$l = \frac{L}{2}$$

This very important relationship, that the horizontal distances from the tangent points to the intersection point are equal, is of considerable use in solving vertical curve problems, and yet is often overlooked.

For the two cases where there is either a true summit or valley i.e. where the gradients are zero, we derive another useful property. In Figure 11.18, D denotes the distance to the highest point of the curve (or the lowest if the curve is a valley). (It should be noted that this will not coincide with I unless $p = -q$.) Remembering that the gradient at the highest (or lowest) point is zero and making use of the definition that the rate of change of gradient (which is constant) is given by the change of gradient divided by the distance in which the change takes place

$$\frac{0 - p}{D} = \frac{q - 0}{L - D} = \frac{q - p}{L}$$

from which

$$D = \frac{-p}{q - p} L \quad \text{and} \quad L - D = \frac{q}{q - p} L$$

can be derived.

It will have been seen (Figure 11.16) that a true summit (highest point) or true valley (lowest point) only occurs when there is a change of sign between p and q, so that the term $q - p$ in the above equations will always be given by the numerical sum of the two gradients, and that the signs in the two equations are consistent.

If at a summit, $p = +4$ % and $q = -3$ %, then

$$D = \frac{4}{7} L \quad \text{and} \quad L - D = \frac{3}{7} L$$

and at a valley, $p = -4$ % and $q = +3$ %, then

$$D = \frac{4}{7} L \quad \text{and} \quad L - D = \frac{3}{7} L$$

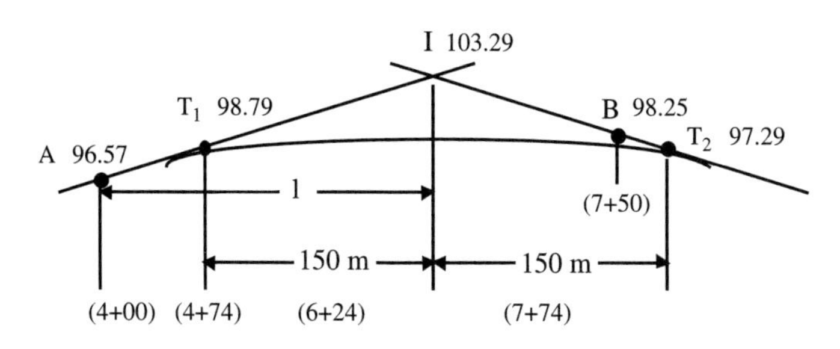

Figure 11.19

Location of tangent points

This is best illustrated by numerical examples; refer to Figure 11.19 in each case.

Example 7

A rising gradient of +3 % and a subsequent falling gradient of –4 % are to be connected by a parabolic vertical curve of horizontal length 300 m. The reduced level of a point A at chainage 4 + 00 on the first gradient is 96.570 m, and the reduced level of a point B at chainage 7 + 50 on the second gradient is 98.250 m. Find the chainage and reduced levels of the tangent points.

Let the horizontal distance from A to the intersection point I be s. Then since the distance AB is 750 – 400 = 350 m, the reduced level of I is given by

$$96.570+\frac{3s}{100}=98.250+4\frac{(350-s)}{100}$$

$$s = 224.0 \text{ and } 350 - s = 126.0$$

Chainage I = 4 + 00 + 224.0 = 6 + 224.0

Chainage T_1 = 6 + 224.0 – $s/2$ = 4 + 74.0

Chainage T_2 = 6 + 224.0 + $s/2$ = 7 + 74.0

$$\text{Reduced level of I}=96.570+\frac{3\times 224.0}{100}=103.290$$

$$\text{Check from } T_2 = 98.250+\frac{4\times 126.0}{100}=103.290$$

$$\text{Reduced level of } T_1 = 103.290-\frac{3\times 150}{100}=98.790$$

$$\text{Reduced level of } T_2 = 103.290-\frac{4\times 150}{100}=97.290$$

Example 8

Points A, B, C and D, data for which are given below, lie on two intersecting gradients which are to be connected by a parabolic vertical curve of horizontal length

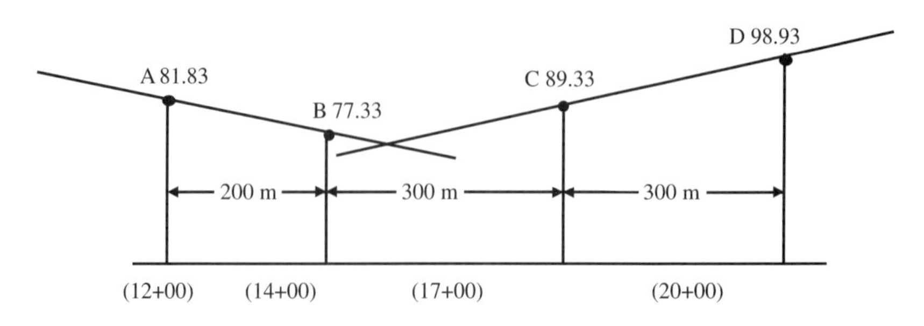

Figure 11.20

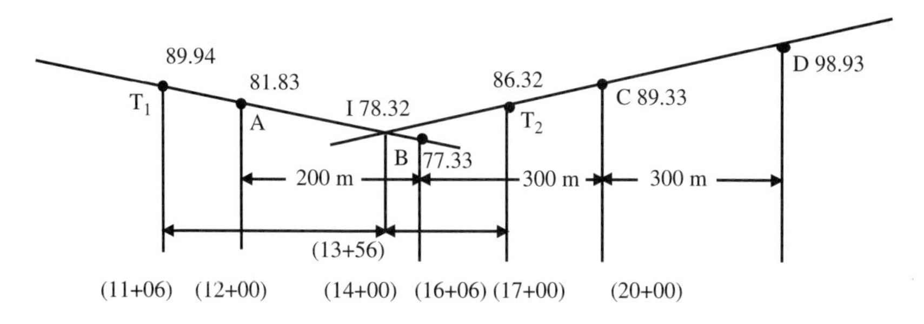

Figure 11.21

500 m (see Figure 11.20). Determine the chainage and reduced level of the tangent points.

Point	*Chainage*	*Reduced level*
A	12 + 00	81.830
B	14 + 00	77.330
C	17 + 00	89.330
D	20 + 00	98.930

Horizontal distances are AB = 200 m, BC = 300 m and CD = 300 m, thus the gradients are AB = –2.25 % and CD = +3.2 %. Let the distance from B to I be s, then the reduced level of I from two points gives

$$77.330 - \frac{2.25s}{100} = 89.330 - \frac{3.2 \times (300 - s)}{100}$$

and

$$s = -44.0 \text{ and } 300 - s = 344.0$$

Since s turns out to be negative the point I lies to the left of B as in Figure 11.21.

Chainage I = 14 + 00 – 44.0 = 13 + 56.0

Chainage T_1 = 13 + 56.0 – 250.0 = 11 + 06.0

Chainage T_2 = 13 + 56.0 + 250.0 = 16 + 06.0

Reduced level of I = 77.330 + 2.25 × 44.0 / 100 = 78.320

Reduced level of T_1 = 78.320 + 2.25 × 250.0 / 100 = 83.940

Reduced level of T_2 = 78.320 + 3.2 × 250.0 / 100 = 86.320

11.24 Reduced levels on curve

Once the chainage and reduced level of the tangent points have been determined, it remains to determine the values of the spot levels along the curve. This is normally done for a series of equidistant points, usually referred to as 'chord' points, although the horizontal distances between them are in fact the horizontal components of the chords.

The calculations are simplified if the curve is set out in equal chords, and the total horizontal length of the curve is an exact multiple of the chord length. In general, both these conditions cannot be satisfied if running chainage is to be preserved, but a convenient compromise can be obtained if the length of the curve is chosen to make the final chord only a short chord.

It should be noted that, for a parabola, second differences are constant, and this provides an alternative method of calculating the curve levels. For practical purposes, however, it is usually suitable to calculate levels to the nearest 0.005 m but to achieve this by the difference method it is usually necessary to calculate the second differences to two further decimal places, so that very little time is saved, unless the second difference is exact to 0.005. The calculation of the second differences acts as a valuable check.

Example 9

In this example the running chainage is preserved, and the curve length is not adjusted. The levels are calculated for the 50 m even chainage points for the curve used in the example of Figure 11.19 in which $p = +3$ %, $q = -4$ % and $L = 300$.

Chainage $T_1 = 4 + 74$

Chainage I = 6 + 24

Chainage $T_2 = 7 + 74$

Reduced level of $T_1 = 98.79$

Reduced level of I = 103.29

Reduced level of $T_2 = 97.29$

The equation of the curve is

$$y = ax^2 + bx + c$$

Table 11.2

Chainage	*x*	*ax*^{2}	*bx*	*ax*^{2} + *bx*	*Reduced levels*	*First differences*	*Second differences*
4 + 74	0	0	0	0	98.79		
5 + 00	26	–0.08	0.78	0.70	99.49	0.91	
5 + 50	76	–0.67	2.28	1.61	100.40	0.32	–0.59
6 + 00	126	–1.85	3.78	1.93	100.72	–0.26	–0.56
6 + 50	176	–3.61	5.28	1.67	100.46	–0.85	–0.59
7 + 00	226	–5.96	6.78	0.82	99.61	–1.43	–0.58
7 + 50	276	–8.89	8.28	–0.61	98.18		
7 + 74	300	–10.50	0.00	–1.50	97.29		

where

$$a = \frac{q-p}{2L} = \frac{-4-3}{200 \times 300} = 1.166 \times 10^{-4}$$

and

$$b = p = \frac{3}{100} \quad \text{and} \quad c = H = 98.79$$

The levels of the curve are tabulated in Table 11.2, where columns 6, 7 and 8 represent the reduced levels of points on the curve and the first and second differences between these levels respectively.

11.25 Choice of design constants for vertical curves

It has been shown that once the value of the two intersecting gradients and the (vertical and horizontal) position of their intersection point have been fixed, only one further property is required to define a vertical curve uniquely. For the purposes of comparison, the most convenient additional property to fix is the (horizontal) length. The determination of a suitable length for a particular vertical curve is the responsibility of the road engineer, but the principles influencing the choice are included here. It will be appreciated that when the length of the curve is derived from traffic considerations, these will be minimum lengths, and the actual curves will usually be longer even if only to obtain a whole number of chords or to preserve running chainage. Consequently, any calculations made to assist in the choice of length need only be approximate.

Safety and free flow of traffic are the first considerations in traffic engineering. With this in view, comfort and prevention of stress on vehicle and driver are the main objectives for valley curves. While these are still of importance for summits, visibility over the crest must also be taken into account. In most practical cases a longer curve will be required, where visibility is the overriding criterion.

Design by limitation of vertical acceleration

A rational method of design takes into account the design speed of the road i.e. the 85 percentile speed or the speed that is not exceeded by 85 % of the vehicles using the road. For safety and comfort the vertical acceleration must be limited to a reasonable value.

Design by vision distance

For summits, in addition to applying other criteria, the question of visibility over the brow of the hill must also be considered as mentioned above. The *vision distance D* is defined as being the length of the sight line between two points at the driver's eye level h above the road (see Figure 11.22), h normally assumed to be 1.143 m (2.75 ft). Assuming a circular arc for the vertical curve we have its radius R limited to a minimum value given by:

$$h \approx \frac{0.5D^2}{2R}$$

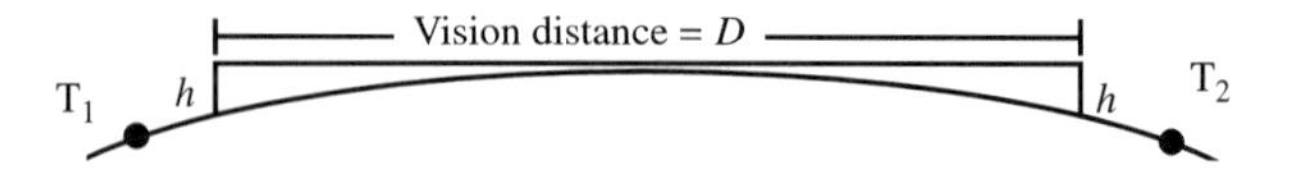

Figure 11.22

therefore

$$\frac{p-q}{L}=\frac{1}{R}=\frac{8h}{D^2} \quad \text{or} \quad L=\frac{D^2(p-q)}{8h}$$

L cannot be less than the vision distance D for an acceptable curve at a given design speed. The required length of this sight line for two-way roads (single carriageway) is based on the minimum distance required for overtaking, and for one-way roads (dual carriageway) on the minimum stopping distance, both of these being dependent on the design speed. Typical values are given in Table 11.3.

Curve controlled by fixed level

In certain cases none of the above criteria will apply and the curve may have to be designed to pass above or below some point of fixed level, for example, under or over a bridge. A numerical example can demonstrate this problem (see Figure 11.23).

Table 11.3 *Vision distance*

Design speed		*Two-way*		*One-way*	
mph	kph	ft	m	ft	m
40	65	950	290	300	91
50	80	1200	366	425	129
60	96	1400	427	650	198
70	113	1650	503	950	290

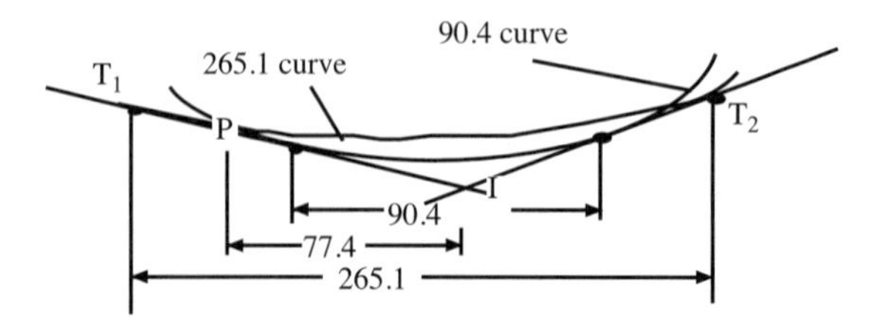

Figure 11.23

Example 10

A falling gradient of 1.8 % is followed by a rising gradient of 2.5 %, their intersection occurring at chainage 8 + 73.2 and RL 72.56. In order to provide sufficient clearance under an existing bridge, the RL at chainage 7 + 95.8 on the curve must not be higher than 74.20.

Find the greatest length of simple parabolic vertical curve that can be used. Let P be the control point under the bridge. Distance PI is given by

$$\text{PI} = (8 + 73.2) - (7 + 95.8) = 77.4$$

$$\text{RL of } T_1 = 72.56 + \frac{1.8L}{200} = H$$

The equation of curve is

$$y = H - \frac{1.8x}{100} + \frac{(2.5+1.8)x^2}{200L}$$

$$\text{RL of P} = 74.20 = 72.56 + 0.009L - 0.018(0.5L - 77.4) + 0.0215\frac{(0.5L-77.4)^2}{L}$$

Hence

$$L^2 - 355.5L + 23963 = 0$$

which yields the solution

$$L = 265.1 \text{ or } 90.4$$

Since 0.5 × 90.4 is less than 77.4, the second solution is impracticable. In problems of this kind a quadratic equation will always occur, producing two solutions as shown in Figure 11.23. Common sense, aided where necessary by a rough sketch, will indicate which solution is applicable.

11.26 Excavations for setting out curves

Obviously, curves to be set out to a design that does not follow the existing ground will require the ground to be excavated or filled as part of the process. The detail of formations and their establishment on the ground are based on the shapes of horizontal and vertical curves.

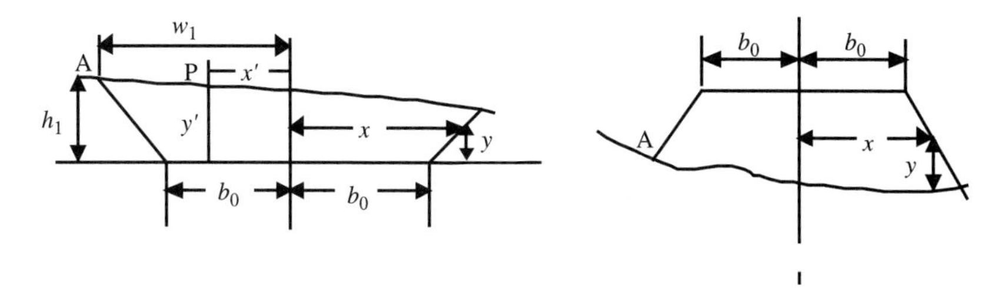

Figure 11.24

Slope staking

Slope staking is a trial-and-error process for locating and staking, in the field, the points at which the proposed side slopes of cuttings or embankments will intersect the existing ground surface. It will be seen that the depths of cut and fill at these points, as well as the cross-section half-breadths, are calculated as part of the procedure. The limits of the earthworks are not only defined on the ground, but data for the calculation of the earthwork volumes is provided. It will be obvious that the proposed formation levels, the formation breadth, and the values of the side slopes must have been settled before any work can be done.

The alternative to slope staking is to set out points from coordinates, itself an iterative process, made easier by the ability of the prism holder to read the setting out measurements and data at their end of the line. Since slope staking can be carried out with simple equipment, it may be a fall back position if other equipment fails.

Principle

In Figure 11.24, any point on the side slope will be defined by the equation

$$x = b_0 + ny$$

Consequently, at A the values of $x = w_1$ and $y = h_1$ will also satisfy this equation.

The field procedure consists of placing the staff at a series of points on the cross-section such as P, determining the values of x' and y' at P (by taping and levelling) and finding out whether they satisfy the equation. When they do, P is at the point A, and w_1 and h_1 are found. A is marked by a peg or batter rail (see following section).

This procedure may sound laborious, but in fact, after a few hours' experience, very few trials are required to locate the correct point. It must be appreciated that even on fairly smooth ground, a correspondence of levels of 0.01 m is all that can be expected, which at a ground slope 1/10 is equivalent to a variation in x of 0.1 m. By evolving a suitable system for the fieldwork, calculation and booking, the work can be made to proceed quite rapidly.

Calculation

Once the level has been set up (Figure 11.25) and the height of the horizontal plane of collimation HPC has been found, then G, known as the *grade staff reading*, can be calculated from the formation level FL as follows.

1. For cuttings, G = HPC – FL
2. For embankments, G = FL – HPC

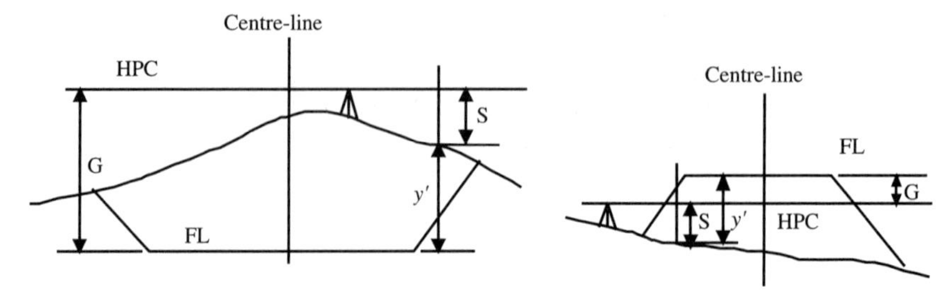

Figure 11.25

This is a constant for one cross-section and one instrument position. From this, the values of y' can be found corresponding to the various values of S, the staff reading.

1. For cuttings, $y' = G - S$
2. For embankments, $y' = G + S$

If the ground is horizontal, the half-breadth will be given by

$$w = b_0 + nh$$

(where h is the depth at central-line), and it is convenient to calculate this for each cross-section, as a starting point for estimating w_1 which will be greater or less than this value according to whether the ground slopes upwards or downwards away from the centre-line (vice versa for embankments).

11.27 Field work and batter rails

The cross-sections are usually taken at equidistant intervals along the centre-line of the route, often at the round-number chainage points, and the work will be easier if the ground levels on the centre-line have been determined first. Since slope staking cannot be carried out until the formation levels have been decided, the centre-line ground levels will almost certainly have been found at some previous time. A plentiful supply of temporary benchmarks is also desirable, and the chainage pegs can be conveniently used for this purpose.

If the values of w are all expected to be less than 30 m, the level can be set up at a convenient point for reading the staff on several cross-sections. The distance from the centre-line is measured with a tape, the staff holder calling out the distance x' for

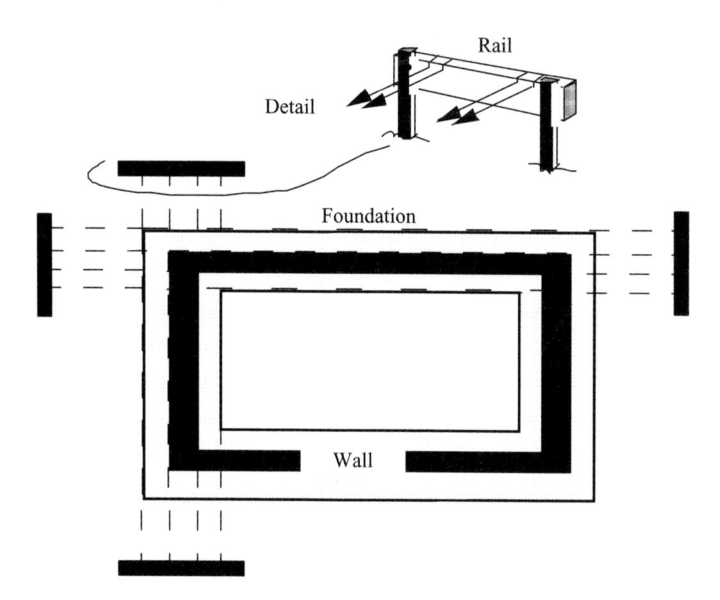

Figure 11.26

each point where the staff is set. If the values of w are greater than 30 m, it will often be more convenient to measure x' tacheometrically using the stadia hairs or by EDM although this, of course, requires the levelling instrument to be set up on the centre-line for each cross-section.

Clearly if pegs are left in the ground at the toe of the slope, they are immediately lost at the beginning of excavation. Therefore *repere marks* in the form of *batter rails* are erected for each cross-section following the slope staking or other form of setting out.

Setting out markers or reperes

The points of the design are usually set out by trial and error. Because these key positions are likely to be disturbed from the very nature of the task, a complex system of repere marks is adopted to recover them throughout the lifetime of the task. The technique is used extensively in setting out works to control such operations as the excavation of trenches for the footings of buildings, pipe laying, pile driving, tunnel or alignment.

A typical arrangement of reperes to control trench digging for the footings of a building is shown in Figure 11.26. The cross rails are marked by saw-cuts which are easy to position, and difficult to remove. When needed, the lines are established by strings between the boards. Excavation depths are also controlled from these strings, although this may be controlled separately by levelling if the ground is steep. All dimensions are measured by tape, giving many checks.

Much use is also made of reperes in a vertical plane to control the slopes of cuttings and embankments in road works, and the fall of a pipe. The latter has to be particularly accurate to preserve the direction of flow. Typical arrangements using fixtures, variously called batter boards, sight rails, boning rods and travellers, are shown in Figures 11.27 and 11.28. The sight rails define the slope of the pipe and its trench, which is different from the ground slope. The traveller acts as a template to

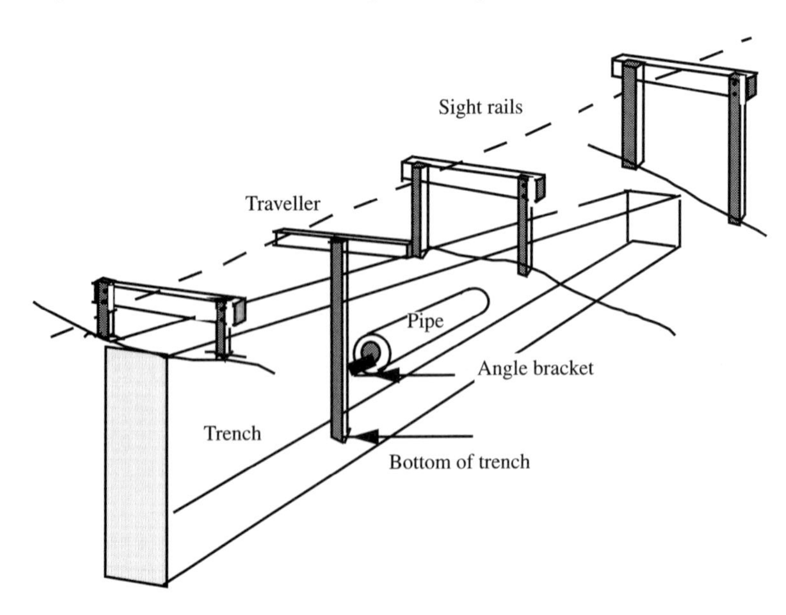

Figure 11.27

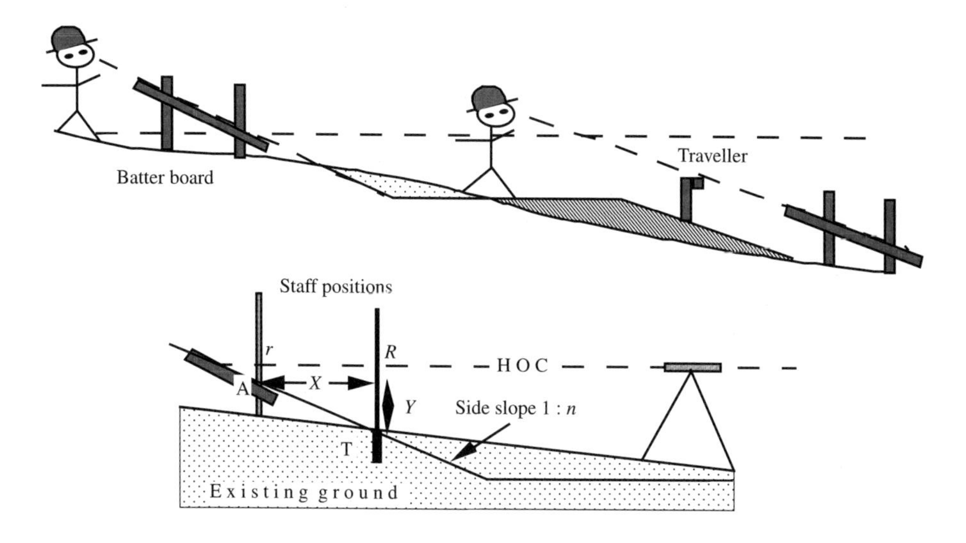

Figure 11.28

control the depths of the trench and of the inside of the bottom of the pipe. The rails are set out with a level using a height of collimation carried in from a benchmark.

Where practical, more than the minimum number of reperes should be used so that the work can be verified as it proceeds. As they can easily be disturbed, reference to semi-permanent control marks is essential. Such marks will be concrete blocks cast *in situ*, or positioned after casting. Important primary blocks are best guarded on site by protective fences. It is not always easy to persuade a site engineer to co-operate in their construction.

To control slopes, the arrangements of Figures 11.27 and 11.28 are typical. The upper diagram shows how the slope is controlled during the excavation of construction. The required line of sight is established by a rail on which the traveller is lined. The lower diagram of Figure 11.28 indicates how the rails are set in position using a level. The holding nails should be hammered through the sloping member before trying to locate them on the uprights.

Lines may also be established by visible lasers or by invisible systems using special detectors. Laser safety on site needs to be considered when using these devices. A similar procedure arises in industrial surveying in which the required alignment is established by a collimator, or components are located on-line by auto-collimation or auto-reflection in a mirror mounted on a component to be positioned. See Chapter 12 for more details. The floor plans of buildings have to be correlated vertically using any of the vertical alignment methods described in Chapter 12.

Comment

It is vital that thorough records are maintained describing the location of points and the nature of any reperes used in a construction process. These will enable work to be checked independently, or to be presented as evidence in the case of dispute or litigation.

Chapter 12
Industrial and Engineering Surveying

12.1 The objectives of industrial and engineering surveying

The fabrication and assembly of complex engineering structures such as oil rigs and ships, often from parts made elsewhere and brought together for assembly, poses severe problems of dimension control. Tolerances vary from a few micrometers in industrial work to a few millimetres in civil engineering structures. For example: drive shafts have to be carefully aligned if engine life is to be adequate; magnets of large particle research accelerators have to be positioned to provide a smooth path of, say, 1 mm over distances in excess of 20 km; the antennae of radio telescopes require to be constructed to sub-millimetre tolerances; and nuclear reactor boilers, weighing hundreds of tonnes, have to be placed in position to within centimetres.

Often this demanding work has to be carried out in a hostile and difficult environment. To avoid extra costs, speed may also be required, so that corrective measures may be taken with the minimum delay. This demand for immediate results sometimes rules out some systems of measurement, such as photogrammetry, although it and television systems are now being used to effect.

It is common practice to base problems of many dimensions on a numerical coordinate system developed and processed by computer software, and to present results in digital form for future use. Methods of working have become very versatile as a result. It is not possible to describe particular systems here. We deal only with the essential principles which enable the surveyor to apply common sense to a working medium dominated by computers and which enjoys a close relationship with the construction industry.

Although many principles are common to both industrial and engineering surveying (with the exception of the tolerances demanded) the working environments differ, as do the instruments required and the terminology used. The *site* and *setting-out* of civil engineering become the *workpiece* and *building* of large-scale metrology.

For various reasons, metrology requires immediate results, which can only be obtained with direct measurements. These require to be almost free of systematic error, and must give real-time results without the need for complex numerical processing. The tooling telescope typifies this principle, in contrast with the standard theodolite. This situation is changing considerably with the advent of on-line processing.

On completion of the site plan for an engineering survey, the design work has to be completed. Eventually the design has to be implemented in the field. If possible, the same control points are used to maintain consistency and design fidelity.

Before beginning the setting out, a field check is essential. In particular it is vital to see that new works such as road alignments match up with the existing terrain. For instance, the alignment of an existing road has to be established. This is done by marking several centres of the road over a distance of about 100 m, then visually checking a best mean fit line to the centre marks. Similar techniques are used for off-sets from walls and directions of power lines, etc. These alignments are compared with the setting out data and plans for accuracy.

For historical reasons the foot (0.3048 m) is still much used in industrial work, although the international metre is fast becoming the standard unit of length.

12.2 Measurement and positioning

The reader should also refer to Section 11.1 for a discussion of the general aspects of setting out works, which also apply to industrial surveying.

Problems generally take one of two forms:

1. Measuring or dimensioning existing components;
2. Positioning components to a designed schedule.

In engineering surveying, structures are set out with reference to fixed points such as bench marks and pegs. The setting out is made by trial and error using total stations. In large-scale metrology, the key reference points are usually some holes in a *jig* or *fixture* to accept a *tool,* from which the assembly of components is controlled. The *tools* are specially designed fittings within which dimensioned parts must fit. For example, a cube may be set at some attitude as defined by the four points A, B, C and D of Figure 12.1, which gives a schematic idea of the layout.

The jig is a large stable frame (not all of which is shown) which creates the reference system from which all components are assembled using the tools attached to the jig. These tools, custom made for the job in hand, are fitted to the jig by a unique arrangement of holes and dowels so that the required point, such as A, is correctly positioned to eventually accept the cube. The component itself is not likely to be a cube, but some machinery with reference points on it. The cube can be thought of as a local Cartesian coordinate system describing the component.

The process does not end there, because the 'cube' shown may be used to fix further cubes within it. Thus the assembly can consist of placing one reference system within another until all components are in their correct positions within the machine. The process consists of a series of coordinate transformations based on physical reference points on each 'cube'.

12.3 Building tools

To effect these transformations, which in practice may amount to moving large heavy objects through very small distances, special *building tools* are constructed. These tools also need to be located in pre-arranged places, with their movement axes also aligned correctly. Figure 12.2, a schematic diagram of some building tools, illustrates some basic concepts.

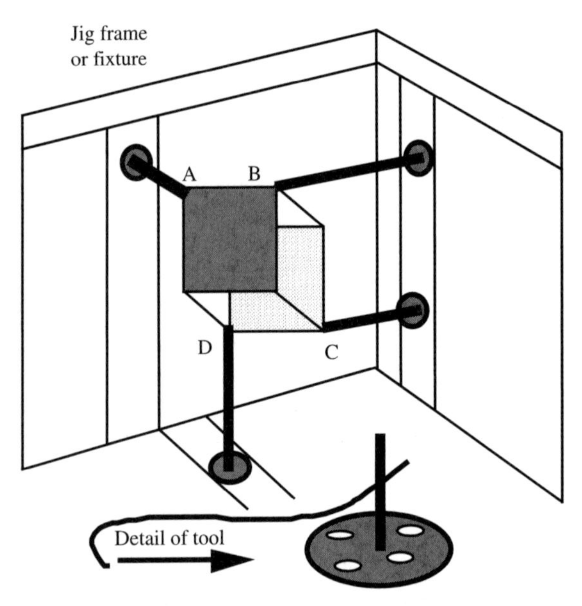

Figure 12.1

The kinematic mount for three balls, mounted below the tool, gives a very stable location. At A the cup-and-ball establishes position, the slot-and-ball at B prevents any rotation about a z-axis through A perpendicular to the plane ABC and the flat-and-ball at C allows rotation about the axis through AB by a fine screw pressing downwards on the base plate. The rotation version of the mount also allows small rotations about the z-axis through A.

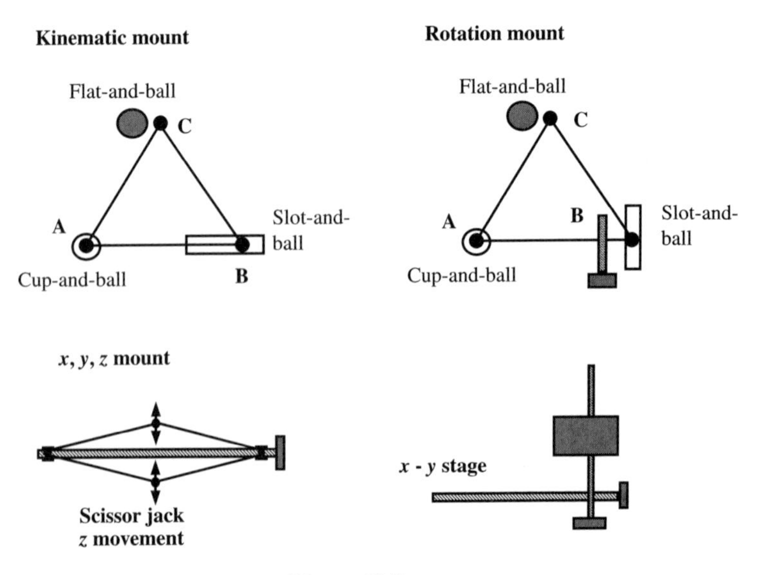

Figure 12.2

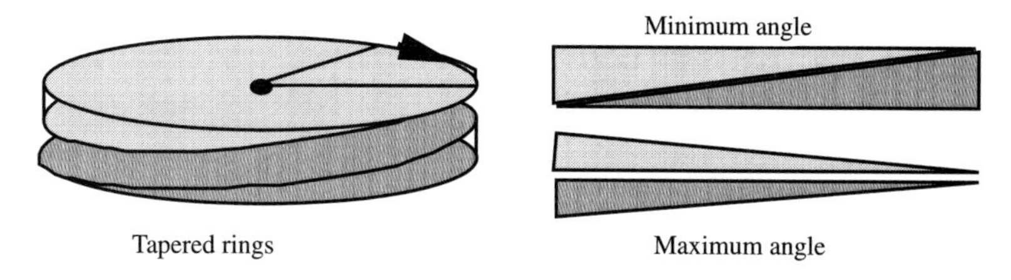

Figure 12.3

The cartesian or (x, y, z) mount consists of a scissor jack for z movement, mounted on a pair of perpendicular arms for (x, y) movement, all operated by fine screws. Sometimes it is better to mount the (x, y) movement on the z movement.

Apart from the standard three foot-screw system common to most theodolites and other fixtures, a tapered wedge system is commonly used since it can accept great weights.

A practical method of introducing a tilt at a required azimuth is by two tapered rings (Figure 12.3) first set to a required angle, which are rotated together to a calculated orientation to produce the required tilt and final orientation. Calculation of the correct settings is achieved by spherical trigonometry.

In practice, various combinations of these basic tools are used according to need. In some cases the whole construction process has to be programmed by computer to optimize efficiency, and records kept of all assembly stages for future dismantling. In civil engineering construction, often a skilled blow from a hammer or the use of wedges and jacks is sufficient to move a component into position.

12.4 Alignment and positioning

At the outset it is most important to stress that *alignment* should be considered separately from *positioning*. Often accurate position is not required whereas alignment may be vital. For example, the rollers of a paper mill or a conveyor belt need careful relative alignment to reduce unnecessary wear on bearings, but their absolute positions are not so stringently required.

Alignment

Alignment is achieved via the establishment of straight lines, either mechanically or optically. It is practicable to establish a straight line mechanically to a tolerance of 0.01 mm by straight edge or tooling bar up to a length of two or three metres. Beyond this length, nylon fishing line (of diameter 0.25 mm) stretched nearly to breaking point has been use profitably. Offsets from these lines are measured in various ways.

The advantage of mechanical systems is that they do not suffer from the effects of refraction, due to the presence of temperature gradients. Many survey control networks have been strengthened by alignments from nylon lines. The use of such lines can be a useful device for carrying bearings from one room to another in the surveys of interiors.

Optical alignment uses either normal (incoherent) light viewed in a telescope, or laser (coherent) light viewed indirectly on a screen or sensor. Optical systems suffer

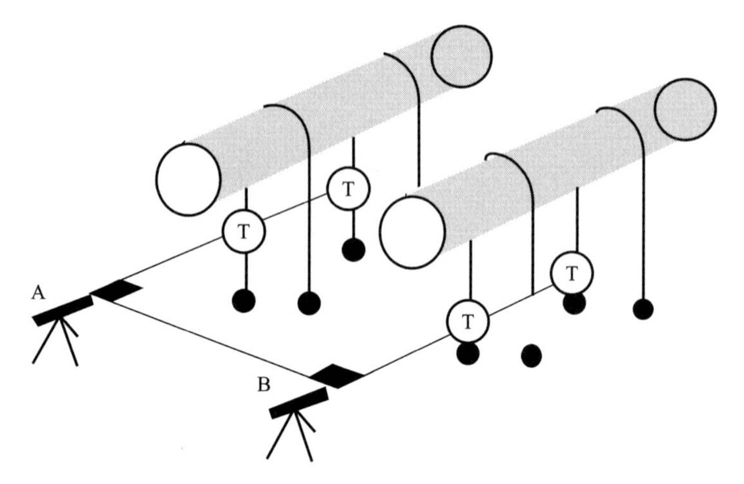

Figure 12.4

from the effects of shimmer and refraction mainly due to temperature gradients close to heating systems. For greatest accuracy, a three-component system is used involving a light source, a lens or zone plate and a target screen.

An important stage in any form of works is the transference of the alignment of a component to the line of sight of a telescope. This line of sight may then transfer direction to others as required, either by right angles created by pentagonal prisms, or by any angle set out by theodolite. The first stage may be achieved by co-planing, Weisbach triangle, autocollimation or autoflexion.

Co-planing or bucking in

The techniques of co-planing or 'bucking in' are used frequently for applications such as the alignment of rollers and engine shafts as shown in Figure 12.4.

Two tooling or theodolite telescopes are co-planed at A and B on the targets T or on the plumb lines directly. This means that the directions of the roller axes are transferred to the telescopes.

The right angles between telescopes are established either by pentagonal prisms or angle measurement. If many rollers have to be aligned, the datum direction AB is established as a semi-permanent feature. This is usually by a collimator, plane mirror, planizer, datum marks or wires. See below for further explanations.

It is essential to change face in theodolite work, because of the alteration to collimation with focus. If telescopes are mutually pointed, the unique pointing routine must be adopted. Another method of transferring the axis of rotation of a roller is by autocollimation or autoreflection from a mirror attached to the end face of the roller.

Technique

The technique of co-planing requires that the telescope can be separately translated sideways, and also rotated about its primary axis. Refer to Figure 12.5 showing the plan view of operations.

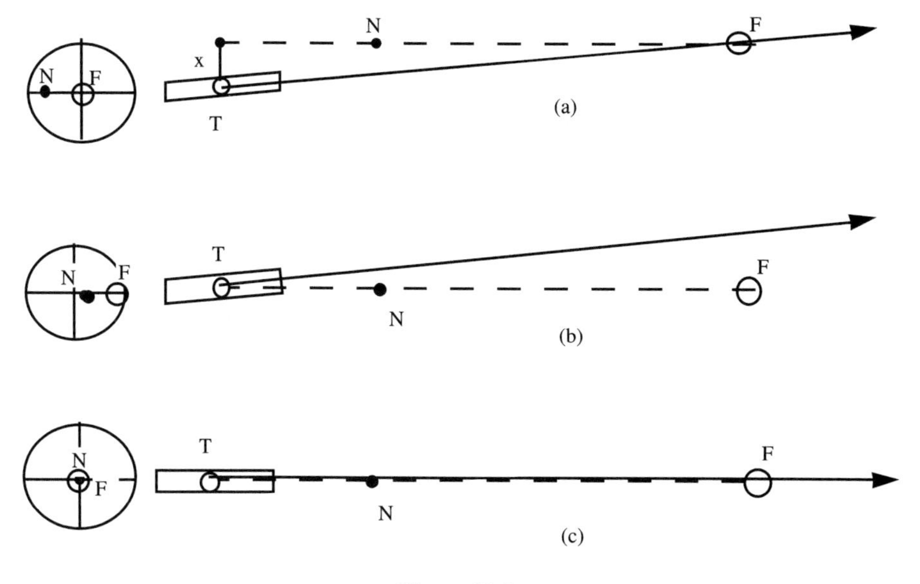

Figure 12.5

The telescope T has to be pointed so that its optical axis lies in the plane of the wires N (near) and F (far) as shown in Figure 12.5(c). An initial position is adopted, as shown in Figure 12.5(a). Both targets will not be in sharp focus at the same time. The line of sight is pointed at F with N to its left side.

The whole instrument is racked through a distance x until target N is a little to the right of the cross-hairs. Rotation of the telescope to point on F should bring N nearly on line. Iteration is required to establish perfect alignment.

It should be remembered that alignment is made on the far target and shift on the near target. Although the amount of the overshoot to be set when sighting N is in proportion to TN/TF, this can only be estimated because both targets cannot be seen simultaneously.

It should also be clear that, if collimation alters with focus, perfect alignment cannot be achieved unless the procedure is reversed. In theodolite work the whole operation of transferring a direction to another line must be repeated on the other face, and a mean taken. In contrast with most theodolite telescopes, tooling telescopes exhibit very little collimation change with focus (2").

The Weisbach triangle

An alternative to co-planing is the Weisbach triangle (Figure 12.6). In this method, the theodolite T is placed slightly off the line NF either outside NF (in case (a)) or within it (in case (b)). In the latter case, the points N and F are often reference targets on walls inside a building from which a bearing is required.

Assume the direction of the line NF in Figure 12.6(a) is known, and its bearing with respect to the x-axis is α. We require the bearing β of the line TR, where T is the theodolite and R the reference object.

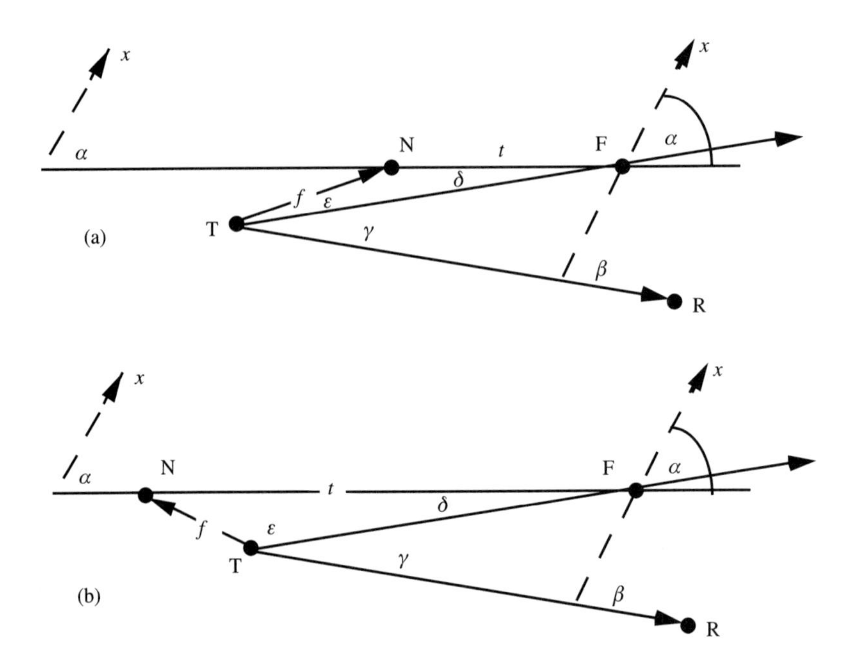

Figure 12.6

From Figure 12.6,

$$\beta = \alpha - \delta + \gamma$$

The technique is to measure the angle γ and calculate the angle δ from other measurements. No special equipment is needed. The angle δ is obtained from

$$\sin \delta = \frac{f \sin \varepsilon}{t} \tag{12.1}$$

The angle ε is observed at the same time as γ, and the distances f and t are taped. Provided the angles ε and δ are kept small (< 1° or 179°) no great accuracy is required for the distances. An error analysis shows this clearly. Differentiating Equation (12.1) partially we have

$$\cot \delta \, \mathrm{d}\delta = \frac{\mathrm{d}f}{f} - \frac{\mathrm{d}t}{t} + \cot \varepsilon \, \mathrm{d}\varepsilon$$

Since the angles are small, it is sufficient to use for error analysis

$$\frac{\mathrm{d}\delta}{\delta} = \frac{\mathrm{d}f}{f} - \frac{\mathrm{d}t}{t} + \frac{\mathrm{d}\varepsilon}{\varepsilon} \tag{12.2}$$

Suppose f = 1.5 m, t = 3.5 m and ε = 00° 40'00" = 2400", then

$$\delta = 00°17'09" = 1029"$$

Suppose the limit of dδ is 10", then we have

$$\frac{1}{102.9} = \frac{\mathrm{d}f}{1.5} - \frac{\mathrm{d}t}{3.5} + \frac{\mathrm{d}\varepsilon}{2400}$$

Considering each measurement independently, we obtain the error bounds of

$\mathrm{d}f = 14.5$ mm; $\mathrm{d}t = 34$ mm; $\mathrm{d}\varepsilon = 23''$

The angle γ must be observed to the 10" tolerance required of the angle δ. The bearing transfer is the combination of three angles obtained to within 10" tolerance, giving the final error bound as 10√3". It must be remembered that the angle γ may have to be observed several times to meet the required specification, and that the bearing should be carried through the longer of the two directions TF. The inter-target distance NF will depend on the limitations of the site. In tunnel work, this is the diameter of the shaft down which the wires are hung.

12.5 Collimators

A *collimator* is a telescope focused at infinity to bring parallel light to a focus at its cross-hairs; or, conversely, a light source placed at the cross-hairs will project a parallel beam from the telescope. A collimator set to point at a reference direction may be used throughout an alignment task, as depicted in Figure 12.7.

To check that the collimator remains undisturbed, a distant fixed target is established on the datum line. Because the roving instrument is also focused at infinity, it can view the collimator cross-hairs from zero distance. Notice this arrangement produces alignment only.

Alignment problems do not require the roving instrument to be collinear with the collimator axis, only parallel to it. This can be achieved anywhere within the aperture size of the telescopes.

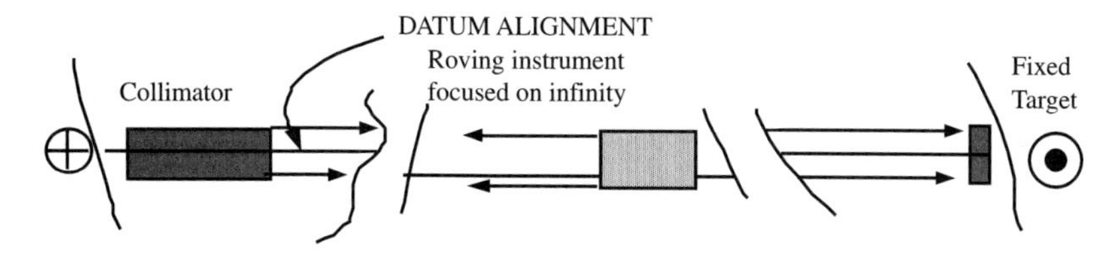

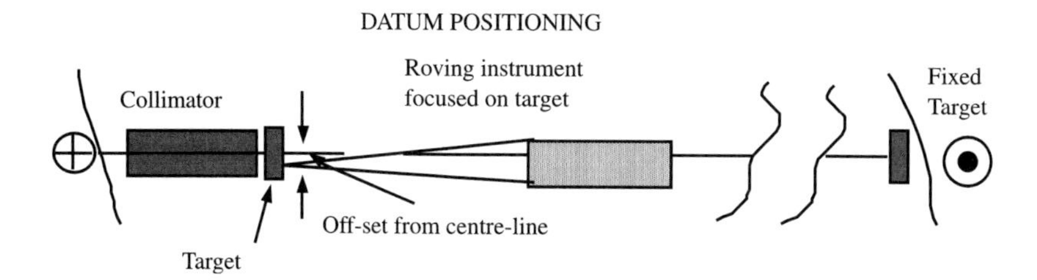

Figure 12.7

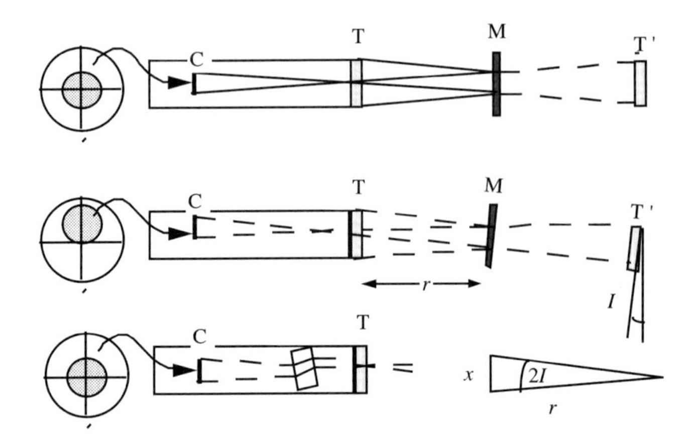

Figure 12.8

If, on the other hand, the roving telescope has to be collinear with the reference axis, the lateral off-set can be seen by focusing on a graticule or target marked with concentric dimensioned rings. The telescope may be then be moved laterally to centre. The alignment has to be checked looking at the collimator with infinity focus. Distance along the line is not always required to any great accuracy, if at all.

Auto-reflection

When we look at ourselves in a mirror we are auto-reflecting. To avoid complexities and double images, a front-silvered plane mirror is used in technical work. Figure 12.8 shows the optical arrangement of auto-reflection.

The top diagram shows the telescope with its line of sight normal to the surface of the plane mirror M at a distance r. With the telescope focused on a target T mounted on the front of the telescope, an image appears to be at T', a distance r behind the mirror. It is central to the cross-hairs. If the mirror is tilted through an angle I the reflected ray moves through $2I$ and no longer appears centrally, as shown in the middle diagram.

If the telescope is fitted with a parallel plate micrometer, the apparent lateral shift x can be corrected and measured. Thus the tilt I in radians is given by

$$I = \frac{x}{2r}$$

Since the parallel plate micrometer of a tooling telescope may operate in two orthogonal directions, orthogonal components of tilt can be measured separately. Thus the mirror can be aligned, or its misalignment allowed for in calculations.

In ordinary theodolite work, no front target is needed. The mid-image position is obtained from the mean of four tangential observations of the aperture image. The process of auto-reflection only ensures that the collimator axis is *parallel* to the mirror normal. If position is also required, the mirror is fitted with scales, dimensioned rings

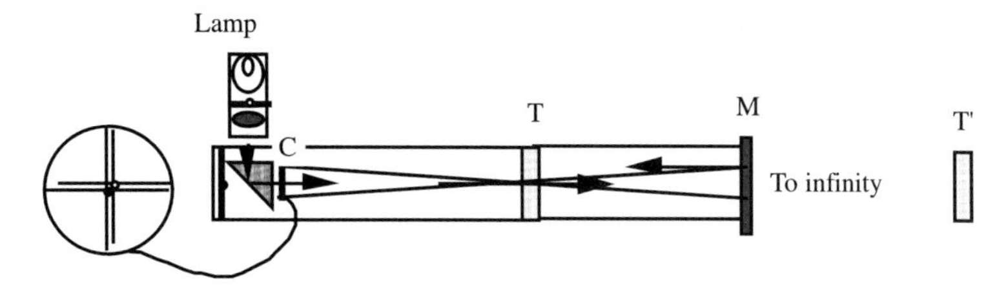

Figure 12.9

or a grid, so that the *position* of the line of sight can be seen from scale readings when focusing directly on the mirror face.

One important application of this technique is in the placing of an industrial component with the aid of a collimator, mirror and tape. If the collimator is set at the desired attitude, the component can be put in position by trial and error until auto-reflection is achieved at the correct distance from the collimator.

Auto-collimation

A similar system involves viewing the *projected* image of the telescope cross-hairs in the mirror, instead of a front target. Greater accuracy is possible at short distances which can be zero because the telescope is focused on infinity. A typical layout is shown in Figure 12.9.

Light from the lamp projects an image of the cross-hairs C to the mirror M and back. The observer sees two sets of cross-hairs in the field of view. One set is steady, while the other reflected set moves as a result of vibration and air turbulence. When the telescope is auto-collimated the two sets of cross-hairs coincide. To distinguish the reflected image from the direct image, the hand can be passed over the aperture to obscure the reflected image momentarily.

The application of auto-collimation is similar to auto-reflection, except that it cannot be used with a parallel plate micrometer to measure small offsets or mirror tilts. The reason for this limitation is that a ray of light passing through the plate is also deviated back to its original position on returning.

Reference mirrors or planizer

A reference line may be established by a plane mirror from which a roving telescope auto-reflects or auto-collimates. The mirror may also carry scales for offset measurements. To give a wider scope for viewing, use is made of two orthogonal mirrors acting like a Porro prism, which may rotate about a horizontal axis. Such devices, shown in Figure 12.10, are obtainable commercially or can be constructed in-house.

The circular graticule or auto-reflected target is viewed after the two reflections which return the incident ray of light parallel to the original direction, provided they all lie in a plane perpendicular to the mirror axis. The device establishes a series of reference vertical planes whose distances apart can be measured on the scale printed

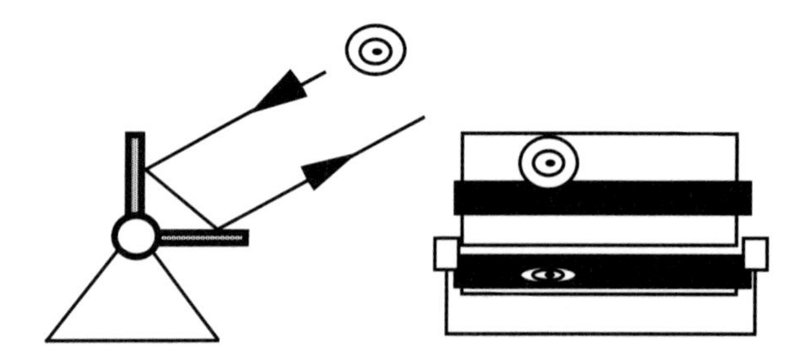

Figure 12.10

on the mirrors. Again, a distant fixed reference target is used to ensure that the mirrors are not disturbed during the measurements.

A plane mirror mounted on the front of a theodolite telescope has many uses, which arise from the fact that the telescope can be pre-set to any required direction in space. Auto-viewing the mirrors from a laser is also convenient.

12.6 Mutual pointing of telescopes

When two theodolite telescopes are pointed at each other, a good practical solution is obtained by sighting a centrally placed target in front of each objective lens in turn. Transparent targets make mutual pointing even quicker.

In metrology, however, greater accuracy at close range is obtained if the fine cross-hairs of each instrument graticule are used as targets. A sharp image can be obtained in each of three circumstances:

1. when both telescopes are focused on a common intermediate plane between them;
2. when both are focused at infinity i.e. collimated;
3. when both are focused on a common plane behind one of them.

The optics of the three cases are illustrated in Figure 12.11 which shows a plan view of the two theodolites whose primary axes of rotation are situated at B and C marked in the middle diagram.

In each diagram, points A and D are the cross-hairs of the telescopes whose fixed objective and sliding internal lenses are shown diagrammatically.

To assist in establishing the first case, the beginner should view a piece of bright card at some intermediate point E, not necessarily midway. This means that the separation b between the instruments must be greater than twice their minimum focusing distance i.e. usually about 3 m. Some slight alteration of the focus of one instrument is needed to obtain a clear view of the other's cross-hairs, illuminated by a bright card held outside the eyepiece. After some practice, the preliminary card at E is not required.

The second case is easy to obtain once the position of infinity focus for each instrument has been noted. This is usually about two turns of the focusing screw back from its maximum position.

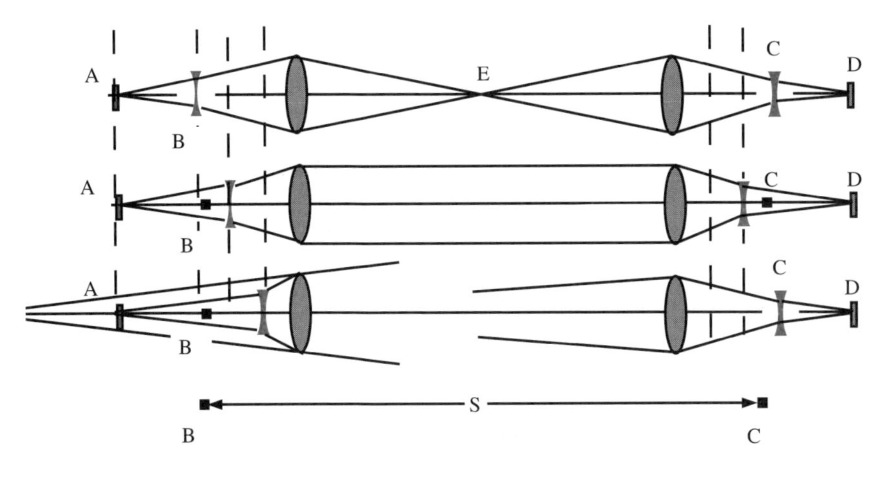

Figure 12.11

The third case may seem somewhat surprising. In most theodolites the internal lens can go beyond the position needed for infinity focus. One telescope is set to this position. The other needs to be focused on the point F, which is easy to do if the backgound is bright.

If one of the telescopes is kept fixed as a reference collimator, the infinity case is used. But if both telescopes are rotated as in theodolite intersection, the original position cannot be recovered because it is not unique. The mutual pointing merely establishes the lines of sight parallel to each other.

Pointing procedure

A good initial pointing is made by iterative pointing at the centre of each telescope (Figure 12.12). Consider the usual case with a front focus point.

By front focus point Consider the case when an intermediate point is used as shown in diagrams (a) and (b). After the lines of sight are made parallel, they are equally inclined to the line BC by an angle β as shown in diagram (a).

Both telescopes are brought to focus on real images of the crosshairs at E_C and E_B. Telescope B is rotated through β to bring it into the required direction BC (diagram (b)). Both telescopes are then set to infinity focus, and keeping telescope B fixed, telescope C is rotated into direction CB. Usually the process has to be repeated because the initial rotation of B is incorrect.

Notice that the angle γ of diagram (b) can be observed, and if the distances BC $= b$ and $BE_C = d$ are measured, the angle β can be calculated from

$$\beta = \frac{d}{b}\gamma$$

Since a mid-point is usually chosen for E_C, $\beta = \gamma / 2$, and no calculation is normally required. The procedure can be very fast with two observers accustomed to the work, and no actual plane is placed at E.

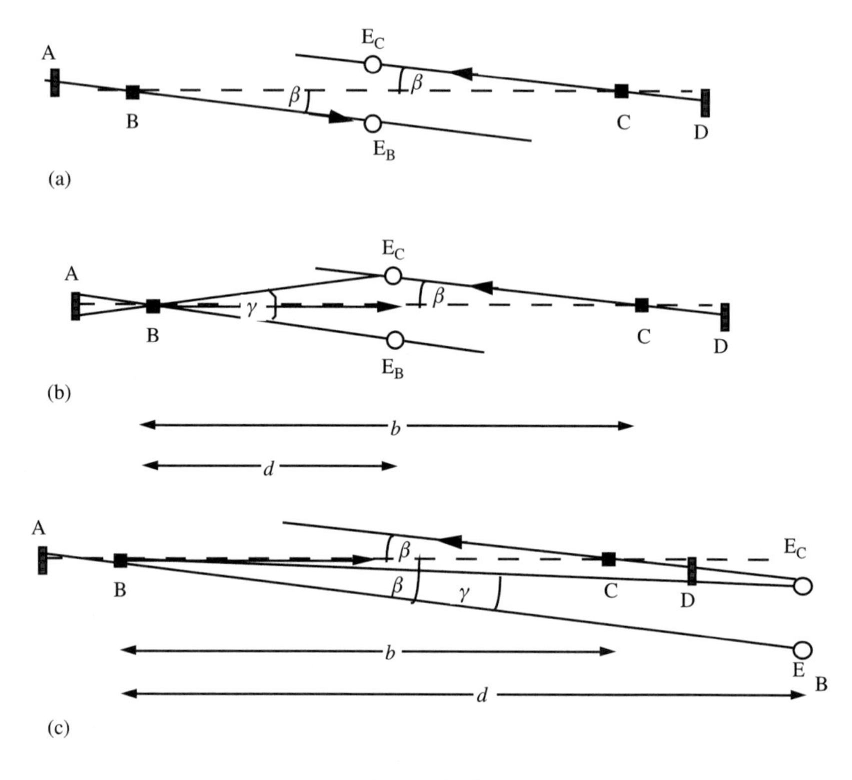

Figure 12.12

By back focus point When the telescopes are too close together to focus on an intermediate point, the second case of diagram (c) has to be adopted. In this case, a virtual image of cross-hairs A appears at E_B which lies outside the line BC.

In this case, a trial and error procedure is more difficult, because the telescope B has to overshoot the image E_C ($\beta > \gamma$). Often both theodolites are so close together that one observer can manipulate them both more easily.

12.7 Laser alignment

Similar to the optical telescope which uses incoherent light, the laser beam of monochromatic (or coherent) light is much used for alignment when the atmosphere is stable, and refraction either predictable or absent. Work in such places as tunnels can give poor results because asymmetric temperature gradients cause serious bending of the beam over long distances. Over shorter distances of, say, less than 50 m, good work can be done.

As with collimators and mirrors, the alignment of the laser line should always be controlled by a reference target. Best results are obtained from the three-point system, consisting of laser source, lens or zone plate and target. The roving lens or zone plate is shown in Figure 12.13 in a common application aligning a shaft or bearings at a power station. The long vertical stand is kept vertical by a precise level.

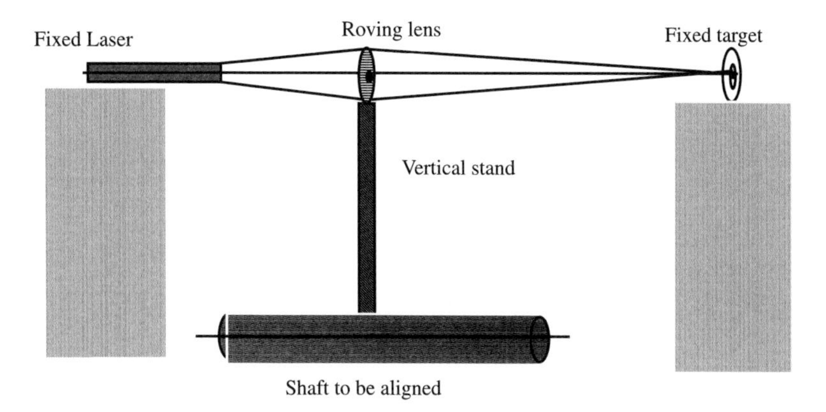

Figure 12.13

12.8 Mechanical alignment by piano wire or nylon line

When optical conditions are unsuitable or inconvenient, alignment in the horizontal plane may also be done by nylon fishing line under tension to near breaking point. Offsets from the line are measured by a travelling microscope or by special electro-optical sensor which uses the line as a reflector.

Piano wires are also much used to align lift shafts and escalators in a typical arrangement such as Figure 12.14. The much stronger piano wire is much more difficult and dangerous to work with than nylon fishing line.

Alignment by inverted plumb line: Rylance method

An extension of the inverted pendulum to permit the transfer of alignment down a shaft was developed by the late K. Rylance in the UK. The arrangement is depicted in Figure 12.15.

Unlike the inverted plumbline, in this arrangement the foot of the downward line is free to move laterally, but so too is the top. Thus the 'vertical' vector can move its

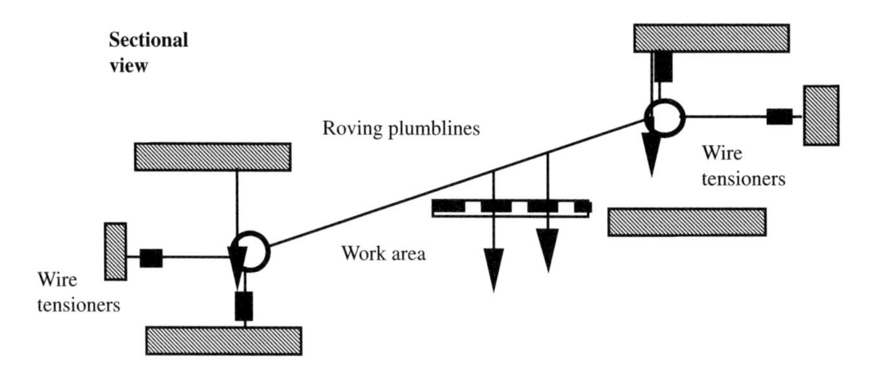

Figure 12.14

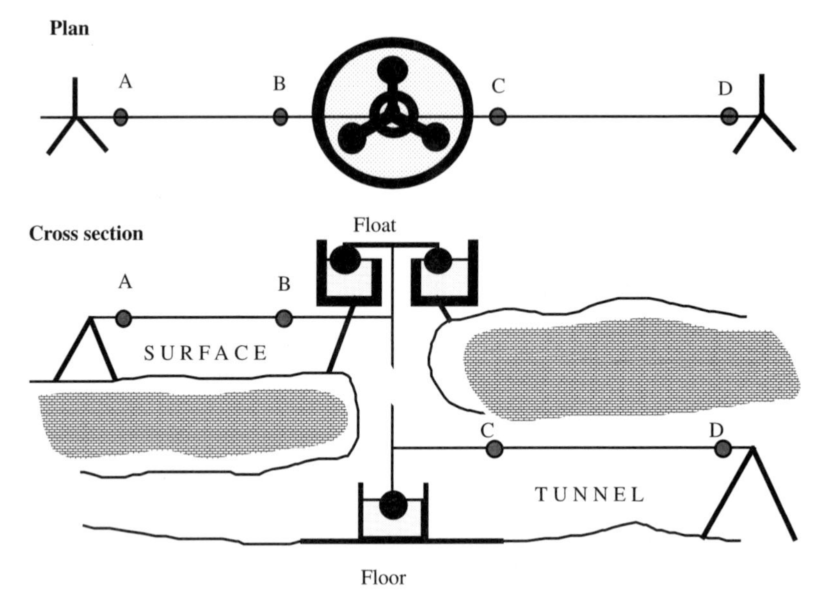

Figure 12.15

plan position within the range of the hole in the water container. As the two lines AB and CD pull against each other with a slight tension, all three lines become coplanar after a short while. The surface bearing is transferred underground using the targets A, B, C and D by two Weisbach triangles, one on the surface, the other in the tunnel, not necessarily simultaneously. The Rylance method does not require the wide shaft of the two-wire method, but only a small pipe sufficient to carry the slightly inclined downward line lying in the common vertical plane of the system. Vertical drops of up to 50 m give a transfer accuracy of 3". The method is cheap and very robust.

12.9 Bearing transfer by gyro-theodolite

If a north-seeking gyroscope is attached to a theodolite, the direction of astronomical north can be determined to an accuracy of about 3" from a series of careful observations.

One version of the gyro-theodolite incorporates a rapidly spinning wheel W suspended by tape T above a theodolite, as indicated in Figure 12.16. A pointer attached to the cage carrying the wheel may be read against a scale to which theodolite horizontal circle readings are related when the two units are clamped together.

The rapid spin of the wheel and the spin of the Earth cause the gyro to oscillate about north. This oscillation is observed on the scale. The turning points P of the oscillation are recorded on the scale, as are the corresponding theodolite horizontal circle readings C_P.

From a series of oscillations, with M approximately first to the left and then to the right of north, the estimated values of M are converted to angular corrections E to

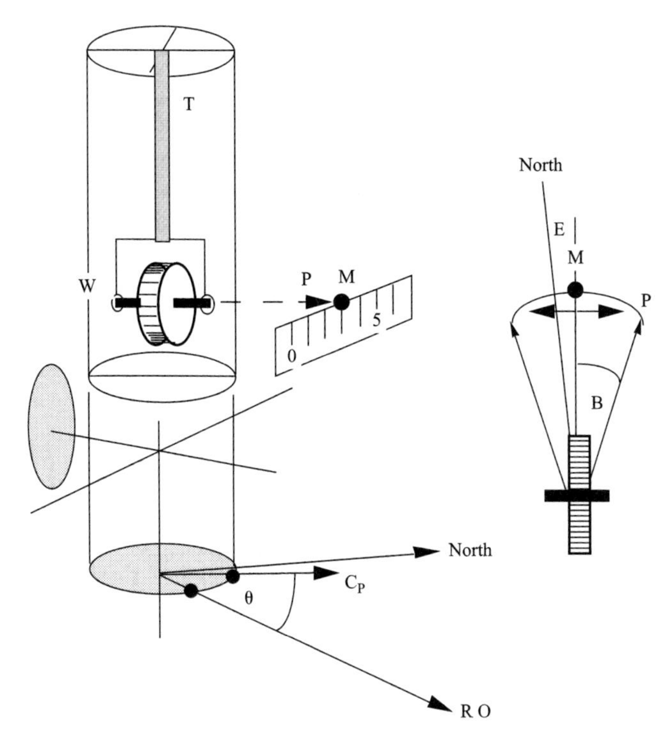

Figure 12.16

yield a circle reading on north. This reading is connected to the RO by angle θ giving the azimuth of the required reference line.

The instrument has to be calibrated for *tape-zero* at every set-up, and for *index* by comparison against a known azimuth, determined astronomically as in Appendix 7.

A complete observation, including calibration, takes about two hours if the best accuracy (3") is required. If a projection bearing is required, the projection convergence is applied (see Figure 8.10 in Appendix 8 and related text).

Although the system has greatest use in mining and tunnelling work, it is occasionally required for industrial applications when wire and laser methods are unsuitable or impractical, for example to calibrate weapon guidance systems in the bowels of a ship. Because the gyroscope needs a stable platform for operation, the technique does not function well on oil platforms where it would be most useful. Readers are referred to Thomas (1982) for more details on north-seeking gyroscopes.

12.10 Vertical alignment

Mechanical and optical (visual and laser) methods are employed in vertical alignment. It has the disadvantage that a distant reference point usually cannot be incorporated into the on-going measurement process, such as the erection of a tall

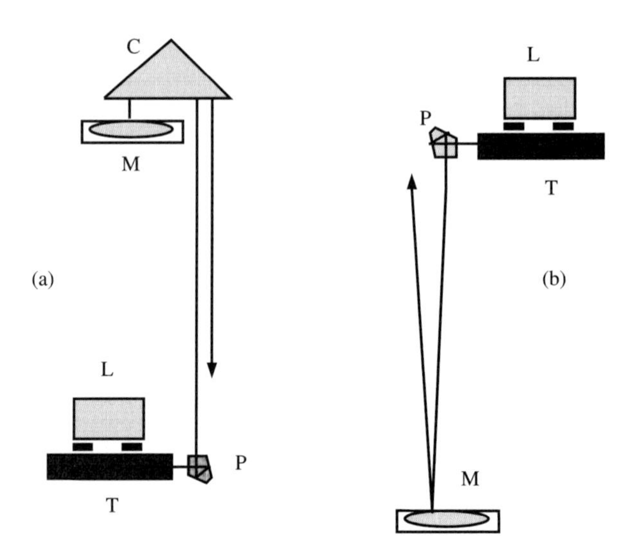

Figure 12.17

building. Dependence has to be placed on the level sensor. Therefore this should be checked against a level line using a pentagonal prism to deviate the beam horizontally.

The *vertical* or horizontal datum is established by bubble tube, pendulous compensator, liquid surface or simple plumb line. Special clinometers with electronic read-out giving accuracies of 0.05" are available: a mercury pool surface is level to better than 1". Refraction is a special problem with optical methods, while wind affects mechanical systems if not shielded. A long transparent plastic tube makes an efficient cheap screen for a plumb line.

The pentagonal prism is much used to transfer a horizontal line through a right angle to look vertically upwards or downwards. Telescopes can be fitted with a prism which rotates about the line of sight within acceptable tolerances. This can be checked like the optical plummet (see Chapter 13). An alterative arrangement is to mount two fixed telescopes above each other, one looking upwards and the other downwards as in the auto-plumb.

To check a system, auto-reflection from a mercury pool can be used to advantage. An upward-looking system, which utilizes a corner cube reflector to view the mercury pool, is checked as in Figure 12.17(a) and a downward-looking system by the arrangement of (b).

In both systems, the line of sight of the telescope, set horizontally with the aid of the precise levelling mechanism L, is directed vertically by the pentagonal prism P, and reflected back parallel to itself by the horizontal mercury surface M. Any departure from verticality, as in (b), is seen at once in the field of view. This arrangement is used to calibrate and test the system which is used without the mercury pool.

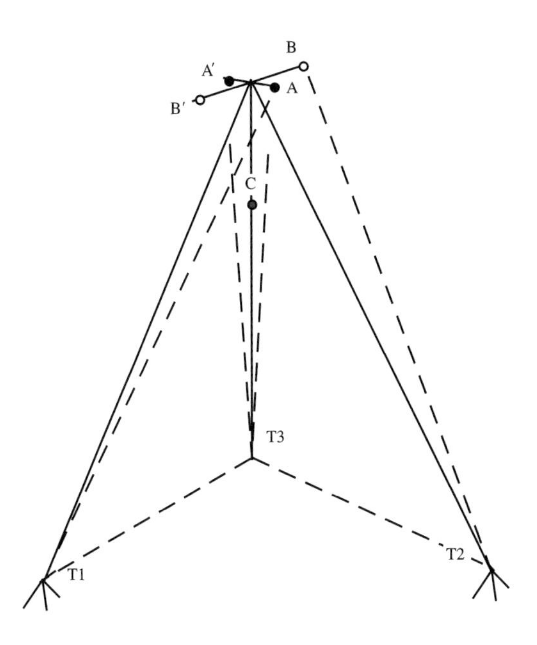

Figure 12.18

Theodolite methods

The theodolite telescope can point vertically to the near-zenith for upward alignment, provided the collimation circle is small (see Figure 12.18). Rotation of the alidade about the primary axis of theodolite T3 causes the line of sight to sweep out a collimation cone into which the target C can be centred. The vertical circle of the theodolite is used for levelling to about 0.5" accuracy.

Downward plumbing can be achieved with a pentagonal prism attachment as in Figure 12.17(b). Some old instruments were capable of pointing vertically downwards through a hollow primary axis; others were fitted with an auxiliary side telescope for this purpose.

Alternatively, if the site permits, a vertical line can be set out by two theodolites T1 and T2, set at right angles to each other as indicated in Figure 12.18. The process is repeated on both faces to give positions A, A', B and B' (greatly exaggerated here), the mean of which gives the correct solution. The instruments are very carefully levelled using the vertical circle or compensator, and the process is repeated to ensure that the result is within tolerance.

Laser systems

The laser has the huge advantage that a projected dot can be thrown on a screen for easy detection. However, the laser suffers from all the errors of any other optical system (collimation, verticality and refraction), and must be calibrated at frequent intervals against the theodolite or mercury pool systems described above.

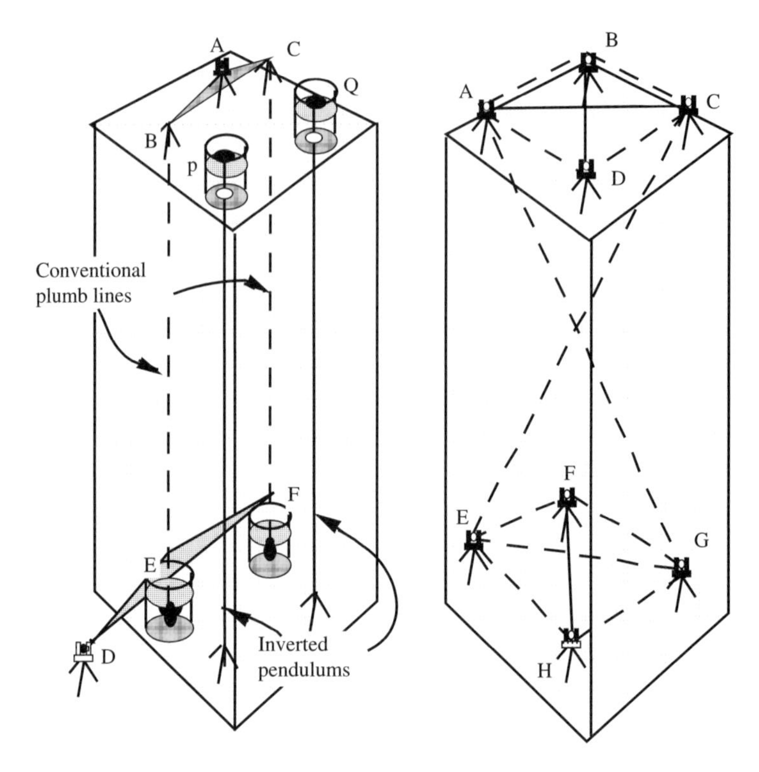

Figure 12.19

Modern systems using invisible lasers operate with special detection devices mounted on x–y rails. Systems using photographic or television recording of visible spot movements are also available to monitor building vibrations.

12.11 Vertical alignment and positioning

In many applications, both alignment and position are required to be transferred vertically. Two methods are commonly used for vertical drops of up to 100 m. Drops greater than this are handled by a series of shorter repeated sections, or by gyro-theodolite for bearing transfer and plumbing for position.

Two-wire method

The method involves two wires suspended from the ground to a tunnel or from the roof of a building to the ground. The alignment of the wire plane relative to others is determined at top and bottom, by simultaneous co-planing or Weisbach triangles. The oscillation of the wires is damped by immersion of the bobs in cans of water or oil. The wires need not be suspended down the same hole. Within reason, the further they are apart the greater the accuracy of alignment. Figure 12.19 depicts the general arrangement.

Triangulation method

A three-dimensional triangulation can produce satisfactory results provided no angles are less than 10° and the theodolites are very carefully levelled and several observations made on both faces. As shown on the right of Figure 12.19, a complete transfer of height bearing and length can be made by this method. It is commonly used to check the other direct methods described in this chapter.

12.12 Length measurement

Length is determined directly by scale, stick micrometer, steel tape or invar wires; it is determined indirectly by subtense measurement or electromagnetic waves. Standards often need to be traced back to a scientific laboratory, especially when components are assembled from various manufacturing sources. An accuracy of from 0.1 mm to 0.001 mm is normally required in large-scale metrology.

The method used depends on the lengths to be measured and the accuracy required. Stick micrometers and clock gauges are excellent for lengths up to 2 m; tooling bars made from heavy steel girders may be used up to 10 m; beyond this steel and invar tapes or wires are most effective.

The most accurate method of length measurement is by laser interferometer, which can achieve results to an accuracy of 0.0001 mm, provided refraction is also known adequately. Distance measurement is by laser interferometer equipped with servo-controlled pointing, reflecting light from special retrospheres (see Reuger 2003).

Carbon fibre rods, although expensive, make very stable length standards, used as subtense bars with theodolite systems. The ends are registered by a typical cup-and-ball system of targets.

Conventional surveyors' EDM systems are not accurate enough as a rule, although developments will no doubt alter this situation.

Tape standardization

As with EDM, all tapes must be calibrated against a known length standard, which is ultimately traceable to an international standard. This is especially necessary for industrial surveying in which components made in different factories and different countries have to conform in the final construction. A common procedure is to set up a temporary base against which working tapes are calibrated. This base is measured with a calibrated invar tape used only for this purpose. The following example is typical of the process.

Example

A 100 ft invar band is calibrated at a standards laboratory for use in catenary with 20 lb pull at 68°F (20°C). The accepted length is 100.000 ft. A site base has to be set up to calibrate 30 m steel tapes.

Three collinear marks A, B and C are placed on a solid concrete floor of a workshop, such that AB $\approx$ 30 m and AC $\approx$ 100 ft $\approx$ 30.48 m, as depicted in Figure 12.20. The standard tape is used to measure AC on the flat under a tension of 20 lb at temperature 8.5°C giving AC = 99.9660 ft.

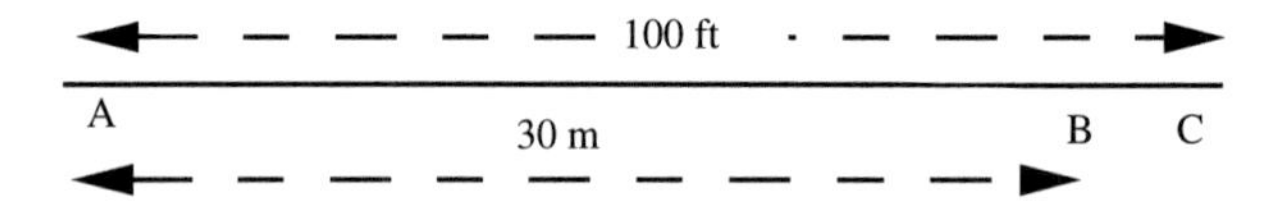

Figure 12.20

The temperature correction is given by

$$(T - T_S)\ C_1 = (8.5 - 20) \times 9 \times 10^{-7} \times 100 = -0.0010 \text{ ft}$$

The coefficient of linear expansion $C = 9 \times 10^{-7}$ for invar.

Because the tape has been standardized in catenary and is used on the flat, a correction has to be applied amounting to

$$\frac{W^2 l^3}{24T^2} = 0.0319 \text{ ft}$$

The weight of the tape (found by weighing the reel with and without the tape) is $W = 1.75$ lb. The taped base length is therefore

$$99.9660 - 0.0010 + 0.0319 = 99.9969 \text{ ft} = 30.47906 \text{ m}$$

The short length BC, measured with a precise calibrated scale, is 0.4820 m, giving the site base AB as 29.9971 m. A working tape gave a mean reading of 29.995 m for the base. The calibration scale factor F is therefore

$$F = \frac{29.9971}{29.995} = 1.000070$$

All lengths measured by this working tape have to be scaled by F. For example, a length of 246.993 has to be corrected to

$$246.993 \times F = 247.010 \text{ m}$$

12.13 Vertical distance measurement

Vertical distance measurement arises in the construction of buildings, towers and dams, etc. Three methods, shown in Figure 12.21 are available for this work: levelling, taping and EDM. Triangulation is also used as a backup to the others.

Taping

In Figure 12.21(a), a steel tape is suspended vertically under the tension T at which it is standardized on the flat. One level is kept at A to read the tape throughout the measurements, while the other, B, makes readings at various levels up or down through the building or excavation.

Small corrections may have to be applied for the calibration scale factor of the tape and for the added extension to the different lengths of tape involved at each level.

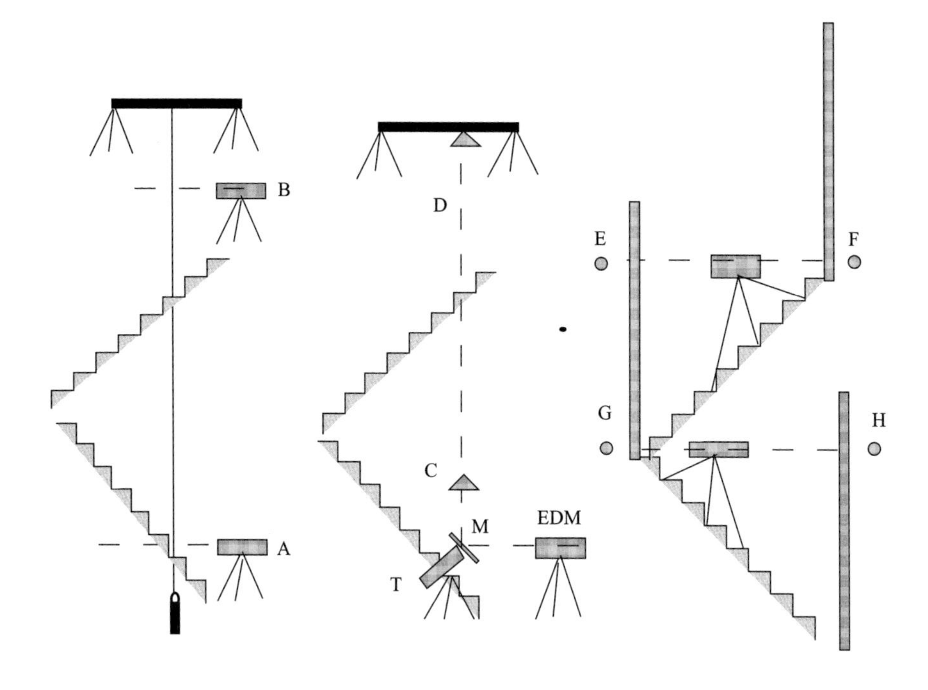

Figure 12.21

The calibration factor is proportional to the length s used for reading the tape whose total length is L. The extension factor for the stretch of the tape ds is given by

$$ds = s\frac{T - T_s}{AE} + (L - 0.5s)\frac{T}{AE}$$

where the field tension is T and the standard tension T_s. A is the cross sectional area of the tape and E is Young's modulus of elasticity. If these tensions are equal, the correction is

$$ds = (L - 0.5s)\frac{T}{AE}$$

EDM measurement

In the EDM method depicted in Figure 12.21(b), the differential distance CD is found by subtracting the two distances folded by the mirror M. The reflector points C and D are tied into others by levelling. The method is particularly practical in large engineering structures and deep shafts.

Levelling

Figure 12.21(c) shows levelling up a stairwell is a common way of transferring heights from floor to floor in building surveys and construction. The height of collimation method of booking is usually preferred so that other height data for floors and ceilings can be recorded. Often, temporary bench marks such as E, F, G and H, are set out during the process for further construction use.

12.14 Positioning and targets

The main problem which arises in close range observations of very high accuracy (of the order 10 μm) is the establishment of the physical position of a telescope or the intersection of the three axes of a theodolite to this tolerance. Targets too have to be centred to this accuracy. The most satisfactory way to proceed is to adopt a cup-and-ball system. The Rank Taylor Hobson Company has established the world standard for this system with a sphere of 2.25 inches in diameter to which most tooling telescopes, fittings and mechanical mounts conform.

Tooling theodolites are designed with a two-foot screw levelling system which enables the instrument to pivot about a small ball resting in a cup, so that the height of instrument is not altered when levelling. The older Brunson Jig Transit had four-foot screws for this purpose.

The centring of a target within a sphere is achieved by means of a shadow graph in which an image of the target is projected on a screen. When the sphere is rotated slightly, the image remains fixed if the target is central. The same procedure can be adopted by viewing the target through a telescope and rotating the sphere in its cup.

Tooling balls are expensive; a cheaper alternative is the ordinary snooker ball, whose centre can be machined out quite easily. The differing colours are helpful too. The cup-and-ball system can also be adapted to accept instruments with a standard thread. The ball is eventually locked in place by a holding ring or cage.

To enable small radial offsets from the line of sight to be measured directly, accurate circular rulings are indispensable, such as on the target shown in Figure 12.22, located inside a tooling ball mounted on a cup. The target can be illuminated by light guided into the ball by a rod of Perspex.

Another useful idea is to project a light from a telescope to a large steel ball bearing to view the small reflected light. Small reflective corner cubes can also be inserted into the ball for use with EDM.

Instrument stands and fixtures

Industrial surveying requires such accuracies that conventional surveying tripods are unsuitable as they lack rigidity; instruments need to be set up in unusual positions and

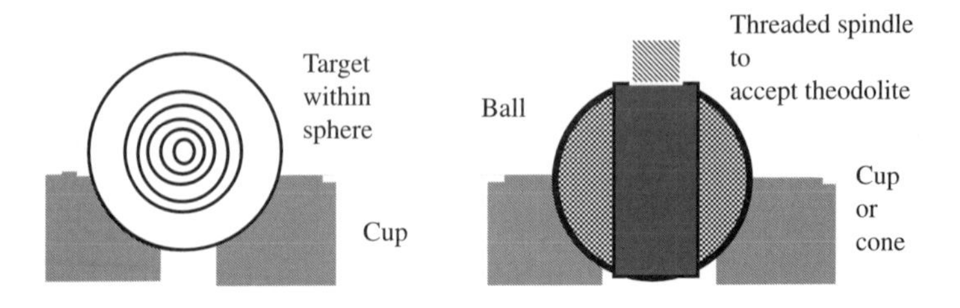

Figure 12.22

so special stands and attachments are required. A typical instrument stand has great rigidity and a head capable of fine x, y and z movements. Stands may also need to be very high, attached to structures or hung from roofs. Instruments may also need to be mounted inside pipes and other awkward places. Targets too need special fixtures. For most purposes, the custom-designed tooling equipment made by Rank Taylor Hobson in the UK is sufficient in most situations. The compatibility and versatility of telescopes, ball and cone mounts, pentagonal prisms, targets, mirrors and stands, etc. is remarkable.

12.15 Example of alignment and distance in sports

An interesting example of the proper use of a scale to measure distance but not straightness, combined with a line of sight for alignment, is the arrangement shown in Figure 12.23 to judge the distance jumped by an athlete. The *alignment* is provided optically by the collimator viewed via the pentagonal prism while the observer views the footprint through the beam splitter by sliding the telescope along the bar which carries a scale on which the *distance* jumped is recorded. Thus the bar need not be very straight or aligned well, but the collimator's direction has to be maintained.

A similar arrangement is used in a *tooling bay*, where two identical systems assembled perpendicular to each other can read off or set out Cartesian coordinates (x, y) directly, as with a map grid. Thus a pattern, such as an aircraft wing shape, can be moulded to shape under the control of the two telescopes. Height is supplied by levelling.

12.16 Theodolite systems

The development of electronic theodolites with on-line real-time computer processing turned a conventional field surveying system into a valuable and practical addition to industrial surveying instrumentation. Either in its own right, or as a back-up to more conventional systems, the theodolite-based provision of coordinated points can deal with the location of fiducial marks on fittings, in the same way as a jig. In addition, it is easier to quantify the accuracies attained. Theodolite methods, however, have limitations, and need to be used in conjunction with other methods. Although laser

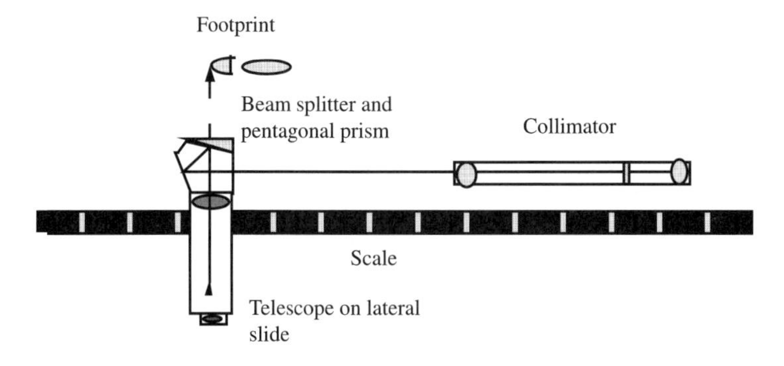

Figure 12.23

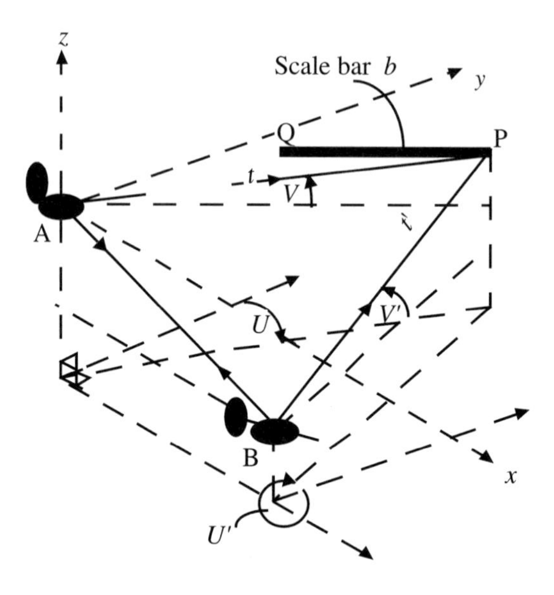

Figure 12.24

trackers are now used in a similar way to total stations to fix points by radiation, they are expensive compared to theodolites.

The principles of theodolite intersection in three dimensions

The treatment of intersection in three dimensions allows for the mismatch of the two rays (vectors) as they intersect a target. Usually this effect is ignored, but in industrial work it has to be considered. The simplest way to find the distance between two theodolites, placed under a few metres apart, is indirectly from observations to a scale bar.

Figure 12.24 shows the minimum configuration with two theodolites and a scale bar. The two theodolites are mutually pointed (see Section 12.6) for relative orientation, and the terminals of the scale bar are observed. Thus a three-dimensional right-handed coordinate system (x, y, z) is set up with one theodolite A as the origin. The axes are aligned with z to the zenith, and the x-axis is in a plane containing the other theodolite's primary axis, also set vertically at B.

It is sufficient to ignore the 10 μm effect of Earth curvature up to distances of 10 m, and refraction effects are usually ignored. The scale bar need not be horizontal, nor aligned in any special way.

At each theodolite A and B, the respective bearings and vertical angles (U, V) and (U', V') to point P are measured. From these the respective direction cosines (l, m, n) for AP and (l', m', n') for BP are found from

$$l = \cos V \sin U$$

$$m = \cos V \cos U$$

$$n = \sin V$$

and

$$l' = \cos V' \sin U'$$
$$m' = \cos V' \cos U'$$
$$n' = \sin V'$$

We assume the length AB = 1 and from the slope angle α between A and B, compute the coordinates of B from

$$x_B = \cos\alpha \quad \text{and} \quad z_B = \sin\alpha$$

Thus we now have the direction cosines of AP (l, m, n) and BP (l', m' and n') and coordinates of A (0, 0, 0) and B (x_B, 0, z_B).

If we assume that the rays AP and BP intersect, which they will do subject to small errors of pointing, we solve for the coordinates of P as follows:

Let AP = t and BP = t', then the coordinates of P are given by

$$x_P = 0 + lt$$
$$y_P = 0 + mt$$
$$z_P = 0 + nt$$

and

$$x_P = x_B + l't'$$
$$y_P = y_B + m't' \qquad (12.3)$$
$$z_P = z_B + n't'$$

Therefore, by eliminating t from the above equations, we have

$$t' = \frac{m\,x_B}{l\,m' - m\,l'}$$

Substitution in Equations (12.3) gives the required coordinates of P. Similarly we obtain the coordinates of Q, from observations of Q from A and B, and calculate the length of the scale bar based on the initial assumption that AB = 1. The coordinate system then has to be scaled to fit the correct length of the scale bar. Before giving a worked example, we consider the matter of quality control on intersection.

Quality control on intersection

The above solution assumes that the two observed rays from A and B intersect at P. In fact they will not do so due to a small error of mispointing or by a large amount if there has been a misidentification of the target. To check for this mismatch, a more rigorous mathematical solution is needed, described as follows.

We now consider that the rays do not intersect and that their common normal PP' has to be found (see Figure 12.25). The angle C between the rays AP and BP is given by

$$\cos C = ll' + mm' + nn'$$

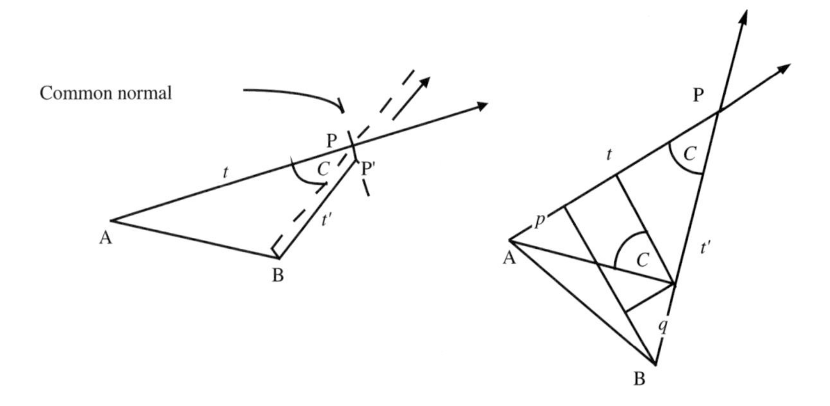

Figure 12.25

Hence we may calculate sin C. The projection p of the ray AB on the ray AP is given by

$$p = l(x_B - x_A) + m(y_B - y_A) + n(z_B - z_A)$$

and the projection q of the ray BA on BP by

$$-q = l'(x_B - x_A) + m'(y_B - y_A) + n'(z_B - z_A)$$

Thus from Figure 12.25,

$$AP = t = \frac{pq \cos C}{\sin^2 C}$$

Hence we have the coordinates of P. Since

$$t' = qt \cos C$$

from Equations (12.3) we have the coordinates of P'. The length of this common normal PP' is a measure of the quality of the intersection, and may be used to alert the observer to a bad measurement which should be repeated. This is practicable because the theodolites are usually linked directly to a computer in the field.

Practical example

Table 12.1 gives the necessary data to set up a coordinate system as described in the above section. The invar scale bar b was found to be 2.0000 m long when calibrated against a laser interferometer.

First compute the value of z_B from tan (–00°08'31"), giving the unscaled coordinates of A and B as

Point	*x*	*y*	*z*
A	0	0	0
B	1.0	0	–0.002 4776

Table 12.1

Direction	*Bearing U*	*Vertical V*
AB	90°00'00"	–00°08'32"
AP	9°25'41.4"	–5°06'33.0"
AQ	30°59'02.1"	–4°27'15.0"
BA	270°00'00"	+00°08'30"
BP'	333°25'48.2"	–4°32'59.5"
BQ'	356°30'33.1"	–5°05'18.2"

Table 12.2 gives the salient stages in the calculations described in the above section. The final results are listed in Table 12.3.

In the calculations below it will be seen that the ray mismatches Δ are very small: of the order a few micrometers. These figures are quite typical of theodolite intersections at this range, provided carefully selected targets are used and the theodolites are placed on stable stands and not on conventional tripods.

To assist data handling by computer, the two theodolites are oriented with respect to the axes systems as a preliminary process to identify the four quadrants for the software. This is usually by pointing at a wall and setting the horizontal circle to zero.

Table 12.2

Line	*Direction cosines*					
AP	*l*	0.163 15986	C	35.870 144 90		
	m	0.982 57229	t	1.527 764 51		
	n	–0.089 05370	t'	1.683 703 26		
BP'	*l'*	–0.445 88030	x	0.249 269 88	*x'*	0.249 269 88
	m'	0.891 57055	y	1.501 139 07	*y'*	1.501 139 07
	n'	–0.079 32660	z	–0.136 053 00	*z'*	–0.136 053 00
		Normal	Δ =	1.3038×10^{-5}		
AQ	*l*	0.513 24281	C	34.357 234 20		
	m	0.854 72246	t	1.768 714 19		
	n	–0.077 66160	t'	1.520 564 25		
BQ'	*l'*	–0.060 64860	x	0.907 779 85	*x'*	0.907 779 90
	m'	0.994 21095	y	1.511 759 75	*y'*	1.511 761 63
	n'	–0.088 69240	z	– 0.137 361 20	*z'*	–0.137 340 10
		Normal	Δ =	2.11020×10^{-5}		

Table 12.3

	Mean Coordinates		
	P		Q
x	0.249 269 86	*x*	0.907 779 87
y	1.501 139 66	*y*	1.511 760 69
z	–0.136 046 50	*z*	–0.137 350 70

Distance PQ = 0.658 596 95

Scale factor = 3.036 758 67

	Final scaled coordinates		
	P		Q
x	0.756 97	*x*	2.75671
y	4.558 60	*y*	4.59085
z	–0.413 14	z	–0.41710

Using industrial software, on-line information is supplied about the quality of intersection and coordination from Least Squares processing, so that observations can be repeated if not to standard.

12.17 Hidden points

It often happens that a point to be surveyed cannot be seen from either or one of the total stations used for radiation. Then indirect methods, such as the following, are applicable:

1. hidden points bar;
2. fitting a sphere;
3. reflection in a mirror.

Combinations of these can also be used to check the work. In all cases it is essential that the bars or mirrors are kept in a stable position during the measurement. Hydraulic articulated arms with clamp attachments, although expensive, provide a good solution. An old theodolite used with telescope-mounted devices also makes a stable, versatile and controllable holding device.

Hidden points bar

The points A and B visible either from a total station or two theodolites, as depicted in Figure 12.26, are coordinated with C held in contact with the hidden point. From the coordinates of A and B and the known distance BC we compute the coordinates of C. Because the length of AB is also known, a partial check on the coordinates of A and B can be made. The formulae are:

$$l = \frac{x_B - x_A}{AB}; \quad m = \frac{y_B - y_A}{AB}; \quad n = \frac{z_B - z_A}{AB}$$

$$x_C = x_B + l\,BC; \quad y_C = y_B + m\,BC; \quad z_C = z_B + n\,BC$$

In practice it is vital that the bar is kept still during the intersection. Various forms of clamping arms are available for this purpose.

If the hidden point is at B, with A and C intersected, greater accuracy is achieved. Sometimes the hidden points bars have to be shaped specially to enable them to be fed into inaccessible parts of machinery. The hidden end is located automatically in position by a cup and ball.

Hidden points by fitting a sphere

A simpler practical alternative is to use a bar of known length r, with one target. The target is placed in at least three different attitudes with one end located at the hidden point C. From the calculated coordinates of the points D, E and F of Figure 12.26, a sphere of radius r is fitted to the points. The coordinates of its centre are the required coordinates of C. At least five bar positions are used as a check.

The solution is by variation of coordinates and Least Squares if more than three positions of the bar are observed. Approximate coordinates of C are estimated. Computer software may take the mean coordinates of the observed points, since this places the provisional centre on the correct side of the sphere. The computed radii to each observed point yield the absolute terms in observation equations, which are of the form:

$$-l\,dx_C - m\,dy_C - n\,dz_C = r - r^* + v$$

l, m, and n are the direction cosines of the typical ray such as CD. There is no index parameter for r because its length is known. The solution and error estimation follow standard procedures explained in Appendix 5. Experience proves that this method is

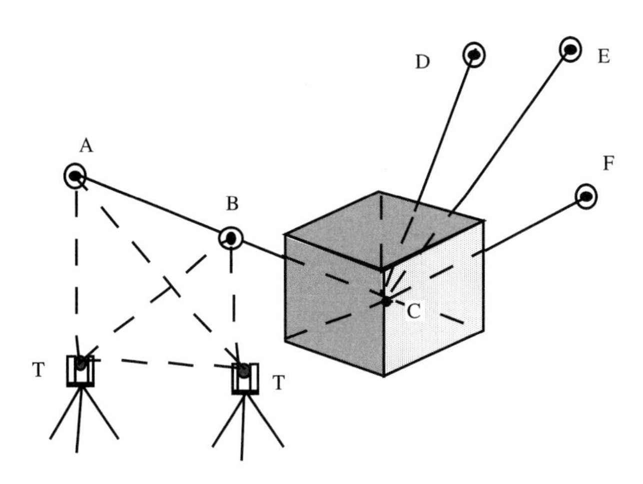

Figure 12.26

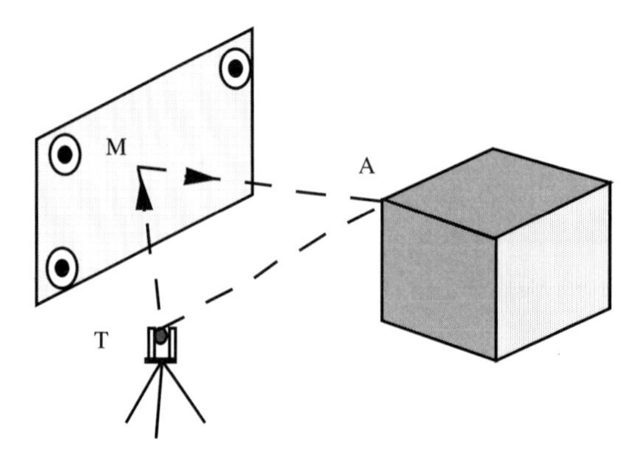

Figure 12.27

not only more practical, but also more accurate than the hidden-points bar method. Where there is limited access to the point C, and a suitably well-conditioned sphere cannot be described, the other method is preferred.

Mirror observations

Sometimes it is convenient to observe behind an object by means of a ray or rays reflected from a plane mirror whose position has been established by intersecting points on its surface. The technique works well provided the mirror is not at a great distance from the point to be fixed. In Figure 12.27, which illustrates the method of fixation of a point from one theodolite only, the distance MA is exaggerated simply to make the principle clear. The method is practicable within a tooling bay where the mirror or mirrors are permanently in place.

Three targets on the mirror face are intersected to obtain the equation of the plane in the form:

$$l_1x + m_1y + n_1z = p$$

where l_1, m_1 and n_1 are the direction cosines of the normal to the plane from the origin of coordinates and p its length. The mirror point M is calculated from the intersection of the vector TM with this plane.

If the known direction cosines of the ray TM are l_2, m_2 and n_2 the required direction cosines l_3, m_3 and n_3 of the ray MA are calculated from

$$\cos\theta = l_1l_2 + m_1m_2 + n_1n_2$$

where θ is the half-angle between the rays MT and MA. Thus θ is known and we have:

$$\cos\theta = l_1l_3 + m_1m_3 + n_1n_3$$

$$\cos 2\theta = l_2l_3 + m_2m_3 + n_2n_3$$

Since the three vectors TM, MA and the mirror normal are coplanar we have a third equation:

$$\begin{vmatrix} l_1 & m_1 & n_1 \\ l_2 & m_2 & n_2 \\ l_3 & m_3 & n_3 \end{vmatrix} = 0$$

The last three equations are solved to give the direction cosines of the ray MA. Finally, the point A is computed by intersection from the theodolite T and the reflected ray MA. If a total station is used, the distance TM + MA can also be measured as a check, although the distance measurement may not be as accurate as the angular intersection.

12.18 Setting out and building

On completion of the site plan for an engineering survey, the design work has to be finalized. Eventually the design has to be implemented in the workshop. If possible, the same control points are used to maintain consistency and design fidelity.

In the factory or workshop a reference system also has to be set up to control construction. This often takes the form of a special tooling bay in which parts are constructed from full-scale design drawings mounted on walls. In both industrial and engineering surveys, reference markers or *reperes* are commonly used to control day-to-day work.

The tooling bay

Figure 12.28 shows the general layout of a typical tooling bay in which the sectional scale drawings, mounted on the tooling bay walls, are used to locate positions from the sighting telescopes T.

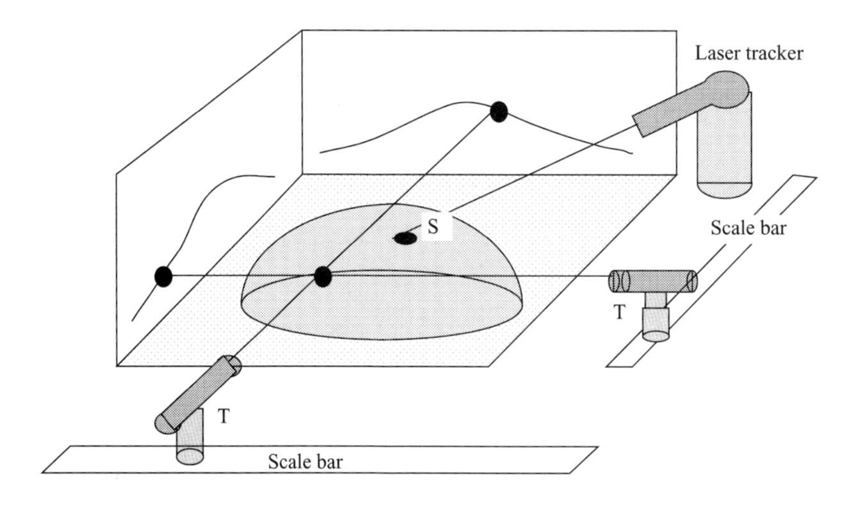

Figure 12.28

These are used in turn to control the pattern maker as he works the scale prototype, say of an aircraft wing section. The same technique is used to control jig fixtures. Computer driven telescopes are also used to create a large-scale three-dimensional coordinatograph.

Polar methods using very costly laser-controlled angle and distance measurers or laser trackers capable of accuracies of a few micrometers over ranges of tens of metres are also to be found. Distance measurement is by laser interferometer, equipped with servo-controlled pointing reflecting light from special retrospheres (Reuger 2003).

Note that the field of industrial and engineering surveying is subject to rapid change as research into sophisticated automated systems progresses. Much progress has been made towards the design and construction of a fully automated robotic system to carry out surveys of hostile environments such as nuclear reactors. On-line computer processing of theodolite and total station data, and the rapid processing of photogrammetric and laser point cloud data, herald even more advances in this branch of surveying. Reference should be made to the websites listed in Appendix 10, especially: http://spectrum-metrology.co.uk for up-to-date information about instruments available for large-scale metrology.

Chapter 13
Instrumentation

13.1 Introduction

In this chapter, we outline some instrumentation commonly used in ground surveying. The objective is to provide a basic understanding of the physical and geometrical principles involved, and of those factors which limit system performance. Because the current manifestation of these fundamental principles into instrumental form depends on the current state of rapidly changing technology, the treatment is kept to a schematic minimum. Readers should refer to the websites listed in Appendix 10 for up to date information.

Ground-based instruments

Most instruments employ optical, mechanical and electronic devices to measure length, angle, time or attitude. They each have a point of reference which has to be related to some physical mark on the ground or structure, albeit only temporarily. Instruments are imperfect measurers which suffer from systematic and random errors. Most instrument designs and procedural use are based on efforts to reduce the systematic errors to a minimum, usually by some form of reversal or balancing. Although instruments depend less and less on observation skills to achieve their best performance, their output measurements have to be incorporated into a properly designed framework. This usually involves the basic principle of working from 'the whole to the part', or always attempting to *interpolate* within an accurate framework and only resorting to *extrapolation* when all else fails. An example of this last case is an open-ended traverse running along the inside of an underwater pipeline leading out to sea.

13.2 Practical Issues

Instrument stands

The most common instrument stand is a tripod with telescopic legs. The feet have to be pressed firmly into the ground, or are set in concrete blocks after greasing, or they may be stuck down with resin glue to a steel surface; stability is essential. Tripod screws and fittings must be tight. The tolerances of industrial work require the use of concrete pillars, or other special metal heavy stands, often fitted with wheels and jacks. Such stands usually cost ten times as much as a tripod.

Some ingenuity is often needed to erect an instrument over or under a mark or close to a wall, etc. Telescopic legs are essential in such circumstances. Sometimes a

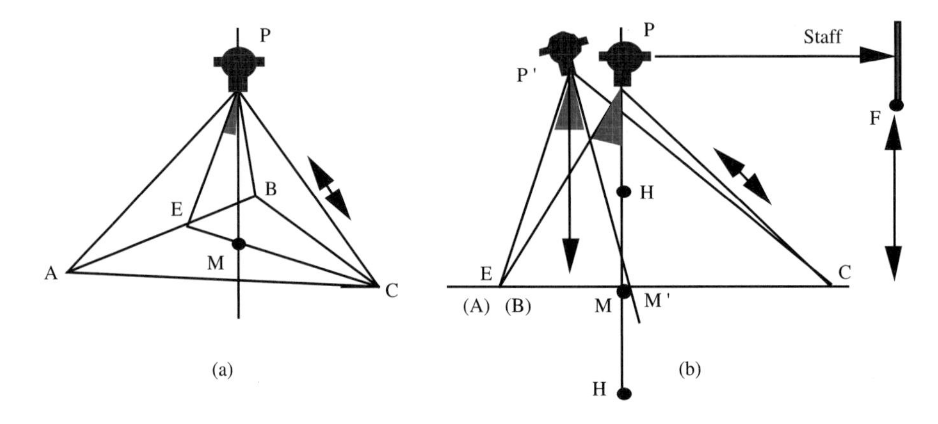

Figure 13.1

theodolite has to be placed directly on the ground for two observers to operate from a prone position, or has to be suspended upside down from a roof. Industrial applications employ special stands clamped to scaffolding.

Footscrews

For a number of reasons older instruments are normally fitted with three footscrews. Two (or four) footscrews have the advantage of preserving the instrument height during the levelling. The two-footscrew instruments usually rest on a central ball about which the instrument pivots, thus preserving the height of instrument. They are now commonly used.

Centring and levelling the instrument on an ordinary tripod are usually achieved together with the aid of an optical plummet. This should be mounted on the alidade of the instrument to allow for collimation error. Instruments with no plummet are centred by plumb line or plumbing rod.

Centring by optical plummet

It is not straightforward to explain this topic; practical demonstration is required. The theodolite P, resting on the tripod with telescopic legs whose feet are at A, B and C as depicted in Figure 13.1(a), has to be centred over the mark M shown at ground level with the primary axis PM vertical.

To set up the instrument begin by placing it in an approximate position P' so that the observer can see the mark through the plummet. With the tripod feet firmly in place, the footscrews are used to centre M from P'. Next, one tripod leg, say PC, is shortened as in Figure 13.1(b), to level the theodolite approximately, with mark M still visible. Since this levelling is in only one direction, it has to be repeated using another leg. It is quite remarkable to the beginner to see how far P' and P can be off-centre while the line of sight of the plummet still passes through M. For level ground, the mismatching distance MM' is given by

$$2\text{EM}\sin^2\frac{e}{2}$$

where the primary axis P'M' is off the vertical by an angle e.

Suppose $e = 5°$ and EM $= 0.8$ m then

MM' is only 3 mm, when PP' $= 1.2 \sin e = 0.1$ m.

Clearly the plummet alone gives no indication of centring accuracy without a carefully levelled instrument. An instrument height of 1.2 m infers levelling to better than 3' for 1 mm centring accuracy. Industrial work to 0.1 mm requires greater attention (18" arc) to levelling. When the mark M is not at ground level, being either in a hole at H or on a tall peg at H, the mismatch of the plummet line of sight gives a much better indication of dislevelment.

The final process is to level by the footscrews and shift the instrument laterally for centring, trying not to rotate its base relative to the tripod top, which is probably not level. Industrial stands are fitted with a precise *xy* stage for lateral movement to be finely tuned.

Height of instrument

It is not easy to measure the height of instrument directly to 0.1 mm. It can be done in two stages using an auxiliary point F whose height above M has been levelled before setting up the instrument. The height of collimation can be read on a small scale at F when the line of sight is horizontal.

Centring under a roof mark

Centring under a roof mark is carried out in a similar manner: either by plumb line, or using the main telescope as a plummet to look vertically upwards if this is possible and if a diagonal eyepiece is available. Collimation error is eliminated by rotating the alidade in azimuth so that the roof mark appears at the centre of the collimation circle. Some surveyors use a pentagonal prism fitted to the objective to achieve this centring, or to measure the amount the instrument is eccentric to the mark. Another technique is to read off an eccentric position from a local coordinate system established by graph paper about the roof point.

Centring from auxiliary point

When centring has to be assured in one direction, such as in EDM calibration, a theodolite placed approximately perpendicular to the line is used to sweep up to the target, as shown in Figure 13.2. To assure against collimation error, the centring is repeated on both faces and a mean taken.

Forced centring systems

Once the tripod has been centred over the mark, the mechanical relationship between theodolite, target, EDM, lamp and mirror, can be preserved to a high order (about 0.2 mm) by a forced centring system of locating pins and matching holes. All components are manufactured to the same height and all fit above the levelling system, or tribrach containing the footscrews. The system is illustrated in Figure 13.3.

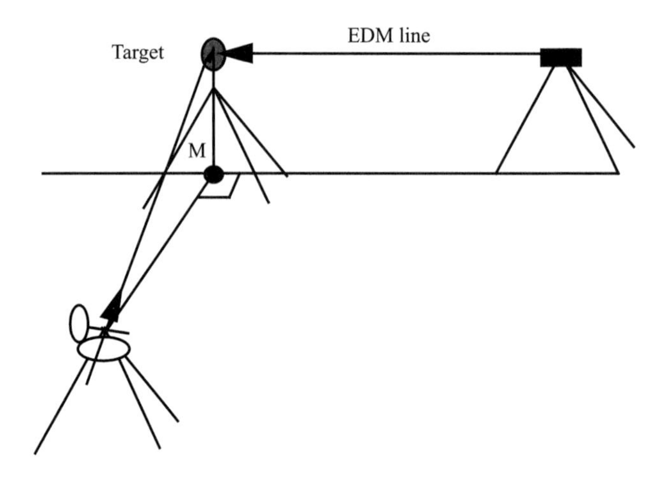

Figure 13.2

Targets

Targets to be observed take many forms. Of paramount importance is the accuracy with which the observation point, or reflection point in the case of EDM, is related to the physical mark being coordinated. Also important is symmetry, so that the appearance is not direction dependent. Various colours are employed for use in differing lighting conditions. Illumination also needs care.

Spheres are much-used as targets. These can be supplied by manufacturers, or devised on site by the surveyor. We have used tennis balls, snooker balls, golf practice balls and ping-pong balls (the last illuminated inside by a torch bulb) with great success.

Reflective materials such as bicycle reflectors, etc. are commonly used. A stiff mixture of flour and water makes an excellent temporary adhesive for a cardboard target on a dusty concrete surface. The list is endless.

Jigs and fixtures

It is not always possible to measure to a mark directly, such the centre line of a rail. Hence a jig carrying a target mark that fits snugly to the rail profile has to be manufactured to assist with pointing. Ordinary mirror screws, for fitting a mirror to a wall, make good hemispherical reference points.

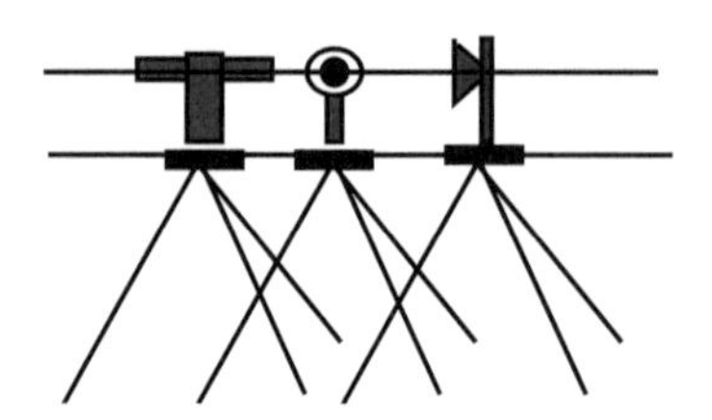

Figure 13.3

Markers

Temporary or permanent marks call for some ingenuity if they are not to be disturbed by accident or design. It is usually wise to tie in a station with measurement from buildings, trees or concreted blocks as a safeguard.

Many types of screw-in site markers are available off the shelf, as are various types of nails either hammered in or shot from a gun. In open ground a buried mark can usefully be recovered if some brightly coloured plastic tape is attached with an end close to the surface.

13.3 Measuring Instruments

The total station

This instrument combines the functions of a traditional theodolite with a distance measurer. Recently some models also incorporate two digital cameras replacing the optical telescope. These instruments are usually operated manually to view individual points, although motor drives are also available to programme initial pointing.

The laser tracker

Automated laser trackers also combine angle measurements with rapid scanning reflectorless distance measurement to build up a three-dimensional picture of a continuously scanned area. They have to be linked into control points so that data from each separate station can be stitched together, in a similar manner to graphic images.

It is convenient to describe the above instruments in terms of their separate functions:

1. angle measurers: the theodolite (goniometer);
2. distance measurer: EDM (electronic distance measurer).

13.4 Angle Measurers

The Theodolite

In Figure 13.4 the essential design features of a theodolite are depicted. The instrument has three axes mounted orthogonally in paired sequence.

1. The ***tertiary axis***, or line of sight PT, is orthogonal to the secondary axis PS, about which it can rotate.
2. The ***secondary axis*** PS is orthogonal to the primary axis PZ, about which it can rotate. Rigidly attached to the tertiary axis is a circle for measuring angles V. These angles are usually measured relative to the *vertical*, giving *vertical angles*. The vertical is established by an independent bubble or compensator to which is attached a measuring index. Some instruments rely solely on the primary axis to define the vertical.
3. The *primary axis* PZ, about which the other two axes can rotate, is usually set vertically, thus allowing *horizontal angles* to be measured on a circle. Because there is no natural reference line equivalent to the vertical, a horizontal *angle* is always the relative difference between two *directions*.

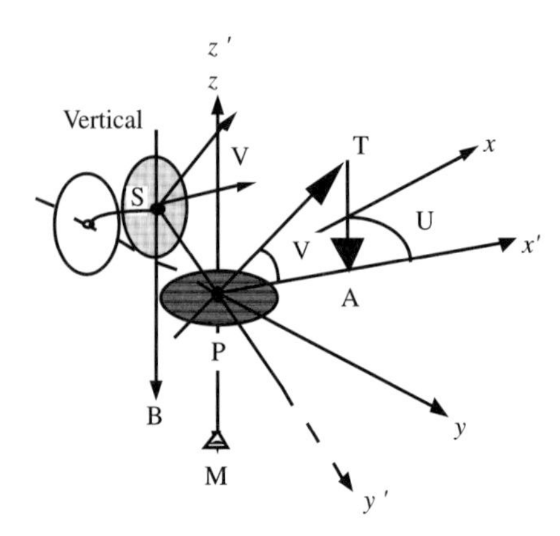

Figure 13.4

The terms *primary*, *secondary* and *tertiary* are to be preferred to all others since they convey the correct information even when the theodolite is stowed in its box. They also describe the rotational hierarchy of each axis. The part of the instrument carrying the secondary and tertiary axes, mounted about the primary axis, is called the *alidade.* Occasionally the horizontal circle is fitted with a magnetic compass to define the *magnetic north* direction, to about 10'. This contrasts with the *vertical* which can be accurate to 0.5".

When the vertical circle is to the observer's left, as in Figure 13.4, the theodolite is said to be on *face left*. After the telescope is transitted through the zenith and the alidade rotated through 180° to sight back on an object, the theodolite is then on *face right*. This *change of face* or reversal almost completely eliminates most errors of construction. Because some cheaper instruments have no independent bubble or compensator, the angle V is measured with respect to the primary axis PZ.

Note that the vertical lines through different points on the Earth's surface are not parallel to each other; for two points a distance of 30 m apart in latitude, the verticals differ by an angle of 1", or 1 in 200 000. This can be significant even in engineering work. Sometimes at industrial sites or on floating platforms or ships, the vertical direction is not useful, and the primary axis is set relative to an arbitrary direction which moves with the structure. In this use the compensator must be locked.

Modern theodolites are capable of accuracies of 0.5" or 1:400 000, a remarkable achievement for circles of 150 mm diameter. This accuracy is achieved by optical Moire fringes or by electrostatic-field techniques. Ordinary optical circle instruments are accurate to 10". Note that these instrument accuracies can only be achieved if the observer employs proper observational procedures.

13.5 Establishment of the vertical direction

The vertical direction is established by a plumb line, pendulum, bubble, or by the normal to a free liquid surface such as mercury. Visual methods of detection are capable of 0.5" accuracy while electronic systems can reach 0.01".

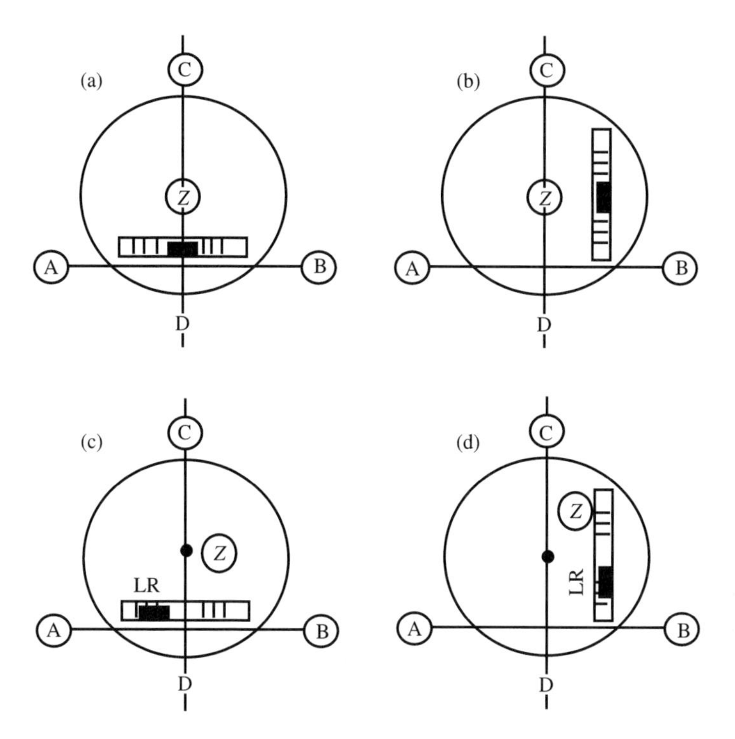

Figure 13.5

Although modern instruments do not have any directly visible bubble or compensator to be viewed by the observer, but use instead a computer-generated picture of a bubble, the principles of the levelling process are best understood by reference to an actual bubble.

The bubble attached to the alidade (see Figure 13.5) is typical of all systems. Figures 13.6(a) and (b) show the bubble LR inside its glass vial when the primary axis Z is vertical. Notice that the bubble is not usually central to the tube. When the alidade is rotated about the *Z*-axis the bubble does not move from its levelling position, provided the primary axis is vertical. Notice also that the ends of the bubble are always labelled L and R *as viewed from the observer's position*.

To set up a theodolite, the first part of the process is to determine this bubble levelling position from the situation shown in Figure 13.6(c) and (d), where the primary axis is tilted from the vertical. Diagram (c) shows the bubble LR with the alidade at any direction; diagram (d) shows it again with the alidade rotated through 180°. The mean of these two bubble positions is the required levelling position. When this is known, the *Z*-axis is tilted by means of the footscrews until the bubble takes up this position; that is until it assumes the circumstances of diagrams (a) or (b).

If the bubble levelling position is too far off-centre for convenience, the vial-holding screw can be adjusted to bring the position more central. There is no point wasting time trying to achieve exact centrality.

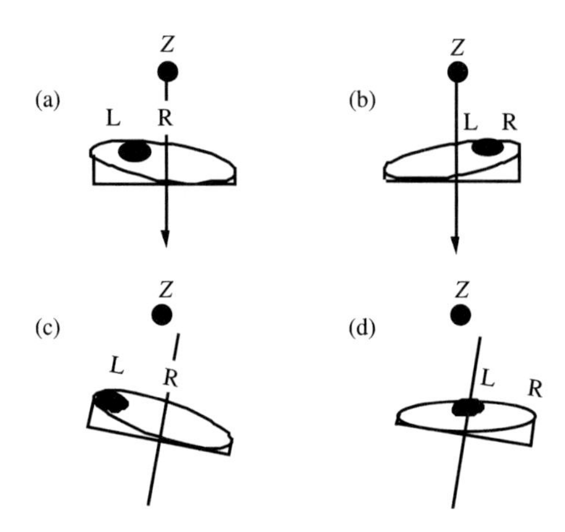

Figure 13.6

Levelling procedure

It is assumed that the instrument is fitted with three footscrews for levelling. The procedure for levelling is illustrated in Figures 13.5 which show the three footscrews A, B and C, the alidade bubble and the *Z*-axis pointing out of the page. For simplicity, the bubble is assumed to be in perfect adjustment. In position (a) the bubble has been aligned parallel to any two footscrews A and B, and in (b) in a direction at 90° to this. In (c) the *Z*-axis has been tilted over to the right so that the bubble moves to the left. In (d) it has been further tilted up the page with the bubble moving down. Bubble vials are graduated in angular terms to enable these tilts to be recorded if necessary.

The process of levelling is the reverse of that just described. The alidade is placed first in position (c) i.e. with the bubble parallel to foot screws AB. It is then rotated through 180° to discover the levelling position, which is then used in position (d) to achieve position (b) by turning footscrew C only. Finally position (a) is achieved by moving footscrews A and B in opposition.

If very precise levelling is needed, the vertical bubble or compensator can be used instead of the alidade bubble. Additionally, if the glass vial containing the bubble is graduated in angular terms, the tilt I can be measured. Modern optical sensors detect this bubble movement, and record it digitally for application by hard-wired software within the instrument, or off-line as required. Such devices require the theodolite to be set vertically by conventional bubbles to within a few minutes of arc, to allow the compensators to operate.

Determination of primary axis tilt *I*

It must be stressed that there is no substitute for proper levelling of the instrument, which should be mounted on a stable stand. However there are cases when the tilt of the axis has to be determined during observations, which are then corrected for its effect. Of particular significance is the correction to horizontal directions for the effect of axis tilt. The most accurate way to determine tilt is to use the vertical circle

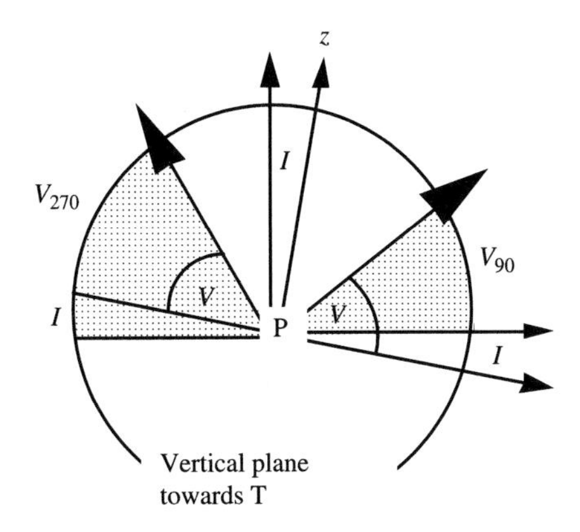

Figure 13.7

bubble or compensator, instead of the less precise alidade bubble. Consider Figure 13.7 which shows the primary axis PZ tilted to the right by an angle I.

With the tertiary axis kept at any fixed angle V, the alidade is turned through 90°, the independent vertical bubble levelled or compensator allowed to operate, and the vertical angle V_{90} recorded. The alidade is then turned through a further 180°, the bubble re-levelled and V_{270} recorded. It can be seen by inspection that the required tilt angle I is given in magnitude and sign by

$$2I = V_{270} - V_{90}$$

The angle V can be any value, but is best not close to zero.

Effect of tilt *I* on a horizontal direction *U*

When the line of sight points to an object at a vertical angle V the effect dU of a cross-tilt I on the horizontal reading U is given by

$$dU = I \tan V$$

For example if $V = 45°$ and $I = 20''$, then $dU = 20''$: a serious amount in precise work.

We will derive this important result and other similar formulae by treating the theodolite as a solid body rotated within a coordinate system.

From Figure 13.4 we can write down the coordinates of point T with respect to axes Px', Py' and Pz' assuming PT = 1, as

$$x' = \cos V$$
$$y' = 0$$
$$z' = \sin V$$

Tilt I clockwise about the axis Px' produces new coordinates given by

$$\begin{bmatrix} x'' \\ y'' \\ z'' \end{bmatrix} = \begin{bmatrix} 1 & 0 & 0 \\ 0 & \cos I & -\sin I \\ 0 & \sin I & \cos I \end{bmatrix} \begin{bmatrix} x' \\ y' \\ z' \end{bmatrix}$$

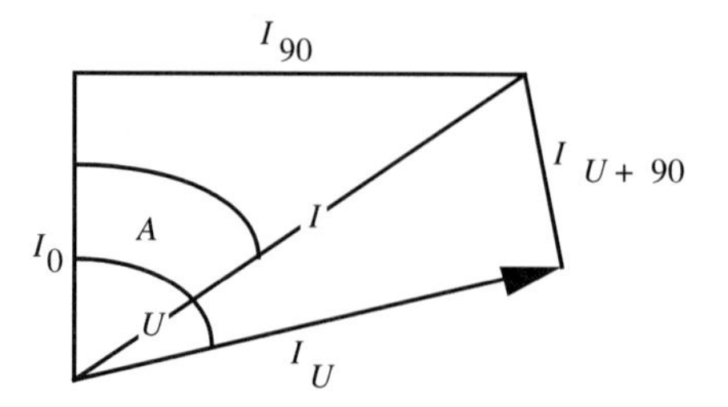

Figure 13.8

The horizontal direction dU which results from this tilt is given by

$$\tan \mathrm{d}U = \frac{y''}{x''} = \frac{y'\cos I - z'\sin I}{x'}$$

Since I is usually quite small, as normally is dU, we can say cos I = 1 and sin I = I, giving approximately

$$\mathrm{d}U = -\frac{I \sin V}{\cos V} = -I \tan V$$

In a practical case, the reading from an already tilted theodolite is too small by dU. Hence the correction to give the reading as if from an untilted instrument is $+I$ tan V. Notice that the sign of I is given already from $2I = V_{270} - V_{90}$. There is, however, no need to measure the tilt I for every pointing, because it can be calculated if the magnitude I and direction A of the maximum tilt are known. Consider Figure 13.8, which shows the maximum tilt I at direction A. Its resolved small components along the axes are I_0 and I_{90} given by

$$I_0 = I \cos A \qquad \text{and} \qquad I_{90} = I \sin A$$

The tilt I_U at any direction U is given by

$$IU = I \cos (U - A) = I_0 \cos U + I_{90} \sin U$$

and in the direction at right angles to U, i.e. at direction $U + 90$, is

$$IU + 90 = -I_0 \sin U + I_{90} \cos U$$

Thus the correction dU to a direction U with vertical pointing V is

$$\mathrm{d}U = (-I_0 \sin U + I_{90} \cos U) \tan V$$

It must be stressed that this is a correction to a *horizontal* reading on a tilted theodolite, not to a vertical angle. The equation is used whenever high accuracy work with steep sights is involved. The effect of this *primary axis tilt* is not cancelled by changing face. However, an inclination of the *secondary axis* to the primary axis is cancelled by changing face. In precise astronomy a special bubble, the *striding level*, is hung directly from the secondary axis to allow for both sources of tilt.

Effect of collimation *c* on a horizontal angle reading *U*

In a similar way to a tilted axis, the component c of the collimation in the horizontal plane also affects the direction U. The change $\mathrm{d}U$ is given by

$$\mathrm{d}U = c \sec V$$

The collimation error c means that the tertiary axis lies along Px' instead of Px (see Figure 13.9). When the telescope shows $V = 0$ or the line of sight lies in the plane Pxy, the point T has coordinates given by:

$$x = \cos c$$
$$y = \sin c$$
$$z = 0$$

or, since c is small,

$$x = 1$$
$$y = c$$
$$z = 0$$

Pointing the telescope at an angle V is achieved by a rotation of the point T about the y-axis by an angle V, giving new coordinates

$$\begin{bmatrix} x'' \\ y'' \\ z'' \end{bmatrix} = \begin{bmatrix} \cos V & 0 & \sin V \\ 0 & 1 & 0 \\ -\sin V & 0 & \cos V \end{bmatrix} \begin{bmatrix} x \\ y \\ z \end{bmatrix}$$

Thus the change in direction is given by

$$\tan \mathrm{d}U = \frac{y''}{x''} = \frac{y}{\cos V} = c \sec V$$

When the telescope is changed to the other face, c changes sign while V does not. The mean of the two directions is therefore free of error. This is illustrated in Figure 13.9, which shows the case when $V = 0$.

However, when a moving target such as the Sun is observed, V cannot be exactly balanced, thus leaving as a residual effect the difference of the two corrections.

Since sec V tends to infinity as V tends to 90°, the effect is very great on steep sights. It also means that when the primary axis is vertical, the line of sight cannot point to the zenith. If the alidade is rotated, the tertiary axis traces out a cone of

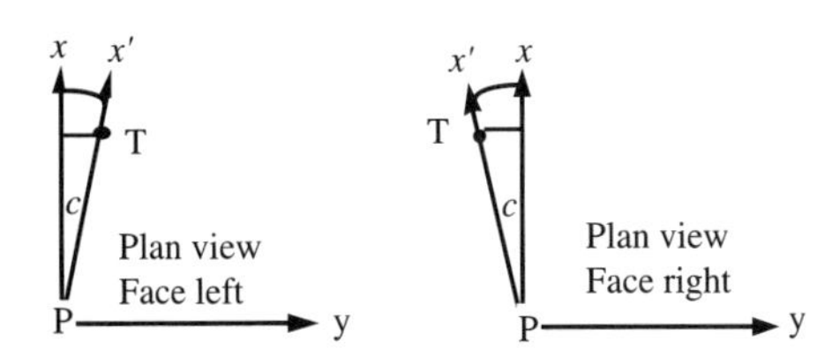

Figure 13.9

collimation centred on the vertical. Hence the observer can locate the zenith at the centre of the circle traced out when the alidade is rotated in azimuth (Figure 13.10).

Eccentricity of circles

Although the accuracy of instrument construction is phenomenal, it is impossible to centre a circle relative to its corresponding axis of rotation without some error. Figure 13.11(a) shows the main eccentricity problem. The line of sight is centred on its axis B while the centre of circle graduations is at A such that AB = d. The angle between the telescope pointings to T and T' is ϕ but the angle recorded is θ_L, indicating the face left position. The error is given by

$$\varepsilon = \phi - \theta_L$$

When the telescope is transitted and the alidade rotated through 180° on change of face, Figure 13.11(b) results. The centre of graduations at A is now 180° from its first position. Thus the error is now given by

$$\varepsilon = \phi - \theta_R$$

Thus we obtain

$$\phi = \frac{q_R + q_L}{2}$$

Therefore the mean of a face left and a face right reading eliminates the error ε. This procedure also applies to the vertical readings where

$$V = \frac{V_R + V_L}{2}$$

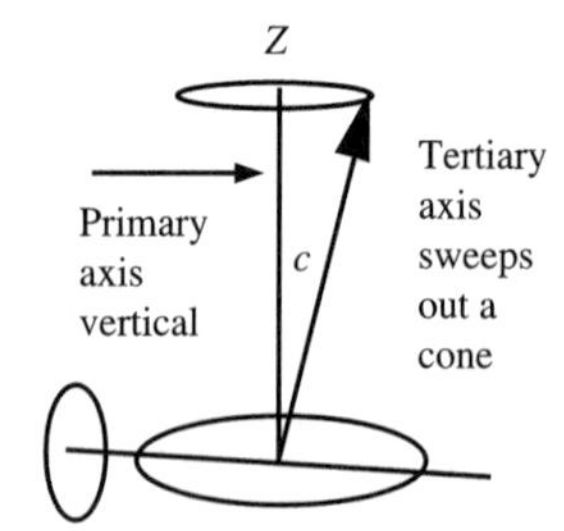

Figure 13.10

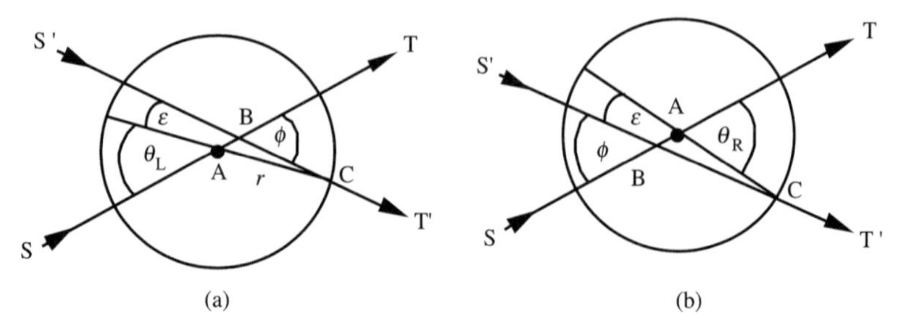

Figure 13.11

In addition, most modern instruments combine readings from both sides of the circle to reduce the eccentricity error. One way to do this is to create Moire fringes by combining images from diametrically opposite parts of a finely ruled circular grating.

For the record we can express the eccentricity error as

$$\sin \varepsilon = \frac{d}{r} \sin \phi$$

This formula occurs in a different context when a theodolite is set up eccentrically to a survey mark, or an eccentric target is bisected. In practice the difference between a face left and a face right horizontal circle reading is the net effect of the collimation, eccentricity and other minor errors. Taking the mean reduces their systematic effects to random noise, not including, of course, the effect of primary axis tilt.

13.6 Instruments for levelling

Instruments for levelling are of two types:

1. *Tilting levels*, in which the line of sight may be tilted separately from the primary axis as in a normal theodolite.
2. *Dumpy and laser sweep levels*, in which the line of sight is fixed rigidly at right angles to the primary axis.

Figure 13.12 shows the principal elements of the tilting and automatic levels. To make a reading, the line of sight has to be horizontal. This is achieved in the tilting level by moving the telescope in response to the bubble about its secondary axis at the pivot, which incidentally does not usually lie on the primary axis.

In the *automatic level,* a horizontal line through the objective centre at O is directed by a pendulous compensator to pass through the centre of the cross-hairs at C. The compensator will only operate in this fashion if the instrument is approximately levelled by a crude circular bubble B.

In the dumpy and sweep laser levels (Figure 13.13), the primary axis has to be set vertically, so that the tertiary axis sweeps out a horizontal plane when the alidade is rotated about the primary axis. In the sweep laser instrument this rotation is motorized. Although convenient on a construction site, the ability to sweep out such a plane is unnecessary for line levelling.

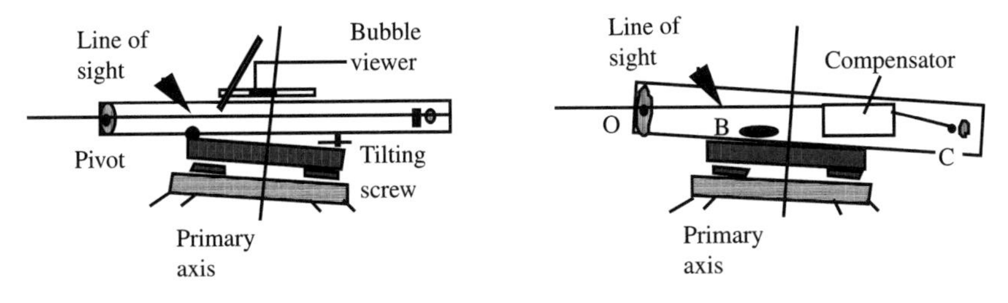

Figure 13.12

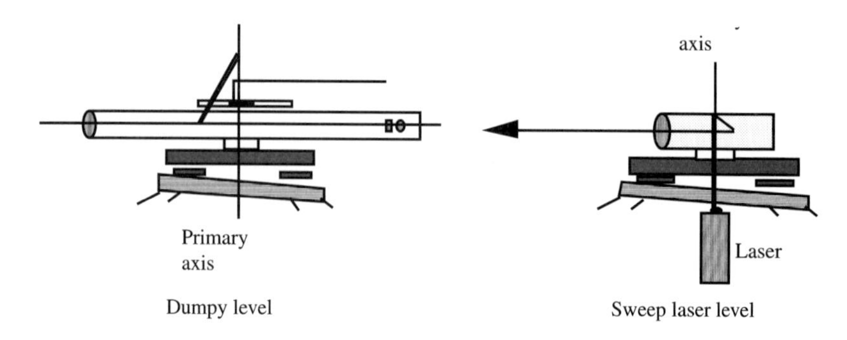

Figure 13.13

All systems are subject to index error, which has to be found by the *two peg test,* and either allowed for or adjusted.

Two peg test for levels

All levels, including sweep lasers, are subject to index error. Index error is the amount by which the system produces a collimation effect. The first step is to determine the magnitude of this error to see if it is significant. This procedure should be carried out before all fieldwork begins, and from time to time during work, especially if the instrument has been bumped in transit. We describe the test for a conventional tilting bubble-level. The test is normally carried out using only one levelling staff, because it too can have an index error.

The level is set up at C, as in Figure 13.14(a), midway between two points A and B located on solid ground about 50 m apart and the staff readings taken in

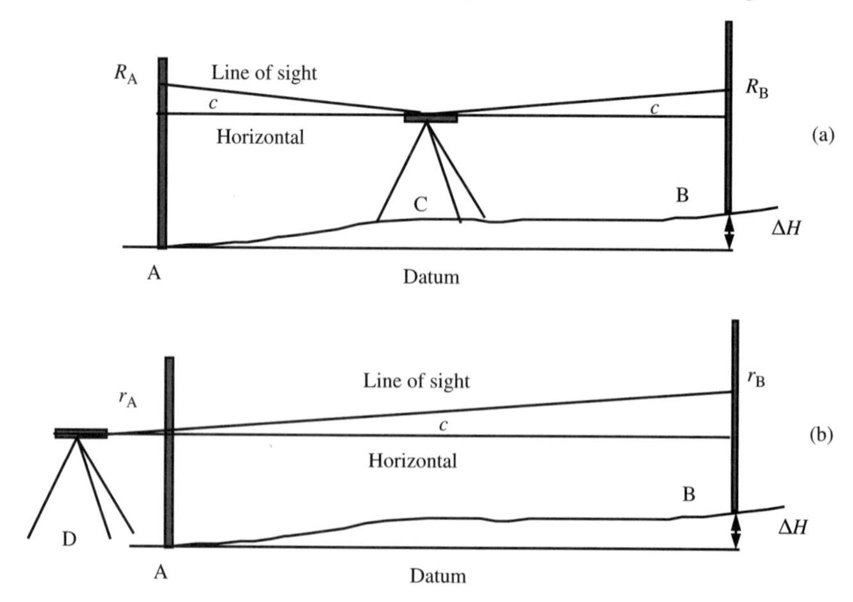

Figure 13.14

succession. Because the index error c cancels out, the true difference in height is given by

$$\Delta H = R_B - R_A = 1.930, \text{ say}$$

The instrument is moved to D as shown in Figure 13.14(b), as close as possible to the near staff at A as the minimum focus distance will allow, usually about 1.5 m. Readings are taken to the staff in the two successive positions. In this case the height difference, affected by the collimation error, is given by

$$\Delta H' = r_B - r_A = 1.937, \text{ say}$$

The correction over the distance DB is –0.007. This correction can be applied to all readings in proportion to distances. In other than precise levelling, it is better to adjust the level to reduce the effect to negligible proportions. The adjustment differs from instrument to instrument. In the case of a conventional tilting level, it is made by altering the bubble adjustment screw, as follows.

With the instrument still at D, the line of sight is tilted to give the correct height difference from a staff reading at B, 1.950 say. The bubble will then be off-centre. This bubble position could be noted and always used when working with the level, or it can be corrected by altering the bubble setting screws; a procedure which needs patience. After the adjustment, the height difference is checked by readings taken again at D.

Clearly the best practice is always to use the level with equal back and fore sights. Dumpy, automatic and sweep laser levels required the adjustment to be made to the cross-hairs or laser alignment: an even more delicate task.

Clinometers

Clinometers are instruments which measure slope alone. They may be hand-held for approximate work, such as the Abney level (good to 10') or very precise industrial clinometers accurate to 0.01", some using electronic sensors with digital output. They too need to be tested for index by the principle of reversal i.e. measuring over a slope forward and back to check for agreement.

13.7 Optical systems

The principal optical system of many instruments is the telescope. Its function is to gather light, from a target or scale, bring it to form a real image in the plane of a reference mark, such as cross-hairs etched on glass, which together are viewed by the human eye or electronic sensor by means of an eyepiece. Telescopes have three functions in most instruments including the theodolite:

1. as a sighting device to view a distant target;
2. in a folded form as a plummet to centre the instrument; and
3. as part of the angle reading system.

Unlike in photogrammetric instruments and cameras, theodolite telescopes function in the paraxial region i.e. never far from the optical axis. The optical axis, defined by the line joining the effective centre of the lens and a physical mark such as a cross-hair, has to be maintained at a fixed relationship to the mechanical axes, to which circles are attached. It is part of the adjustment process to correct any misalignment of the optical axes, and a feature of the observing procedure to reduce any residual measurement error by reversal of the telescope and reading systems, principally by changing face. An efficient instrument matches the performance of its optical and mechanical components with the precision of its reading system to achieve consistent accuracy for its purpose.

We now consider the various factors involved in the optical performance of an instrument.

Resolving power and magnification

The detection system of rods and cones in the human eye is able to resolve an angle of about 30" arc sexagesimal, or 1:7000. This means that to resolve an angle of 1" a telescope with 30X magnification is required to aid the eye, and that the diameter of the circular aperture must be at least 50 mm.

The eyepiece contributes to the magnification, while the resolving power depends on the diameter d of the aperture. The formula to give the minimum resolved angle α, Rayleigh's criterion for a circular aperture, is

$$\alpha = \frac{1.22\,\lambda}{d}$$

where λ is the wavelength of the electromagnetic radiation, and the resolved angle α is in radians. Since visible light has a wavelength of about 0.5 μm or 0.0005 mm, it will be seen that a normal telescope can resolve about 1". To exploit this accuracy the telescope needs to be held still on some form of rigid stand. Conversely, the magnification limit for hand-held binoculars is about 8X.

The telescope and its use

Figure 13.15 shows the essentials of a classical telescope in use by the surveyor. Tracing two incident rays is sufficient to explain the nature of the transformations involved. One of these rays passes through the optical centres, while the other, initially parallel to the axis, passes through the foci of lenses, such as F.

Light from the sun or an artificial source, such as a lamp, is required to illuminate the target A. Some of the reflected light travels towards the telescope, passing through its lenses until it emerges through a small aperture at E, the *exit pupil* of the telescope. The observer places the eye so that its *entrance pupil* coincides with E, and thus light

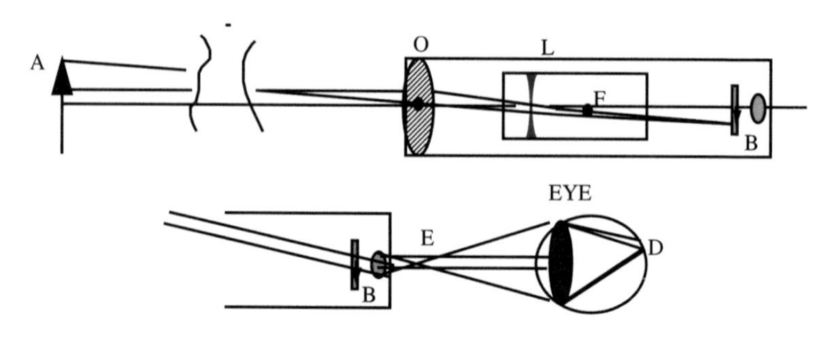

Figure 13.15

passes into the eye and is focused on the retina at D with a resolution of about 30". To achieve a sharp image, the eyepiece lens at B has to focus a virtual or apparent image of the target and cross-hairs on the retina. This image of the target has in turn been focused on the glass plate in front of B by the internal lens L.

It is important to see a sharp image of the cross-hairs at B when first looking into the telescope. If this is not so, eyestrain and headaches will result. The eyepiece focus is unique to each person.

For each target distance AO the internal lens L has to be moved forward or back to achieve a sharp image on the glass plate at B. If this is not so the observer will see a relative movement of the target against the cross-hairs: an effect called *parallax*, which if not removed causes bad observation. Most modern telescopes use an extra lens to present an erect image to the observer; which improves convenience at the expense of optical performance.

It is most important to preserve a fixed relationship between the line of sight, and the secondary axis about which the telescope rotates. In a perfect instrument, the angle between these should be right. The small departure from 90° is called the collimation angle c. It has components in and perpendicular to the plane formed by the tertiary and secondary axes.

For example, Figure 13.16 could represent either the horizontal view shown, or a vertical view if we interchange the primary and secondary axes labels. The component of collimation in the vertical plane is combined with its bubble or sensor error into the vertical index E. The detail diagram of the cross-hairs shows the opposing screws which can position the cross-hairs central to the telescope tube.

In a good telescope the collimation angle will not change by more than 20" at different focusing distances; that is from the minimum, usually about 1.5 m, to infinity. To overcome its effect, the telescope is reversed in attitude. Special expensive tooling telescopes preserve collimation, from zero to infinity distances, to within 2". Should the collimation prove to be too large, the cross-hairs can be moved laterally by opposing screws to reduce its effect: a task requiring patience and luck.

Diagonal eyepiece

Steep sights, possibly vertical, are viewed with the aid of a diagonal eyepiece as shown in Figure 13.17, which bends the light through a right angle.

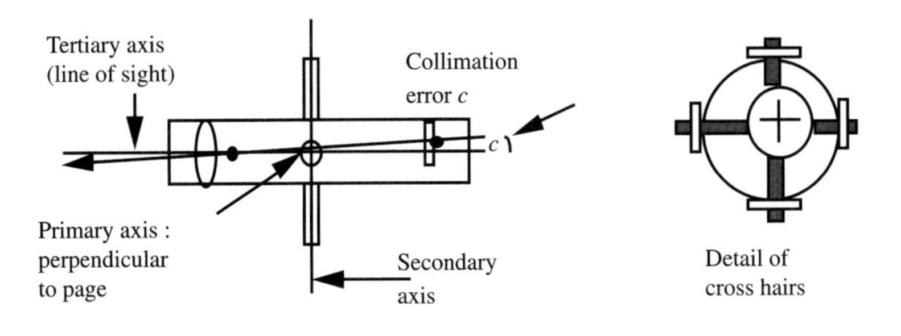

Figure 13.16

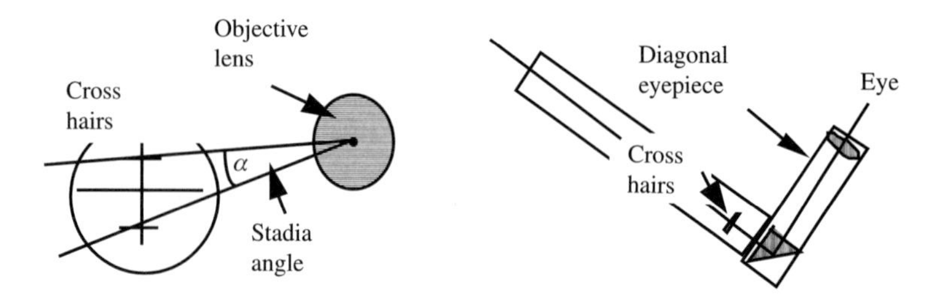

Figure 13.17

Optical plummet

Figure 13.18 shows the optical arrangement of the plummet which consists of a folded telescope. Some have full focusing of both the objective and the eyepiece. Most just have the latter.

Plummets are best mounted in the alidade as shown in Figure 13.18. When the alidade is rotated, the ground mark should not appear to move. If it does, the telescope has a collimation error which can be reduced by adjusting the circular reference mark. Ultimately, the small residual error is eliminated by centring the theodolite within the collimation cone. Recollect that the centring is only as good as the ability to level the instrument.

However, most forced centring instruments do not have the plummet mounted on the alidade. They therefore cannot be checked by rotation at each set up. Instead, the following field check can be carried out.

Draw round the triangular outline of the footscrew base (the tribrach) on the tripod top. Centre and level the instrument with the plummet over a dot marked on a piece of white card stuck to a floor. Then rotate the whole tribrach through 120° to fit the triangular drawing on the tripod top, level up and view again through the plummet to see if the dot is still central. If it is not, mark another dot on the card in the apparently central position. Repeat for the third position. The centre of the triangle formed by the three dots on the card is the correct position vertically below instrument. Now use the

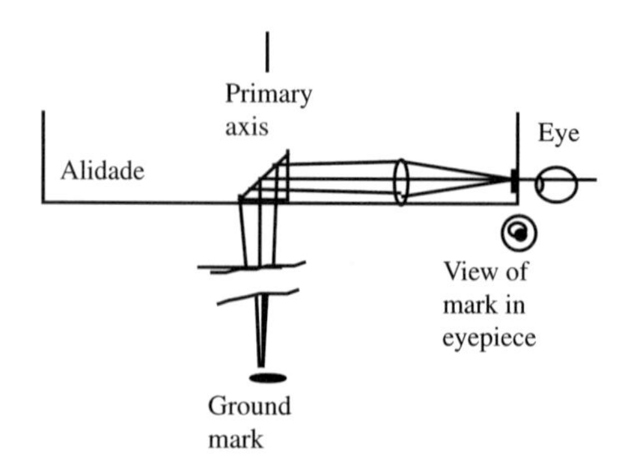

Figure 13.18

adjusting screws of the plummet eyepiece to view this centre through the plummet. The result is checked from all three positions 120° apart. The same technique can be used when centring the tripod in the field.

In the office, the plummet can be tested by clamping the theodolite alidade to the edge of a bench, with the plummet free to rotate. By viewing a wall through the plummet it can be seen if the line of sight describes a cone or not. (If no clamp is available, a bag of sand acts as a useful weight to keep the theodolite in place.) Where the tribrachs can be separated from the alidade, as with most forced centring systems, all available plummets can be tested in this way at one session.

13.8 Ancillary optical components

The following optical components augment the performance of a theodolite or level, and have uses in their own right:

1. Parallel plate;
2. Pentagonal prism;
3. Optical wedge;
4. Corner cube.

Parallel plate

The glass plate with parallel plane sides shown in Figure 13.19 is used to measure small offsets d (up to 5 mm) from a line of sight. The offset d is calibrated in terms of the thickness of the plate t, its refractive index n and the rotation angle β by the expression

$$d = t\beta\frac{n-1}{n}$$

The parallel plate is used within instruments to interpolate scale readings, and externally as a telescope attachment, usually to a level telescope, to interpolate staff readings. Notice that, unlike the wedge, the line of sight is shifted sideways. In the tooling telescope, the plate can be tilted both horizontally and vertically to measure offsets.

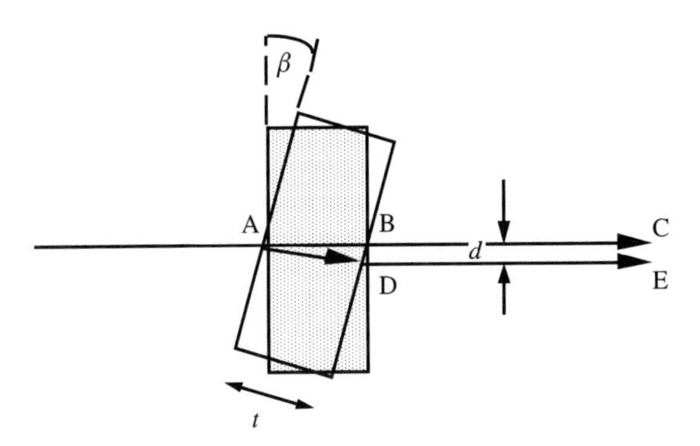

Figure 13.19

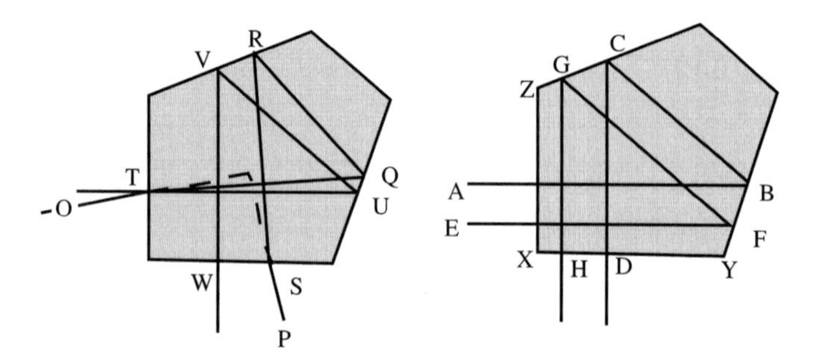

Figure 13.20

Pentagonal prism

The pentagonal prism depicted in Figure 13.20 deviates a line of sight through a right angle irrespective of the attitude of the prism, provided the light rays and principal prism section all lie in the same plane, as shown.

Ray OT, meeting a face at an angle of more than 45°, is refracted by the prism to Q, ultimately emerging as SP perpendicular to OT. The optical paths taken by orthogonal rays from A and E are each of the same length L given where d = XZ = XY by

$$L = (2 + \sqrt{2})d$$

The optical path length, as measured by EDM, is given, by

$$nL = (2 + \sqrt{2})nd$$

where n is the refractive index of the glass for the carrier wavelength of the EDM. It is in the region of 1.5. The prism has many uses: as an optical square; as an attachment to a telescope for special steep sights; as well as in instruments such as the laser level and tunnel profiling equipment to sweep out planes. Like all instruments its accuracy can be checked by the reversal principle, as indicated in Figure 13.21.

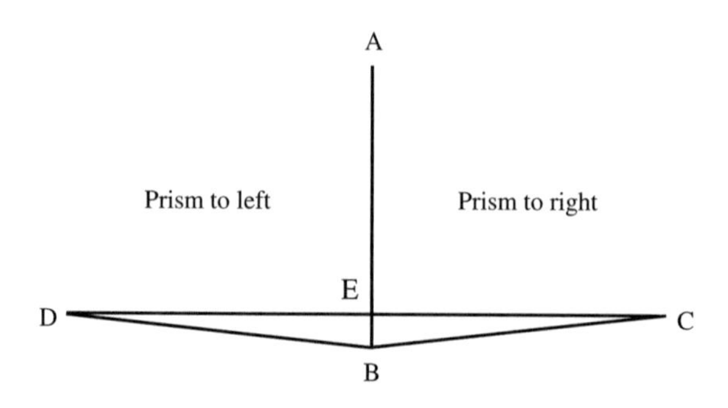

Figure 13.21

From A the point C is located looking through the prism at B, and D located similarly. If the prism angle is 90°, points CBD are collinear. If not, the angular error c is given by

$$\tan c = \frac{\text{EB}}{\text{EC}}$$

The optical square

The optical square is a small hand-held instrument with two pentagonal prisms mounted at right angles to each other. The observer's field of vision is large enough to see through both prisms and straight ahead at the same time. If the observer's eye in Figure 13.21 is at point E all three points A, D and C can be seen together, and thus is on the line DC at right angles to EA. In this way position is fixed by resection using only right angles. The device is of great value in detail surveying and setting out small works.

Optical wedge

The optical wedge is commonly used with another to form a matched pair capable of rotation relative to each other, as shown in Figure 13.22. There are two reasons for the use of a pair of wedges.

1. The deviation angle may be varied.
2. The deviation angle is confined to only one plane.

The wedge is used within and externally to instruments. Fitted to the telescope of a level it can establish an inclined line of sight needed to set out pipework, etc.

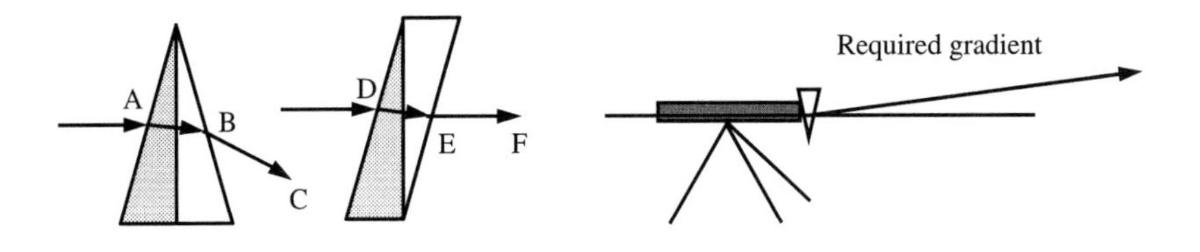

Figure 13.22

Corner cube prism

The prism formed by cutting away the corner of a perfect glass cube, to form a prism of equal side length d = AD (Figure 13.23), is of great importance as a reflector in electro-optical distance measurement.

The prism has the property of returning an incident ray parallel to the original direction. This can be demonstrated with a pocket laser or proved mathematically by considering the reflections of a vector, with direction cosines L, M and N, at three orthogonal surfaces. The returned vector has direction cosines $-L$, $-M$ and $-N$ and also undergoes a displacement, unless it is the central ray. The path is illustrated in Figure 13.23(a).

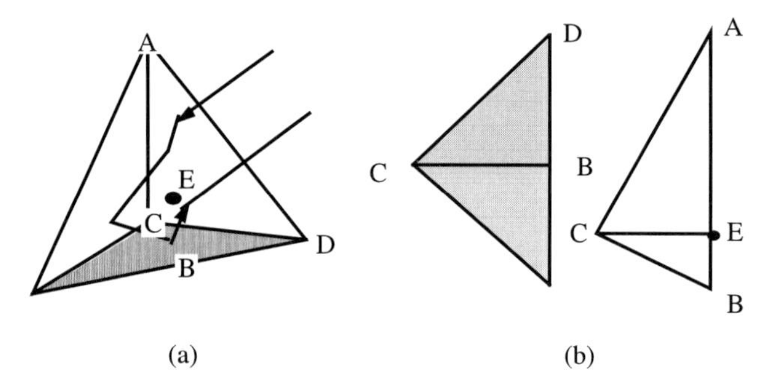

Figure 13.23

The *geometrical* path length p = CE of a ray normal to the front face is given by

$$p = \mathrm{CE} = \frac{d}{\sqrt{6}}$$

and the *optical* path length of interest to EDM is given by

$$n\,p = \frac{n\,d}{\sqrt{6}}$$

For a typical corner cube with refractive index $n = 1.5270$ for a carrier wavelength of 0.910 µm, $p = 41.5$ mm implying $np = 63.4$ mm.

If the ray makes an angle θ with the normal, its optical path length is increased to $p \sec \theta$. These results are evident from Figure 13.23(b) by considering the direct central ray. The proof for a general ray is lengthier. This information is required when considering the geometrical requirements of electro-optical distance measurers capable of 0.1 mm accuracy. For example, this same corner cube misaligned by 5° increases the path length by $np\ (\sec \theta - 1) = 0.24$ mm.

For use in the field, the corner cube is mounted in a housing which fits into a tribrach, usually interchangeable with the total station or theodolite.

It can also be mounted on a telescopic rod provided with a bubble for detail surveying. This device is often called a *pogo stick.* Modern pogo sticks also incorporate a level sensor and a speech link to the total station, which in one case is motorised and telemetered to track the pogo stick as the surveyor moves around to fix detail points.

13.9 Distance Measurers

Simple distance measurement

Before the advent of EDM, all distances had to be measured by tape or wire or optical methods. Any distance less than a 30 m tape length is still usually measured by tape of that length, although longer tapes are still in use. Hand-held laser rangers are now also common (see Rueger 2003).

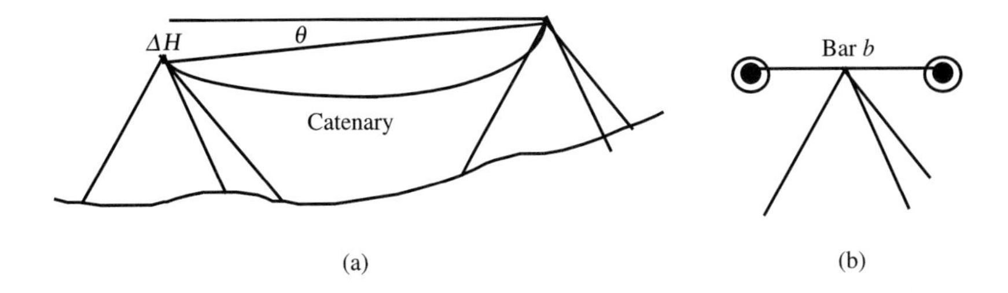

Figure 13.24

Tapes, rods and bars

Tapes, wires, rods and bars are very varied. Each has its use according to need. Attainable accuracy varies from 1 cm with a cloth tape to 0.02 mm with a tooling bar. Long 50 m steel and invar tapes are seldom used today. Rods used in levelling generally conform to the 'E' pattern of marking. Bar-coded staves for automatic reading by pattern recognition are also common.

Tapes

Tapes are used in a straight line along a surface such as a railway line or suspended in catenary (Figure 13.24). In the latter case, care has to be taken to apply the correct tension, noting that the correction dL to catenary length L is given by

$$\mathrm{d}L = -\frac{w^2 L^3}{24T^2}\cos^2\theta$$

The tension T can be applied by a spring balance, a special tension handle, a weight suspended over a pulley or by a tensioning lever device. The weight per unit length w has to be determined by weighing the tape; the slope θ is measured by clinometer or obtained from the difference in height ΔH.

Corrections to catenary and line measurements are also applied for changes from standard temperature t from the standard tension T and for slope, as summarized below.

1. Temperature t: $L_t = L_0 + \alpha L\,(t_F - t_0)$

where the coefficient of linear expansion of the tape material is α.

2. Tension T: $L' = L + \dfrac{L(T_F - T_0)}{AE}$

where A is the tape cross-sectional area and E is Young's modulus of elasticity.

3. Slope: $L_S = L - \dfrac{\Delta H^2}{2L}$

Good practice is to standardize the equipment empirically against a laser interferometer or high precision EDM. Special industrial systems using wires cut to size, automatic tensioning devices and laser interferometer calibration can readily achieve accuracies of 0.025 mm up to 60 m in length.

Subtense bars

Bars made of invar or carbon fibre with end targets for use in subtense methods have applications in industrial surveying (Figure 13.24(b)). They also need to be calibrated against a standard before use.

Distance is obtained from the subtense bar by a miniature triangulation method. The bar, usually 2 m long, forms one side of a triangle, from which the other sides are computed. To simplify the process, the bar can be fitted with a sighting vane enabling it to be aligned normal to a theodolite, at which a horizontal angle is read. The observed angle B subtended by the bar *b*, enables the horizontal distance S to be computed from

$$S = \frac{b}{2} \cot \frac{1}{2} B$$

The method is still much used in precise industrial surveying, at close range up to 20 m with a two-metre bar. Modern levels use the technique by pattern matching a bar-coded staff to give the distance at which a precise difference of level is also read off.

13.10 Electronic Distance Measurement (EDM)

The most accurate measurement of distance using the electromagnetic spectrum is by *interferometry*. In this system the carrier waves themselves are used directly in an incremental counting system. The technique is employed in the laboratory and factory at optical wavelengths, with radio frequencies at sea, from ground to satellite in differential global position fixing by satellites (GPS), and from ground to stars in Very Long Baseline Interferometry (VLBI).

Most other practical instruments use modulated waves to measure distance by timing or phase difference measurement. In the modulation process, a high frequency carrier wave is varied to create a more manageable, lower frequency.

The sonar systems of hydrographic and seismic surveying do not use the EM spectrum. They depend on shock waves which need a physical medium such as water or rock to transmit energy, and travel at a speed depending on this medium e.g. 300 $m\ s^{-1}$ in air and 1500 $m\ s^{-1}$ in water, etc. By contrast, EM waves travel through a vacuum at about $0.3 \times 10^{-9}\ m\ s^{-1}$.

Since the introduction of Bergstrand's visible-light Geodimeter in 1946, Wadley's microwave radio Tellurometer in 1956 and Holscher's near infra-red MA 100 Tellurometer in 1962, the use of electromagnetic waves for distance and range difference measurement has expanded into all branches of surveying and geodesy, especially into satellite systems.

13.11 Basic Principles of EDM

Far from being distance measurers, EDM instruments are really *time measurers*, making use of the basic equation relating distance <*s*>, speed of waves *c* and transit time tim:

$$<s> = c\ \mathrm{tim} \qquad (13.1)$$

To achieve useful results, time difference has to be measured to an accuracy of better than 10^{-9} sec (a nanosecond). Time differences are measured either

1. directly, by time of flight measurement, or
2. indirectly, by phase comparison.

The speed of EM waves, 299 792.458 m s^{-1}, is accurate to 10^{-9}, but the refractive index, about 1.0003 for air, is difficult to obtain to 10^{-6}, a figure of 0.5×10^{-6} being more practicable. This seriously limits the accuracy of EDM work.

In satellite to ground propagation, the limits are even worse. It does not make much sense to talk about absolute accuracies of 0.1 mm over a distance greater than 100 m unless very special equipment is available to measure the density and humidity of the air. Improvements to range differences can be achieved by the method of *length ratios*, which is of some value in monitoring *changes* in length between different epochs.

One way to improve matters is to use two different carrier waves for the measurement, because refractive index is wavelength-dependent. The dual frequency technique is standard for radio instruments, especially in satellite systems, where the additional problems of ionospheric refraction are also tackled.

Direct measurement

This system has two synchronized clocks, one at each end of the line to be measured. The time difference between the transmitted and received signal, suitably marked for identification, is recorded as the difference of the two clock readings, tim. Such a simple concept depends on very accurate time keeping and clock stability, such as is found in the atomic clocks used by GPS satellites.

The receivers, having short-term consistency, are able to determine their error on the GPS time scale by matching a pseudo-random code transmitted by the satellite with one held in their database. This can be done to a linear equivalent accuracy on the C/A code to about a few metres.

If there is a clock error $E = d$ tim, the distances will be in error by

$$ds = cd\,\text{tim} = cE$$

and the proportion error is given by

$$\frac{\mathrm{d}s}{s} = \frac{d\,\text{tim}}{\text{tim}}$$

Apart from a clock error, the precision of timing can be to 10^{-10}, which means that, for a GPS satellite some 26×10^6 m away from Earth, a relative precision of 0.0026 m is possible. However the actual speed of the signal is known only to about 10^{-6} due to refraction uncertainty, thus debasing the result to about 26 m. Ranging by this direct method is called *pseudo-ranging* because it still incorporates clock errors. Differential phase techniques can achieve the maximum accuracy of 2 mm under certain conditions, thus yielding line vectors on the Earth to this accuracy.

Indirect measurement

In this method the carrier wave is modulated to a longer wavelength on which a phase comparison is made. Typically the modulation produces an effective wavelength of 10 m. The modulation is produced by one of three methods.

1. *Amplitude modulation* (AM), in which the *intensity* of the signal is varied sinusoidally, is adopted by most infra-red instruments.
2. *Frequency modulation* (FM), in which the *frequency* is varied sinusoidally, is used by radio instruments such as the MRA 7 Tellurometer (now obsolete) and GPS satellites.
3. *Elliptical polarization*, in which an optical signal is polarized for phase comparison, is used by the two Froome-derived instruments, Mekometer and Geomensor.

In modern electro-optical instruments, the infra-red or laser light sources are modulated directly at frequencies, usually in the region of 15 MHz, giving an effective 10 m wavelength. GPS uses two carriers at the 1575.42 and 1227.6 MHz frequencies.

When distance is varying with time, as in satellite or hydrographic applications, the signal is sampled over a selected period and allowance made for the Doppler effect. Ground-based instruments can also operate in a *tracking* mode provided signal lock is maintained. This feature is useful in setting out works.

13.12 Instruments

Microwave radio instruments, such as GPS with special omni-directional aerials, generally require active receiver-transmitters at each line terminal. Electro-optical instruments are varied in design. Whatever the arrangement, all require a pointing telescope, a transmitting optic and a receiving optic.

Generally, all distances are obtained by subtracting a measurement to an internal reference point from the distance measured to the external reflector. This subtraction is to allow for the time drift between the effective electrical centre and the physical mounting of the instrument, and gives rise to the so-called *constant* index E.

Corner cubes are most commonly used as reflectors, although cat's eyes and plane mirrors have their applications. Some laser interferometers use expensive omni-directional hemispherical reflectors. Laser rangers can operate over limited distances without a reflector but with a drop in accuracy.

Ground-based microwave radio instruments (now obsolete) required two active units, one at each end of the line, and usually incorporated duplex telephones or other telemetric links.

Coaxial instruments

All current instruments have coaxial optics with the sighting telescope incorporated into the coaxial design. This feature is particularly useful in engineering work, because alignment can be carried out simultaneously with distance measurement.

An electro-optical distance measurer, illustrated in Figure 13.25 (not to scale), shows the instrument reference point at G close to a physical mark A. A modulated

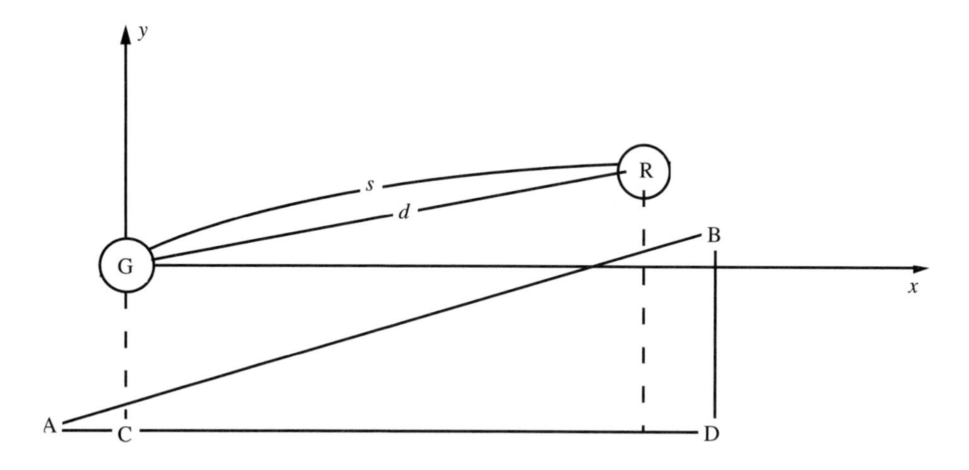

Figure 13.25

signal (light) is transmitted along some path s until it is reflected back to G via a reflector whose reference point R is close to another physical mark B. The surveyor wishes to know the length of the line AB between the physical marks, which can be related to the ground.

The EODM instrument measures the time interval tim which elapses between the transmission of the signal from G and its return along the double path via R. This time interval can be expressed in linear units as a distance $<s>$ called the *optical path length* defined by

$$<s> = \frac{1}{2} c \text{ tim}$$

where c is the speed of light in vacuo, in which the refractive index $n = 1$ by definition. If the average refractive index for the air mass through which the signal passes is n_A and if the recorded distance, the *geometrical path length*, is s_A, Fermatt's principle of stationary time means that

$$n_A s_A = <s> = \frac{1}{2} c \text{ tim} = \text{constant, K}$$

Thus

$$s_A = \frac{<s>}{n_A} = \frac{\text{K}}{n_A}$$

The refractive index is calculated from readings of air pressure, temperature and humidity. To find the chord length d and ultimately the distance AB requires further small corrections.

Frequency standard

The unit of length stems from the fundamental modulation frequency used by the instrument. This must be calibrated to guard against scale error. A laboratory check against a frequency counter controlled by a fundamental radio standard is quite easy to carry out. Some manufacturers provide equipment for this purpose. Experience shows that frequencies are now very stable to at least 10^{-6} over periods of years. In precise work, direct calibration should be carried out at least against a known distance, if the frequency is not tested.

Range

Although the range of microwave radio systems (MDM) is virtually limitless provided there is radio line-of-sight, electro-optical (EODM) systems suffer from *attenuation* or absorption of the signal, which limits their operational distance. The reduction dI in signal of intensity I due to absorption can be expressed by Lambert's Law:

$$\frac{\mathrm{d}I}{I} = -K\,\mathrm{d}x$$

where K is an absorption coefficient and x is the range. K varies with weather conditions and the carrier wavelength selected to utilize windows in the absorption spectrum. A wavelength of 0.93 μm is much used by infra-red instruments. From empirical data, a typical infra-red instrument gives the following equation relating range x in km achievable from a number N of standard Aga circular prisms

$$N = \exp(1.22(x - 0.7))$$

If $N = 1, x = 1.5$; if $N = 3, x = 1.6$, which shows the severe practical limitations on range. A realistic maximum number of prisms is 6.

Range can be extended by increasing the energy transmitted and by narrowing the beam width, for example by pulsed laser instruments, which can reach 15 km in clear weather.

13.13 Propagation and refraction

The speed of EM waves in a medium varies with the carrier wavelength. The phenomenon is known as *dispersion*. For normal dispersion i.e. if dn/dλ is negative, the refractive index n can be expressed by Cauchy's equation

$$n = A + \frac{B}{l^2} + \frac{C}{l^4} + \ldots \tag{13.2}$$

For air under the conditions 0°C, 760 mm pressure and 0.3% carbon dioxide content, the refractive index for optical wavelengths is given by

$$N = (n-1)\times 10^{-6} = 287.604 + \frac{1.6288}{l^2} + \frac{0.0136}{l^4}$$

For a typical visible-light instrument, $\lambda = 0.56$ giving $N = 292.9$ and for a typical infra-red instrument, $\lambda = 0.92$ giving $N = 289.5$.

The dispersion effect is used to great advantage in two-colour instruments to measure the refraction effect with high precision.

Dual Frequency measurement

Further corrections for the effects of refraction are made by a dual frequency technique. At optical wavelengths the troposphere affects the phase velocity of the signal. At microwave frequencies, the effect is most pronounced in the ionosphere.

Suppose the distances measured by a synchronous phase instrument at two frequencies, called red and blue for convenience, are s_R and s_B and the refractive indices are n_R and n_B, we have

$$\frac{<s>}{l} = \frac{s_R}{n_R} = \frac{s_B}{n_B}$$

$$\frac{s_R - s_B}{n_R - n_B} = \frac{<s> - s_B}{n_B}$$

which on rearranging gives

$$<s> = s_B + \frac{l - n_B}{n_R - n_B}(s_R - s_B)$$

$$<s> = s_B + \frac{l - n_B}{n_R - n_B}(s_R - s_B)$$

Using the figures for the visible and infra-red wavelengths mentioned above, we have A = 86. The Terrameter actually uses blue and red light, and has A = 55. This means that the difference $(s_R - s_B)$ has to be measured to very high precision (0.1 mm) to give a required length $<s>$ of 5 km to a precision of 10^{-6}.

A similar dual frequency principle is involved in improving the microwave radio measurements used in satellite systems to correct for ionosperic refraction affected by group velocity.

Group velocity

Because the distance is achieved on a modulation pattern, the speed of this modulation is required, not that of the carrier. This speed is obtained from the group refractive index n_G defined as

$$n_G = n - \lambda \frac{dn}{dl}$$

given by differentiating Equation (13.2)

$$\frac{dn}{dl} = -\frac{2B}{l^3} - \frac{4C}{l^5}$$

The correction is of the order of 10^{-5}.

Correction for refractive index

Built into every EDM system is a standard refractive index n_o, which enables it to give results s_o in length units, from the equation

$$s_o = \frac{c}{n_o}\text{tim}$$

We need the actual distance s_A using the actual refractive index n_A from the equation

$$s_A = \frac{c}{n_A}\text{tim}$$

Thus we have

$$s_A = \frac{n_o s_o}{n_A} \tag{13.3}$$

Optical wavelengths

The refractive index can be expressed in terms of meteorological parameters T (K, +273°C), atmospheric pressure P and water vapour pressure e (both in mm of Mercury), and the instrumental standard refractive index n_S by the equation (Barrel and Sears, 1969)

$$n_A - 1 = (n_S - 1)\frac{273}{T} \times \frac{P}{760} + \frac{15.02e}{T} \times 10^{-6} \tag{13.4}$$

The standard refractive index, which depends on the carrier wavelength, is provided by the manufacturer. Since $n = 1.0003$ approximately, the quantity $N = n \times 10^{-6} = 300$ is often used for simplicity. The following results, obtained by differentiation of Equation (13.4), are significant

$$\frac{dN}{dt} = -1; \quad \frac{dN}{dP} = 0.4; \quad \frac{dN}{de} = -0.05 \tag{13.5}$$

From these equations, it can be seen that the humidity effect can be neglected in all but the most accurate work and only temperature and pressure need be measured, to 1°C and 3 mm Mercury (4 mb), respectively. Thermometers and barometers need to be calibrated for reliable results.

Microwave radio corrections

The important difference between radio and optical waves is that the former are seriously affected by humidity. Therefore the relative humidity must be measured by wet and dry bulb thermometers. The corresponding refractive index formula (Froome and Essen, 1969) is

$$N = 103.49\frac{P}{T} - 17.23\frac{e}{T} + 49.59\frac{e}{T^2} \times 10^4 \tag{13.6}$$

The partial pressure of water vapour e is obtained from

$$e = e' - C\frac{t-t'}{755}$$

where $C = 0.5$ for a water-covered thermometer bulb and $C = 0.43$ for an ice-covered bulb. The saturation vapour pressure is obtained from Marvin's or Smithsonian tables, or estimated by the formula

$$e' \approx \exp(1.526 + 0.07243t' - 0.2945t'^2 \times 10^3)$$

Example

If P = 725 mm, t = 9°C and t' = 8°C, we find e' = 8.211; e = 8.210 and N = 317.

Differentiation of Equation (13.7) gives

$$\frac{dN}{dT} = -1; \quad \frac{dN}{dP} = 0.4; \quad \frac{dN}{de} = 6 \qquad (13.7)$$

The need to obtain e to an accuracy of 0.2 mm causes problems with psychrometers using wet and dry bulb thermometers, especially in hot, dry climates or in very cold climates.

13.14 Meteorological readings

It is customary to make pressure readings with a suitable barometer which must be calibrated against a mercury standard. If a surveyor's altimeter is used, it must be remembered that an index has been applied to the scale, to avoid negative height readings. The conversion from altimetric heights to equivalent pressure is made with the aid of the Smithsonian Table (number 51; see Appendix 1).

Temperature readings are made with traditional glass or solid-state thermometers which should also be calibrated. Taking wet bulb readings with a hand whirling psychrometer needs great care if systematically biased readings, and therefore lengths, are to be avoided. The water should be pure, distilled, and the sleeve covered. Rapid readings are essential after stopping the rotary motion, which should be vigorous, because the whole theory depends on a fast minimum speed being exceeded.

Care should be taken to ensure that thermometers are recording genuine air temperatures and that they have not been lying in sunlight.

At about 0°C the wet bulb, coated with super-cooled water, will give erroneous results. The water should be triggered to form ice by touching, before making readings.

13.15 Reduction of slant ranges to datum

In contrast with GPS and industrial measurements, it is unusual to calculate topographical coordinates in a three-dimensional system. The two-plus-one system of plan and height is usual. Therefore lines measured by EDM have to be reduced to a datum for incorporation into a survey network. The most elaborate case is the reduction of long lines to the ellipsoid which will be dealt with first.

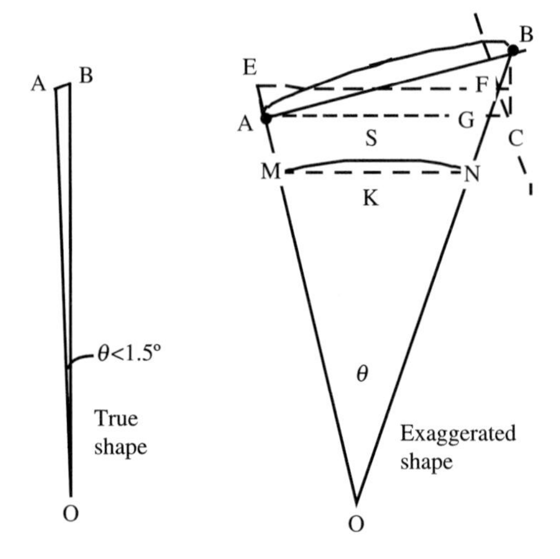

Figure 13.26

Corrections have to be applied for the following factors:

1. curvature of signal path;
2. second refraction correction;
3. chord to arc;
4. slope C_1;
5. height C_2.

The first three of these corrections, small even for long lines measured by obsolete microwave systems, are combined into one correction C_3. For details, see Allan (1997) and the example below. With lines under a few kilometres, only C_1 and C_2 are significant.

Slope correction C_1

In Figure 13.26, the chord AB of length D is to be reduced to the datum chord K = MN. The heights of A and B are, respectively, H_A and H_B and their difference ΔH. Since the arc subtended at the centre of the Earth is never more than 2° it is quite permissible to set

$$\Delta H = BG = BC.$$

Then let

$$AC = D_1 = \sqrt{D^2 - \Delta H^2} = D - (\beta + \frac{b^2}{2D} + \ldots) \tag{13.8}$$

by binomial expansion, where

$$\beta = \frac{\Delta H^2}{2D}$$

or

$$D_1 = D - C_1, \text{ say.}$$

Height correction C_2

We are considering the ellipsoidal heights of A and B. In Figure 13.26 since AC = EF, the chord length K is obtained by reducing D_1 from the mean height

$$H_M = \frac{1}{2}(H_A + H_B)$$

thus

$$K = \frac{RD_1}{R + H_M} = D_1 - \frac{D_1 H_M}{R} = D_1 - C_2 \quad (13.9)$$

with ample accuracy. Strictly R is the radius of curvature of the ellipsoid at latitude ϕ and azimuth α of the line AB given by Euler's theorem (see Appendix 4.7),

$$\frac{1}{R} = \frac{\cos^2 \alpha}{\rho} + \frac{\sin^2 \alpha}{\nu}$$

Typically the quantity $(2R)^{-1} \approx 7.85 \times 10^{-8}$ at latitude 30° and azimuth 45°, which gives the correction

$$C_2 \approx 7.85 D_1 (H_A + H_B) \times 10^{-8}$$

Summary

The final ellipsoidal arc S is given by

$$S = D - C_1 - C_2 + C_3 \quad (13.10)$$

Note there may still be a scale factor to apply if coordinates are treated on a projection. Note also that if the coordinates are being estimated by Least Squares in a three-dimensional system, the EDM line may only need reduction to the slant chord distance D, but with the inclusion of the first two small corrections only.

Example

A microwave radio MDM instrument with a standard refractive index of 1.000 275 gives the line AB to be 41 998.26 m. The actual refractive index calculated from meteorological data is 1.000 3182. The heights of A and B are respectively 90 m and 46 m above the ellipsoid, at a mean latitude of 51° and an azimuth of 66°. We calculate the ellipsoidal length S.

$$D = \frac{1.000275 \times 41998.26}{1.0003182} = 41996.45$$

$$C_1 = 0.02; \; C_2 = 0.45; \; C_3 = 0.04$$

$$S = D - C_1 - C_2 + C_3 = 41\ 996.02$$

Alternative formulae

When reducing very long lines, the following two alternative formulae should be considered, provided precision of computation is sufficient.

1. The arc S can be obtained from the angle θ subtended at the centre curvature of the ellipsoid from $S = R\,\theta$ where

$$\sin\frac{\theta}{2} = \sqrt{\frac{(s-a)(s-b)}{ab}}$$

and

$$a = R + H_A; \quad b = R + H_B; \quad 2s = a + b + D$$

2. The chord K can be computed from

$$K^2 = \frac{(D - \Delta H)(D + \Delta H)}{\left(1 + \frac{H_A}{R}\right)\left(1 + \frac{H_B}{R}\right)} \tag{13.11}$$

In both formulae the correct ellipsoidal value for R has to be used.

Reduction of short lines

In the reduction of short distances of less than 1 km, the small corrections can usually be neglected. For example, if $D = 45.678$ m and the slope angle $V = 20°$ then

$$D_1 = D \cos V = 42.923 \text{ m}$$

Attention is however drawn to the need for care with steep slope reduction, and the accurate measurement of slopes or height differences. If a precision of 1 mm is sought, the accuracy with which slope angles have to be measured is given by

$$\mathrm{d}V'' = \frac{206.265}{D \sin V} = 14$$

or the difference in height to about 3 mm.

13.16 Satellite ranging corrections

The troposphere and ionosphere both seriously degrade ranging to satellites, the former by as much as 10 m depending on the angle of elevation, and the latter by as much as 150 m depending on sunspot activity. Refraction models for dual frequency systems and tropospheric effects are complex. However, as most systems are able to be employed in differential modes, many of these effects are removed when measuring relatively short (up to 20 km) ground distances. In fact, there is evidence to support the view that attempts to apply tropospheric corrections for lines in excess of 20 km debases, rather than enhances, results.

Integer counts

A distance S can be expressed as

$$S = MU + D$$

where U is the unit of measurement, D is a fractional part of this unit and M is an integral number of such units. For example, we can express the length 123.45 m as $123 \times 1 + 0.45$; here $M = 123$, $U = 1$ m and $D = 0.45$ m.

In most systems (but not all) the fractional part is obtained from a reliable phase measurement. The integer count M is derived by most systems by introducing different frequencies (or wavelengths) to yield different values for the fractional part of the distance D on each frequency. These are subtracted and scaled to give M. The principle is like a Vernier scale used in reverse (Allan, 1997).

For various reasons it is possible to miscount M and therefore make an error of some magnitude depending on the value of U. In satellite systems such as GPS, U is of the order of 20 cm. Hence just one integer miscount (or cycle slip) introduces this error. Much care is needed in processing to eliminate these possible effects (see Chapter 7, Section 7.3).

13.17 Calibration and testing

Calibration and testing form part of the much wider topic of quality assessment, in which the characteristics of an instrument are established and verified. General characteristics can include reliability, robustness, range and accuracy of an instrument. Instrument parameters are assessed with a view to improving performance by the application of corrections. For example, a prism-instrument index will be quantified and applied to all distances measured with the instrument.

In testing we simply determine if the instrument is performing to a specification, usually claimed by the manufacturer. If it does not meet the criteria, then it may be possible to calibrate it so that it does. This latter process is usually the task of the manufacturer.

Scale checks

The scale of an instrument can be wrong for two reasons:

1. the refractive index n_A is incorrect, or has been incorrectly applied, and
2. the fundamental frequency of the oscillator, from which the effective wavelength λ_o is derived, is out of specification.

Bearing in mind that $n \approx 1.0003$, a specification of 10^{-6} implies that, for an accuracy of 0.1 mm, a distance of 100 m lies at the limit of our competence. Since modern surveying instruments have this high performance, this matter is not entirely theoretical. At the accuracy level of 5 mm the problems of refraction are solvable, provided due care is taken in measuring the meteorological parameters.

There are two ways to check the frequency:

1. directly by measuring the fundamental frequency of the generator producing the wavelength, and
2. indirectly by comparing a measured distance against a better-known value, usually obtained from a laser interferometer or a more accurate EDM, or by taping.

Frequency calibration

Suppose the modulation frequency used for the accurate distance measurement is 100 MHz, and the instrument is provided with a port for comparison with a frequency counter, which in turn is regulated by an 'off air' radio standard. Typical comparison figures might show accuracies of 1 in 10^{10} for a new instrument. Ageing of crystals causes frequency drift with time so annual checks are advised, thus avoiding scale errors.

Field calibration and testing

In field-testing or calibration we make measurements of a number of lines, whose lengths are selected to sample the effective pattern wavelength of the modulation at stepped intervals. There are advantages and disadvantages in making these steps equal. It is assumed that the EDM is always properly operated, slope corrections applied, and that refractive indices are set to the standard used by the manufacturer. Meteorological readings are taken to allow the actual refractive index to be applied later. In this way the work can be checked. Stability of instrument stands is essential, pillars being ideal provided they are properly constructed and monitored for movement. Other marks are suitable, such as ground or wall marks on well-established concrete or stone. With such marks, the transfer of centres has to be done with great care. Reflectors mounted on theodolite telescopes looking at the zenith are ideal for this purpose (Reuger, 1996).

13.18 Graphic surveying

Still in use today are two graphical surveying methods. One uses a hand held reflectorless laser ranger, together with alignment techniques to revise existing maps. These simple measurements are processed on site in a specially designed laptop computer (see Section 10.3). The other uses a small portable plane table to draw maps directly in the field, again assisted by a laser ranger mounted on an alidade. The device still has its professional uses and is much valued in teaching. The surveyor sights through a telescope or open sight to which a parallel rule is attached. The direction is drawn directly on to the plotting surface, which is either paper or plastic. The alidade may also incorporate a small EDM and clinometer to measure distances and slopes.

Both systems are cheap and very versatile and able to be used for any ground survey technique: intersection, resection, contouring and radiation, etc. They suffer from the disadvantage of simplicity and lack of technological image. In the hands of an expert it is a most efficient and cost-effective tool with which to carry out small surveys.

13.19 Future vision and automated systems

The most obvious vision system used in surveying is the human eye together with its linkage to the human brain. This system has the ability to interpret what is seen as well as to resolve an image. Much effort is being made to copy this process by electronic detectors coupled to a computer as part of measurement systems to be used by automated levels, motorised total stations and robotic survey instruments. Consideration of the human vision system tells us what automated systems must do. Now two digital cameras are used to copy the human vision system: one to deal with paraxial precision and the other to deal with peripheral vision. The automated systems have the advantage of direct links to computer processing and the potential to achieve better accuracy. In such systems, the observer does not look through the telescope at all, viewing a small visual display instead.

The total automation of the surveying process is unlikely, if only on the grounds of costs. For many applications, however, such as in the de-commissioning of nuclear reactors, it is the only way ahead.

Appendix 1
Useful Data

Fractions and multiples of units

power of 10	*prefix*	*symbol*
– 1	deci	d
– 2	centi	c
– 3	milli	m
– 6	micro	μ
– 9	nano	n
– 12	pico	p
– 15	femto	f
– 18	atto	a
+ 1	deka	da
+ 2	hecto	h
+ 3	kilo	k
+ 6	mega	M
+ 9	giga	G
+12	tera	T

Units of length

1 metre (trad) = length of a quadrant of the Earth $\times 10^{-7}$
$= 0.5\ \pi\ \mathrm{R}\ 10^{-7}$ where R is the radius of the Earth

Radius of the Earth $\mathrm{R} \approx 6.4 \times 10^{6}$ m

1 metre = distance travelled by light in vacuo during a period of 1/ 299 792 458 th of a second

1 British foot = 0.3048 metre (m) exactly (one inch= 254 mm)

1 British statute mile = 5280 feet (ft)
= 1760 yards (yd)
= 1609.344 m
$\approx \frac{8}{5}$ km

1 nautical mile = average distance subtended by a sexagesimal minute of arc on the surface of the Earth
= 1.853 18 km

1 ft = 12 inches (in)
1 inch = 25.4 mm exactly
1 yard = 3 ft
1 chain = 22 yards = 66 feet

Area

1 are (a) = 10 square metres (m^2)
1 hectare (ha) = 10^2 ares = 10^4 m^2
= twice the area of the average football pitch

1 acre = 10 square chains ≈ 0.405 ha
= about the area of a football pitch

Angle

1 radian (rad) = the angle subtended by the arc of a circle equal to its radius
1 radian = $\frac{180}{\pi}$ sexagesimal degrees (°)

≈ 57.3° ≈ 3438' ≈ 206265"
= $\frac{200}{\pi}$ centesimal degrees (gon) (g)

2 π radians = one cycle = 360 sexagesimal degrees (°)
= 400 centesimal degrees (g)

60 sexagesimal minutes of arc (') = one sexagesimal degree
60 sexagesimal seconds of arc (") = one sexagesimal minute

100 centesimal minutes (c) = one centesimal degree (g)
100 centesimal seconds (cc) = one centesimal minute (c)

one sexagesimal minute ≈ 0.003 rad
one sexagesimal second ≈ 3 centesimal seconds (cc)
≈ 0.000 05 rad

Temperature conversion

Fahrenheit to Celsius

$$C = \frac{5}{9}(F - 32)$$

Equivalent pressure units

The conversion of barometer readings in inches (in) or millimetres (mm) of Mercury to millibars (mb) is given in the following table. In accurate EDM work the barometer has to be calibrated against a mercury (Fortin) or other accurate barometer.

Most altimeters reading in feet (ft) or metres (m) require the scale zero index of 1000 ft (or other) to be subtracted from readings before using the table.

in	*mb*	*mm*	*mb*	*ft*	*mb*	*m*	*mb*
19	643	490	653	– 1000	1050	– 250	1044
20	677	515	687	0	1013	0	1013
21	711	540	720	1000	976	250	982
22	745	565	753	2000	941	500	953
23	779	590	787	3000	907	750	925
24	813	615	820	4000	874	1000	897
25	847	640	853	5000	842	1250	871
26	880	665	887	6000	812	1500	845
27	914	690	920	7000	783	1750	820
28	948	715	953	8000	754	2000	795
29	982	740	987	9000	727	2500	749
30	1016	765	1020	10 000	701	3000	705
31	1050	790	1053	12 000	651	3500	664

The Greek alphabet

α	alpha	ν	nu
β	beta	ξ	xi
γ	gamma	ο	omicron
δ	delta	π	pi
ε	epsilon	ρ	rho
ζ	zeta	σ	sigma
η	eta	τ	tau
θ	theta	υ	upsilon
ι	iota	ϕ	phi
κ	kappa	χ	chi
λ	lambda	φ	psi
μ	mu	ω	omega

Appendix A2
Spherical Trigonometry

Three basic formulae

In this appendix we derive the three basic formulae of spherical trigonometry, and quote several others which may be derived from them. Napier's rules for the relationships between the circular parts of a right-angled spherical triangle are also given.

The formula for the spherical excess of such a triangle is derived.

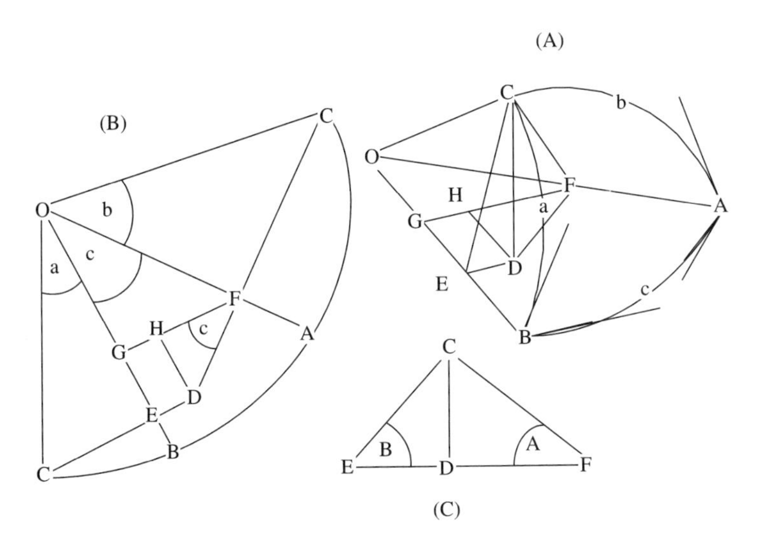

Figure A2.1

A2.1 The spherical triangle

The triangle ABC formed by three great circles of a sphere, centre O and unit radius, is shown in Figure A2.1(A). If the sphere is *developed* along the radius OC it can be presented as the sector of a unit circle shown in Figure A2.1(B). The figure CEDF of Figure A2.1(A), in which CD is perpendicular to the plane OAB, is developed as Figure A2.1(C).

The *angles* of a spherical triangle are the angles between the tangents to the sphere at each apex point.

The *"sides"* are the *angles* subtended by the linear arcs at the centre of the sphere. This definition of sides means that the size of the sphere is of no importance. Thus we can treat a unit sphere as being quite general.

If the reader has difficulty in visualizing the three-dimensional nature of Figure A2.1(A), Figure A2.1(B) should be redrawn at a larger scale, cut out and folded along OA and OB to create a three-dimensional model, with CD represented by an elastic band.

A2.2 Sine formula

In triangles OCE, OCF, CED and CFD of Figure A2.1(A), we have respectively

$$CE = \sin a : \quad CF = \sin b : \quad CD = \sin a \sin B \quad : \quad CD = \sin b \sin A$$

Equating the two expressions for CD gives the sine rule

$$\frac{\sin a}{\sin A} = \frac{\sin b}{\sin B}$$

Similarly

$$\frac{\sin a}{\sin A} = \frac{\sin c}{\sin C}$$

therefore

$$\frac{\sin a}{\sin A} = \frac{\sin b}{\sin B} = \frac{\sin c}{\sin C} \qquad \text{A2(1)}$$

This formula links *sides* with their opposite *angles*. Although a convenient formula, care has to be taken with signs, because the sine is positive in both the first and second quadrants. For this reason, a more apparently complicated alternative formula is often preferred.

A2.3 Cosine formula

Consider triangles FDH, CDF and OCF of Figure A2.1(A) in turn to give

$$HD = DF \sin c = \sin b \cos A \sin c$$

From triangles OCE, OFG and OCF in turn we have also

$$HD = GE = OE - OG = \cos a - \cos b \cos c$$

Equating values of HD gives the cosine formula:

$$\cos a = \cos b \cos c + \sin b \sin c \cos A \qquad \text{A2(2)}$$

This is the most useful formula of spherical trigonometry. It relates two sides and their included angle to the side opposite. There is no sign ambiguity between the first and second quadrants.

A2.4 Cotangent formula

Consider triangles HFD, CDF and OCF of Figure A2.1 (A) to give

$$\text{HF} = \text{DF} \cos c = \sin b \cos A \cos c$$

From triangles OFG, OCF, CED, and OCE we have

$$\text{HF} = \text{GF} - \text{GH} = \text{GF} - \text{ED} = \cos b \sin c - \sin a \cos B$$

Equating values of HF gives

$$\sin b \cos A \cos c = \cos b \sin c - \sin a \cos B$$

Dividing by sin b gives

$$\cos A \cos c = \cot b \sin c - \frac{\sin a \cos B}{\sin b}$$

$$\cos A \cos c = \cot b \sin c - \sin A \cot B \qquad \text{A2(3)}$$

This cotangent formula relates four adjacent parts of the triangle.

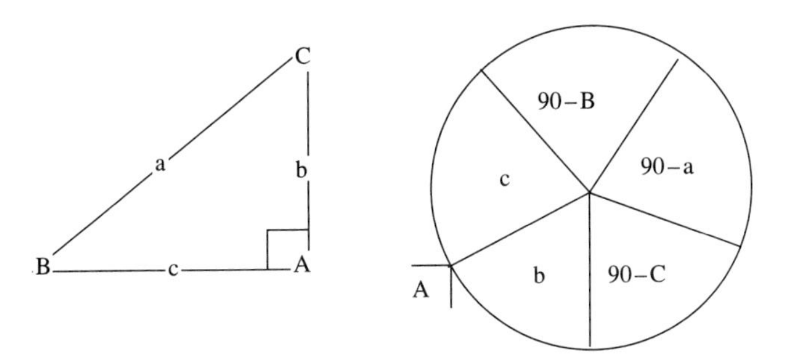

Figure A2.2

A2.5 Napier's rules for a right-angled triangle

If one angle of a spherical triangle is 90° the various formula simplify considerably. For example from equation (1), if angle A is right, we have

$$\sin a \sin B = \sin b$$

The Scottish mathematician, Napier, showed that thirty formulae for a right angle spherical triangle can be found by applying the simple rules which bear his name.

Consider the triangle ABC right-angled at A. A circle with five sectors is drawn as shown in Figure A2.2. The right angle A is marked outside the circle as indicated. Into the sectors are written the five remaining parts of the triangle, according to the following scheme.

Parts adjacent to the right angle are written as they stand, but the complements of the others are inserted in correct order as shown.

Ten formulae for the right angle at A are obtained by applying the following rules.

s**I**ne of the m**I**ddle part = product of the t**A**ngents of **A**djacent parts

= product of the c**O**sines of the **O**pposite parts

The *middle part* is the parameter in any of the five sectors. This selection defines the parts *adjacent* and *opposite* to it.

For example, if we select 90 – a as the middle part we obtain the formulae

$$\sin(90 - a) = \cos a = \tan(90 - B)\tan(90 - C) = \cot B \cot C$$

and

$$\cos a = \cos b \cos c$$

Or if we select b as the middle part we have, as before,

$$\sin b = \cos(90 - B)\cos(90 - A) = \sin B \sin A$$

and

$$\sin b = \tan c \cot C$$

There are ten similar formulae for right angles at B and at C.

A2.6 Spherical excess

The amount by which the sum of the angles of a spherical triangle exceeds 180° is called the spherical excess ε: in other words

$$A + B + C = 180° + \varepsilon$$

On a sphere of radius R, the spherical excess in seconds of arc of a triangle, whose area is Δ, is given by

$$\varepsilon'' = \frac{\Delta}{R^2} \times 206\,265 \qquad \text{A2(4)}$$

In spheroidal work $R = \sqrt{\rho v}$ (See Appendix A4.10)

Thus a terrestrial triangle, such as an equilateral triangle of 20 km side, whose area is about 180 km^2 as a spherical excess of one second of arc.

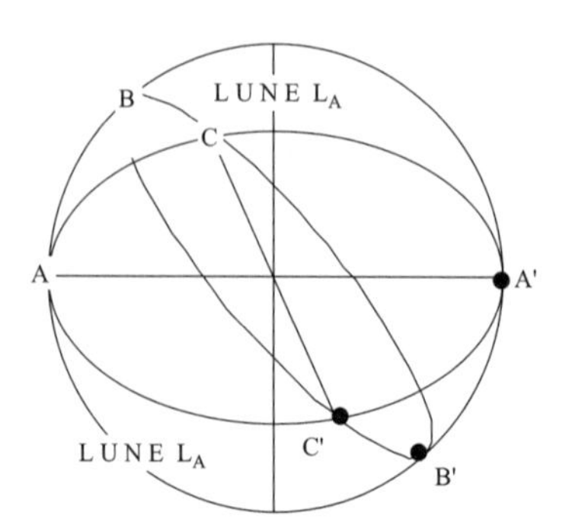

Figure A2.3

To derive equation A2(4) refer to Figure A2.3. The beginner may care to use a table tennis ball as a model on which to draw lunes to follow the arguments used in the derivation.

The surface area of a complete sphere can be thought of as the area swept out by a great circle as it rotates 180°, or π radians, about a diameter. For any intermediate angle A the surface area of the *lune* swept out, L_A, is proportional to A, thus we have

$$\frac{A}{\pi} = \frac{L_A}{4\pi R^2}$$

The formula for the spherical excess follows by considering the surface areas of the lunes L_A, L_B and L_C formed by the three angles A, B and C. Adding these areas we have

$$\frac{A}{\pi} + \frac{B}{\pi} + \frac{C}{\pi} = \frac{L_A}{4\pi R^2} + \frac{L_B}{4\pi R^2} + \frac{L_C}{4\pi R^2}$$

$$A + B + C = \frac{1}{4R^2}(L_A + L_B + L_C)$$

But $L_A + L_B + L_C =$ the surface area of the sphere $+ 4\Delta = 4\pi R^2 + 4\Delta$ therefore

$$A + B + C = \pi + \frac{\Delta}{R^2}$$

Thus

$$\varepsilon = \frac{\Delta}{R^2}$$

Some additional formulae

The following additional formulae of spherical trigonometry are sometimes of value in surveying.

A2.7 Half angle formulae

If the perimeter of the spherical triangle ABC is denoted by

$$2s = a + b + c$$

we have

$$\sin^2 \tfrac{1}{2}A = \frac{\sin(s-b)\sin(s-c)}{\sin b \sin c} \qquad \text{A2(5)}$$

$$\cos^2 \tfrac{1}{2}A = \frac{\sin s \sin(s-a)}{\sin b \sin c} \qquad \text{A2(6)}$$

$$\tan^2 \tfrac{1}{2}A = \frac{\sin(s-b)\sin(s-c)}{\sin s \sin(s-a)} \qquad \text{A2(7)}$$

When computing angles from sides, formula A2(7) is the safest to use in all cases.

A2.8 Delambre's formulae

The following fomulae are useful to derive meridian convergence.

$$\tan\tfrac{1}{2}(A+B)\ \tan\tfrac{1}{2}C = \frac{\cos\tfrac{1}{2}(a-b)}{\cos\tfrac{1}{2}(a+b)} \qquad \text{A2(8)}$$

$$\tan\tfrac{1}{2}(A-B)\ \tan\tfrac{1}{2}C = \frac{\sin\tfrac{1}{2}(a-b)}{\sin\tfrac{1}{2}(a+b)} \qquad \text{A2(9)}$$

Appendix A3
General Least Squares

Observations, conditions and constraints

In this appendix we derive the principal equations used in the Least Squares process. In this treatment we assume that the reader is familiar with the basic rules of matrix algebra. Worked examples are to be found throughout the book where they arise in appropriate applications. In this appendix the most important results are inscribed within a box for easy location.

The various practical matrix operations may seem complicated, but when considered in the light of modern computer software, and spreadsheet systems, small problems of up to 40×40 size are easily solved by direct matrix operations. It is important to note that because matrices **N, W**, and their inverses are symmetric, they are unaffected by transposition.

A3.1 Mathematical models and notation

In Least Squares estimation problems two types of mathematical models are used: functional models and statistical models.

The first type concern the surveying processes which relate measurements, or *observed* parameters, to the required quantities, or *unobserved* parameters. An example of a functional model is an equation which relates the distance between two points to their coordinate differences. The treatment of these models will be written in ordinary type such as

$$s^2 - \Delta x^2 - \Delta y^2 = 0$$

Partial differentiation of the above equation is also written in ordinary type as follows

$$2s\,\delta s - 2\,\Delta x\,\delta\Delta x - 2\,\Delta y\,\delta\Delta y = 0$$

$$s\,\delta s - \Delta x\,\delta\Delta x - \Delta y\,\delta\Delta y = 0 \qquad \text{A3(1)}$$

In this equation the observed quantity may be the distance s, with Δx and Δy being the parameters. However, the measured values, say from a map or air photo, could easily be the coordinate differences Δx and Δy, with s, the distance derived from them, being the parameter.

Ultimately we reduce the problem to a standard mathematical form to be treated in matrix terms expressed in heavy type. Then the parameters will be expressed as the vector **x** and the observations as the vector **s**. This change of notation may confuse the beginner. To avoid some possible confusion we prefer to develop the full equations in ordinary algebra normally associated with a problem, before converting to matrix notation. For example;

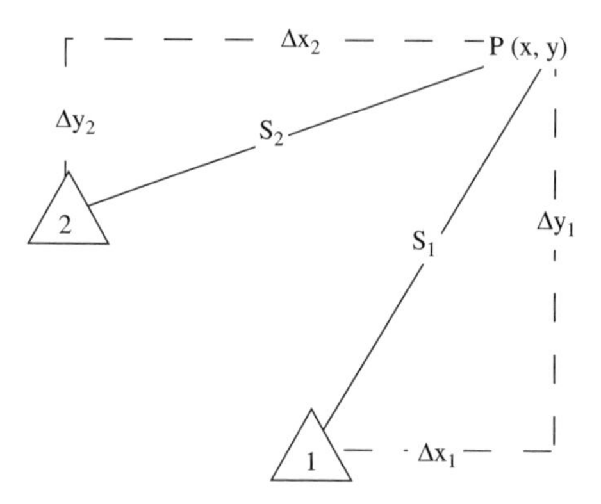

Figure A3.1

Refer to Figure A3.1. Consider the two equations, in two parameters (x, y), which describe the fixation of a new point P by two distances s_1 and s_2 measured from two fixed points 1 and 2. That is

$$s_1^2 = \Delta x_1^2 + \Delta y_1^2$$
$$s_2^2 = \Delta x_2^2 + \Delta y_2^2$$

The observation equations are

$$s_1\, ds_1 - \Delta x_1\, dx - \Delta y_1\, dy = 0$$
$$s_2\, ds_2 - \Delta x_2\, dx - \Delta y_2\, dy = 0$$

In matrix form these equations are written in full as

$$\begin{bmatrix} -\Delta x_1 & -\Delta y_1 \\ -\Delta x_2 & -\Delta y_2 \end{bmatrix} \begin{bmatrix} dx \\ dy \end{bmatrix} + \begin{bmatrix} s_1 & 0 \\ 0 & s_2 \end{bmatrix} \begin{bmatrix} ds_1 \\ ds_2 \end{bmatrix} = \begin{bmatrix} 0 \\ 0 \end{bmatrix}$$

and in condensed matrix notation as

$$\mathbf{A\,x} + \mathbf{C\,s} = \mathbf{0} \qquad \text{A3(2)}$$

However, if the *observations* are the *coordinates*, we write

$$\begin{bmatrix} s_1 & 0 \\ 0 & s_2 \end{bmatrix} \begin{bmatrix} ds_1 \\ ds_2 \end{bmatrix} + \begin{bmatrix} -\Delta x_1 & -\Delta y_1 \\ -\Delta x_2 & -\Delta y_2 \end{bmatrix} \begin{bmatrix} dx \\ dy \end{bmatrix} = \begin{bmatrix} 0 \\ 0 \end{bmatrix}$$

To be consistent, keeping the parameters as **x**, the condensed equations *are still written*

$$\mathbf{A\,x} + \mathbf{C\,s} = \mathbf{0}$$

Such changes in notation between the two types of mathematical model can be confusing to the unwary. The reason for adopting a standard notation for the matrix algebra is to enable all problems to be treated in the same way once the matrix model has been formed. Thus we will derive the observation equations first in terms of the mathematical notation common to a particular branch of surveying, before recasting them in the standard matrix form of equation A3(2).

As explained in the next section, the vector **s** is usually expressed as the sum of two vectors **L** and **v,** giving the standard equations the following form

$$\boxed{\mathbf{A\,x} + \mathbf{C\,v} + \mathbf{C\,L} = \mathbf{0}} \qquad \text{A3(3)}$$

Notation

In discussing statistical models a rather complex notation is required to distinguish between *observed, provisional* and *estimated* versions of a parameter. Once the model has been cast into a standard form, the notation can be simplified without confusion.

Figure A3.2 illustrates the notation used to described the various statistical values used in error analysis.

The reader should remember that a point can have three positions: a *provisional position*, a *best position* estimated from the sample observations, and a *population position*, sometimes referred to as the true position. Generally the observed parameters will not fit any of these positions. Therefore, for statistical purposes the notation followed in this book is as follows.

Observed parameters or "observations" for short	$\overset{o}{s}$
Selected provisional values of these observations	$\overset{*}{s}$
Best estimates of these observed parameters	$\hat{s}$
Population values of the observed parameters	$\bar{s}$
Errors in best estimates	$\delta s = \bar{s} - \hat{s}$
Sample residuals	$v = \hat{s} - \overset{o}{s}$
Population residuals	$V = \bar{s} - \overset{o}{s}$
Differential changes	$ds = \hat{s} - \overset{*}{s} = \overset{o}{s} - \overset{*}{s} + v = L + v$

$$\text{Where } L = \overset{o}{s} - \overset{*}{s}$$

The objective is to find the *best estimates* of all parameters. The best estimates are invariably the best linear unbiased estimates (BLUE) obtained by applying the Least Squares principle to sampled known values of L, i.e. from the observations.

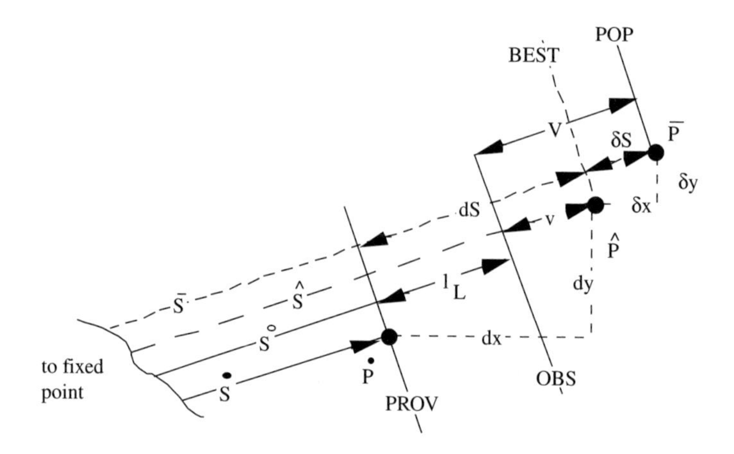

Figure A3.2

In addition to the observed parameters we often have to estimate additional parameters for which no observations are available. The notation for these is

Best estimates of the unobserved parameters	$\hat{x}$
Population values of the unobserved parameters	$\bar{x}$
Errors in best estimates	$\delta x = \bar{x} - \hat{x}$
Selected provisional values of these parameters	$\overset{*}{x}$
Differential changes	$dx = \hat{x} - \overset{*}{x}$

The key relationships are summarized as follows:

Observations

$$\hat{s} = \overset{*}{s} + ds$$

thus

$$ds = \overset{o}{s} - \overset{*}{s} + v$$

$$ds = L + v$$

Parameters

$$\hat{x} = \overset{*}{x} + ds$$

Notice that the difference between these relationships is that there are no residuals attached to the unobserved parameters, and that when written in matrix algebra the small changes δs and δx become **s** and **x** respectively.

The significance of the "^" is that *Least Squares estimates* of the parameters **x** are obtained. Where no confusion arises we omit the hat from the text.

In the following sections the population values are not considered until the matter of dispersion matrices is discussed.

A3.2 Mathematical models for the general case

Ordinary algebra

To be consistent, the best estimates and the provisional values must satisfy the mathematical models exactly, thus

$$F(\hat{x} : \hat{s}) = O \quad \text{and} \quad F(\overset{*}{x} : \overset{*}{s}) = O \qquad \text{A3(4)}$$

Expressing the model in terms of small differential changes we have

$$F(\hat{x} : \hat{s}) = F(\overset{*}{x} : \overset{*}{s}) + \left(\frac{\partial F}{\partial \overset{*}{x}}\right) dx + \left(\frac{\partial F}{\partial \overset{*}{s}}\right) ds$$

therefore

$$\left(\frac{\partial F}{\partial \overset{*}{x}}\right) dx + \left(\frac{\partial F}{\partial \overset{*}{s}}\right) ds = O$$

and since

$$ds = L + v$$

we have

$$\left(\frac{\partial F}{\partial \overset{*}{x}}\right) dx + \left(\frac{\partial F}{\partial \overset{*}{s}}\right) v + \left(\frac{\partial F}{\partial \overset{*}{s}}\right) L = O \qquad \text{A3(5)}$$

The partial differentials are gathered together into a matrix, called the Jacobian matrix, when expressed in matrix algebra.

Matrix algebra

We adopt the convention that there are m observations, having a weight matrix **W,** in n parameters linked together by r equations of type A3(5), which in matrix form, with their dimensions, are

$$\underset{(r\times n)}{\mathbf{A}}\ \underset{(n\times 1)}{\mathbf{x}} + \underset{(r\times m)}{\mathbf{C}}\ \underset{(m\times 1)}{\mathbf{v}} + \underset{(r\times m)}{\mathbf{C}}\ \underset{(m\times 1)}{\mathbf{L}} = \underset{(r\times 1)}{\mathbf{0}} \qquad \text{A3(6)}$$

Because $L = \overset{o}{s} - \overset{*}{s}$ are both known, and the design matrices **A** and **C** are also known and the product **CL** is also known, and we let

$$\mathbf{C\ L} = -\mathbf{b}$$

therefore equation A3(6) becomes

$$\boxed{\mathbf{A\ x} + \mathbf{C\ v} = \mathbf{b}} \qquad \text{A3(7)}$$

To obtain a unique solution for $\mathbf{x}$ we invoke the principle of Least Squares, making the sum of squares of the weighted residuals ($\mathbf{v}^T \mathbf{W} \mathbf{v}$) a minimum. For convenience we denote the expression ($\mathbf{v}^T \mathbf{W} \mathbf{v}$) by Ω.

Introducing a vector λ of Lagrangian multipliers, of dimension ($r \times 1$), we may put

$$\Omega = \mathbf{v}^T \mathbf{W} \mathbf{v} - 2 \lambda^T (\mathbf{A}\mathbf{x} + \mathbf{C}\mathbf{v} - \mathbf{b}) \qquad \text{A3(8)}$$

This vector $\boldsymbol{\lambda}$ of *correlates* or *undetermined multipliers* is a device invented by the French mathematician Lagrange to treat problems of conditioned minima. Equation A3(8) is valid because the second term is zero.

Notice that the dimension of Ω is (1×1) i.e. a single number. The $(m \times m)$ weight matrix $\mathbf{W}$ is formed from the inverse of the dispersion matrix of the observations estimated in some way. (See chapter 5.)

We now partially differentiate the function Ω with respect to the variables $\mathbf{x}$ and $\mathbf{s}$, and equate them to zero. Before doing so, however, we will establish that differentiation with respect to the observed parameter $\mathbf{s}$ is the same as differentiation with respect to $\mathbf{v}$.

$$\left(\frac{\partial\Omega}{\partial\mathbf{s}}\right) = \left(\frac{\partial\Omega}{\partial\mathbf{v}}\right)\left(\frac{\partial\mathbf{v}}{\partial\mathbf{s}}\right)$$

Because $\mathbf{s} = \mathbf{L} + \mathbf{v}$, and L is a constant,

$$\left(\frac{\partial\mathbf{v}}{\partial\mathbf{s}}\right) = \mathbf{I}$$

and thus

$$\left(\frac{\partial\Omega}{\partial\mathbf{s}}\right) = \left(\frac{\partial\Omega}{\partial\mathbf{v}}\right)$$

For a minimum value of Ω we have

$$\left(\frac{\partial\Omega}{\partial\mathbf{v}}\right) = \mathbf{O} \text{ and } \left(\frac{\partial\Omega}{\partial\mathbf{x}}\right) = \mathbf{O}$$

Applying these expressions to equation A3(8) gives

$$\left(\frac{\partial\Omega}{\partial\mathbf{v}}\right) = \mathbf{O} = 2\mathbf{v}^T\mathbf{W} - 2\lambda^T\mathbf{C}$$

Since $\mathbf{W}$ is symmetric

$$\mathbf{W}^T \mathbf{v} = \mathbf{C}^T \lambda = \mathbf{W}\mathbf{v}$$

Thus we have a most important equation

$$\boxed{\mathbf{v} = \mathbf{W}^{-1} \mathbf{C}^T \lambda} \qquad \text{A3(9)}$$

And again

$$\left(\frac{\partial\Omega}{\partial\mathbf{x}}\right) = \mathbf{O} = 2\lambda^T\mathbf{A}$$

therefore

$$\boxed{\mathbf{A}^T \lambda = \mathbf{0}} \qquad \text{A3(10)}$$

Substituting for **v** from A3(9) in A3(7) gives

$$\mathbf{A\,x} + \mathbf{C}\ \mathbf{W}^{-1}\mathbf{C}^T \lambda = \mathbf{b}$$

Putting $\mathbf{N} = \mathbf{C}\ \mathbf{W}^{-1}\mathbf{C}^T$

we have $\mathbf{A\,x} + \mathbf{N}\ \lambda = \mathbf{b}$

and

$$\lambda = \mathbf{N}^{-1}(\mathbf{b} - \mathbf{Ax}) \qquad \text{A3(11)}$$

Substituting for λ from A3(11) in equation A3(10) gives

$$\mathbf{A}^T \mathbf{N}^{-1}(\mathbf{b} - \mathbf{Ax}) = \mathbf{0} \qquad \text{A3(12)}$$

$$\mathbf{A}^T \mathbf{N}^{-1}\mathbf{A\,x} = \mathbf{A}^T \mathbf{N}^{-1}\mathbf{b}$$

And putting $\mathbf{N}_1 = \mathbf{A}^T \mathbf{N}^{-1}\mathbf{A}$ and $\mathbf{b}_1 = \mathbf{A}^T \mathbf{N}^{-1}\mathbf{b}$

we have the final set of equations

$$\boxed{\mathbf{N}_1 \mathbf{x} = \mathbf{b}_1} \qquad \text{A3(13)}$$

The matrix $\mathbf{N}_1$ is square and symmetric. Traditionally the equations A3(13) were called 'normal equations'. Their solution gives the required parameters.

From A3(7) and A3(12) we obtain another very important result

$$\boxed{\mathbf{A}^T \mathbf{N}^{-1}\mathbf{C}\ \mathbf{v} = \mathbf{0}} \qquad \text{A3(14)}$$

It is so important that we remind readers that

$$\mathbf{N} = \mathbf{C}\ \mathbf{W}^{-1}\mathbf{C}^T$$

A3.3 Mathematical model with added constraints

It sometimes happens that there are certain constraints amongst the parameters which need to be allowed for. For example we may need to hold a length fixed between two points whose positions may be allowed to vary. A constraint is written as a linear equation which has to be satisfied exactly, i.e. as an equation, which has no residual, of the form

$$\mathbf{E}\,\mathrm{x} - \mathbf{d} = \mathbf{0} \qquad \text{A3(15)}$$

It is often possible, and thoroughly desirable for simplicity, to eliminate one parameter for each constraint, right at the outset. Not only does this simplify the solution but it reduces the size of the matrix to be solved. The simplest application of this procedure is to eliminate the variables for fixed points in a network.

Another practical way to solve the problem is to assign a very high weight to the constraint equations, which are then treated as observations, thus ensuring that the constraints are almost satisfied.

A theoretically correct and standard procedure is to employ further Lagrangian multipliers **k,** one for each constraint equation, and proceed as before, and as follows.

$$\text{Let} \quad \Omega = \mathbf{v}^T \mathbf{W} \mathbf{v} - 2\,\boldsymbol{\lambda}^T (\mathbf{A}\mathbf{x} + \mathbf{C}\mathbf{v} - \mathbf{b}) - 2\,\mathbf{k}^T (\mathbf{E}\mathbf{x} - \mathbf{d})$$

Putting

$$\left(\frac{\partial \Omega}{\partial \mathbf{x}}\right) = \mathbf{O}$$

gives

$$\mathbf{O} = 2\,\boldsymbol{\lambda}^T \mathbf{A} + 2\,\mathbf{k}^T \mathbf{E}$$

therefore

$$\mathbf{A}^T \boldsymbol{\lambda} + \mathbf{E}^T \mathbf{k} = \mathbf{O}$$

Since λ is the same as before, we obtain

$$\mathbf{N}_1 \mathbf{x} + \mathbf{E}^T \mathbf{k} = \mathbf{b}_1 \qquad \text{A3(16)}$$

But

$$\mathbf{E}\mathbf{x} - \mathbf{d} = \mathbf{O} \qquad \text{A3(15)}$$

Sets of equations A3(15) and A3(16) may be combined into a *hypermatrix* as follows

$$\begin{bmatrix} \mathbf{N}_1 & \mathbf{E}^T \\ \mathbf{E} & \mathbf{O} \end{bmatrix} \begin{bmatrix} \mathbf{x} \\ \mathbf{k} \end{bmatrix} = \begin{bmatrix} \mathbf{b}_1 \\ \mathbf{d} \end{bmatrix}$$

(Note: These may be solved as they stand, although if Cholesky's method is used, the null matrix causes negative square roots which have to be flagged during the procedure to identify negative squares when they arise. Gauss's method of solution has no problems in its general form.)

This hypermatrix form is useful when making design studies to add and subtract constraint equations.

If an explicit solution for **x** only is obtained, a smaller set of equations has to be solved. The explicit solution is obtained as follows: from A3(16)

$$\mathbf{x} = \mathbf{N}_1^{-1}(\mathbf{b}_1 - \mathbf{E}^T\mathbf{k})$$

and from A3(15)

$$\mathbf{E}\mathbf{N}_1^{-1}(\mathbf{b}_1 - \mathbf{E}^T\mathbf{k}) = \mathbf{d}$$

or

$$\mathbf{E}\mathbf{N}_1^{-1}\,\mathbf{E}^T\mathbf{k} = \mathbf{E}\mathbf{N}_1^{-1}\,\mathbf{b}_1 - \mathbf{d}$$

or

$$\mathbf{N}_2\,\mathbf{k} = \mathbf{b}_2$$

and

$$\mathbf{k} = \mathbf{N}_2^{-1}\,\mathbf{b}_2$$

Substituting for **k** in A3(16) finally gives

$$\mathbf{N}_1 x = \mathbf{b}_1 - \mathbf{E}^T \mathbf{N}_2^{-1} \mathbf{b}_2$$

or

$$\boxed{\mathbf{N}_1 \, x = \mathbf{b}_3} \qquad \text{A3(17)}$$

Thus the problem once again reduces to the solution of a symmetric set of equations of the form

$$\mathbf{N}\,\mathbf{x} = \mathbf{b}$$

A common simplification

Many problems are simplified at the outset so that they do not contain any constraint equations, for example by the elimination of fixed parameters when forming up the observation equations.

Also in many cases it is possible to select a mathematical model which has only one observation in each equation, giving an observation equation which contains only one residual. Thus the **C** matrix reduces to the unit matrix, or alternatively it has an inverse.

In such cases, equations A3(7) may be simplified by premultiplying by this inverse giving

$$\mathbf{C}^{-1}\mathbf{A}\ \mathbf{x} + \mathbf{C}^{-1}\mathbf{C}\ \mathbf{v} = \mathbf{C}^{-1}\mathbf{b}$$

which is of the form

$$\mathbf{A}_1\,\mathbf{x} = \mathbf{L}_1 + \mathbf{v}$$

giving directly the normal equations

$$\mathbf{A}_1^T\mathbf{W}\,\mathbf{A}_1\mathbf{x} = \mathbf{A}_1^T\mathbf{W}\,\mathbf{L}_1$$

or

$$\mathbf{N}_3\,\mathbf{x} = \mathbf{b}_4 \qquad \text{A3(18)}$$

These equations A3(18) are commonly found in many surveying applications. For example equations A3(2) can be written

$$\mathbf{C}^{-1}\mathbf{A}\,\mathbf{x} + \mathbf{C}^{-1}\,\mathbf{C}\,\mathbf{s} = \mathbf{0}$$

$$\mathbf{A}_2\,\mathbf{x} + \mathbf{s} = \mathbf{0}$$

or

$$\mathbf{A}_2\,\mathbf{x} = \mathbf{L} + \mathbf{v}$$

where $\mathbf{A}_2$ is in full

$$\begin{bmatrix} -\dfrac{\Delta x_1}{s_1} & -\dfrac{\Delta y_1}{s_1} \\ -\dfrac{\Delta x_2}{s_2} & -\dfrac{\Delta y_2}{s_2} \end{bmatrix}$$

This is the most common form of observation equations.

A3.4 Figural approach by conditions

As an alternative to the parametric approach to Least Squares problems and which is used generally throughout this book, a once-popular figural approach may be adopted instead. In certain cases, where the figural conditions among the observed parameters are linear as in levelling networks, this is by far the simpler of the two methods.

Consider the example of the level network of Chapter 6.5. Because the **C** matrix is the unit matrix, each observed height difference is written as a simple observation equation of the form

$$\mathbf{A}\,\mathbf{x} = \mathbf{L} + \mathbf{v}$$

The **A** matrix is also the unit matrix, and the **L** vector made zero by selecting the provisional values to be the observed values. Thus

$$\mathbf{x} = \mathbf{v}$$

Thus there are no unobserved parameters in the problem. Conditions among the observed parameters, to ensure that the loops of levelling must close, give constraint equations of the form

$$\mathbf{E}\mathbf{x} = \mathbf{d}$$

Following the same procedure as in A3.3 incorporating Lagrangian multipliers (or correlatives) **k** to minimize the sum of squares of residuals, leads to

$$\mathbf{E}^{T}\,\mathbf{k} = \mathbf{W}\,\mathbf{x}$$

and

$$\mathbf{x} = \mathbf{W}^{-1}\,\mathbf{E}^{T}\,\mathbf{k} \qquad \text{A3(19)}$$

Substituting for **x** in the constraint equations gives

$$\mathbf{E}\,\mathbf{W}^{-1}\,\mathbf{E}^{T}\,\mathbf{k} = \mathbf{d} \qquad \text{A3(20)}$$

Therefore

$$\mathbf{k} = (\mathbf{E}\,\mathbf{W}^{-1}\,\mathbf{E}^{T})^{-1}\,\mathbf{d}$$

The equations A3(20), called the *correlative normal equations,* are solved to give the vector of correlatives **k**, and then substituted in A3(19) to give the desired solution for **x** (**v**).

This approach, well suited to simple constraint equations as in levelling where all the coefficients are unity, becomes complex when non-linear equations are involved. Also, the mechanism for forming constraint equations is by no means straightforward. Being purely figural, however, the method does avoid the datum problems associated with coordinates, and is useful in theoretical studies and in blunder detection. Since there are no examples of this method elsewhere in this book, one seems useful here. The weight matrix is taken to be the unit matrix.

Example

The data of Table 6.1 give the constraint equations for each loop of levels as

$$\begin{bmatrix} 1 & 0 & -1 & 0 & 1 & 0 \\ 0 & -1 & 1 & 0 & 0 & -1 \\ 0 & 0 & 0 & 1 & -1 & 1 \end{bmatrix} \mathbf{x} = \begin{bmatrix} 0.11 \\ 0.04 \\ -0.08 \end{bmatrix}$$

and the correlative normals as

$$\begin{bmatrix} 3 & -1 & -1 \\ -1 & 3 & -1 \\ -1 & -1 & 3 \end{bmatrix} \mathbf{k} = \begin{bmatrix} 0.11 \\ 0.04 \\ -0.08 \end{bmatrix}$$

The solution for $\mathbf{k}^T$ is

$$0.045 \qquad 0.0275 \qquad -0.0025$$

giving the solution for $\mathbf{x}^T$ as

$$0.045 \qquad -0.028 \qquad -0.018 \qquad 0.002 \qquad 0.048 \qquad -0.030$$

Summary

In the above sections A3.1, A3.2 and A3.3, the various formulae needed to obtain solutions of the required parameters are given. Often not all parameters in a problem are required, since only a few may be of interest, whilst the others are either of no concern or are not actually recoverable in a physical sense. In such cases only a partial solution of the normal equations is needed. Also, in design problems or in real-time problems, not all the observation equations will be used at the same time, and a sequential approach is adopted by forming the normal equations by the method of contributions.

We now proceed to discuss the statistical information which can be obtained from these solutions.

A3.5 Dispersion matrices

An important feature of the Least Squares estimation process is the ability to estimate *quality.* This quality is expressed by dispersion matrices.These symmetric square matrices contain the variances of the parameters along the diagonals and the co-variances as off-diagonal terms. When the correlations are zero, these matrices are diagonal and therefore easy to invert.

We denote the dispersion matrix of the observed parameters by $\mathbf{D_0}$ and of the derived parameters by $\mathbf{D_x}$. Because the dispersion matrix of the observed parameters can only be estimated from some *a priori* data such as from previous experience, we make allowance for its error by introducing an unknown scaling factor, σ_0^2, called the variance of an observation of unit weight. We relate the inverse of the weight matrix $\mathbf{W}^{-1}$ to the *a priori* estimated dispersion matrix $\mathbf{D_0}$ by

$$\sigma_0^2 \ \mathbf{W}^{-1} = \mathbf{D}_0$$

The data of any Least Squares calculation will be used to *estimate* this scaling factor from the sample residuals, and thus gain a better value for the dispersion matrix of the observations. If the calculated value of σ_o^2 is equal to 1 then the original estimated weight matrix is correct and

$$\mathbf{W}^{-1} = \mathbf{D_o}$$

Population parameters and sample statistics

We now relate the best estimates from a sample to their theoretical population from which the sample has been drawn.

If $\bar{s}$ is the population value of the observed parameter, then the population residual V is given by

$$\bar{s} = \overset{o}{s} + V$$

Correspondingly the vectors are written in bold notation

$$\bar{\mathbf{s}} = \overset{o}{\mathbf{s}} + \mathbf{V}$$

This compares with the vector of data from the sample

$$\hat{\mathbf{s}} = \overset{o}{\mathbf{s}} + \mathbf{v}$$

Thus

$$\delta\mathbf{s} = \bar{\mathbf{s}} - \hat{\mathbf{s}} = \mathbf{V} - \mathbf{v}$$

where $\delta\mathbf{s}$ is the difference between the population value and the sample estimate.

Applying similar ideas to the mathematical model A3(3) we have the respective sample and population models

$$\mathbf{A\,x} + \mathbf{C\,v} + \mathbf{C\,L} = \mathbf{O}$$

and $$\mathbf{A\bar{x}} + \mathbf{CV} + \mathbf{CL} = \mathbf{O}$$

Subtracting gives $$\mathbf{A(\bar{x} - x) + C(V - v) = O}$$

$$\mathbf{A\,\delta\,x} + \mathbf{C(V - v) = O} \qquad \text{A3(21)}$$

where $\delta\,\mathbf{x}$ is the vector of the differences between the population parameters and their corresponding values estimated from the sample by Least Squares.

Now we know from equation A3(14) that

$$\mathbf{A^T N^{-1} C\,v} = \mathbf{O}$$

thus by premultipling equation A3(21) by $\mathbf{A^T N^{-1}}$ we obtain

$$\mathbf{A^T N^{-1} A\,\delta x} + \mathbf{A^T N^{-1} C\,V} - \mathbf{A^T N^{-1} C\,v} = \mathbf{O}$$

or $$\mathbf{A^T N^{-1} A\,\delta x} + \mathbf{A^T N^{-1} C\,V} = \mathbf{O}$$

or say

$$\mathbf{N}_1\ \delta\mathbf{x} = \mathbf{B\ V}$$

where

$$\mathbf{B} = -\mathbf{A}^{\mathrm{T}}\ \mathbf{N}^{-1}\ \mathbf{C}$$

$$\mathbf{N}_1 = \mathbf{A}^{\mathrm{T}}\ \mathbf{N}^{-1}\ \mathbf{A}$$

Now

$$\mathbf{N}_1\ \delta\mathbf{x}\ (\mathbf{N}_1\ \delta\mathbf{x})^{\mathrm{T}} = \mathbf{B\ V}\ (\mathbf{B\ V})^{\mathrm{T}}$$

therefore because $\mathbf{N}_1^{\mathrm{T}} = \mathbf{N}_1$

$$\mathbf{N}_1\ \delta\mathbf{x}\ \ \delta\mathbf{x}^{\mathrm{T}}\ \mathbf{N}_1 = \mathbf{B\ V\ \ V}^{\mathrm{T}}\ \mathbf{B}^{\mathrm{T}}$$

Taking expectations, and remembering that

$$\mathrm{E}\ (\delta\mathbf{x}\ \ \delta\mathbf{x}^{\mathrm{T}}) = \mathbf{Dx} \text{ and } \mathrm{E}\ (\mathbf{V\ \ V}^{\mathrm{T}}) = \mathbf{D_o}$$

we have $\mathbf{N}_1\ \mathbf{D_x}\ \mathbf{N}_1 = \mathbf{B\ \ D_o\ B}^{\mathrm{T}}$

Pre and post multiplying both sides by $\mathbf{N}_1^{-1}$ gives

$$\mathbf{N}_1^{-1}\ \mathbf{N}_1\ \mathbf{D_x}\ \mathbf{N}_1\ \mathbf{N}_1^{-1} = \mathbf{N}_1^{-1}\ \mathbf{B\ D_o\ B}^{\mathrm{T}}\mathbf{N}_1^{-1}$$

or

$$\mathbf{D_x} = \mathbf{N}_1^{-1}\ \mathbf{B\ D_o\ B}^{\mathrm{T}}\mathbf{N}_1^{-1}$$

but

$$\mathbf{B\ \ D_o\ B}^{\mathrm{T}} = \mathbf{A}^{\mathrm{T}}\ \mathbf{N}^{-1}\ \mathbf{C\ D_o}\ (\mathbf{A}^{\mathrm{T}}\ \mathbf{N}^{-1}\ \mathbf{C})^{\mathrm{T}}$$

$$= \mathbf{A}^{\mathrm{T}}\ \mathbf{N}^{-1}\ \mathbf{C}\ \ \mathbf{D_o}\ \mathbf{C}^{\mathrm{T}}\ \mathbf{N}^{-1}\ \mathbf{A}$$

and

$$\mathbf{C\ D_o C}^{\mathrm{T}} = \mathbf{C}\sigma_0^2\ \mathbf{W}^{-1}\ \mathbf{C}^{\mathrm{T}}$$

$$= \sigma_0^2\ \mathbf{N}$$

$$\mathbf{B\ D_o B}^{\mathrm{T}} = \sigma_{\mathbf{o}}^2\ \mathbf{N}_1$$

giving

and finally

$$\boxed{\mathbf{D_x} = \sigma_0^2\ \mathbf{N}_1^{-1}} \qquad \text{A3(22)}$$

Thus the dispersion matrix of the required parameters is the scaled inverse of the normal equations. Hence, although this inverse may not be needed to obtain the solution, it is needed for this statistical information.

Equation A3(22) gives the expression for $\mathbf{D_x}$ for the general case of observations with conditions **C**.

Special case

When $\mathbf{C} = \mathbf{I}$ the unit matrix, the result is of the same form, namely

$$\boxed{\mathbf{D_x} = \sigma_0^2\ \mathbf{N}^{-1}} \qquad \text{A3(23)}$$

because in full

$$\mathbf{D_x} = \sigma_o^2 \, (\mathbf{A}^T \, (\mathbf{C} \, \mathbf{W}^{-1} \, \mathbf{C}^T)^{-1} \, \mathbf{A}^{-1}$$

which reduces to equation A3(23).

A3.6 Estimation of σ_o^2

We now show how σ_o^2 is estimated from the sample itself, as a bi-product of the Least Squares computation, using the expression:

$$\boxed{\sigma_o^2 = \text{Exp}\frac{\mathbf{v}^T \mathbf{W} \mathbf{v}}{m - n}} \qquad \text{A3(24)}$$

For economy of space, the derivations are not given in full, with some work left to the reader, who is once again reminded of the symmetric nature of many of the matrices.

From equation A3(11) we have

$$\boldsymbol{\lambda} = \mathbf{N}^{-1} \, (\mathbf{b} - \mathbf{A}\mathbf{x}) \qquad \text{A3(11)}$$

and from equation A3(9)

$$\mathbf{v} = \mathbf{W}^{-1} \, \mathbf{C}^T \, \boldsymbol{\lambda} \qquad \text{A3(9)}$$

Thus we have the sample residuals given by

$$\mathbf{v} = \mathbf{W}^{-1} \, \mathbf{C}^T \, \mathbf{N}^{-1} \, (\mathbf{b} - \mathbf{A}\mathbf{x}) \qquad \text{A3(25)}$$

The population residuals are therefore given by

$$\mathbf{V} = \mathbf{W}^{-1}\mathbf{C}^T\mathbf{N}^{-1}(\mathbf{b} - \mathbf{A}\bar{\mathbf{x}}) \qquad \text{A3(26)}$$

giving by subtraction

$$\begin{aligned}
\mathbf{V} - \mathbf{v} &= \mathbf{W}^{-1}\mathbf{C}^T\mathbf{N}^{-1}(\mathbf{b} - \mathbf{A}\bar{\mathbf{x}} - \mathbf{b} + \mathbf{A}\mathbf{x}) \\
&= -\mathbf{W}^{-1}\mathbf{C}^T\mathbf{N}^{-1}\mathbf{A}(\bar{\mathbf{x}} - \mathbf{x}) \\
&= -\mathbf{W}^{-1}\mathbf{C}^T\mathbf{N}^{-1}\mathbf{A}\delta\mathbf{x}
\end{aligned}$$

$$\mathbf{V} = \mathbf{v} - \mathbf{W}^{-1} \, \mathbf{C}^T \, \mathbf{N}^{-1} \, \mathbf{A} \;\; \delta \, \mathbf{x}$$

$$\mathbf{V} = \mathbf{v} - \mathbf{K} \, \delta \, \mathbf{x}$$

where

$$\boxed{\mathbf{K} = \mathbf{W}^{-1} \, \mathbf{C}^T \, \mathbf{N}^{-1} \, \mathbf{A}} \qquad \text{A3(27)}$$

Now from equation A3(14)

$$\mathbf{A}^T \, \mathbf{N}^{-1} \, \mathbf{C} \;\; \mathbf{v} = \mathbf{O} \qquad \text{A3(14)}$$

Thus we have the important result

$$\boxed{\mathbf{K}^T \, \mathbf{W} \, \mathbf{v} = \mathbf{O}} \qquad \text{A3(28)}$$

Now

$$\mathbf{V}^T\mathbf{W}\ \mathbf{V} = (\mathbf{v} - \mathbf{K}\,\delta\,\mathbf{x})^T\mathbf{W}\ (\mathbf{v} - \mathbf{K}\,\delta\,\mathbf{x})$$

$$\mathbf{V}^T\mathbf{W}\ \mathbf{V} = \mathbf{v}^T\mathbf{W}\ \mathbf{v} + \delta\,\mathbf{x}^T\mathbf{K}^T\mathbf{W}\,\mathbf{K}\ \delta\,\mathbf{x} \qquad \text{A3(29)}$$

because $\mathbf{K}^T\mathbf{W}\ \mathbf{v} = \mathbf{0}$ and $\mathbf{v}^T\mathbf{W}\ \mathbf{K} = \mathbf{O}$

Now, remembering that the *trace* of a matrix is the sum of its diagonal terms, we have

$$\mathbf{V}^T\mathbf{W}\ \mathbf{V} = \text{trace}\,(\mathbf{V}\,\mathbf{V}^T\mathbf{W})$$

$$\delta\,\mathbf{x}^T\mathbf{K}^T\mathbf{W}\,\mathbf{K}\ \ \delta\,\mathbf{x} = \text{trace}\,(\delta\,\mathbf{x}\ \ \delta\,\mathbf{x}^T\mathbf{K}^T\mathbf{W}\,\mathbf{K})$$

Thus equation A3(29) becomes

$$\text{trace}\,(\mathbf{V}\,\mathbf{V}^T\mathbf{W}) = \mathbf{v}^T\mathbf{W}\ \mathbf{v} + \text{trace}\,(\delta\,\mathbf{x}\ \ \delta\,\mathbf{x}^T\mathbf{K}^T\mathbf{W}\,\mathbf{K})$$

and taking expectations of both sides we have

$$\text{Exp trace}\,(\mathbf{V}\,\mathbf{V}^T)\ \mathbf{W} = \text{Exp}\,(\mathbf{v}^T\mathbf{W}\ \mathbf{v}) + \text{Exp trace}\,(\delta\,\mathbf{x}\ \ \delta\,\mathbf{x}^T)\,\mathbf{K}^T\mathbf{W}\,\mathbf{K}$$

Now

$$\text{Exp}\,\{\,\delta\mathbf{x}\,\delta\mathbf{x}^T\} = \mathbf{D}_\mathbf{x} = \sigma_o^2\ \ \mathbf{N}_1^{-1}$$

Also

$$\text{Exp trace}\,\{\,\mathbf{V}\,\mathbf{V}^T\}\,\mathbf{W} = \text{trace}\,\{\,\mathbf{D}_\mathbf{o}\ \mathbf{W}\} = \text{trace}\,\{\,\sigma_o^2\ \mathbf{W}^{-1}\ \mathbf{W}\,\}$$

Therefore

$$\sigma_o^2\ \text{trace}\ \mathbf{I}_m = \text{Exp}\,\{\,\mathbf{v}^T\ \mathbf{W}\,\mathbf{v}\,\} + \text{trace}\ \sigma_o^2\ \mathbf{N}_1^{-1}\ \mathbf{K}^T\ \mathbf{W}\,\mathbf{K}$$

$$= \text{Exp}\,\{\,\mathbf{v}^T\ \mathbf{W}\,\mathbf{v}\,\} + \sigma_o^2\ \text{trace}\ \mathbf{I}_n$$

because

$$\mathbf{K}^T\mathbf{W}\,\mathbf{K} = \mathbf{N}_1$$

Now $\mathbf{I}_m$ and $\mathbf{I}_n$ are unit matrices of dimensions m and n, thus

$$m\,\sigma_o^2 = \text{Exp}\ \{\,\mathbf{v}^T\ \mathbf{W}\,\mathbf{v}\,\} = n\,\sigma_o^2$$

and finally

$$\sigma_o^2 = \text{Exp}\,\frac{\mathbf{v}^T\ \mathbf{W}\,\mathbf{v}}{m-n}$$

If we put

$$S^2 = \frac{\mathbf{v}^T\ \mathbf{W}\,\mathbf{v}}{m-n}$$

then

$$\text{Expectation}\ \ S^2 = \sigma_o^2$$

and we say that

S^2 is an *unbiased estimator* of σ_o^2

The denominator $m - n$ is the number of degrees of freedom in the problem, or the number of redundant observations which are not essential for a solution to be obtained.

Special case of one parameter

In the special case of one parameter estimated from n observations the above expression for S^2 reduces to

$$\boxed{\begin{aligned} \text{Exp}\, S^2 &= \text{Exp}\, \frac{\mathbf{v}^T\, \mathbf{W}\, \mathbf{v}}{n-1} \\ &= \sigma_o^2 \end{aligned}} \qquad \text{A3(30)}$$

and

$$S^2 = \frac{n}{(n-1)}\, s^2 = \frac{n}{(n-1)}\, \frac{\mathbf{v}^T\, \mathbf{W}\, \mathbf{v}}{n}$$

or in words

$$S^2 = \frac{n}{(n-1)} \times \text{ sample variance}$$

is an *unbiased estimator* of the population variance.

A3.7 Dispersion matrix of the sample residuals D_v

Although the quality of the derived parameters, as expressed by their dispersion matrix $\mathbf{D}_x$ in equation A3(22), is generally of most interest, there are several statistical reasons for calculating the quality of the *residuals* derived in the Least Squares process.

From equation A3(25) we have

$$\mathbf{v} = \mathbf{W}^{-1}\mathbf{C}^T\, \mathbf{N}^{-1}\, (\mathbf{b} - \mathbf{A}\mathbf{x}) \qquad \text{A3(25)}$$

$$\mathbf{v} = -\ \mathbf{W}^{-1}\, \mathbf{C}^T\, \mathbf{N}^{-1}\, \mathbf{C}\mathbf{L}\ -\ \mathbf{W}^{-1}\, \mathbf{C}^T\, \mathbf{N}^{-1}\, \mathbf{A}\ \text{x}$$

$$\mathbf{v} = -\ \mathbf{G}\,\mathbf{L}\ -\ \mathbf{K}\,\mathbf{x}$$

where

$$\mathbf{G} = \mathbf{W}^{-1}\, \mathbf{C}^T\, \mathbf{N}^{-1}\, \mathbf{C}$$

$$\mathbf{K} = \mathbf{W}^{-1}\, \mathbf{C}^T\, \mathbf{N}^{-1}\, \mathbf{A}$$

Differentiating gives

$$\delta\,\mathbf{v} = -\ \mathbf{G}\,\delta\,\mathbf{l}\ -\ \mathbf{K}\,\delta\text{x} \qquad \text{A3(31)}$$

$$\delta\,\mathbf{v}\ \ \delta\,\mathbf{v}^T = (\mathbf{G}\,\delta\,\mathbf{l} + \mathbf{K}\,\delta\text{x})\ (\mathbf{G}\,\delta\,\mathbf{l} + \mathbf{K}\,\delta\text{x})^T$$

Now

$$\mathbf{L} = \overset{o}{\mathbf{s}} - \overset{*}{\mathbf{s}}$$

therefore

$$\delta\mathbf{L} = \delta\overset{o}{\mathbf{s}} - \delta\overset{*}{\mathbf{s}}$$

but $\delta\overset{o}{\mathbf{s}}$ is the error vector of the observed parameters, that is $\delta\overset{o}{\mathbf{s}} = \mathbf{V}$ the population residuals, and $\delta\overset{*}{\mathbf{s}} = \mathbf{O}$, thus

$$\delta\,\mathbf{v}\ \delta\,\mathbf{v}^{T} = (\mathbf{G\,V} + \mathbf{K}\,\delta x)\ (\mathbf{G\,V} + \mathbf{K}\,\delta x)^{T}$$

$$\delta\,\mathbf{v}\ \delta\,\mathbf{v}^{T} = \mathbf{G\,V\,V}^{T}\,\mathbf{G}^{T} + \mathbf{K}\,\delta x\ \ \delta x^{T}\,\mathbf{K}^{T}$$

$$+\ \mathbf{GV}\ \ \delta x^{T}\,\mathbf{K}^{T} + \mathbf{K}\,\delta x\ \ \mathbf{V}^{T}\,\mathbf{G}^{T}$$

Taking expectations we have

$$D_{v} = G\ \mathbf{D_{o}}\,G^{T} + \mathbf{K}\ \mathbf{D_{x}}\,\mathbf{K}^{T} + \mathbf{G}\,\exp(\mathbf{V}\ \ \delta x^{T})\,K^{T}$$

$$+\ \mathbf{K}\ \exp\,(\delta x\ \ \mathbf{V}^{T})\ \ \mathbf{G}^{T}$$

The third and fourth terms of this expression are both equal to

$$-\,K\,\mathbf{D_{x}}\,\mathbf{K}^{T}$$

giving the result

$$\boxed{\mathbf{D}_{v} = \mathbf{G}\ \mathbf{D_{o}}\,\mathbf{G}^{T} - \mathbf{K}\ \mathbf{D_{x}}\,\mathbf{K}^{T}} \qquad \text{A3(32)}$$

We give the outline proof that terms three and four above are both equal to

$$-\,K\,\mathbf{D_{x}}\,\mathbf{K}^{T}$$

From A3(13)

$$x = \mathbf{N}_{1}^{-1}\,\mathbf{b}_{1}$$

therefore

$$\delta\mathbf{x} = \mathbf{N}_{1}^{-1}\,\delta\,\mathbf{b}_{1}$$

$$\delta\mathbf{x} = -\mathbf{N}_{1}^{-1}\,\mathbf{A}^{T}\,\mathbf{N}^{-1}\,\mathbf{CV}$$

The third term then is

$$\mathbf{G}\exp\,(\mathbf{V}\ \delta\mathbf{x}^{T})\,\mathbf{K}^{T} = \mathbf{G}\exp\,(\mathbf{V}\,(-\mathbf{N}_{1}^{-1}\,\mathbf{A}^{T}\,\mathbf{N}^{-1}\,\mathbf{CV})^{T})\,\mathbf{K}^{T}$$

$$\mathbf{G}\ \exp\,(\mathbf{V}\ \ \delta x^{T})\,\mathbf{K}^{T} = \ -\ \mathbf{G}\ \mathbf{D_{o}}\,\mathbf{C}^{T}\,\mathbf{N}^{-1}\,\mathbf{A}\ \mathbf{D_{x}}\,\mathbf{K}^{T}$$

which reduces to

$$-\,\mathbf{K}\ \mathbf{D_{x}}\,\mathbf{K}^{T}$$

This is symmetric and equals the fourth term transposed.

A common simplification

In the case of the simple model in which the **C** matrix is the unit matrix, it is easy to show that the general result in A3(32) reduces to the much simpler expression

$$\boxed{\mathbf{D}_{\mathrm{v}} = \mathbf{D}_{\mathrm{o}} - \mathbf{A}\,\mathbf{D}_{\mathrm{x}}\,\mathbf{A}^{\mathrm{T}}} \qquad \text{A3(33)}$$

A3.8 Dispersion matrix of the estimated observed parameters $\mathbf{D}_s$

The dispersion matrix $\mathbf{D}_s$ of the Least Squares estimates of the observed parameters is given by the simple expression

$$\boxed{\mathbf{D}_{\mathrm{s}} = \mathbf{D}_{\mathrm{o}} - \mathbf{D}_{\mathrm{v}}} \qquad \text{A3(34)}$$

where $\mathbf{D}_v$ is given by expression A3(32) for the conditioned model and by A3(33) for the simple model. In this latter case we have the even simpler expression

$$\boxed{\mathbf{D}_{\mathrm{s}} = \mathbf{A}\,\mathbf{D}_{\mathrm{x}}\,\mathbf{A}^{\mathrm{T}}} \qquad \text{A3(35)}$$

The proof of A3(34) follows similar lines to that for A3(32) in which some heavy, but straightforward, matrix manipulation is involved. The basic steps are as follows.

The best estimates of the observed parameters are given by

$$\mathbf{s} = \mathbf{v} + \overset{\mathbf{o}}{\mathbf{s}}$$

$$\delta\mathbf{s} = \delta\mathbf{v} - \delta\overset{\mathbf{o}}{\mathbf{s}}$$

but from A3(31)

$$\delta\mathbf{v} = -\,\mathbf{G}\delta\mathbf{L} - \mathbf{K}\delta\mathrm{x}$$

therefore

$$\delta\mathbf{s} = -\,\mathbf{G}\delta\mathbf{L} - \mathbf{K}\delta\mathbf{x} + \delta\overset{\mathbf{o}}{\mathbf{s}}$$

but

$$\delta\mathbf{L} = \delta\overset{\mathbf{o}}{\mathbf{s}} = \mathbf{V}$$

therefore

$$\delta\mathbf{s} = (\mathbf{I} - \mathbf{G})\,\mathbf{V} - \mathbf{K}\delta\mathrm{x}$$

Substituting for

$$\delta\mathbf{x} = -\mathbf{N}_1^{-1}\,\mathbf{A}^{\mathrm{T}}\,\mathbf{N}^{-1}\,\mathbf{C}\mathbf{V}$$

gives the expression

$$\delta\mathbf{s} = (\mathbf{I} - \mathbf{G} + \mathbf{K}\mathbf{N}_1^{-1}\,\mathbf{A}^{\mathrm{T}}\,\mathbf{N}^{-1}\,\mathbf{C}\,)\,\mathbf{V}$$

or say

$$\delta\mathbf{s} = \mathbf{F}\,\mathbf{V}$$

where

$$\mathbf{F} = (\mathbf{I} - \mathbf{G} + \mathbf{K}\mathbf{N}_1^{-1}\ \mathbf{A}^{\mathrm{T}}\ \mathbf{N}^{-1}\ \mathbf{C})$$

Now

$$\delta\,\mathbf{s}\ \ \delta\,\mathbf{s}^{\mathrm{T}} = \mathbf{F}\,\mathbf{V}\,\mathbf{V}^{\mathrm{T}}\,\mathbf{F}^{\mathrm{T}}$$

Taking expectations

$$\mathbf{D}_{\mathrm{s}} = \mathbf{F}\,\mathbf{D}_{\mathbf{o}}\,\mathbf{F}^{\mathrm{T}}$$

and substituting for $\mathbf{F}$ we obtain, after some heavy algebra,

$$\mathbf{D}_{\mathbf{s}} = \mathbf{D}_{\mathbf{o}} - \mathbf{G}\ \mathbf{D}_{\mathbf{o}}\,\mathbf{G}^{\mathrm{T}} + \mathbf{K}\ \mathbf{D}_{\mathbf{x}}\,\mathbf{K}^{\mathrm{T}}$$

which is equation A3(34) and equation A3(35) follows.

Comment

Throughout this book we have assumed that we are dealing with properly posed problems which do not give rise to singular normal equation matrices. This means that such matrices have no rank defects due to insufficient information or inadequate datum definition. For example there must be at least two sufficient measurements to fix a point in two dimensions, and a network, described in terms of cartesian coordinates, must have an assigned origin and orientation. This approach is usually sufficient for most problems, other than for deformation studies.

The selection of a datum for coordinates will affect some dispersion matrices and statistical information deduced from them. For a treatment of more general ways to remove rank defects which give unique dispersion matrices, the reader should refer to references (Cooper 1987, Cooper and Cross 1988 and Mikhail 1976).

Appendix A4
The Earth Ellipsoid

For many purposes it is possible to neglect the Earth's curvature, and to treat its surface as plane over a limited area. This greatly simplifies computation. Sometimes it is sufficient to consider the Earth as a sphere and work in terms of spherical trigonometry. In the most refined work, a spheroid or ellipsoid of reference has to be used. The ellipsoid of geodesy is the figure described by the rotation of an ellipse about its minor axis, often called an oblate spheroid.

A4.1 The geoid

If the land masses were covered by a network of canals through which the waters of the oceans were permitted to flow freely under gravity, neglecting tidal effects, the surface of water on the canals and the oceans would form an equipotential surface called the *geoid,* which simply means 'Earth-shaped'. Note this is a surface based on a force field which has no simple *geometrical shape* other than that it approximates to a ellipsoid of revolution.

As a result of various measurements carried out since the 18th century in different parts of the world a number of geodesists have produced estimates of the size of ellipsoid which best fits the geoid for the part of the Earth considered in their calculations. These ellipsoids were adopted in various countries for geodetic computations: e.g. the United Kingdom is computed on Airy's ellipsoid, much of Africa is computed on the Clarke 1880 ellipsoid.

It is often sufficiently accurate to consider the surface of a sphere which closely approximates to the ellipsoid at the particular place in question, and whose radius equals the radius of curvature of the ellipsoid at that point.

A4.2 The meridian ellipse

The ellipse which defines an ellipsoid is called the meridian ellipse. The parallels of latitude are small circles in planes parallel to the equator, which is a great circle. An ellipse is defined in many ways and has a multiplicity of geometrical properties. We shall consider it defined with respect to its semi-major axis a and semi-minor axis b by the equation

$$\frac{x^2}{a^2}+\frac{y^2}{b^2}=1 \qquad \text{A4(1)}$$

The coordinates of a point on the ellipse with respect to the origin at its centre are (x, y). The following properties will also be used:

$$b^2 = a^2 (1 - e^2) \qquad A4(2)$$

where e is the eccentricity of the ellipse, which is about 1/12 for a terrestrial ellipsoid. Also the flattening f given by

$$f = (a - b) / a$$

is about 1/ 300 for the Earth. The area of an ellipse is πab.

Given either of the size parameters a or b, and a shape parameter e or f, the others may be derived. It is usual to define a meridian ellipse, and therefore an ellipsoid, in terms of a and f. For historical reasons, there are many reference ellipsoids recommended for use in different parts of the world, because all previous surveys and maps were based on them.

The parameters of early ellipsoids were calculated from terrestrial arc measurements in specific parts of the Earth. With the advent of artificial Earth satellites, geodesists have been able to use the Earth as a whole to define an ellipsoid. The basic parameters, obtained from orbital analyses, are the semi-major axis a, and the first harmonic of gravitational potential J_2. The latter can be used to derive an equivalent flattening according to a selected mathematical model. Datums, such as the world geodetic reference system WGS 84, are used for satellite work. The subject is one of some complexity, and much influences position fixing, especially off-shore in oil exploration, where geodetic cadastral problems arise. (See Iliffe 2000.)

A4.3 Types of latitude

Since the reference ellipsoid will not fit the geoid exactly at all points, two main types of latitude may be distinguished:

(i) *geodetic latitude,* ϕ_G, is the angle between the normal to the ellipsoid at a point and the plane of the equator (see Figure A4.1). The plane of the equator is perpendicular to the spin axis of the Earth.

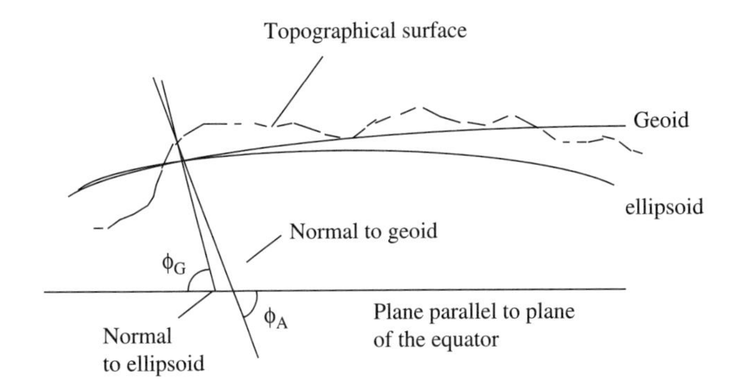

Figure A4.1

(ii) *astronomical latitude*, ϕ_A, is the angle between the equator and the meridian component of the normal to the geoid at that point. The normal to the geoid is the direction of the vertical and will not usually lie in the plane of the meridian ellipse. The angle between the direction of the plumb line and the normal to the ellipsoid is the deviation of the vertical. This depends on the ellipsoid chosen and the point or points at which the ellipsoid and geoid are related to each other. Latitude is positive north of the Equator.

Other types of latitude, such as *geocentric, reduced* and *isometric* latitude used in projection theory, will not concern us here.

A4.4 Types of longitude

In a like manner to latitude, we distinguish two types of longitude which arise out of the lack of coincidence between the direction of the plumb line, or the vertical, and the normal to the ellipsoid.

(i) Geodetic longitude, λ_G, is the angle measured along the equator between the meridian of Greenwich and the meridian ellipse of the place.
(ii) Astronomical longitude, λ_A, is the angle measured along the equator between the astronomical meridian of Greenwich and the astronomical meridian of the place: the astronomical meridian of a place being defined as the plane containing the Earth's axis and the astronomical zenith of the place.

Longitude is reckoned positive East of Greenwich.

A4.5 Types of azimuth: Laplace azimuth equation

Since there are two meridians at any place, the astronomical and the geodetic, directions referred to these meridians, azimuths, will be different, i.e. we distinguish astronomical azimuth α_A and geodetic azimuth α_G. Since computation is carried out in terms of geodetic azimuth, and only astronomical azimuth can be observed to control a survey, the geodetic value corresponding to an observed astronomical azimuth must be obtained. The connection between the two is the *Laplace azimuth equation.* Its use enables a reference ellipsoid to be aligned with the spin axis of the Earth.

Azimuth is positive clockwise from north.

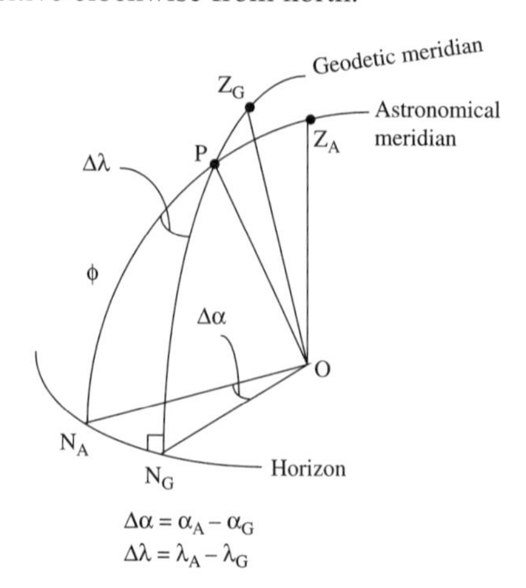

Figure A4.2

Let the geodetic and astronomical longitudes and azimuths be respectively

$$\lambda_G, \lambda_A, \alpha_G, \text{ and } \alpha_A$$

Then the Laplace azimuth equation connecting them is

$$\alpha_A - \alpha_G = (\lambda_A - \lambda_G) \sin \phi \qquad \text{A4(3)}$$

This may be proved by reference to Figure A4.2 in which the following are shown: P the pole, O the centre of the Earth; Z_A and Z_G the respective astronomical and geodetic zeniths, N_A and N_G the astronomical and geodetic north points on the horizon; and astronomical and geodetic azimuth and longitude differences $\Delta\alpha$ and $\Delta\lambda$.

It should be noted that, because there is only one pole, the effect is to align the minor axis of the ellipsoid parallel to the spin axis of the Earth.

In the triangle P N_AN_G, arc $N_AN_G = \Delta\alpha$, and angle PN_GN_A is a right angle if N_AN_GO is the geodetic horizon. Then

$$\sin \Delta\alpha = \sin \Delta\lambda \sin \phi$$

and, since both $\Delta\alpha$ and $\Delta\lambda$ are small angles,

$$\Delta\alpha = \Delta\lambda \sin \phi \qquad \text{A4(4)}$$

It makes no difference which latitude is used in this equation.

Traditionally the Laplace equation was used to align the ellipsoid parallel to the Earth's spin axis and to control the swing of traverses and networks. It required a very precise and tedious determination of astronomical longitude.

Black showed that a *geodetic* azimuth may be determined directly from *astronomical* observations without the need to observe astronomical longitude. Black's method is outlined in Allan 1997.

A4.6 Curves on the surface of an ellipsoid – Normal sections and geodesics

The shortest curve traced out along the ellipsoid surface between two points is a geodesic. It can be envisaged by thinking of a string on the surface pulled tight between the two points P_1 and P_2 of Figure A4.3 The curves traced out along the ellipsoid surface by the normal sections from P_1 and P_2 and the shortest line along the surface, called the *geodesic*, are shown in Figure A4.3. The very small angles between them are divided in the ratio of 1/3 to 2/3. Little use is made of the geodesic today.

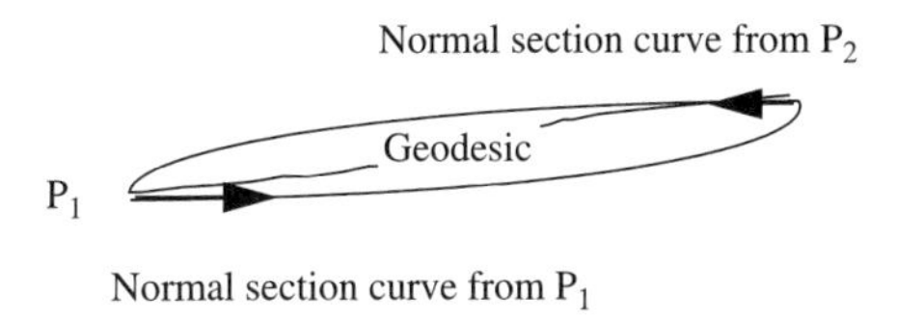

Figure A4.3

A4.7 Radii of curvature of the ellipsoid

The double curvature of the surface of the ellipsoid is usually resolved into two components; one in the plane of the meridian ellipse, and the other, in a plane at right angles to the meridian: the plane of the *prime vertical*. The respective radii of curvature in these two directions, denoted by the Greek letters ρ and ν, are given by

$$\rho = a^2 b^2 / p^3$$

$$\nu = a^2 / p$$

where

$$p^2 = a^2 \cos^2 \phi + b^2 \sin^2 \phi \qquad \text{A4(5)}$$

where ϕ is the latitude of the place. For any given ellipsoid for which the values of a and e are defined, both radii are functions of latitude alone. The distance p is the pedal distance to the ellipse.

The radius of curvature R at latitude ϕ at any azimuth α is given by

$$\frac{1}{R} = \frac{\cos^2 \alpha}{\rho} + \frac{\sin^2 \alpha}{\nu} \qquad \text{A4(6)}$$

This is Euler's theorem, which is used to calculate R for computations involving survey lines and their reduction to the ellipsoid. (See Allan 1997.)

The mean value of the radius of curvature at a point is the geometrical mean of the principal radii of curvature, i.e.

$$\sqrt{\rho \nu}$$

This expression is used when areas are considered, e.g. in the computation of spherical excess. The following section gives outline proofs of these expressions.

A4.8 Principal radii of curvature of the ellipsoid

Consider Figure A4.4 showing the meridian ellipse.

In Allan 2004 it is shown that

$$\nu = PS = a^2 / p$$

and that

$$x = a^2 \cos \phi / p$$

therefore

$$x = \frac{a \cos \phi}{(1 - e^2 \sin^2 \phi)^{\frac{1}{2}}} \qquad \text{A4(7)}$$

and

$$\nu = x \sec \phi = \frac{a \cos \phi \sec \phi}{(1 - e^2 \sin^2 \phi)^{\frac{1}{2}}} = \frac{a}{(1 - e^2 \sin^2 \phi)^{\frac{1}{2}}} \qquad \text{A4(8)}$$

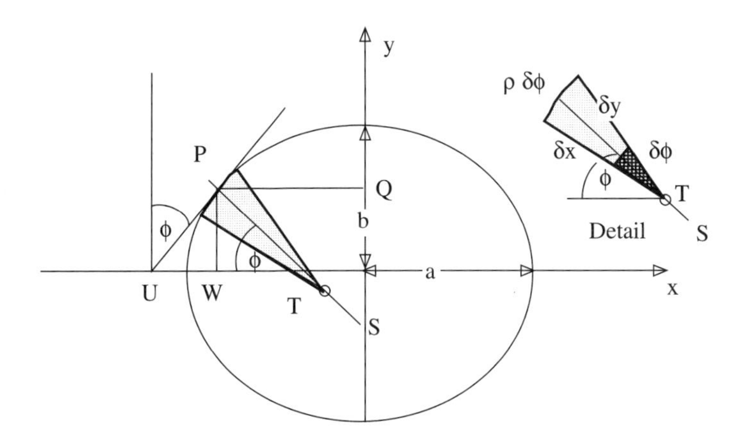

Figure A4.4

To derive the expression for ρ we use the fact that

$$\rho\ \delta\phi = \delta x \operatorname{cosec} \phi$$

or

$$\rho = \frac{\delta x}{\delta \phi} \operatorname{cosec} \phi \qquad \text{A4(9)}$$

which on differentiation of x in A4(7) with respect to φ and substitution in A4(9) gives

$$\rho = \frac{a(1-e^2)}{(1-e^2 \sin^2 \phi)^{\frac{3}{2}}}$$

or alternatively

$$\rho = a^2 b^2 / p^3 \qquad \text{A4 (10)}$$

A4.9 Euler's Theorem

Consider a point P on the surface of the ellipsoid touched by a tangent plane AA' (see Figures A4.5). If BB' is a plane parallel to AA' at a small distance z below it, it will cut the ellipsoid whose surface will describe an ellipse in plane BB'. This ellipse is called the Tissot indicatrix.

Now consider the normal section through P at azimuth α and let P' be a point on the indicatrix at this azimuth a distance s from P. Let the semi-major and semi-minor axes of the indicatrix be m and n respectively. Then from Figure A4.5

$$z = s^2 / 2R \qquad z = m^2 / 2\nu \qquad z = n^2 / 2\rho$$

And the equation of the indicatrix is

$$\frac{x^2}{m^2} + \frac{y^2}{n^2} = 1$$

or

$$s^2 \sin^2 \alpha / m^2 + s^2 \cos^2 \alpha / n^2 = 1$$

giving, from the equations in z,

$$\frac{1}{R} = \frac{\cos^2\alpha}{\rho} + \frac{\sin^2\alpha}{\nu}$$

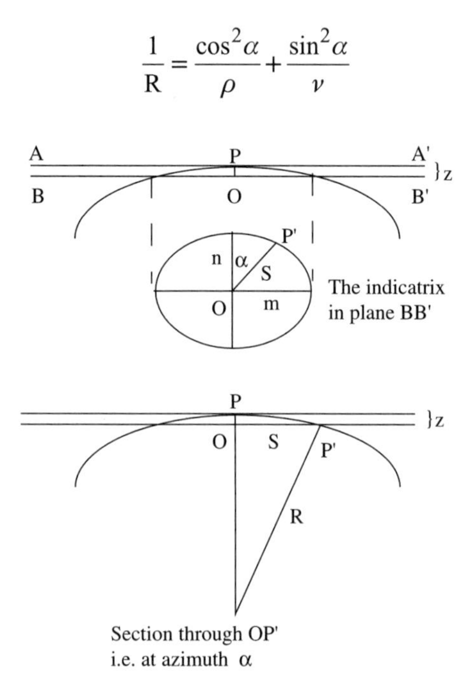

Figure A4.5

A4.10 The mean radius of curvature

The mean value of R is given by

$$\frac{1}{2\pi}\int_0^{2\pi} R \, d\alpha$$

and from the indicatrix this is

$$\frac{1}{2\pi}\int_0^{2\pi} \frac{s^2}{n^2} \rho \, d\alpha$$

But

$$\frac{1}{2\pi}\int_0^{2\pi} s^2 \, d\alpha$$

is the area of the ellipse, i.e. πmn whence the

$$\text{mean radius} = m\,\rho / n = \rho\sqrt{\frac{\rho}{\nu}} = \sqrt{\rho\nu}$$

A4.11 Distances along a meridian

The length of a small arc of the meridian in the vicinity of the point P whose latitude is ϕ is given by

$$ds = \rho\, d\,\phi$$

where ds is the arc length subtended by $d\phi$. This expression is often used in practice for approximate computations when tables of meridional distances are not available. For longer arcs, e.g. between latitudes ϕ_1 and ϕ_2, the length of the arc is given by

$$s = \int_{\phi_1}^{\phi_2} \rho\, d\,\phi$$

$$s = a(1-e^2)\int_{\phi_1}^{\phi_2} \frac{d\phi}{(1-e^2\sin^2\phi)^{\frac{3}{2}}} \qquad \text{A4(11)}$$

To evaluate this elliptical integral we expand the elliptical part as a power series and integrate term by term to the accuracy required, obtaining a result of the form

$$s = a(1-e^2)[A(\phi_2-\phi_1) - \tfrac{1}{2}B(\sin 2\,\phi_2 - \sin 2\,\phi_1) + \tfrac{1}{4}C(\sin 4\,\phi_2 - \sin 4\,\phi_1)]$$

where A, B, and C are constants for the particular ellipsoid, being themselves power series of the eccentricity of the meridian ellipse. For example with the Clarke 1866 ellipsoid these coefficients have the following values: A= 1.005 1093; B= 0.005 1202; C= 0.000 0108. (See Allan 1997.)

An alternative is to carry out the integration by the numerical addition of many small arcs, $ds = \rho\, d\,\phi$, evaluating ρ for each mid arc latitude.

A4.12 Length of a parallel of latitude

The length of a parallel is simply found, because the parallel is a circular arc of radius $\nu \cos\phi$. Hence for a difference in longitude of $\Delta\lambda''$ the corresponding arc length Δs is given by

$$\Delta s = \nu \cos\phi\, \Delta\lambda'' \sin 1''$$

A4.13 Geodetic computations

Traditionally the complex formulae of geodesy were most arduous to handle before the advent of electronic computers. Much use was made of special tables laboriously calculated by hand methods. Although tables may still have some uses for cartographic work they have been entirely superseded by computer programs. Because of the very large numbers involved, care was needed in retaining the necessary accuracy throughout. It is important to remember that one second of arc subtends about 30 m on the Earth's surface. Modern practice is to work in three dimensional plane geometry as described in Chapter 3.10.

A4.14 Location of geographical boundaries

Sometimes the surveyor has to set out or relocate a geographical boundary defined by a meridian, or a great circle between two points, or a parallel of latitude. The first two cases are comparatively simple because the theodolite can be made to point along a great circle, for all practical purposes. The parallel of latitude however has to be set out from a great circle by offsets, as with a railway curve.

In all three operations a point close to the required line has to be located on the correct datum. This may be carried out in many possible ways using any recognized survey method, now mostly by GPS. The basic principle is to locate a point as close to the required position as possible by reconnaissance surveys or aerial photography, then establish its exact position and finally to compute a small offset to the required boundary,which can then be checked by GPS.

Suppose that x is the provisional position whose coordinates can be calculated through the survey. Then assuming that the point y lies on the required meridian or parallel, the azimuth and distance of y from x are computed by the mid-latitude formulae (see Allan 1997) and the required point is located on the ground, making due allowance for height above sea-level, slope, etc.

Points are then set out along the required route, with checks applied from time to time by tying-in to control surveys or making GPS determinations. Such surveys can often involve the greatest of ingenuity on the part of the surveyor to overcome obstacles, though EDM and satellite sytems have made the work easier. The main problem will be to establish that the work is on the correct datum.

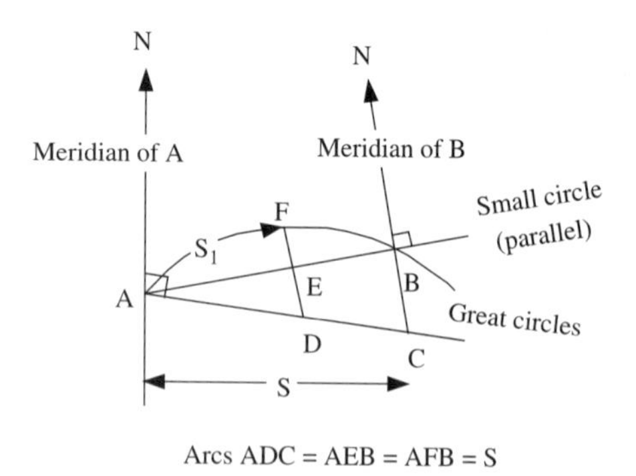

Figure A4.6

A4.15 Setting out a parallel of latitude

This operation may be carried out from key points by offsets from a tangent to the parallel, or by offsets from a chord. Refer to Figure A4.6 in which AC is the tangent to the parallel AEB at A, and AFB is the chord of this parallel. Both arcs AFB and AC are great circles, and AEB is a small circle. The small angles of Figure A4.6 have been exaggerated. The length CB is the offset from the tangent at a distance ADC equal to s. Since BC is small it is sufficiently accurate to take the lengths AFB, AEB, and ADC to be equal to s. The diagram is grossly exaggerated for clarity.

Offset from the tangent AC

Let $\Delta\alpha$ be the difference between the forward and back azimuth from A to C, neglecting 180°. The angle ACB is $90 - \Delta\alpha$ and the offset BC is given by

$$BC = \frac{s^2 \tan \phi}{2\nu}$$

For, remembering that $\alpha = 90°$,

$$\begin{aligned} BC &= s \cos (90 + \tfrac{1}{2} \Delta \alpha) \\ &= s \sin \tfrac{1}{2} \Delta \alpha = s \tfrac{1}{2} \Delta \alpha = s \tfrac{1}{2} \Delta \lambda \sin \phi \\ &= s \tfrac{1}{2} \sin \phi \; s/\nu \cos \phi \end{aligned}$$

$$BC = \frac{s^2 \tan \phi}{2\nu}$$

After positioning the total station at C by radiation from A, the angle ACB = $90° - \Delta\alpha$ and distance BC are used to set out B.

Offset from the chord AFB

To locate B directly along the chord from A, sight along AFB at an azimuth of $90 - \frac{1}{2} \Delta \alpha$, and set out B at a distance s. Intermediate points along the parallel are often required, say at intervals of one km. The offset FE from an intermediate point F is given by

$$FE = \frac{s_1 (s - s_1) \tan \phi}{2\nu}$$

For

$$\begin{aligned} FE = FD - ED &= \frac{s_1 \Delta \alpha}{2} - \frac{s_1^2 \tan \phi}{2\nu} \\ &= \frac{s_1 BC}{2s} - \frac{s_1^2 \tan \phi}{2\nu} \end{aligned}$$

$$FE = \frac{s_1 (s - s_1) \tan \phi}{2\nu}$$

When point B is reached, the next stage of the setting out is repeated as from A, and so on. In thickly wooded country the method of offsetting from a chord is preferable since less cutting is required for the shorter offsets. The survey points are often marked by huge concrete pillars so that warring factions cannot plead ignorance of the boundary location.

Appendix 5 Photogrammetry

A5.1 Photogrammetric mapping — surveying control

Photogrammetry At present, in the first decade of the 21st century, most mapping employs air survey. Photogrammetry is a technique for obtaining information about the position, size and shape of an object by measuring images of the object instead of measuring the object directly. By far the most important field of application of photogrammetry is topographical and other mapping. The surveyor's role is to provide the geometric connection of mapping from the air to the national, or other, ground survey coordinate system. In order to understand the requirements placed on the surveyor it is necessary to enter into a brief discussion about photogrammetry.

Photogrammetry is based on an extremely simple idea: in the thought experiment outlined below, at least two, and often more, cameras are employed, photographing the object of interest from different positions and directions. The surface of an object may be thought of, and described, as consisting of a multiplicity of points. We may define the surface numerically by determining the three-dimensional coordinates of these points. The more irregular the surface and the more accurately we wish to represent it, the closer the spacing and hence the greater the number of the points required. Measuring the shape, size, position and orientation of an object reduces to the measurement of the three dimensional coordinates of such individual points.

In Figure A5.1, if an object is photographed by two or more cameras, light travels from each point, I, on the object through the lenses which focus it on the recording surface (film or electronic sensor) where an image, i, is created in each camera.

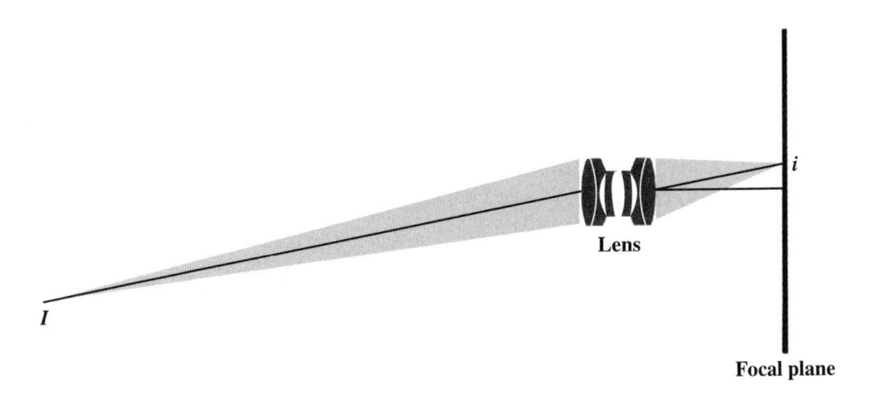

Figure A5.1 Camera

If the rays of light could be immediately reversed in direction so that they travelled back out through the lenses, they would follow identical (reversed) paths in space; that is to say, the cameras would now function as projectors; at every point, *I*, on the object, sets of rays from all the cameras would intersect. Suppose the object were then to be removed; the sets of projected rays would continue to intersect at points occupying exactly the same positions in space as those previously occupied by the points on the surface of the object. An exact, but insubstantial, model of the visible surface of the object would have been created by this re-projection of the rays. We call this a photogrammetric model.

If the points at which corresponding groups of rays intersect could be determined one by one, then the model could be measured in three dimensions point by point. The idea is wonderfully simple. We recreate the directions in space of all the rays by sending them back through the lenses of all the cameras where they will follow exactly the reverse paths previously followed. Points in space are then determined by intersection. What has been described above could be achieved if each camera were rigidly fixed in position, whilst the photographs were being taken and afterwards. Nothing, including the rays, would have been displaced from its original position; the model of the object would have been formed at a scale of 1:1 and it would coincide exactly in position with the original object.

The essence of photogrammetry as a means of three-dimensional measurement is thus seen to be extremely simple; the rest is technology and mathematics. The transmutation of this idea into practical systems for air survey encounters significant difficulties, not least of which stems from the fact that it would be difficult, to say the least, to maintain two or more cameras fixed in position and orientation, perhaps 1000 m above the Earth's surface. Furthermore, the model created is at a scale of 1:1 and it coincides with the original surface. One might as well bypass photogrammetry and measure the actual Earth's surface. Clearly, the whole system has to be modelled and, in today's world, that means mathematical modelling and computation.

If we can convert each real camera into a mathematical projector generating a direction vector for each point of interest, then all we need to know is the position and

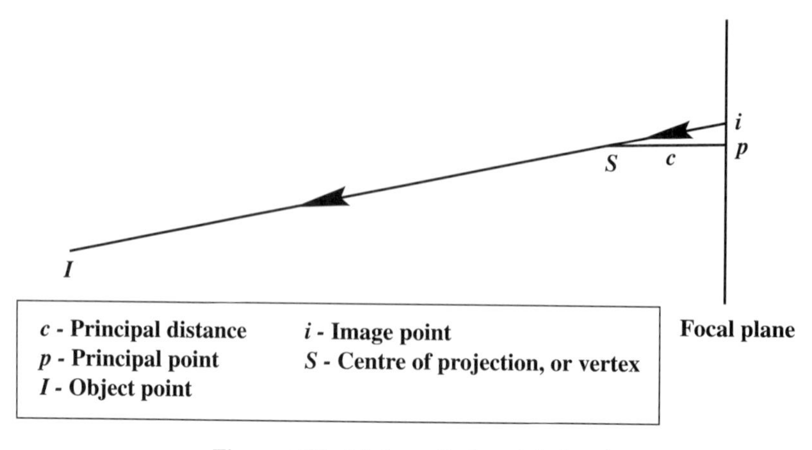

Figure A5.2 Mathematical model of projector

orientation of each projector in order to compute the positions of ground points by intersection.

Until very recently, despite much inventive effort, no satisfactory methods had been developed for the direct determination of the position and orientation of the camera at each instant when a photograph is exposed in the air. Airborne global positioning systems (such as GPS) and inertial measurement systems (IMS) can partially, or even fully in some cases, provide this information. In general, however, indirect methods must be used for the computation of position and orientation. Providing the link between photographs and the ground survey coordinates is the role of the surveyor: the surveyor may provide what are termed control points, or ground control points (GCP). A ground control point consists of a well defined point on the ground whose image is unambiguously identifiable in the photograph(s) and whose survey coordinates are known. For small and medium scale mapping the surveyor in the field, with a paper print of the aerial photograph in hand and perhaps a magnifying glass or a pocket stereoscope, will select suitable points on the ground and survey them. At such scales as these, suitable points might be an isolated bush, the intersection of two paths, a fence corner. For engineering or other large scale mapping, it is seldom possible to find sufficient already existing points whose position can be defined with the accuracy required. In this case GCPs may be "targets", or "signals", of appropriate dimensions, fixed on the ground before the air photo mission is flown. While a great deal could be written about the practical aspects of choosing GCPs, one point is paramount. There is always the likelihood of misidentification of the point in the photogrammetric process; it is even possible that the surveyor makes a mistake in his determination of coordinates. Either case is possible. As in all surveying, redundancy is vital for the discovery and rejection of erroneous information and for estimation of precision. Always select and fix more GCPs than the theoretical minimum. For reasons that will not be analysed here, some GCPs will be chosen for plan control only, others for height control only and some will be used in all three dimensions.

A5.2 Resection of the camera station(s)

A surveyor may set up a theodolite at an arbitrarily chosen position and record directions to three suitably located control points. (See Chapter 8.12.) From the recorded data and the known coordinates of the control points it is possible to compute the position of the theodolite station as well as an angular datum; this permits further work to be done at that station in the coordinate system of the control points. Except in special cases, such as in workshop metrology in which a resection is sometimes performed in three dimensions, the theodolite observations and the computations are in two dimensions only. The mathematical model of the camera as a projector, emitting rays to all points of interest, implies that a camera can be considered as an instrument for the determination of directions in three dimensions, just as a theodolite records directions in three dimensions. In plane resections, using three control points and hence six known coordinates, one computes three unknowns; these are the plane coordinates of the unknown survey station and a correction for bearings. In the resection of a single photograph the minimum control is three points, each in three dimensions, and hence nine known coordinates. Six unknowns are computed; these are the three-dimensional coordinates of the air station and three parameters defining the orientation, or angular attitude, of the camera. These latter three parameters must be carefully specified and may be defined in several ways. For the present simply think of them as a rotation of the camera about three coordinate axes parallel to each of the ground coordinate axes. The six unknowns are known collectively by the rather strange name of "exterior orientation". While the resection of a single camera station is very frequently used in close range photogrammetry, it is seldom used in air survey. As will be described in a later section, two point resection (See also Chapter 2) is a standard technique in aerial photogrammetry, as also is the simultaneous resection of large numbers of photographs, a technique usually known as block adjustment.

Before going further it is necessary to consider in more detail the geometry of the camera and of the photographs.

A5.3 The camera

Technology changes constantly and at the time of publication analogue imaging and processing techniques in photogrammetry, as elsewhere, are being replaced by fully digital methods. The vast majority of existing aerial photographs, certainly those taken in the second half of the 20th century, have been taken with analogue cameras, that is to say film cameras, and it is not necessary for the purposes of this section on survey control for photogrammetry to consider digital cameras.

Analogue air survey cameras are frame cameras; in a frame camera the film is held flat against an opening lying in the focal plane of the lens. An ordinary 35 mm film camera is a frame camera. The format of an analogue aerial photograph is 230 mm × 230 mm; an example of such a photograph is shown below, at reduced scale. The

Figure A5.3 Aerial photograph of part of Brisbane, Australia
With acknowledgement to the Queensland Government Department of Natural Resources, Mines and Energy

projector must be modelled mathematically to match the internal geometry of the original. The parameters describing the internal geometry of a camera are referred to collectively as the "inner orientation" of the camera. The inner orientation is usually defined in terms of a simple "idealized" model (Figures A5.2 & A5.5) and of departures from this model. The simple model assumes that each ray passes without deviation through a point *S*. A line, *Sp*, perpendicular to the focal plane meets the focal plane in *p* which is defined as the *principal point* (Figure A5.4). The length of *Sp* is known as the *principal distance* (signified here by the symbol *c* as in Figure A5.2). The principal distance is often loosely, but incorrectly, referred to as the *focal length* of the camera; in many books and papers the symbol *f* is used for the principal distance rather than *c*. The focal length is a property of the lens; the principal distance is a property of the camera, and will vary if the camera is refocused. It is important to distinguish between the two.

In normal English, the orientation of an object implies a direction or angular attitude. Photogrammetric usage, probably deriving from German usage, applies the word to groups of camera parameters. Interior orientation parameters, in their simplest form, consist of a distance and two plane coordinates and imply no angular values. The use of the terminology is, however, firmly established in photogrammetric usage. The parameters of exterior orientation, a term which was used above, specify the spatial position and orientation (used in the English sense of angular attitude) of the camera in a global coordinate system.

Negative and positive

Figure A5.4 shows a (positive) photograph of a laboratory object. A *coordinate system*, *pxy*, in the focal plane has its origin at the principal point, *p*; the axes *px* and *py* are usually taken approximately parallel to the sides of the picture. Then *p*S forms the third axis of a rectangular Cartesian coordinate system; *pS* is known as *the camera axis* and the coordinate axes *Sxyz*, parallel to *pxyz*, are known as *the camera coordinate axes*. Figure A5.5 shows the image in both the *positive* and the *negative* positions; we most usually look at and measure pictures in the *positive* position.

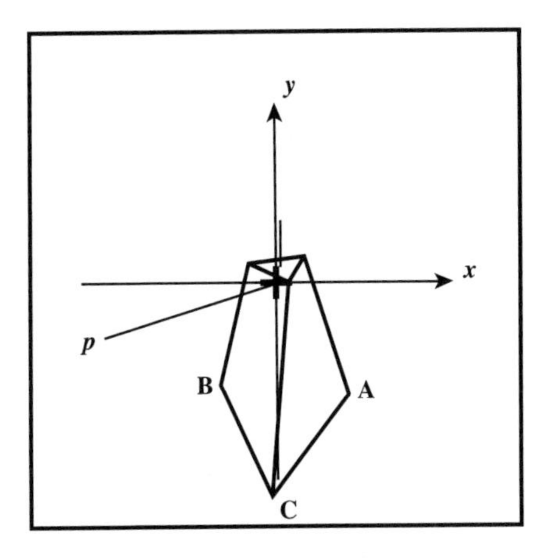

Figure A5.4 Positive photograph

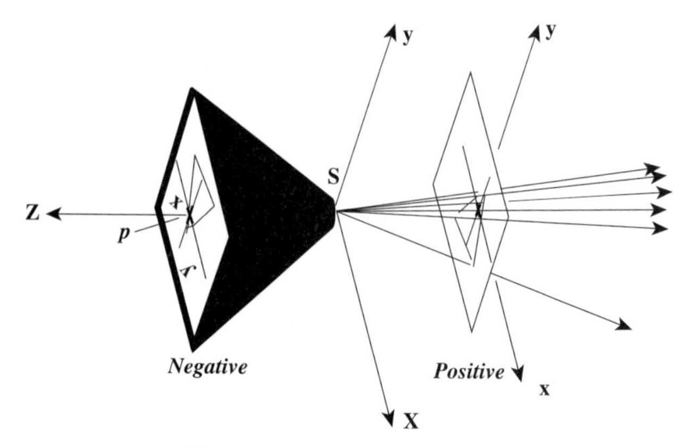

Figure A5.5 Coordinate systems

In a metric camera using film, images of *fiducial marks* are transferred from the camera to each photograph. The fiducial marks are in known and fixed positions in the camera; when their positions in a photograph are measured in the same coordinate system as used for measuring the image coordinates the latter may be transformed into the camera coordinate system. In an air survey camera there are typically four fiducial marks; Fig A5.6 shows one of the fiducial marks from the Brisbane photograph, (Fig A5.3). In our idealized model the inner orientation consists of the principal distance, c, and the coordinates, x_0, y_0, of the principal point in the image coordinate system. A real camera departs from the idealized model and refinement of the mathematical model is required in order to fully define the inner orientation.

Air survey cameras are manufactured and adjusted so that the point of intersection of the lines joining pairs of fiducial marks in diagonally opposite corners intersect at the principal point, to a moderate degree of accuracy. For accurate work, the true position of the principal point must be found by calibration.

For mapping, vertical photographs are used; that is to say photographs in which the camera axis is nominally vertical. Without modern methods of stabilization,

Figure A5.6 Typical fiducial mark on an air survey photograph

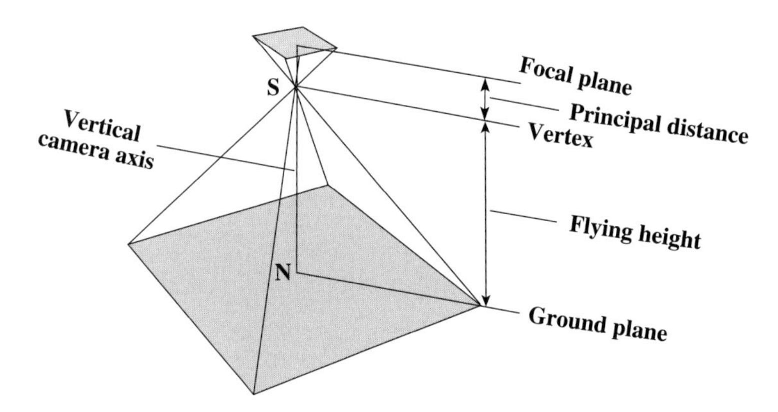

Figure A5.7 Vertical photographs

random variations of up to 2° or 3° are common. Figure A5.7 shows the idealized case of a truly vertical photograph taken over flat level ground. The focal plane and the ground are therefore parallel. From simple geometry it is clear that the portion of ground covered is square, that the scale of the photograph is uniform over this area and that the scale is given by the relationship:

$$\text{photoscale} = 1/m = c/H$$

where c is the principal distance and H is the flying height above ground

If *either* the ground is not flat and level *or* the photograph is not vertical the scale varies throughout the photograph. Nevertheless, in common parlance we speak of the ratio c/H as "the" photoscale. For special purpose mapping, for engineering for example, the scale may be as large as 1: 2 000; for small scale mapping the photoscale is limited by both the ceiling on the flying height and by the fact that at scales below, say, 1:100 000 it is not possible to interpret detail. For a very large majority of cameras the value of c is 152.4 mm or 6 inches. These are known as 'wide-angle cameras".

Vertical photography is very seldom in the form of a single photograph; more usually it is in the form of "runs" or "strips", which may be straight or curved, but are generally straight. For systematic mapping the photography consists of a number of runs forming a "block", as illustrated in Figure A5.8. The runs should be as straight

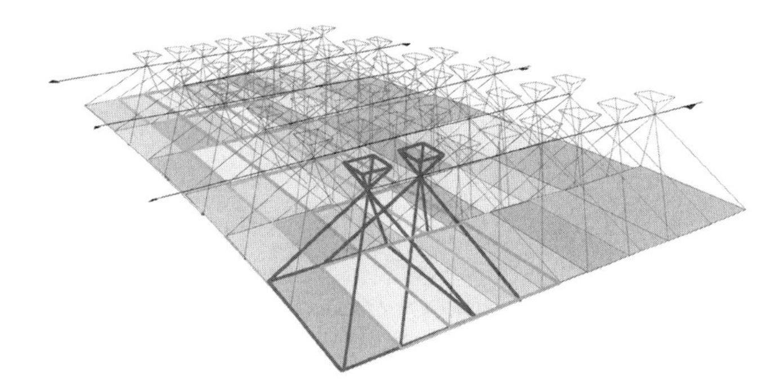

Figure A5.8 A block of photographs

and nearly parallel as possible; they are usually flown in an East-West direction (or West-East). The altitude should be maintained constant as nearly as possible. Satisfactory and economical mapping depends greatly on good flying.

Successive photographs along the strip must overlap and the strips themselves must overlap laterally; the amount of overlap is very important, especially along the strip. Specifications generally call for:

- longitudinal overlap of 55% to 65% with 60% as optimum;
- and a lateral overlap of 15% to 30%.

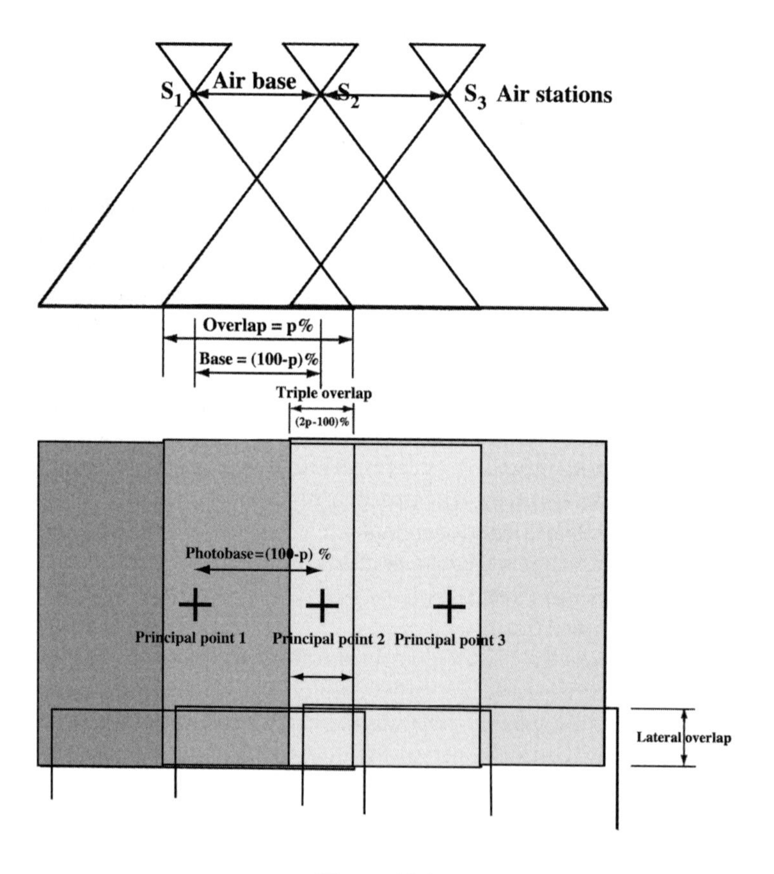

Figure A5.9

The upper part of Figure A5.9 represents the formation of three successive aerial photographs in a strip, shown in elevation. The lower part of the diagram may be thought of the ground areas covered by these photographs, in which case the points marked "Principal point *n*" represent the nadir points, or they may be thought of as the resulting photographs laid out on a table in such a way that corresponding parts of the images overlap. The percentages relate either to the side of the ground area covered by each photograph or to the side of each photograph itself. It will be seen that each photograph contains a point corresponding to the principal point of the

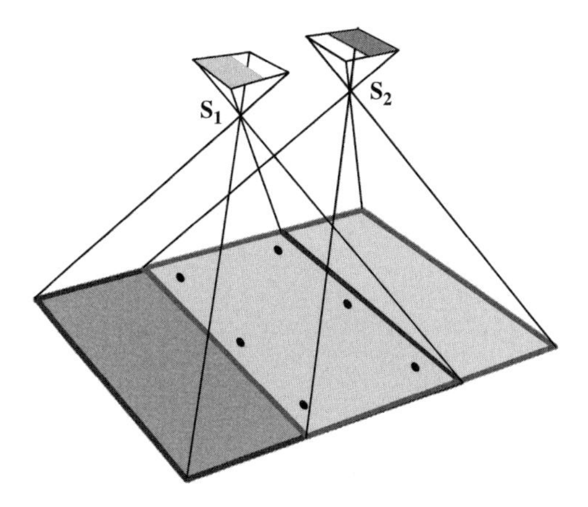

Figure A5.10 Two adjacent photographs forming a stereopair

adjacent photograph. The distance between this point and that photograph's own principal point is called the photobase.

A5.4 Formation of the photogrammetric model

Two adjacent photographs in a strip make up a "stereopair". It is to be assumed that we have no information about the photographs beyond the inner orientation of the camera; nothing is known of the exterior orientation of either picture. The aim is to determine the coordinates of points in the ground coordinate system whose images fall in both photographs. In processing a stereopair, at least six points common to the two pictures are initially chosen, more or less in the positions shown in the above diagram as black dots; their image coordinates are measured in each photograph. The procedure has similarities to two point resection in plan; in two-point resection in plan (see Figure A5.11) angles are measured at two unknown survey stations, A and B, to two control points, P1 and P2.

The length and bearing of the base AB are unknown; using arbitrary values for these it would be possible to compute the shape of the figure ABP2P1. This figure could then be scaled, oriented and shifted to fit the control points. This may not be the way the computation would be done but it describes the conclusion.

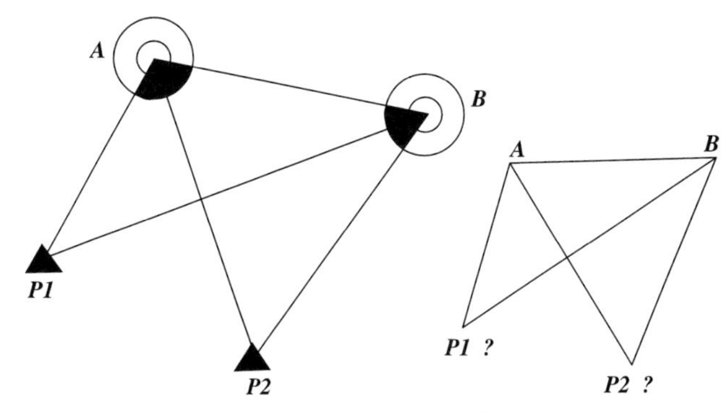

Figure A5.11 Two point resection in plan

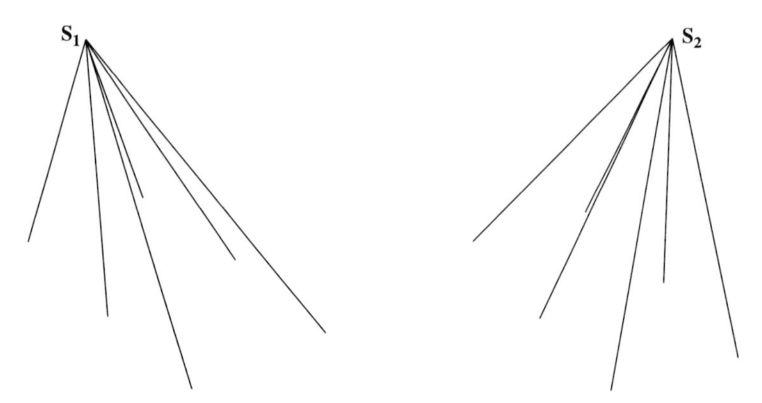

Figure A5.12 Bundles of rays computed from photographs at S1 and S2

The first stage in the photogrammetric processing is "relative orientation", the outcome of which is the formation of a correctly shaped model, but a model whose scale, orientation and position are unknown.

The directions of projected rays from each camera/projector could be computed, as in Figure A5.12. It would not be possible, as it would be in the two dimensional case in which the orientation of the bundles of rays at A and B relative to each other is already known, to go immediately to the computation of points of intersection of corresponding rays. Surprisingly, it is possible in the photogrammetric case, by applying the algebraic condition that all pairs of corresponding rays must intersect, to compute the relative orientation knowing only the shapes of the two bundles. The coordinates of the points of intersection may be found, using arbitrary values for scale and orientation. These coordinates together with those of the air stations, S1 and S2, comprise a "model", as illustrated in Figure A5.13 below. The shape of this model is correct but the scale, orientation and position are unknown in relation to the ground coordinate system; these must be found by the process of absolute orientation.

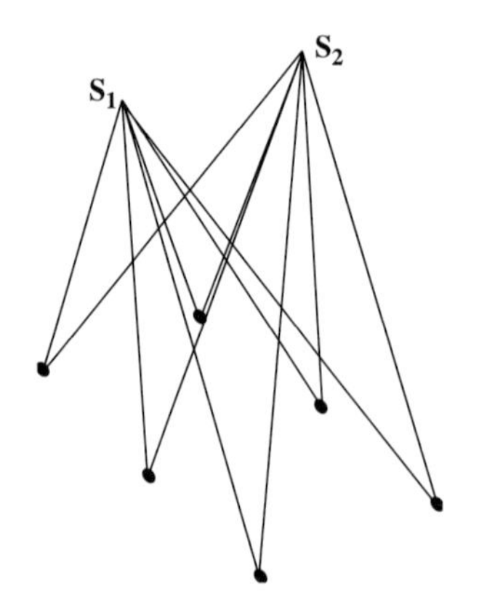

Figure A5.13 The uncontrolled model, consisting of the points of intersection of corresponding rays and the two camera stations. Scale and orientation are unknown.

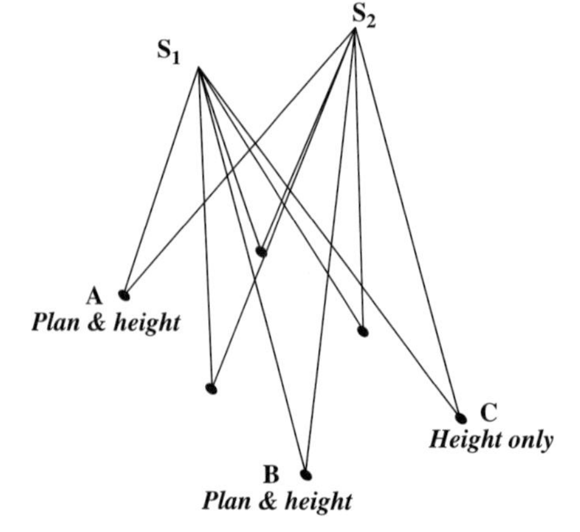

Figure A5.14 The model has been transformed to fit (minimal) control, using points A, B and C.

A5.5 Absolute orientation

"Absolute orientation" is used to describe not only the orientation but also the scale and position of the model. Let us assume, for purposes of illustration, that a surveyor has found the ground survey coordinates of points A and B in both plan and height and has also found the height of point C. This information is sufficient. The surveyor has found seven coordinates and there are seven unknowns to be found: scale, three parameters of orientation and three of position. For maximum accuracy it is advisable to have plan control points widely separated; a line of known length, AB, is used to scale the model and the bearing of the line AB controls the azimuth of the model. It is necessary for height control points to be well separated in both the direction of the air base, the X direction, and at right angles to this. Imagine Figure A5.14 to represent points on a rigid body; this body will sit firmly on the ground on the three height control points A, B and C. Should A, B and C be collinear, or nearly so, the "levelling" of the model will be unstable. The surveyor must bear these points in mind as well as the need for redundancy. Without redundant points, an error in any one of the seven coordinates considered above will go undetected.

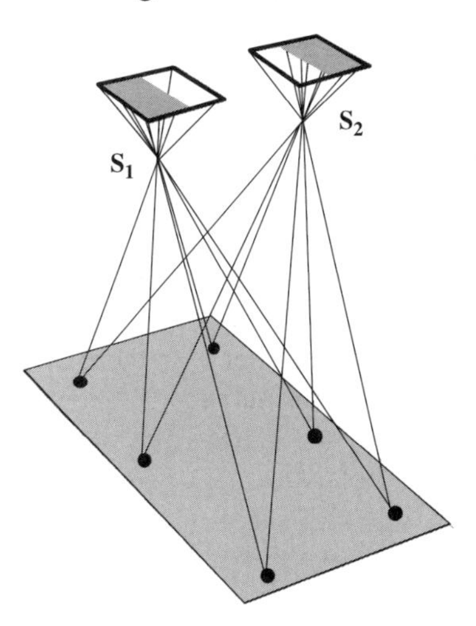

Figure A5.15 The "absolutely oriented" model. All points in the model may now be mapped.

When scale, orientation and translation have been applied we will have a model, as represented in Figure A5.15. As many points as necessary may now be found in the ground system, by measuring image coordinates and computing the intersections of projected rays. This is analogous to two point resection in plan: once the resection in plan has been carried out, surveying in the vicinity of the survey stations can continue and the results will have been found in the established ground system.

A5.6 Surveys to control strips and blocks

It is not necessary to supply control in every model of a strip or a block in the manner described above. The technique of block adjustment, better described as photogrammetric triangulation, has already been mentioned, and has been described as the simultaneous resection of large numbers of photographs, all the photographs in a strip or block. Specification of the control required to compute the triangulation of a block is quite complex and will not be discussed in any detail here. In general, control is required only at intervals around the perimeter of the block, with additional lines of height control running across the block. A surveyor undertaking such control surveys can expect to be advised of the specifications in each case.

Appendix 6
Quality Control

A6.1 Introduction

Quality Control (QC) of work forms part of the wider process of Quality Assurance (QA) in which later matters of administration and documentation, as well as technicalities, are involved.

In this book we are concerned mainly with technical aspects; i.e. with quality control.

The use of error analysis is paramount both to the design of surveys to a required specification, and to test the results of observations and derived parameters, to see if they have met the specification. This *before (a priori)* and *after* (*a posteriori*) treatment applies generally.

Figural and parametric methods

Most observations today are treated parametrically by the method of Least Squares. This depends on the selection of some framework of reference, such as a coordinate system, which affects the appearance of any statistical results. Care has to be taken, therefore, with the interpretation of these results. On the other hand, the formerly popular method of figural treatment, which does not suffer from datum bias, has much to commend it.

Norms

Norms other than Least Squares, such as minimum modulus, are more robust at detecting blunders in data, but suffer from excessive computational demands. The whole topic of quality control is one of current research, in which topics like cost effectiveness are not being ignored.

A6.2 Blunder detection

Blunders, or mistakes, are usually located by the careful processes of instrument practice and calibration. The subsequent Least Squares analysis usually assumes that all significant factors have been modelled by the mathematics. If this is not so, statistical tests on data can be quite wrong.

For example, it might be assumed that an EDM index has been applied to all measurements correctly, and therefore no index model parameter was introduced to the analysis. If the index had not been applied, or applied with the wrong sign, the subsequent error distribution is not random.

It might therefore be thought that all possible factors should be modelled. If this is done, the degree of redundancy in the model can be reduced until no statistical analysis is possible. Tests have shown that the cost of modelling every centring error at a theodolite station can easily be offset by the construction of pillars to avoid the problem at the outset.

Thus good instrument practice and calibration turns out to be the simplest and the best solution.

Figural approach

The filtering of observations by common-sense figural methods can be of immense value. Obvious checks are the misclosure of fully observed closed figures such as a triangle or traverse.

In networks, side equations can be used to locate blunders and save much time in wasteful re-observations, or to justify the omission of data from a subsequent parametric treatment.

See Chapter 8 for an example of blunder detection in a traverse.

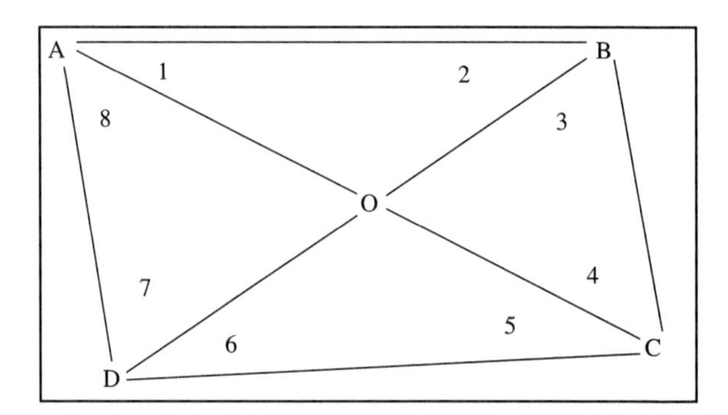

Figure A6.1

Example

Consider the braced quadrilateral ABCD of Figure A6.1. Suppose the triangles ABD and BCD close, while ABC and ADC have large errors of opposite signs. This shows that there is a mistake either in direction AC or in CA, or in both. Using the side equation technique will resolve the matter of location.

Taking the pole at A we have

$$\frac{AB}{AC}\frac{AC}{AD}\frac{AD}{AB}=1$$

therefore

$$\frac{\sin 4}{\sin(2+3)}\frac{\sin(6+7)}{\sin 5}\frac{\sin 2}{\sin 7}=1 \qquad \text{A6(1)}$$

Because equation (1) does not contain the angles at A, if it is satisfied within the tolerance of the observations, the mistake is at A. If not, the mistake is at C or A.

Using a similar equation by taking the pole at C would substantiate whether the error is just at C or at both points. A decision whether to reject or re-observe the faulty data would be taken before the Least Squares analysis by variation of coordinates.

Statistical approach

A blunder may be suspected if a residual is larger than expected. The criterion used may be quite arbitrary, such as three times the standard deviation, or more likely it will be based on a statistical distribution. When a large number of residuals is present (say more than 30) the normal distribution is used. Smaller samples will use the Student's t distribution. This aspect has been treated in Chapter 5. Rejection limits are set usually at the 95% confidence level.

The limit factors of the standard deviation for one, two and three dimensions are respectively

2.0 2.5 2.7

For example, we reject a three-dimensional vector residual greater than 2.7σ at the 95% confidence level, where

$$\sigma^2 = \sigma_x^2 + \sigma_y^2 + \sigma_z^2$$

This test can be applied automatically to sets of data which are never seen by an observer, as for example in a satellite ranging system.

A6.3 Least Squares analysis

In testing for blunders and reliability in the data and results of a Least Squares analysis we are able to use a number of statistical parameters, in addition to figural tests, which can be used to locate really large blunders.

Most of these statistical parameters may also be used to predict the possible results and quality of proposed networks, without making any actual observations. See Chapters 5 and 6 for more information.

Dispersion matrices

Dispersion matrices are made up of the estimated variances and co-variances of observed and unobserved parameters and functions of them.

The *a priori* estimates of the variances and co-variances of observed parameters are made from specific tests of repeated measurements, or from previous Least Squares estimates using similar techniques. The global variance scale factor σ_o^2 is estimated from observations. If the original estimates of the dispersion matrix of the observations is correct, this variance factor should be unity.

The Fisher test is used to judge whether a value of σ_0 is acceptable. It is tested against a theoretical value of unity from an infinitely large population. See Chapter 5 for more details.

If the test fails, a reconsideration of the original dispersion matrix is required. This often means a re-assessment of some types of measurements or instruments. This topic is also the subject of further research.

Error ellipses and ellipsoids

The error vector at a point may be depicted graphically by the pedal curve to an ellipse, or by the pedal surface to an ellipsoid, showing the variances of derived parameters in all directions.

Care is needed to interpret these curves properly. The main point to watch is that positional error ellipses are datum dependent.

A6.4 The error ellipse and its pedal curve

A particularly useful application of an ellipse and its pedal curve is to display some results of an error analysis in two dimensions. The pedal curve which shows graphically the size and direction of an error function in the vicinity of a point is useful in presenting results to clients, and in design studies, especially of networks. In three dimensions, the error ellipsoid and its pedal surface are also of value.

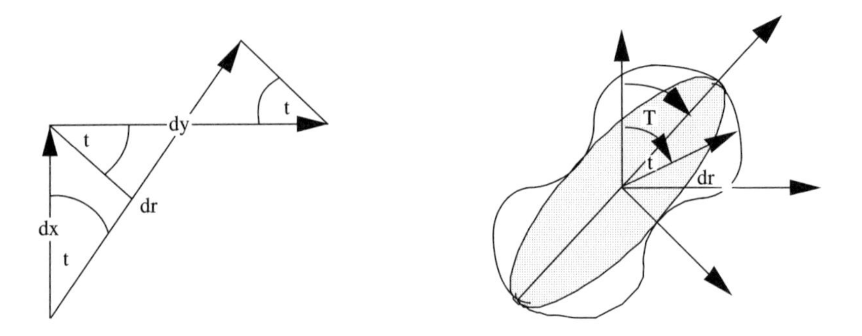

Figure A6.2 **Figure A6.3**

Refer to Figure A6.2 which shows displacements dx and dy and their compounded effect dr at a bearing t. As will be seen shortly these displacements will be represented by variances derived from the dispersion matrix obtained in a Least Squares analysis.

By inspection from the figure we have

$$dr = dx \cos t + dy \sin t$$

therefore

$$dr^2 = dx^2 \cos^2 t + dy^2 \sin^2 t + 2\, dx\, dy \sin t \cos t \qquad A6(2)$$

Taking expectations

$$\sigma_r^2 = \sigma_x^2 \cos^2 t + \sigma_y^2 \sin^2 t + 2\sigma_{xy} \sin t \cos t \qquad A6(3)$$

which may be recast as

$$2\sigma_r^2 = \sigma_x^2 + \sigma_y^2 + (\sigma_x^2 - \sigma_y^2) \cos 2t + 2\sigma_{xy} \sin 2t \qquad A6(4)$$

It is clear from the form of A6(4) with reference to equation A6(2) that the variance at any direction t is the square of the pedal distance to an ellipse. We now establish the orientation and axes of this ellipse by examining the turning values of this variance.

Differentiating A6(4) with respect to t we have, putting $V = 2\sigma_r^2$

$$\frac{dV}{dt} = -2(\sigma_y^2 - \sigma_y^2)\sin 2t + 4\sigma_{xy}\cos 2t \qquad A6(5)$$

The turning values are otained when t = T, and $\frac{dV}{dt} = 0$, i.e. when

$$0 = -2(\sigma_x^2 - \sigma_y^2)\ \sin 2\,T + 4\sigma_{xy} \cos\ 2\,T$$

$$\frac{\sin\ 2T}{\cos\ 2T} = \frac{2\sigma_{xy}}{\sigma_x^2 - \sigma_y^2}$$

$$\frac{\Delta \sin 2T}{\Delta \cos 2T} = \frac{2\sigma_{xy}}{\sigma_x^2 - \sigma_y^2}$$

where Δ is a positive constant. Therefore we may separate the numerator and the denominator as follows

$$\Delta \sin 2\,T = 2\,\sigma_{xy} \qquad A6(6)$$

$$\Delta \cos\ 2\,T = \sigma_x^2 - \sigma_y^2 \qquad A6(7)$$

Substituting from A6(6) and A6(7) in A6(4) we have, after some rearranging,

$$F = 2\sigma_r^2 = \sigma_x^2 + \sigma_v^2 + \Delta \cos 2(t - T) \qquad A6(8)$$

F is a maximum and a minimum respectively when

$$\cos 2\,(\,t - T\,) = +1 \text{ and } -1$$

That is σ_r has maximum and minimum values given by

$$F_{max} = 2\sigma_{max}^2 = \sigma_x^2 + \sigma_y^2 + \Delta \qquad A6(9)$$

$$F_{min} = 2\sigma_{min}^2 = \sigma_x^2 + \sigma_y^2 - \Delta \qquad A6(10)$$

Adding and subtracting A6(9) and A6(10) gives

$$\sigma_{max}^2 + \sigma_{min}^2 = \sigma_x^2 + \sigma_y^2 \qquad A6(11)$$

$$\sigma_{max}^2 - \sigma_{min}^2 = \Delta \qquad A6(12)$$

Thus A6(8) may be expressed in the final form

$$F = 2\sigma_r^2 = \sigma_{max}^2 + \sigma_{min}^2 + (\sigma_{max}^2 - \sigma_{min}^2)\cos\ 2(t - T) \qquad A6(13)$$

This is the equation of the pedal curve to the ellipse whose semi axes are respectively σ_{max} and σ_{min} with its axes oriented with respect to the original axes by the bearing T.

In practice, to avoid potential problems with zero values, the computational procedure is to obtain Δ from equations A6(6) and A6(7) using the positive root of

$$\Delta^2 = (\sigma_x^2 - \sigma_y^2)^2 + (2\sigma_{xy})^2$$

and then T from either A6(6) or A6(7) or from an algorithm using ATAN2 to identify the correct quadrant for T.

The error ellipse may then be drawn together with its pedal curve by hand or by suitable graphics software.

A6.5 Relative error ellipses

Provided the orientation of the coordinate axes is not altered, the error ellipses derived for *pairs of points* are not dependent upon the coordinate datum. These relative error ellipses are obtained from the dispersion matrix of the parameters as follows.

Consider any two points for which variances and co-variances are available. Usually these will come from the dispersion matrix

$$\sigma_0^2 \, N^{-1}$$

Since

$$\Delta x = x_2 - x_1 \quad \text{and} \quad \Delta y = y_2 - y_1$$

$$d\,\Delta x = d\,x_2 - d\,x_1 \qquad \text{and} \qquad d\,\Delta y = d\,y_2 - d\,y_1 \qquad \text{A6(14)}$$

The expectations of

$$(d\,\Delta x)^2, \ (d\,\Delta y)^2 \ \text{and} \ \ d\,\Delta x\, d\,\Delta y$$

are respectively the variances and co-variance

$$\sigma\Delta_x^2, \ \sigma\Delta_y^2 \ \text{and} \ \sigma_{\Delta x\, \Delta y}$$

These are given from A6(14) by

$$\sigma_{\Delta x}^2 = \sigma_{x1}^2 + \sigma_{x2}^2 - 2\,\sigma_{x1\,x2} \qquad \text{A6(15)}$$

$$\sigma_{\Delta y}^2 = \sigma_{y1}^2 + \sigma_{y2}^2 - 2\,\sigma_{y1\,y2} \qquad \text{A6(16)}$$

$$\sigma_{\Delta x\,\Delta y} = \sigma_{x1\,y1} - \sigma_{x1\,y2} - \sigma_{x2\,y1} - \sigma_{x2\,y2} \qquad \text{A6(17)}$$

These expressions are treated as above to give the relative error ellipse parameters for the two points.

$$\frac{\Delta \sin 2T}{\Delta \cos 2T} = \frac{2\sigma_{\Delta x \Delta y}}{\sigma_{\Delta x}^2 - \sigma_{\Delta y}^2} \qquad \text{A6(18)}$$

$$\sigma^2_{\max} = \sigma^2_{\Delta x} + \sigma^2_{\Delta y} + \Delta$$
$$\sigma^2_{\min} = \sigma^2_{\Delta x} + \sigma^2_{\Delta y} - \Delta \qquad \text{A6(19)}$$

It must be remembered that the two points of the network need not have been connected by direct measurement. All that is needed is their variances and co-variances in a common system.

Example

For example, the positional error ellipses, for a straight unclosed traverse of n equal legs starting at a fixed point A with a fixed reference bearing, grow larger from zero at the fixed starting point to a maximum at its end. All directions and lengths are assumed of the same weight. See Figure A6.4.

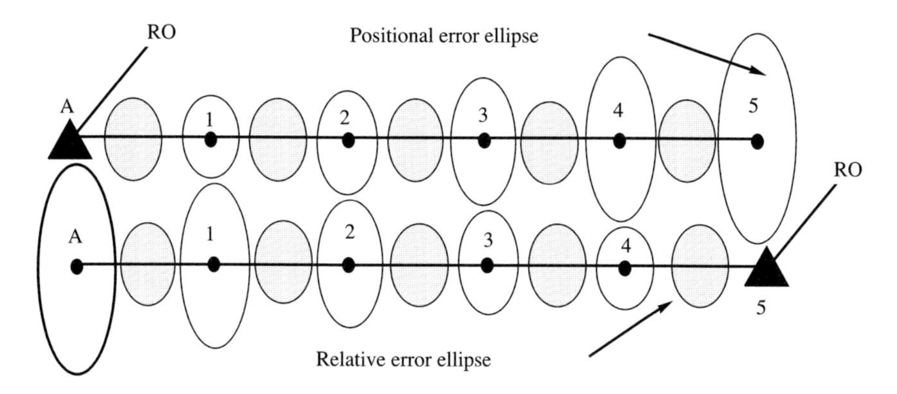

Figure A6.4

The same traverse computed from the other end of the line, i.e. from point 5, would exhibit an identical set of error ellipses, but different values for each point, except the middle one, if there is one. The *positional error ellipses* are merely the *relative error ellipses* between each point and the fixed origin.

A6.6 The error ellipsoid

When the above concepts of the error ellipse are extended into three dimensions, a different approach is required. We repeat some of the above derivations by this different approach as a preliminary to the ellipsoid treatment.

Consider equation A6(3) again :

$$\sigma_r^2 = \sigma_x^2 \cos^2 t + \sigma_y^2 \sin^2 t + 2\,\sigma_{xy} \sin t \cos t \qquad \text{A6(3)}$$

Now we can write

$$\cos t = L \quad \text{and} \quad \sin t = M$$

and equation A6(3) becomes

$$\sigma_r^2 = \sigma_x^2 \, L^2 + \sigma_y^2 \, M^2 + 2 \, \sigma_{xy} \, LM \qquad \text{A6(20)}$$

The quantities L and M are the *direction cosines* of the vector dr with respect to the x and y axes respectively, and these are subject to the constraint

$$L^2 + M^2 - 1 = 0$$

Hence the turning values of the variance are obtained by setting the differentials $\frac{dV}{dL}$ and $\frac{dV}{dM}$ to zero, where V is now the function

$$V = \sigma_r^2 = L^2 \, \sigma_x^2 + M^2 \, \sigma_y^2 + 2 \, L \, M \, \sigma_{xy} - k(L^2 + M^2 - 1) \qquad \text{A6(21)}$$

and k is a Lagrangian multiplier. Thus we have

$$\frac{dV}{dL} = 0 = 2 \, L \, \sigma_x^2 + 2 \, M \, \sigma_{xy} - 2 \, Lk$$

$$\frac{dV}{dL} = 0 = 2 \, M \, \sigma_y^2 + 2 \, L \, \sigma_{xy} - 2 \, Mk$$

which may be written in matrix form as

$$\begin{bmatrix} \sigma_x^2 & \sigma_{xy} \\ \sigma_{xy} & \sigma_y^2 \end{bmatrix} \begin{bmatrix} L \\ M \end{bmatrix} = k \begin{bmatrix} L \\ M \end{bmatrix} \qquad \text{A6(22)}$$

This may written in the form

$$\mathbf{A\,x} = \boldsymbol{\lambda}\,\mathbf{Ix}$$

That is a standard eigenvalue problem, in which **x** is the eigenvector and $\boldsymbol{\lambda}$ an eigenvalue. Subtracting, we obtain the homogeneous equations

$$\begin{bmatrix} \sigma_x^2 - k & \sigma_{xy} \\ \sigma_{xy} & \sigma_y^2 - k \end{bmatrix} \begin{bmatrix} L \\ M \end{bmatrix} = \begin{bmatrix} 0 \\ 0 \end{bmatrix} \qquad \text{A6(23)}$$

They have solutions, other than trivial, if the determinant of $(\mathbf{A} - \boldsymbol{\lambda}\mathbf{I})$ is zero; which leads to the quadratic equation in k, called the characteristic equation of the dispersion matrix **A,** as follows

$$k_2 - (\sigma_x^2 + \sigma_y^2)k + (\sigma_x^2 \, \sigma_y^2 - \sigma_{xy}^2) = 0 \qquad \text{A6(24)}$$

Notice that the coefficient of k and the absolute term are respectively minus the *trace* and plus the *determinant* of the matrix.

The solution of this quadratic gives the two values of k which, the reader may verify, are the maximum and minimum values of the variance in dr found by the previous method, that is

$$k_1 = \sigma_{max}^2 \text{ and } k_2 = \sigma_{min}^2$$

Substituting for k_1 in equations A6(23) we obtain the ratio

$$\frac{L}{M} = B \quad \text{say.}$$

then

$$L = B\,M$$

but

$$L^2 + M^2 - 1 = 0$$

Therefore

$$M^2 = \frac{1}{1+B^2}$$

$$M = \frac{1}{\sqrt{(1+B^2)}} \text{ and } L = \frac{B}{\sqrt{(1+B^2)}}$$

These are the normalized forms of the direction cosines L and M.

The mathematically inclined reader may care to verify that the dispersion matrix **A** may be transformed into a diagonal matrix **R** by the following formula

$$\mathbf{R} = \mathbf{P}^T\mathbf{A}\ \mathbf{P} \qquad \text{A6(25)}$$

where **P** is an orthogonal matrix whose columns are the normalized direction cosines L and M, and the diagonal terms of **R** are the eigenvalues k_1 and k_2. This form is convenient for computer packages dealing with eigenvalue problems. However a direct solution is possible as outlined above.

A6.7 The error ellipsoid continued

The required elements of the dispersion matrix are extracted and presented as

$$\begin{bmatrix} \sigma_x^2 - k & \sigma_{xy} & \sigma_{xz} \\ \sigma_{xy} & \sigma_y^2 - k & \sigma_{yz} \\ \sigma_{xz} & \sigma_{yz} & \sigma_z^2 - k \end{bmatrix} \begin{bmatrix} L \\ M \\ N \end{bmatrix} = \begin{bmatrix} 0 \\ 0 \\ 0 \end{bmatrix}$$

From the determinant of the matrix we obtain the error function in a three-dimensional problem (x, y, z) expressed in terms of three direction cosines L, M and N such that

$$\begin{aligned} V = \sigma_r^2 &= L^2\,\sigma_x^2 + M^2\,\sigma_y^2 + N^2\,\sigma_z^2 \\ &+ 2\,L\,M\,\sigma_{xy} + 2\,L\,N\,\sigma_{xz} + 2\,M\,N\,\sigma_{yz} \\ &- k(L^2 + M^2 + N^2 - 1) \end{aligned} \qquad \text{A6(26)}$$

The process of finding three turning values follows identical lines for two dimensions giving the following cubic characteristic equation

$$a\,k^3 + b\,k^2 + c\,k + d = 0$$

where

$a = 1$

$b = -$ trace of **A**

c = sum of determinants of diagonal minors of **A**
d = – determinant of **A**

A cubic equation may be solved directly by Ferreo's method (see Allan 2004) or iteratively by Newton's method. The three solutions for k are all positive because of the nature of the symmetric dispersion matrix (Positive definite). They equal the squares of the semi-axes (a, b, c) of the error ellipsoid, whose equation is

$$\frac{x^2}{a^2}+\frac{y^2}{b^2}+\frac{z^2}{c^2}=1$$

i.e.

$$k_1 = a^2 \qquad k_2 = b^2 \qquad k_3 = c^2$$

Substitution of these values for k in turn in equations A6(26) yields values for the ratios of the direction cosines

$$\frac{L}{M}=B \qquad \frac{N}{M}=C$$

These are normalized to give

$$L=\frac{1}{\sqrt{(1+B^2+C^2)}} \qquad M=\frac{B}{\sqrt{(1+B^2+C^2)}}$$

$$N=\frac{C}{\sqrt{(1+B^2+C^2)}}$$

The three sets of direction cosines give the direction of the semi-axes of the error ellipsoid relative to the coordinate axes.

Example
To simplify the arithmetic consider the untypically simple elements of a dispersion matrix and its characteristic equation A6(27):

$$\begin{bmatrix} 3 & 1 & -1 \\ 1 & 3 & -1 \\ -1 & -1 & 5 \end{bmatrix}\begin{bmatrix} L \\ M \\ N \end{bmatrix}=\begin{bmatrix} 0 \\ 0 \\ 0 \end{bmatrix} \qquad \text{A6(27)}$$

Consider Figure A6.5. Diagram (A) is a general view of the ellipsoid. Each of the shaded diagrams (B), (C) and (D) show the key planes of interest. Diagram (D) is drawn to a larger scale than the others and it alone shows part of the pedal surface in dots. The beginner is encouraged to make a cardboard model of these diagrams to see their relationships clearly.

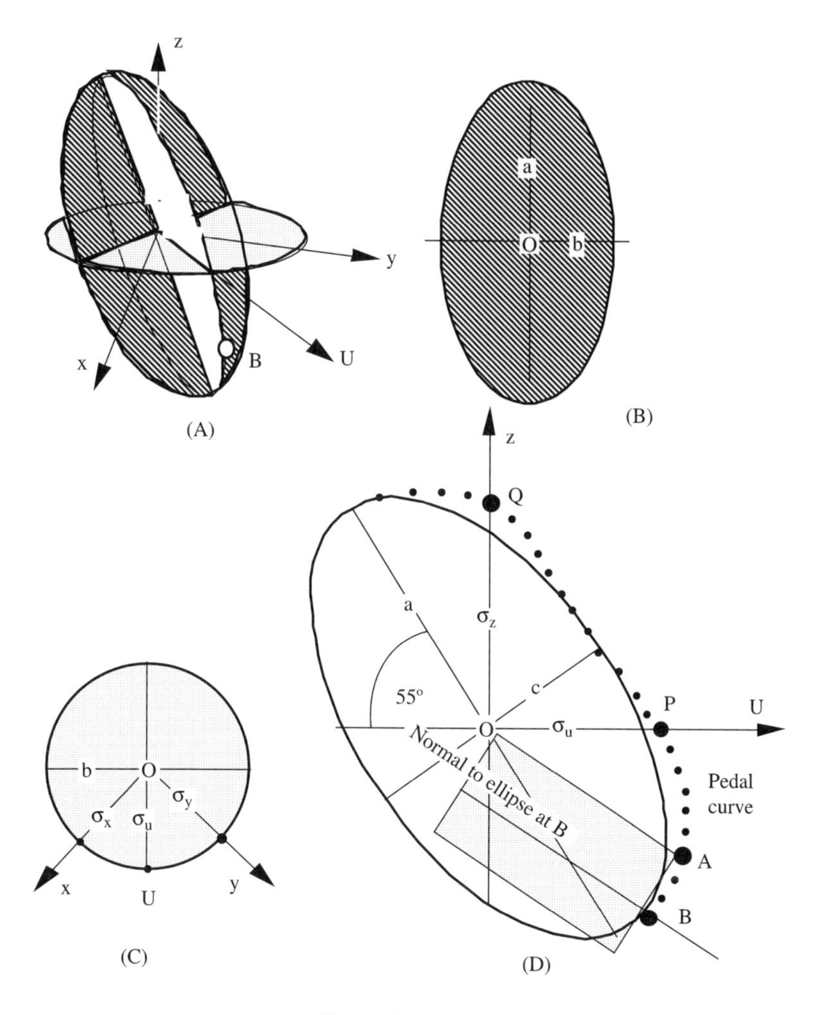

Figure A6.5

The cubic equation is

$$a\,k^3 + b\,k^2 + c\,k + d = 0$$

where $a = 1$ $\quad b = -11$ $\quad c = 36$ $\quad d = -36$

giving the solution

$$k_1 = 6 \qquad k_2 = 3 \qquad k_3 = 2$$

These give the normalized direction cosines as

$$\frac{1}{\sqrt{6}}, \frac{1}{\sqrt{6}}, -\frac{2}{\sqrt{6}} \quad \frac{1}{\sqrt{3}}, \frac{1}{\sqrt{3}}, \frac{1}{\sqrt{3}} \quad \frac{1}{\sqrt{2}}, -\frac{1}{\sqrt{2}}, 0$$

The reader should check that these vectors are orthogonal to each other as expected. Thus we can illustrate the error ellipsoid as in Figure A6.5, whose semi-axes are

$$a = \sqrt{6} \qquad b = \sqrt{3} \qquad c = \sqrt{2}$$

Diagram (D) shows all detail to plot the error ellipse at bearing U. Its semi-axes a and c are used to draw the ellipse. The pedal curve is sketched in using a set square placed tangential to the ellipse at B and marking A such that OA is perpendicular to AB. It is then seen that

$$OQ = \sigma_z = \sqrt{5} \qquad OP = \sigma_U \approx 1.8$$

This checks the value of σ_z from the original dispersion matrix of equation A6(27).

Using σ_U and b we can plot the error ellipse and its pedal curve (both nearly circular) in the plane Oxy illustrated in diagram (C). A check is made on the values of

$$\sigma_x = \sqrt{3} = \sigma_y$$

Finally the three-dimensional drawing is depicted in diagram (A).

The whole graphic process can be programmed for computer if required. Stereoscopic viewing of error ellipsoids adds realism to the presentation of results.

A6.8 DOP factors

In the field of position fixing by satellite, the useful notation has arisen to describe the quality of a position in terms of its Dilution Of Precision. This is a multiplier DOP of the standard error of unit weight σ_o. Thus to express a

height precision we have VDOP $= \sigma_z$

plan precision we have HDOP $= \sqrt{(\sigma_x^2 + \sigma_y^2)}$

positional precision we have PDOP $= \sqrt{(\sigma_x^2 + \sigma_y^2 + \sigma_z^2)}$

These are rough guides only since they do not take correlations into account.

A6.9 Comment

Some error ellipses are easy to predict for simple cases, such as for a radiated point or a straight traverse without bearing closures. In general they are best predicted from the actual dispersion matrix of the task.

Note: the error curve is not the error ellipse but its pedal curve, which may differ seriously from the ellipse, as shown in Figure A6.3.

A design problem of survey networks for general use, such as control for cadastral surveying, is to obtain ellipses which are all circular and of the same size (homogeneous isotropic). This is achieved by a suitable selection of observed data with appropriate weights.

It is also convenient to draw error ellipses or ellipsoids to represent the 95% confidence limits. These will have axes given respectively in two and three dimensions by

$$(2.5\ a, 2.5\ b) \text{ and } (2.7\ a, 2.7\ b, 2.7\ c)$$

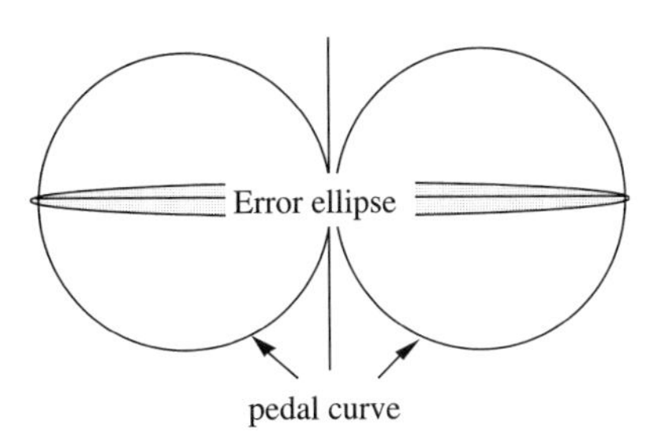

Figure A6.3

The monitoring of structures or land masses to detect movements is also of interest and research. Changes between epochs may be compared if two sets of measurements at different epochs are compared by identical analyses. Often the absolute positions of points are of no interest, only changes between them. There is much to be gained from a figural treatment which does not involve coordinates at all. (See Cooper 1987.)

Appendix A7
Field Astronomy

A7.1 Astronomy theory

The theory of field astronomy is still needed by the surveyor, who may occasionally require to determine azimuth or position from the stars or the Sun. A knowledge of the Sun's movement is also required for architectural studies of shadows, and can be useful to determine the time of day from photographs used in legal cases.

Due to the axial rotation of the Earth, the period of which defines the day, the stars appear to rotate round the Earth. The Earth also revolves around the Sun in an elliptical orbit in a period of time – the year. Since the stars are so distant from the Earth, both the Earth and its orbit can be considered as a point in space, and the stars can be considered as fixed to the inside surface of a huge sphere – the celestial sphere – at whose centre is the Earth and its orbit.

The use of geocentric directions to satellites enables the simple model of the celestial sphere to be applied also to near Earth satellites. Only directions on the Earth have significance.

The *three fundamental directions* in field astronomy are

(a) The *axis* of the Earth.
(b) The direction of the plumb line : the *vertical.*
(c) The *line of nodes*: the intersection of an orbital plane with the plane of the equator.

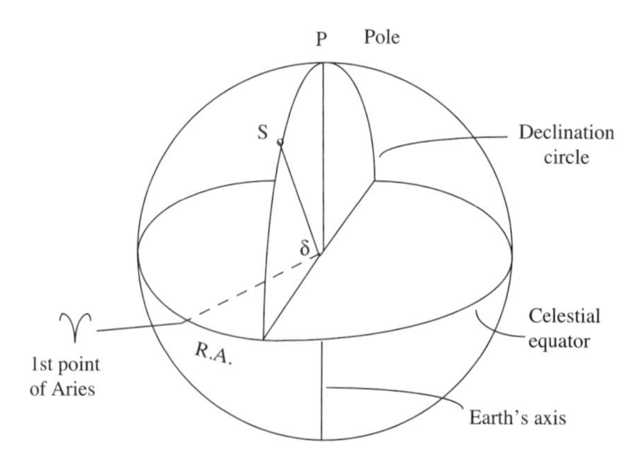

Figure A7.1 shows the celestial sphere as seen from a point outside it and above the North Pole P.

A7.2 System defined by the spin axis of the Earth

Figure A7.1 shows the following information:

(a) The North and South celestial *poles* about which the stars appear to revolve.
(b) The plane of the *celestial Equator* which is perpendicular to the axis and passes through the centre of the Earth.
(c) The *declination* δ of a star (Sun) or satellite - the angle between the equator and the star, measured in the great circle passing through the star and the pole - the *declination circle.* A South declination is negative.
(d) The *right ascension* (RA) of the star or satellite; the angle measured anticlockwise along the celestial equator from a reference point γ – the *first point of Aries* – to the declination circle through the star. The directions clockwise and anticlockwise are as viewed from outside the celestial sphere and above the North Pole. The first point of Aries is also known as the *Equinoctial point.*

The quantities declination δ and right ascension (RA) are tabulated in almanacs from data obtained by direct observations at observatories. The almanac specially designed for land surveyors for all but geodetic work is *The Star Almanac for Land Surveyors,* published annually. Both stellar parameters, declination and right ascension, change very slowly throughout the year. The Sun changes more rapidly than the other stars. Satellite parameters alter very quickly.

The argument used in the SALS for various quantities is generally Universal Time (UT). The UT is obtained from radio time signals or telephone time. In precise work the UT has to be corrected to Atomic time defined with reference to the mean position of the Pole and the non uniform rotation of the Earth. Details are given in SALS.

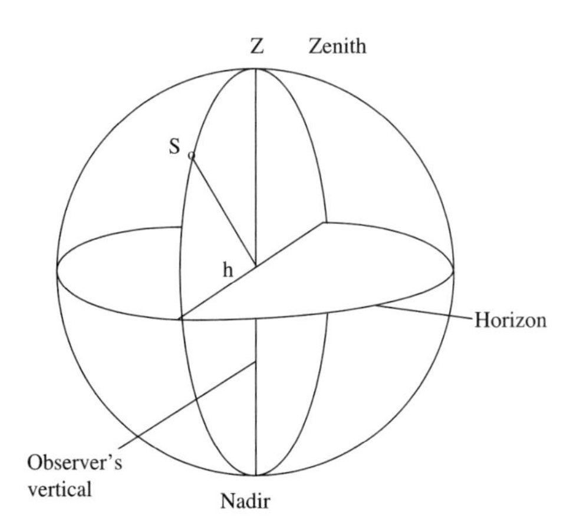

Figure A7.2

A7.3 The system defined by the observer's vertical

Figure A7.2 shows the following information defined by the observer's vertical:

(a) The *zenith* and *nadir* – the points vertically above and below the observer respectively.

(b) The *topocentric horizon* – the plane through the observer perpendicular to the vertical. Sometimes a distinction needs to be made between this and the parallel plane through the Earth's centre – the geocentric horizon.

(c) The *altitude* (h) of a star – the angle between the horizon and the star measured in the great circle through the star and the zenith – i.e. in the *vertical circle* through the star.

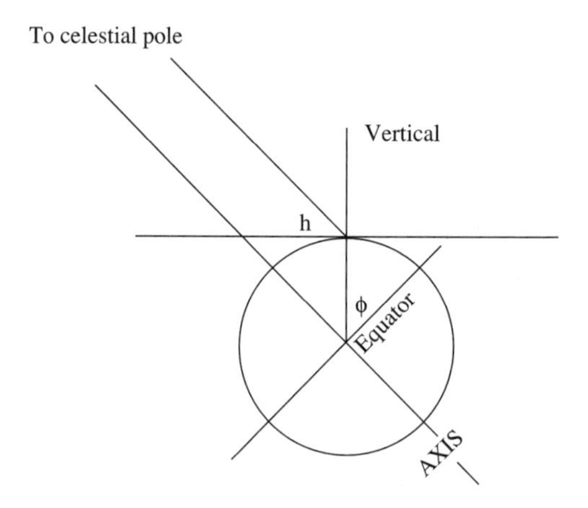

Figure A7.3

A7.4 Combined systems

The two systems defined by the vertical and the Earth's spin axis are combined together via the following:

(a) The great circle containing the zenith and the poles – the *meridian*.

(b) The angle between the Earth's axis and the vertical – the *co-latitude* c. The latitude, $\phi = 90 - c$, is the angle between the vertical and the equator, measured in the meridian (see Figure A7.3). It will be seen that the altitude of the elevated pole equals the observer's latitude.

Figure A7.4 shows the two systems combined together, with the plane of the meridian in the plane of the paper, and the pole to the right of the zenith. If P is the North Pole, the North and South cardinal points are defined by the intersection of the meridian and the horizon, and points east and west are at right angles to this north-south line.

When a star or satellite crosses the meridian it is said to *transit*. Since a star does so twice in one revolution of the Earth, we distinguish *upper transit* when the star transits on the same side of the pole as the zenith, and *lower transit* when the transit is on the opposite side of the pole from the zenith.

A7.5 Time and longitude

The ability of an observer to determine longitude on the surface of the Earth is directly dependent on the accuracy with which *time* can be kept relative to another

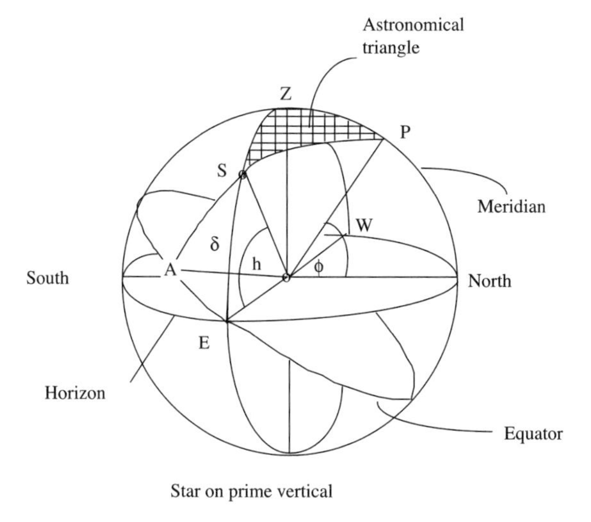

Figure A7.4

place. *Longitude* is the difference of time between the transits of two meridians by the same star. *Latitudes* are much simpler to determine without any need for time keeping.

Early astronomers and navigators could measure intervals of time at one place only by pendulum clocks, timing glasses and the human pulse, but could not measure instants relative to another place. Harrison's chronometers were the first to solve this problem in the 18th century: now radio signals and atomic clocks are capable of accuracies of one in 10^{11}.

Concepts of the day and the year are based on the daily spin of the Earth about its axis and its annual trip around the Sun. Today, the Earth is considered a poor time-keeper in relation to atomic repetitions. The whole subject is one of some complexity which is clearly summarized in SALS (Star Almanac for Land Surveyors. See references). However, we need some elementary ideas to understand the principles of astronomy and satellite position fixing. These may be summarized as follows:

(a) The Earth spins on its axis relative to the Sun in one *solar day* and relative to a distant star in one *sidereal day.* That these are different can be understood when one remembers that the Earth goes round the Sun but not round the other stars.

(b) The real Sun appears to move round its orbit (*the ecliptic)* at an irregular speed, so a harmonized fictitious Sun moving round the Equator *at constant speed* is used to define Greenwich Mean Time or *Universal Time* (UT). The true hour angle of the Sun is given by the expression

$$\text{GHA (Sun)} = \text{UT} + \text{E} \qquad \text{A7(1)}$$

The quantity E is about 12 hours. The exact values are tabulated in the SALS. Also tabulated are coefficients for polynomial interpolation.

(c) One annual trip of the Earth around the Sun defines a *year*. There are 365.2422 *solar* days or 366.2422 *sidereal* days in the year, depending on the reference body used for the count. The Earth's trip around the Sun accounts for the one–day difference which cannot be counted from the Earth. The sidereal day is about four minutes shorter than the solar. The two are in step on about the 22nd September (the *autumnal equinox*) at the Greenwich meridian. The relationship between the two time systems is expressed by the equation

$$GST = UT + R \qquad \text{A7(2)}$$

At the autumnal equinox, R = 0 gradually increasing by about 4 minutes per day throughout the year. The exact values are tabulated in the SALS, which also gives polynomial coefficients for computer use.

(d) Civil time zones and the calendar are an administrative convenience. Observation times need to be corrected to UT for scientific use. The necessary corrections are obtained from time–signals transmitted at various locations: see SALS for details.

(e) Orbit time is based on the period of one trip of a satellite round its orbit. It is usually reckoned from the orbital point closest to the Earth (*perigee)*, which in turn is known relative to the *ascending node* of the orbit (*perigee argument*). See 7.7. For simplified circular orbits the node is used directly as a marker.

A7.6 The astronomical triangle ZPS

The spherical triangle formed by the zenith Z, the pole P, and the star or satellite S is the astronomical triangle ZPS (see Figure A7.5). The angle Z is called the *azimuth angle*, the angle t the *hour angle,* and the angle S the *parallactic angle*. When the azimuth angle Z is 90° the star is said to be on the *prime vertical* (PV) and when the parallactic angle S = 90°, the star is at *elongation.* A star cannot elongate unless

$$PS < PZ \text{ i.e. } 90 - \delta < 90 - \phi \text{ or } \delta > \phi.$$

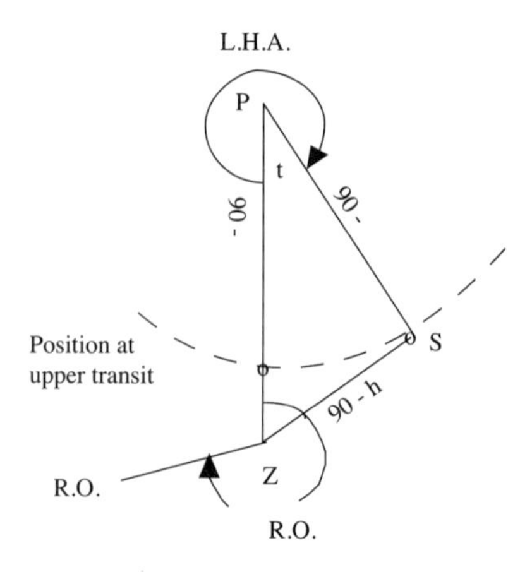

Figure A7.5

The *azimuth* α of a star is the angle between the meridian plane and the plane of the vertical circle through the star, reckoned clockwise from North, even in the southern hemisphere. Hence for an eastern star, $Z = \alpha$ and for a western star, $Z = 360 - \alpha$. The azimuth of a reference object (RO) is found by observing the horizontal angle between the star and the RO at the time of observation.

The Local Hour Angle (LHA) of a star is the angle between the plane of the meridian and the plane of the declination circle through the star, measured clockwise from the star's position at upper transit. Hence, for an eastern star, t = 24 hours – LHA and for a western start = LHA. Hour angles are reckoned in units of time. Computer algorithms will use arguments in radians.

The *sides* of the astronomical triangle are

$$PZ = 90 - \phi \qquad PS = 90 - \delta \qquad ZS = 90 - h$$

The sides and angles are related by the formulae of spherical trigonometry: especially the cosine and cotangent formulae which avoid the sign ambiguities arising from the sine formulae. Napier's rules for right-angled triangles are also useful. See Appendix A2.

A7.7 The determination of azimuth from the Sun and stars

Although the azimuth between two points can be determined from two satellite positions (Chapter 7), or from a North seeking gyroscope (Chapter 12), it is still a viable, and cheaper, alternative to do so from theodolite observations of the Sun or stars. In certain circumstances it may be the only way. An isolated oil platform, for instance, is unsuited to north–seeking gyropscopic methods because of instability, and another platform may not be visible from it.

The process of levelling a theodolite sets the primary axis parallel to the direction of the gravity field. Therefore *astronomical* positions and azimuths result. If it were possible to tilt the theodolite primary axis parallel to the ellipsoidal normal, then *geodetic* parameters would result. For many practical purposes, such as in cadastral surveying, the distinction between astronomical and geodetic azimuths is academic and can be ignored. It is however possible to observe geodetic azimuth by Black's method. (See Allan 1997 for details.)

Azimuth determination

The simplest method is to observe horizontal and vertical angles by theodolite, noting the approximate time only to interpolate data from SALS. This method is not capable of the highest accuracy, but can easily give results to about 5" arc.

The best method is to observe time and horizontal angle, noting the approximate vertical angles for possible i tanh corrections.

It is now practicable to observe all three quantities, horizontal angles, vertical angles and time, with electronic theodolites on–line to a field computer. The author has obtained first order results by this method from Sun observations. (For more details see Allan 1997.)

Azimuth by simultaneous observation of horizontal and vertical angles

Applying the cosine formula of spherical trigonometry to the astronomical triangle ZPS gives:

$$\cos Z = \frac{\sin \delta - \sin \phi \sin h}{\cos \phi \cos h}$$

from which Z and therefore α is derived. The angle θ to the RO is also known, and thence the azimuth of the RO is found.

Azimuth by time and horizontal angles

This method of determining azimuth is potentially more accurate than by altitudes, since the observation involves the bisection of only the vertical cross hair (horizontal angle); vertical refraction does not affect the method; and the stars or Sun may be observed low down close to the horizon thus reducing the errors due to axis tilt.

Theory

From the observed time the hour angle t is calculated: the declination δ is obtained from SALS and the latitude and longitude of the observer are scaled from a map or obtained from the survey coordinates. Applying the cotangent formula to the astronomical triangle gives

$$\cot Z = \frac{\cos \phi \tan \delta - \cos t \sin \phi}{\sin t}$$

The azimuth is obtained from Z and the horizontal angle to the RO.

Warning. The Sun should never be observed directly through a telescope without a dark glass. One quick glance can cause serious damage to the eye. The easiest way is to project the Sun's image on to a piece of white card, which can be observed by the unaided eye.

For detailed examples of Sun azimuth observations the reader should consult Allan 1997.

Appendix A8
Survey Projections

Although the positions of points may be computed on the surface of a reference ellipsoid or in three-dimensional cartesian coordinates, a common end product of surveying, the map, has to be drawn on a plane surface. Hence some system of map projection is required. In the past, hand computation on the curved surface of an ellipsoid or a sphere was more tedious than on a plane surface so preference was given to projection methods. Current practice is still to work on a projection for limited surveys which have to be tied into a national coordinate system. Large survey organizations dealing with national mapping compute in three dimensions or on the ellipsoid and convert to map projections for plotting purposes.

In practice, computation on a projection is usually in two forms.

(1) Control points are converted to two dimensional cartesian coordinates (E and N) or (X and Y) from the three-dimensional measurement system such as WGS84. The conversion formulae are complex for a spheroid. (See Snyder 1987.)

(2) Minor control points or photogrammetric control points are computed on the projection by plane formulae slightly modified or distorted. These modifications are for:

 (a) Grid convergence applied to azimuths;
 (b) Scale factors applied to distances;
 (c) Arc to chord corrections (t – T) to directions.

In this appendix we explain the principles involved in these operations.

A8.1 Map projections

A *map projection* is any orderly system of representing points of the sphere or ellipsoid on a plane. We shall confine our theoretical deliberations to the sphere since the main principles can be more easily understood by this approach. With the exception of the stereographic projection, the projections used by surveyors cannot easily be drawn by direct geometrical perspective, but are plotted from a grid.

The two most commonly used projections are the *transverse Mercator* and *Lambert's conical orthomorphic*, both of which are conformal or orthomorphic projections, i.e. on these projections the scale factor at a point is independent of azimuth.

Although the conversion of geographical or ellipsoidal coordinates to projection coordinates is now carried out by computer software, some use of tables has been retained here to provide data for the examples and to verify computer software.

A8.2 Basic concepts of projection

The following basic concepts are essential for a grasp of the subject.

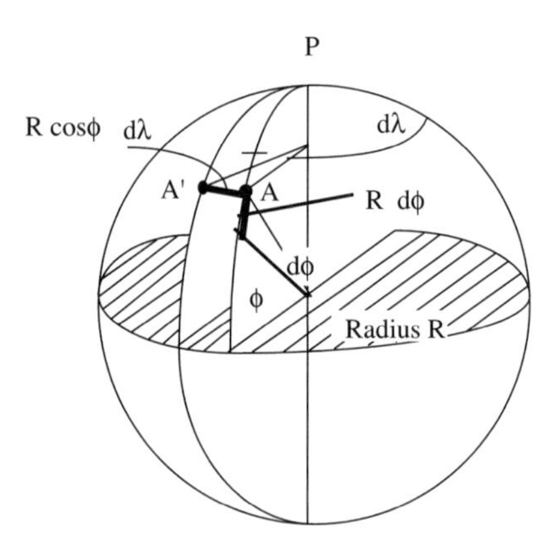

Figure A8.1

A *graticule* is the network of lines formed by the meridians and parallels either on the sphere or on the map, the latter being the projected positions of the former.

A *grid* is a system of squares drawn on a map, from which the map graticule and points of detail are plotted.

A8.3 Transformation formulae

In Figure A8.1, let A be a point on a sphere of radius R representing the Earth, whose North pole is P. The map projection system relates the latitude and longitude (ϕ, λ) of A to its grid coordinates (E,N) by some functional relationship which establishes a unique one-to-one correspondence between a point on the sphere and its plotted position on the map. Expressing this mathematically we write

$$E = F_1(\phi, \lambda) \qquad N = F_2(\phi, \lambda)$$

$$\phi = F_3(E, N) \qquad \lambda = F_4(E, N)$$

These transformation formulae have to be explicitly stated for each projection according to some rules which will preserve some properties of the original surface, but not all of them. *Conformal* or *orthomorphic* projections preserve local small shapes, *equal area* or *orthembadic* projections preserve areas, and so on. Much of the theory is developed around the *scale factor* of the system.

A8.4 Scale factor

The scale factor K is the ratio of a straight line distance on the map to the corresponding distance on the sphere or ellipsoid. The scale factor is often considered under the headings *nominal scale* and *differential scale* or *scale error*. The nominal scale of a map is its representative fraction, say 1: 50 000, stating in a general way the magnitude of the scale factor. The scale factor over the whole map is not constant but varies from the nominal scale by small differential amounts, often called scale errors.

Surveyors usually work implicitly with a nominal scale of unity (at life size); hence the differential scale or scale error is the amount by which the scale factor differs from 1. For example if the scale factor at a point on the projection is 0.9996 the corresponding scale error is – 0.000 4. The subsequent reduction of coordinates from life size to some convenient map scale merely alters the nominal scale of the survey.

The use of the word 'error' is perhaps unfortunate, since it connotes the idea of a mistake, whereas the introduction of a scale factor is deliberate and its effects are predictable. In a sense it is a known *systematic* error.

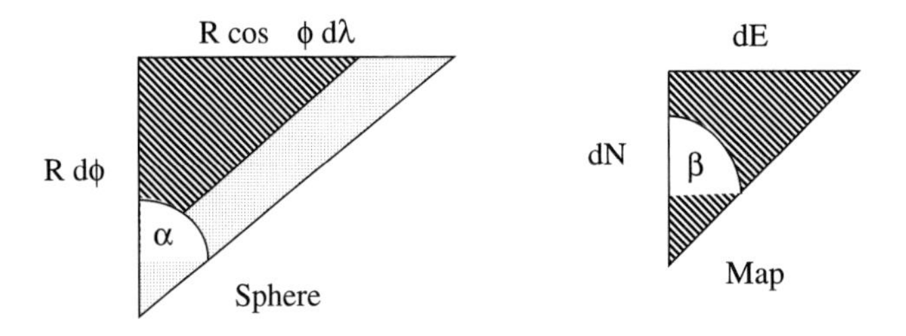

Figure A8.2

The concept of differential scale is explained as follows. Let a small linear element of the meridian of A in the vicinity of A be R d ϕ, and a small element of the parallel of A be R cos ϕ d λ.

(Remember that the meridians are great circles of radius R and the parallels are small circles of radius R cos ϕ)

Then if dN and dE are the elements on the map corresponding to R d ϕ and R cos ϕ d λ, the scales, K_1 in a north-south direction and K_2 in an east-west direction, are given by

$$K_1 = dN / R\, d\, \phi$$

$$K_2 = dE / R \cos \phi\, d\, \lambda$$

(Note: Similar fomulae for a spheroid are

$$K_1 = dN / \rho\, d\, \phi$$

$$K_2 = dE / \nu \cos \phi\, d\, \lambda$$

and since ρ and ν are themselves functions of the latitude ϕ the whole theory becomes much more complex.)

(Note: It is sometimes convenient to define a new variable ψ, called the *isometric latitude*, by putting

$$d\, \psi = \cos \phi\, d\, \phi$$

Isometric latitude is not be employed here.)

The projection is conformal or orthomorphic if $K_1 = K_2$. A corollary to this is that *for a small element*, angles on the surface of the sphere are correctly represented on the map. Thus in Figure A8.2

$$\tan\alpha = R\cos\phi\, d\lambda / R\, d\phi$$
$$= (dE / K_2)\ (K_1 / dN)$$
$$= dE / dN = \tan\beta$$

thus

$$\alpha = \beta$$

In practice a small element will be the theodolite circle used for the angle measurement. For such a small part of the Earth's surface the angle equality of the conformal map can be considered exact.

It is important to note that these properties of a conformal projection hold only for small differential amounts. Since the surface of the sphere cannot be correctly represented on a plane over large areas, the scale factor for long lines will not be independent of azimuth, neither will angles be correctly preserved in conversion from the sphere to the map.

Just what is meant by 'long lines' depends on the ultimate accuracy desired. The angular distortion over a line of 8 kilometres is normally less than 1 second of arc on a conformal projection, hence the projection effect could be neglected for most work under this distance, and the survey computed in simple rectangular coordinates.

In non-conformal projections, angular and differential scale distortions seriously restrict the lengths of lines that may be computed without applying corrections.

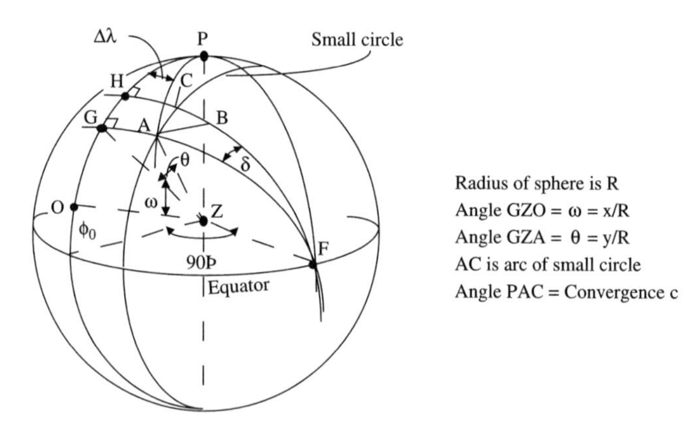

Figure A8.3

A8.5 Rectangular spherical coordinates or Cassini's projection

The simplest form of projection used by surveyors for early mapping is the Cassini projection. Although many old cadastral maps were plotted by this method, it is little used today. However, it does provide a simple way to develop formulae for the more useful conformal projection, the transverse Mercator.

Figure A8.3 shows the terrestrial sphere. P is the pole; OGHP is the central meridian of the projection system; O is the origin of the projection; A and B are two points on the surface of the sphere; G and H are the feet of the perpendiculars from A and B respectively to the central meridian; arc AC is part of a small circle passing through A lying in a plane which is parallel to that of the central meridian; F is the pole of the central meridian; the radius of the sphere is R; latitudes and longitudes of the various points are denoted by (ϕ_A , λ_A), with the suffix referred to the point.

Let the angles subtended at Z, the centre of the sphere, by OG and GA be ω and θ respectively. Let the arc length of OG be x_A and that of GA be y_A (see Figures A8.4 and A8.5).

To construct a projection draw a straight line O'G'H', mark off O'G'= OG = x_A, and G'A' perpendicular to O'G' at G' equal to GA = y_A, the point A' so obtained is the position of A on the the Cassini projection whose origin is at O and whose central meridian is OGH (O'G'H').

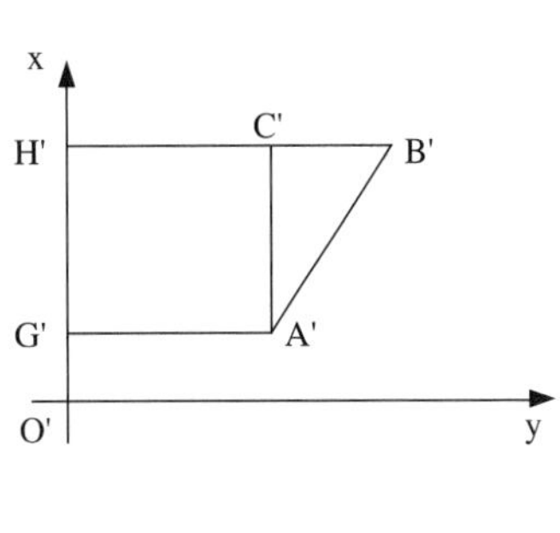

Figure A8.4

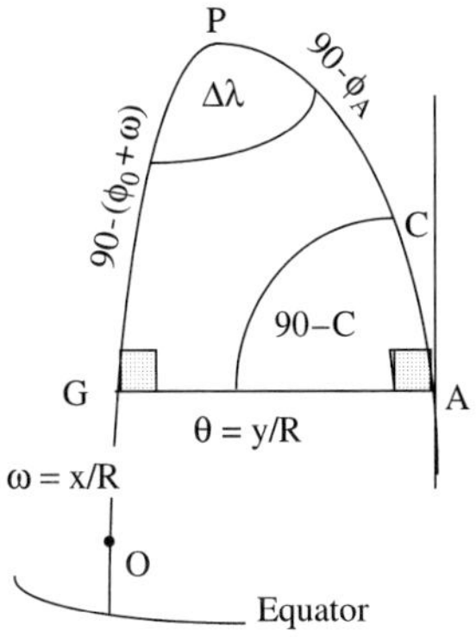

Figure A8.5

A8.6 Transformation formulae for Cassini projection

Projection computation is concerned with the forward and reverse transformation from geographical to rectangular coordinates and vice versa, and the calculation of convergence from either starting point. Also involved are the calculation of graticule-grid cutting points, point and line scale factors, and arc-to-chord corrections.

Forward transformation

Figure A8.5 shows the spherical triangle formed by A, P, and G. The angle at G is right. Since CA is perpendicular to GA and if the angle PAC = C angle PAG is 90 – C.

The line AC is projected as A'C' on the map (Figure A8.4), which defines the direction of grid north. The angle C is therefore the difference between true and grid north. It is called the *map convergence* or simply *the convergence a*t A. Notice this is strictly different from the meridian convergence $\Delta\alpha$ on the ellipsoid although they both are about the same magnitude.

From spherical trigonometry (See Appendix A2)

$$\sin (y/R) = \sin \theta = \sin \Delta \lambda \ \cos \phi_A \qquad A8(1)$$

$$\cot (\phi_0 + x/R) = \cot (\phi_0 + \omega) = \cos \Delta \lambda \cot \phi_A \qquad A8(2)$$

$$\tan C = \tan \Delta\lambda \sin \phi_A \qquad A8(3)$$

where

$$\Delta\lambda = \lambda_A - \lambda_O$$

Given the value of R for the sphere, the coordinates of A' on the map can be calculated from its given geographical coordinates (ϕ_A, λ_A).

Reverse transformation

For the reverse computations, given (x and y) and therefore ω and θ can be derived, the following expressions are used to obtain ϕ_A and λ_A

$$\sin \phi A = \cos \theta \sin (\phi_o + \omega) \qquad A8(4)$$

$$\cot \Delta\lambda = \cos (\phi_o + \omega) \cot \theta \qquad A8(5)$$

A8.7 Scale and angular distortions on the Cassini projection

Figure A8.4 shows points A and B plotted in positions A' and B' on the map. C' lies on H'B' such that H'C' = G'A' = y. Figure A8.6 shows an enlarged view of triangle A'B'C' with the original spherical triangle ABC superimposed upon it. The grid bearing of A'B' is denoted by T, and angle B'A'B' by dT. Since B'C' = BC, the scale east-west is correct.

But the line AC, equal to A'C' = H'G' = HG = R δ, is plotted too long. The angle HFG = δ, or δ is the angle subtended by GH at the centre of the sphere (Figure A8.3).

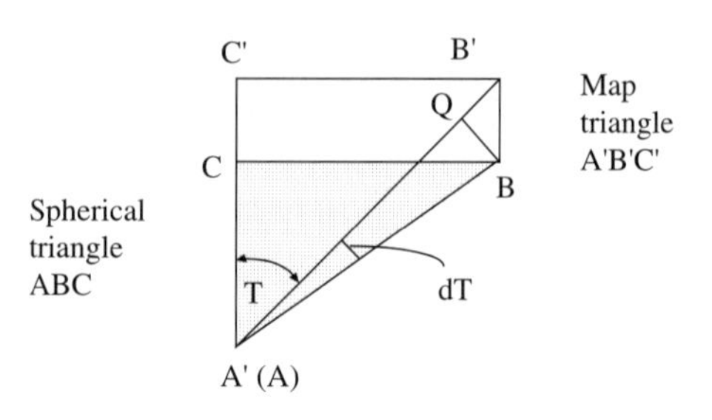

Figure A8.6

In Figure A8.6 we have

$$AC = R\,\delta \cos\theta = GH \cos\theta = A'\,C' \cos\theta$$

therefore the scale factor in a north-south direction at A is

$$A'\,C' / AC = \sec\theta = K \text{ say}$$

then

$$BB' = CC' - AC = A'\,C'\,(1 - \cos\theta)$$

$$= A'C' \frac{1}{2}\theta^2 \text{ approx}$$

$$= \frac{1}{2}\theta^2 \text{ s} \cos T$$

where s is the length of the line AB and T is the grid bearing.

Angular distortion

The angular distortion dT is given approximately as follows:

$$dT'' \sin 1'' = BQ / s = BB' \sin T / s$$

$$= \frac{1}{2}\theta^2 \sin T \cos T$$

$$= \frac{1}{4}\theta^2 \sin 2\,T$$

$$= (y\ 2)/\ 4\ R^2) \sin 2T \qquad \text{A8(6)}$$

Hence dT is zero for a line north to south, and a maximum at azimuths of 45°, 135°, 225°, and 315°. Notice the angular distortion even for a short line is considerable at points away from the central meridian; for example its maximum value when y = 500 km is about 30".

Scale distortion

The line AB becomes A'B' on the map, hence the scale is increased by

$$B'Q = BB' \cos T = \frac{1}{2}\theta^2 \text{ s} \cos^2 T$$

Expressed as a fraction of the length s this scale error is

$$\frac{1}{2}\theta^2 \cos^2 T$$

and the scale factor K is

$$K = 1 + \frac{1}{2}\theta^2 \cos^2 T$$

When $T = 90°$ or $270°$ $K = 1$:

when $T = 0°$ or $180°$ $K = 1 + \frac{1}{2}\theta^2$

Put another way, the scale factor is sec θ in a north-south direction, and unity in an east-west direction.

Because of these excessive angular and differential scale distortions the projection is not well suited to mapping, though it was widely used in the past on account of its computational and graphical simplicity.

A8.8 The transverse Mercator or Gauss-Krüger projection

This projection can be thought of as a Cassini projection modified to be made conformal: i.e. so that the point scale factor is independent of bearing T, with no angular distortion over a very short line.

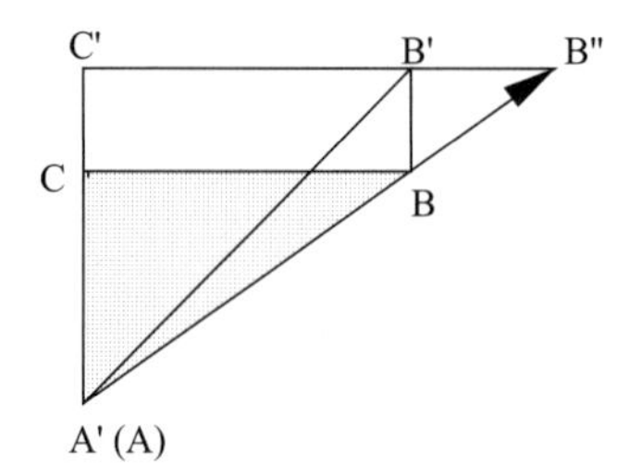

Figure A8.7

The angular distortion at a point, such as A in Figure A8.7, will be eliminated by increasing the projection easting y of point B' until B' lies in position B", i.e. until A'BB" is a straight line. This is achieved if the scale factor for the element BC is made the same as for the element AC, i.e. BC is increased by the value of sec (y/R) = sec θ.

This increase has to be made to all short lines out from the central meridian; that is each element of the arcs G'A' and H'B' of Figure A8.4 will be increased by its appropriate sec θ.

The easting thus obtained is the transverse Mercator easting E which is given by

$$E = \int_{0}^{\theta} R \sec\theta \; d\theta$$

or

$$E = R \ln \tan [\pi/4 + \theta/2] \qquad \text{A8(7)}$$

The transverse Mercator (TM) northing N is the same as x on the Cassini projection.

Note: this is the transverse case of Mercator's projection. In the direct case, the latitude ϕ replaces the angle θ of this projection, and we have

$$N = R \ln \tan [\pi/4 + \phi/2]$$

A8.9 Computation on the transverse Mercator projection

It is easy to apply this formula for E with modern calculators or computer software. Traditionally this was not so and an alternative approach using expansion series was preferred. Since this second method still has its counterpart in ellipsoidal treatment of the projection problem we will give a brief account of it here.

Sec θ is expanded as a power series of θ and integrated term by term. We have

$$E = \int_{o}^{\theta} R \sec \theta \; d\theta$$

$$E = R \int_{o}^{\theta} (1 + \frac{1}{2}\theta^2 + \frac{5}{24}\theta^4 \ldots\ldots) \, d\theta$$

$$= R \, (\theta + \frac{1}{6}\theta^3 + \frac{1}{24}\theta^5 \ldots)$$

The third term has a value of 0.1 m for θ = 3° which is normally the maximum distance from the central meridian used. When a country has a larger extent east-west than this, the area is divided into two or more projection 'belts' or 'zones' with a suitable overlap (see the Universal Transverse Mercator projection).

Reverse case

To obtain θ from E, the method of successive approximation is adopted, putting θ = E/R in the formula instead of the correct θ = y /R. We have

$$E = R \, (\theta + \frac{1}{6}\theta^3 + \frac{1}{24}\theta^5 \ldots)$$

$$= R \, \theta \, (1 + \frac{1}{6}\theta^2 + \frac{1}{24}\theta^4 \ldots)$$

or

$$\theta = (E/R) \, (1 - \frac{1}{6}\frac{E^2}{R^2} \ldots) \text{ approx}$$

The new value of θ is then used in the reverse series until convergence is achieved.

In practice the precise ellipsoidal formulae are programmed for computer computation without much difficulty.

A8.10 Scale factors on the transverse Mercator projection

The scale factor used in the design of the projection was sec θ which was integrated to give the Eastings. Thus for a short line the scale factor is sec θ. This means that all lines away from the central meridian are increased by an amount which can be quite large. To reduce the overall effect, a central scale factor K_0 less than unity is applied to the whole map, making the scale factors at the central meridian and at the designed maximum distance from it to be equal and opposite in size. TM projection zones are usually set at ±3° from the central meridian, i.e. θ max is 3°.

Without the application of K_0 the maximum scale factor sec 3° = 1.0014 approx. If we make $K_0 = 1 - 0007 = 0.9993$, then the combined new scale factor is $K_0\,K = 0.9993$ sec θ for any point. At two eastings either side of the central meridian the net scale factor is 1. These eastings are at angles $\theta = 0.7 \times 3° = 2.1°$, i.e. whose secant is 1.0007.

Line scale factors

Because the scale factors apply only to points, the appropriate value to apply to a line is the average for the points on that line. Sometimes the mid-point value will suffice; for others Simpson's rule for approximate integration is required.

False eastings

To avoid the unnecessary nuisance of positive and negative eastings the origin of a projection is moved to the west to a *False Origin*. This is usually 500 000 metres for a 6° belt, or in the case of the United Kingdom, 400 000 metres. Before the projection formulae can be used, the listed eastings must be reduced to the true eastings with respect to the central meridian.

A8.11 The Universal Transverse Mercator projection (UTM)

The UTM projection has been adopted for mapping by many countries and agencies. Before giving a worked example of its use, we describe its basic characteristics.

Zones

Because the scale error becomes excessive, even with a central scale factor, the practical width of a transverse Mercator projection is limited to about 3° on either side of the central meridian. This limitation also permits relatively simple formulae to be used for the calculation of scale and angular distortions with ample accuracy for most work. To map areas of greater east-west coverage than 6°, two or more belts or zones, consisting of identical transverse Mercator projections, are used.

The Universal Transverse Mercator projection system consists of sixty belts each 6° wide, beginning at 180° west longitude with Zone 1 as its western edge, i.e. its central meridian is 177° west longitude. The zones are then numbered eastwards, e.g. Zone 31 has the Greenwich meridian as its western edge.

UTM projections

Each zone consists of an identical transverse Mercator projection whose characteristics are as follows:

(a) The north-south extent is from 80° N to 80° S latitude.
(b) In the northern hemisphere the origin is on the equator at a point 500 000 m to the west of the central meridian, i.e. the false easting (FE) = 500 000 m.
(c) In the southern hemisphere the origin is 10×10^6 m to the south of the equator and 500 000 m to the west of the central meridian.
(d) The central scale factor $K_o = 0.9996$ exactly.
(e) Various ellipsoids are recommended for different parts of the world, e.g. the Clarke 1880 ellipsoid is suggested for Africa.
(f) Projection tables, for all the recommended ellipsoids, are available for the conversion of geographical coordinates to grid coordinates and vice versa.
(g) Tables are also available for zone-to-zone transformations.
(h) Tables are available giving the grid coordinates to 0.1 m of the intersections of the graticule at intervals of 5 minutes of arc, and at intervals of 7.5 minutes of arc. These are of great use to the cartographer since the edges of various map sheets may be plotted directly from the tables.

A8.12 The computation of a survey on a map projection

As discussed before, latitude and longitude, may be computed on the surface of a reference ellipsoid. These geographical coordinates can then be projected on to a plane surface by the formulae for a particular projection system. Thus a set of plane rectangular coordinates representing the points of a survey is obtained, which will be used as the basis for mapping, for the calculation of areas etc.

As an alternative to computation on the surface of the ellipsoid, it is possible to compute directly in terms of plane trigonometry, provided the ellipsoidal lengths and the measured angles are distorted to suit the particular projection used. This method of computation is still of some relevance for surveys of limited areas. Computer software written for plane coordinate computations by Least Squares is easily adapted for use on the projection by suitably altering the absolute terms **L** of the matrix of observation equations.

The computation of a triangle on the UTM is considered in detail to show the general principles of projection computation.

A8.13 Computation on the UTM projection

The data for triangle ABC used in the ellipsoidal computation will be computed on the Universal Transverse Mercator system Zone 36, central meridian 33 east longitude, origin at 10×10^6 m south, 500 000 m west. The ellipsoidal angles of the triangle are denoted by A', B', and C'; their corresponding projection or grid counterparts are denoted by A, B, and C. The ellipsoidal lengths are denoted by a', b', c'; the corresponding grid lengths are denoted by a, b, and c. Let the angular distortions be dA, dB, and dC and the side distortions be da, db, and dc such that;

$$A = A' + dA \qquad B = B' + dB \qquad C = C' + dC$$

$$a = a' + da \qquad b = b' + db \qquad c = c' + dc$$

If the spherical excess of triangle A'B'C' is ε, then since A+B+C=180°, dA + dB + dC = – ε. The grid angles and sides have been calculated from the grid coordinates of A, B, and C obtained from the geographical coordinates of A', B', and C' by the direct projection formulae A8(2) and A8(7). The corresponding ellipsoidal and grid data are given in Table A8.1

Table A8.1

Angles	*Ellipsoidal*	*Plane*	*Diff*	"
A'	30° 48' 03.96"	30° 48' 10.48"	dA	+06.52
B'	61° 05' 49.07"	61° 05' 57.90"	dB	+08.72
C'	88° 05' 07.74"	88° 05' 51.55"	dC	–16.01
Sides	m	m		m
a'	13 327.628	13 333.512	da	+5.884
b'	22 785.523	22 794.865	db	+9.342
c'	26 013.282	26 023.251	dc	+9.969

Alternatively, from the measured ellipsoidal angles and sides, we could compute the grid coordinates entirely in plane trigonometry once we had the minimum of starting data on the projection, say one point A and a bearing to B. Often this information is given as part of the control from which a survey is begun. We need only compute the distortions given in the fourth column of Table A8.1, apply them to the observed data, and proceed with the ordinary formulae thereafter.

There is the difficulty that the formulae for the angular and scale distortions involve the known coordinates of the starting points and also those of the points required. Thus a process of successive approximation is required, though only one approximation is necessary in most work. The method is first to compute approximate coordinates, by plane trigonometry, to a metre precision using the observed data, or a mixture of observed and grid data, and then to repeat the process with ellipsoidal lengths altered to grid lengths, and ellipsoidal angles to grid angles. Whichever combination of fixation methods is used, a unique result is obtained. For example, an intersected point computed from grid angles will give the same result as the same fixation computed by a linear grid distance and a grid bearing using the same original data. The working formulae are for

(a) the line scale factor; and
(b) the arc-to-chord, or t – T, correction.

A8.14 Line scale factors

Assuming a required accuracy of 0.01 m and distances greater than 1 km, the line scale factor is obtained from the point scale factors, evaluated at its terminals and mid-point, by Simpson's rule. Consider the line AB whose mid-eastings and northings are E_M and N_M, and whose point scale factors are K_A, K_M, K_B. The scale factor for AB, i.e. K_{AB}, is given by Simpson's rule.

$$K_{AB} = \frac{1}{6}(K_A + 4K_M + K_B)$$

Example: If the scale factors for AB at A, its mid-point, and B are found to be respectively

1.000 353 4,   1.000 383 6, and 1.000 414 5

then

$$K_{AB} = 1.000\ 383\ 7.$$

Since the ellipsoidal length of A'B' is 26 013.282, this gives the grid length

$$AB = 26\ 023.263$$

An alternative formula for the line scale factor K is

$$K_{AB} = 1 + \frac{E_A^2 + E_A\ E_B + E_B^2}{6\ R^2} \qquad \text{A8(8)}$$

A8.15 Arc-to-chord correction or t – T correction

To convert the ellipsoidal angle A' to a plane projection angle A we proceed as follows. Refer to Figure A8.8.

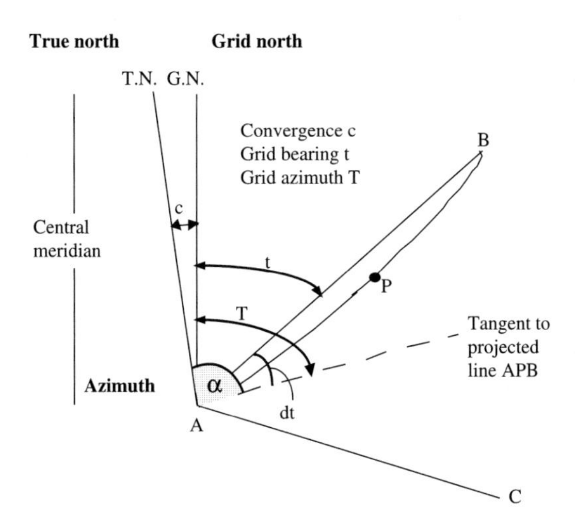

Figure A8.8

The grid angle A is derived from the grid bearings of AC and AB denoted by t_{AC} and t_{AC} respectively, i.e.

$$A = t_{AC} - t_{AB}$$

Let T_{AB} and T_{AC} be the grid *azimuths* of AB and AC respectively. The grid azimuth or spherical bearing is the azimuth α reduced by the convergence C. That is

$$T = \alpha - C$$

Because the projection is orthomorphic, the angle at A between the projected ellipsoidal arcs, or (which is the same thing) the angle between the projected tangents to these arcs, equals the ellipsoidal angle A', i.e. $A' = T_{AC} - T_{AB}$.

Let the differences between the grid azimuths and grid bearings be denoted by dt, then

$$dt_{AC} = t_{AC} - T_{AC}$$

$$dt_{AB} = t_{AB} - T_{AB}$$

Subtracting gives

$$dt_{AC} - dt_{AB} = t_{AC} - t_{AB} - (T_{AC} - T_{AB})$$
$$= A - A'$$

Then

$$A = A' + dt_{AC} - dt_{AB}$$

The distortion factor dt = t – T is usually called the "t – T" correction, or the "arc-to-chord correction".

In practice the values of t – T for each direction are found first by the formula given below. They may be subtracted in pairs to give the distortion to each angle, if computation uses the angles.

The sign of the correction to a direction is obtainable from a strict application of the formulae with due regard to signs; but in practice it is quicker to draw a diagram which shows the signs at once. In Figure A8.8, A and B are the projected positions of A' and B', c = convergence at A. If P' is a point on the ellipsoidal line A'B', on the projection it plots as P on the dotted line which curves away from the line of least scale, i.e. away from the central meridian of the projection. Thus the sign of $t - T_{AB}$ is at once evident from the diagram.

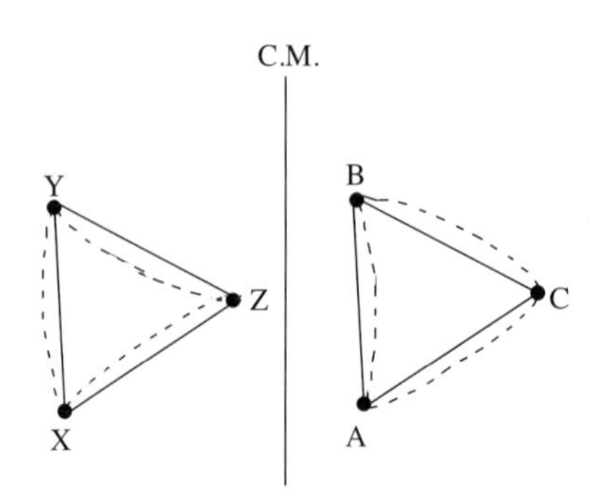

Figure A8.9

Figure A8.9 shows the projected ellipsoidal arcs for the triangle ABC lying to the east of the central meridian, and the projected ellipsoidal arcs for the triangle XYZ lying to the west of the central meridian. Remembering that azimuths and bearings are reckoned clockwise from north, even in the southern hemisphere,

Grid bearing = Grid azimuth + (t – T) with due regard to signs.

Thus the $(t - T)_{ZY}$ is negative and $(t - T)_{YZ}$ is positive. The working formula for the arc-to-chord correction for the bearing AB is

$$(t-T)''_{AB} = -\frac{(2\,E_A + E_B)\,(N_B - N_A)}{6\,\rho\,\nu\,K_o^2\;\sin 1''} \qquad \text{A8(9)}$$

and that for the bearing BA is

$$(t-T)''_{BA} = -\frac{(2\,E_B + E_A)\,(N_A - N_B)}{6\,\rho\,\nu\,K_o^2\;\sin 1''}$$

It will be evident that the (t – T)s at either end of the same line are not identically equal in magnitude, though they will be of the same order. The size of the correction varies with easting, i.e. increases away from the central meridian, and it also increases with the difference in northing. A line running east-west will have no correction.

A8.16 Computation of (t – T) corrections

Since the denominator of the formula changes very slowly with latitude it may be tabulated for particular ellipsoids and central scale factors, or programmed for computer.

Because the algebraic sum of the t – T corrections for the angles of a closed figure must equal the spherical excess of that figure, the spherical excess should always be calculated to check the arithmetic.

Example of solution of triangle ABC

Given the projection coordinates of A and B and the adjusted angles of triangle ABC, the coordinates of C will be derived using the cotangent formula for angles

$$E_P = \frac{E_A \cot B + E_B \cot B - N_A + N_B}{\cot A + \cot B} \qquad \text{A8(10)}$$

$$N_P = \frac{N_A \cot B + N_B \cot B + E_A - E_B}{\cot A + \cot B} \qquad \text{A8(11)}$$

Table A8.2 gives the complete computation. Provisional coordinates of C, E'_C and N'_C are obtained from the ellipsoidal angles, and the coordinates of A and B, using equations A8(10) and A8(11).

From these provisional coordinates and the coordinates of A and B, the six t – T corrections for the directions are derived, remembering to subtract the false easting of 500 000 m.

The factors tabulated in column three are

$$F = \frac{2\,E_1 + E_2}{6\,\rho\,\nu\,K_o^2\ \sin 1''}\,10^3$$

Hence the t – T correction is given by

$$-\,F\,(N_2 - N_2)\ 10^{-3}$$

The reader can easily verify the magnitude of F using R = 6 378 000 m and a central scale factor of 0.9996, and that the t – T is 8.81" by this means, and the spherical excess from

$$\varepsilon = \frac{1}{2}\ a\,b \sin C\ \sin 1''$$

The correct radii of curvature are used in the example.

Table A8.2

True E	*True N 8 000 000 +*	*Ellipsoidal angle*	*t – T*	*Grid angle*	
A	246 793.673	243 690.628	30° 48' 03.96"	+ 6.52"	10.48"
B	256 612.703	267 790.343	61° 05' 49.07"	+ 8.72"	57.79"
C'	264 990.5	257 418.5	88° 05' 67.74"	–16.01"	51.53"
C	264 991.40	257 418.28	180° 00' 00.77"	–00.77"	00.00"

Line	*$2\,E_1 + E_2$*	*Factor*	*N_2–N_1*	*t – T Direction*	*t – T Angle*
AC	758 578	0.64558	– 13 727	– 8.86"	+ 6.52"
AB	750 201	0.63846	– 24 099	– 15.38"	
BA	760 020	0.64681	+ 24 099	+ 15.59"	+ 8.72"
BC	778 216	0.66228	+ 10 372	+ 6.87"	
CB	786 593	0.66940	– 10 372	– 6.94"	– 16.01"
CA	776 774	0.66105	+ 13 727	+ 9.07"	
			Sph excess	– 0.77"	– 0.77"

The t – T correction for each angle is obtained by subtraction of corrections for pairs of directions. The spherical excess of the triangle is found to check the sum of the (t – T)s. The grid angles are then derived by application of the t – T corrections to the adjusted angles. The final coordinates of C are computed using the cotangents of the grid angles. In Least Squares and semi–graphic methods of fixation, the t – T corrections are incorporated into the O–C terms at the stage at which the coordinates are known to within a metre of their correct values.

A8.17 Other fixations on the projection

This procedure of converting observed angles and measured sides into their grid counterparts is used in all methods of fixation, i.e. in traverse, trilateration, etc. The process is similar to that given for the triangle, namely, scale factors and/or t – T corrections are derived from provisional coordinates. The plane trigonometrical formulae are then applied to grid lengths and grid bearings or grid angles as the case may be.

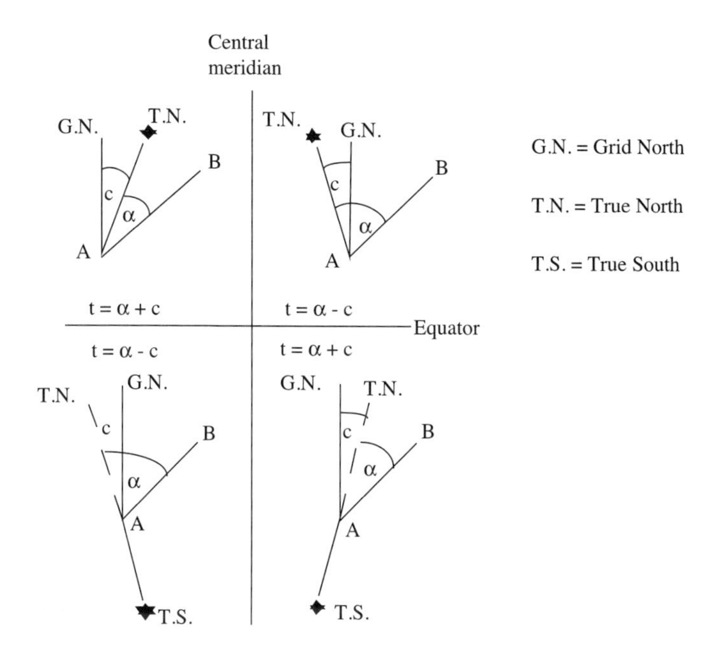

Figure A8.10

Figure A8.10 shows the signs of the convergence c in the four differing positions with respect to the central meridian and the equator.

As has been mentioned before, modern computational strategy is to work entirely in geographical or three dimensional cartesian terms usually by variation of coordinates and Least Squares estimation. Points are then transformed to a projection for mapping. Computation on the projection is now applied to surveys of limited scope only.

A8.18 Proof of formula for (t – T) correction

The following simplified derivation of the arc-to-chord correction on the tranverse Mercator projection assumes a spherical model Earth. The reader who prefers a more rigorous proof should refer to the first edition of this book. Refer to Figure A8.11.

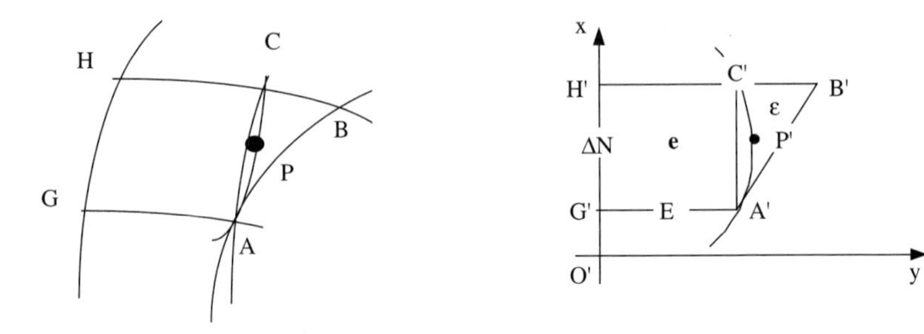

Figure A8.11

Figure ABCHG on the sphere is projected on the map as A'B'C'H'G'. Point C lies due north of A. The arc APC is a great circle whereas AC is a small circle. The spherical excess of Figure APCHG is e, given by

$$e = \text{area}/ R^2 = \Delta N.\,E\ / R^2$$

By inspection and symmetry the small angle CAP = C'A'P' = e/2.

If the spherical excess of triangle ABC is ε then

$$\varepsilon = \Delta N.\,\Delta E / 2 R^2$$

In the spherical triangle we may use Legendre's theorem to compute sides, and therefore deduct ε/3 from angle A for calculations. Since the projection is conformal, the angle A = A' of the map. Hence the bearing of A'B' will be

$$A - \frac{e}{2} - \frac{\varepsilon}{2}$$

which on substitution for e and ε gives the required bearing as

$$A - \Delta N\,(2\,E_A + E_A) / 6R^2$$

or

$$t - T''_{AB} = -\Delta N\,(2\,E_A + E_A) / 6R^2 \sin 1''$$

If a central scale factor of K_0 has been introduced to the projection, this amounts to a reduction of the radii of the ellipsoid model, and we have the final result

$$t - T''_{AB} = -\Delta N(2\,E_A + E_A)/6\rho\nu^2\,K_0^2\ \sin 1''$$

A8.19 The Lambert conical orthomorphic projection

The Lambert conical orthomorphic projection is much used to map areas which have a great extent east-west and a small extent (about 6°) from north to south. For example, about one half of the states of the USA are mapped on this projection. It is easier to construct than the transverse Mercator projection, since its graticule is formed by meridians which project as straight lines, and parallels which plot as circular arcs. Interpolation of coordinates between tabulated portions of the graticule is relatively uncomplicated. One drawback is that the convergence becomes excessive away from the centre of the projection.

A8.20 Simple conical projection

The Lambert projection is formed by adapting a simple conical projection, in a similar way to that in which the transverse Mercator projection is adapted from the Cassini projection.

Simple conical coordinates are produced by developing the surface of a cone which touches the sphere along a parallel of latitude passing through the middle of the area to be mapped. This parallel is called the *standard parallel* defined by the standard latitude ϕ_o. The apex of the cone lies on the prolongation of the axis of the sphere. In Figure A8.12 the cone is shown touching the sphere along the standard parallel.

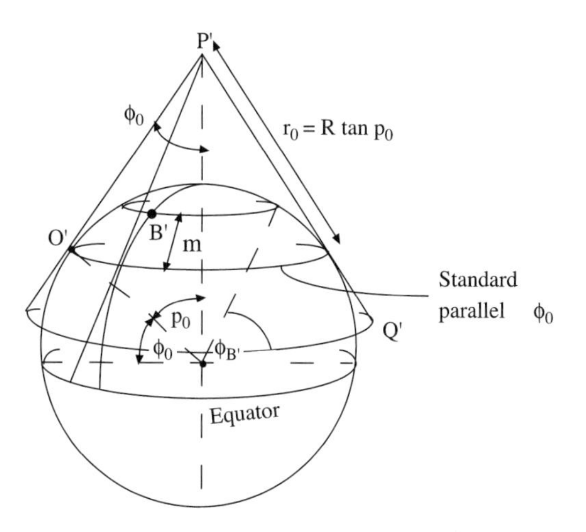

Figure A8.12

A point at the centre of the east-west dimension of the area to be mapped and lying on the standard parallel is chosen to be the origin of the coordinates. The meridian through this origin O is the central meridian. As will be seen later, the character of the projection depends much more on the standard parallel than on the central meridian. The latitudes of points to be mapped are reckoned from the standard parallel, and the longitudes from the central meridian.

If the radius of the sphere is R, the slant length of the cone from the standard parallel, P'O', is given by

$$P'O' = R \cot \phi_o = R \tan p_o$$

where $p_o = 90 - \phi_o$ is the co-latitude of the standard parallel.

The cone is developed by cutting it along a line such as P'O' producing the shape shown in Figure A8.13. The standard parallel on the map is plotted as a circular arc of radius r_o = P'O'' = PO centred on the projected apex of the cone, i.e. at P. The length of this arc equals the length of the standard parallel on the sphere. The length of the standard parallel on the sphere = $2 \pi R \cos \phi_o$, hence

$$\text{arc } Q_1 O Q_2 = 2 \pi R \cos \phi_o$$

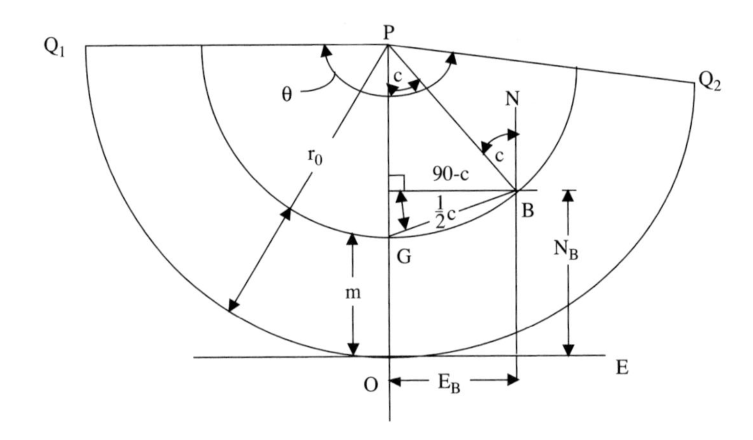

Figure A8.13

The angle $Q_1 P Q_2 = \theta$ is given by

$$\theta = \frac{2 \pi R \cos \phi_o}{R \tan \phi_o} = 2 \pi R \sin \phi_o = 2 \pi n$$

Hence 2π radians of longitude on the sphere are represented by $2 \pi n$ radians on the map; in general, any longitude difference $\Delta\lambda$ on the sphere will be represented by an angle $n \Delta\lambda$ where n, called the *constant of the cone*, defines the spacing of the meridians on the map.

The other parallels are represented by circular arcs centred on P such that their spacing equals the meridian arc distance separating these parallels on the sphere. Consider a point B' on the sphere at latitude ϕ_B. The spherical arc distance $m = R(\phi_B - \phi_O)$ separates the projected parallel of B from the standard parallel. The radius of the projected parallel of B is given by

$$r_B = r_o - m$$

The projected point B lies on a projected meridian which makes an angle c with the central meridian such that

$$c = n(\lambda_B - \lambda_o)$$

where c is the *map convergence* at B. The map coordinates of B are given from Figure A8.13 as follows.

$$E_B = DB - r_B \sin c = (r_o - m) \sin c \qquad \text{A8(12)}$$

$$N_B = OD = m + E_B \tan \frac{1}{2} c \qquad \text{A8(13)}$$

A8.21 Lambert's conical orthomorphic projection

Since the point scale along a meridian does not equal the point scale along a parallel, other than the standard parallel, the projection is not orthomorphic, or it is not conformal.

To make it conformal, the north-south spacing of the parallels is altered to equate the point scale factor along the projected meridian to the point scale factor along a projected parallel.

Let K_1 be the scale factor along a parallel, and K_2 be the scale factor along a meridian, then

$$K_1 = \text{map distance / distance on sphere}$$

$$= r\,d\theta / R \sin p\, d\lambda$$

$$= nr / R \sin p$$

Also

$$K_2 = dr / R\, dp$$

For the projection to be conformal

$$K_1 = K_2$$

$$dr / R\, dp = nr / R \sin p$$

$$\frac{dr}{r} = \frac{dp}{\sin p}$$

Integrating to find r we have

$$\int_{r_o}^{r_r} \frac{d\,r}{r} = \int_{p_o}^{p_r} \frac{dp}{\sin p}$$

$$\ln\left(\frac{r}{r_o}\right) = n \ln\left(\frac{\tan \frac{1}{2} p}{\tan \frac{1}{2} p_o}\right) \qquad \text{A8(14)}$$

$$\frac{r}{r_o} = \left(\frac{\tan \frac{1}{2} p}{\tan \frac{1}{2} p_o}\right)^n$$

$$r = F\left(\tan \frac{1}{2} p\right)^n$$

where $F = r_o / (\tan \frac{1}{2} p_o)^n$ is a scale constant defined by the latitude of the standard parallel

The computation of the grid coordinates (E, N) is identical to that for the simple conical projection with the exception that the radius of the parallel for latitude is found by equation A8(14).

Because the radii of the projected parallels are very large it is simpler to work in terms of corrections to the standard radius. If

$$r = r_o + m'$$

it can be shown by application of Taylor's theorem to expression (14) that

$$m' = m + \frac{m^3}{6R^3} - \frac{m^4}{24R^3} \tan \phi_o \ldots \qquad \text{A8(15)}$$

where the meridional arc distance m is given by

$$m = R(\phi - \phi_o)$$

The distance m is reckoned from the standard parallel, positive southwards, i.e. with increasing r, and negative northwards. The values of m' are tabulated in projection tables or derived in software.

The character of the Lambert projection is very similar to that of the transverse Mercator projection, since

$$m' \approx m + \frac{m^3}{6R^3}$$

which compares with the easting on the transverse Mercator of

$$E \approx y + \frac{y^3}{6R^3}$$

The scale variation on Lambert depends on the distance of a point north or south of the standard parallel. In fact, the transverse Mercator projection is merely a special case of Lambert's conformal projection, as is the Mercator projection itself.

Scale factor

The scale factor K by analogy is is given approximately by

$$K \approx 1 + \frac{m^3}{2R^3}$$

A central scale factor K_o is introduced in the same way as for the transverse Mercator projection, which produces a negative net scale along the standard parallel and reduces the scale error at the maximum extent north or south to a similar numerical amount.

The arc-to-chord correction on the Lambert projection

By analogy with the transverse Mercator projection, the first term of the formula for the t – T correction is

$$t - T''_{AB} = -\Delta E(2N_A + N_A)/6\rho\nu^2 K_o^2 \ \sin 1''$$

Northings are interchanged with eastings by comparison with the transverse Mercator projection.

A8.22 Oblique or skew projections

In Figure A8.14 a projection system was derived from the point P and the great circle PHGO. In the cases of the Cassini and transverse Mercator projections, P is the Earth's geographical pole, and PHGO is a meridian.

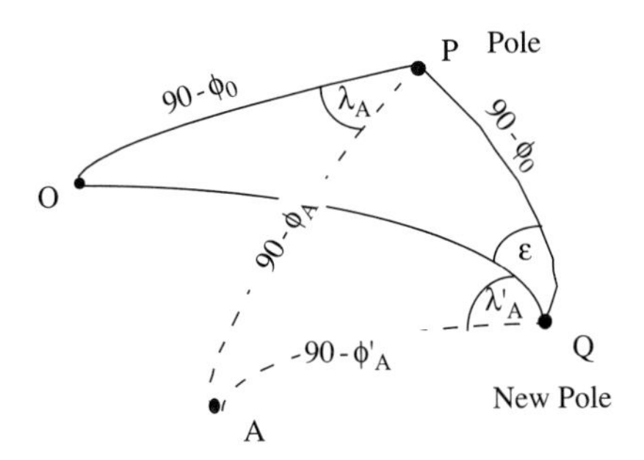

Figure A8.14

However, the same formulae can be used to obtain similar projection systems, if P is any point on the surface of the sphere, and PHGO is any great circle; provided the coordinates analogous to 'latitude' and 'longitude' with reference to this 'pole' and this 'meridian' can be calculated.

In Figure A8.14 Q is the new 'pole' and the great circle QO is the new 'central meridian' of a projection whose origin is at O. A is any point to be projected. The geographical latitudes and longitudes (ϕ, λ) of Q, O, and A respectively can be used to derive new 'latitudes' and 'longitudes' (ϕ', λ'), with respect to the pole at Q, from spherical triangle OPQ.

Also the new latitiudes and longitudes can be found on solution of the spherical triangle PQA. Thence a Cassini type or transverse Mercator type projection can be computed, based on pole Q and meridian QO and new coordinates of points such as A.

The general case is an *oblique* or *skew* form of projection. The skew projection is suited to a country which is orientated at some azimuth intermediate between 0° and 90°. For example Malaysia and Borneo use oblique Mercator projections.

Although the projection character and formulae are based on a skew 'meridian', the coordinates obtained are finally rotated on to a north–south grid axis to accord with convention and reduce the convergence between the final grid and graticule.

The name transverse Mercator derives from the fact that this projection is the transverse case of the Mercator projection in the sense described here. The Mercator projection is based on the equator as the 'central meridian' and was developed historically before the transverse case and before the Gauss-Kruger projection as the latter is also called. Alternatively the Mercator projection could be called the 'transverse Gauss-Kruger' projection.

A8.23 Comment

Many map projections can be considered as conical projections. For example, if the standard parallel of the Lambert projection is the equator, Mercator's projection is produced, and thus the Gauss-Kruger projection is a special transverse case of Lambert's conical orthomorphic projection. If the standard parallel is at latitude 90° i.e. at the pole, an azimuthal projection is obtained. A full treatment of map projection is beyond the scope of this book. For more information the reader should refer to Snyder 1987.

For simple treatment of the Stereographic projection see Allan 2004. Chapter 11.12

Appendix 9
Satellite Surveying

The following is a brief outline of the past and current (2006) roles played by satellite technology in geospatial surveying. Remote sensing and other systems are omitted.

History

Prior to the 1990s, terrestrial positions and azimuths were obtained from precise Optical Astronomical Methods (Torge 2001).

1958 – 1992 US Navy's Doppler Transit system
1960's Satellite Laser Ranging(SLR) and VLBI (Very Long Base lime Interferometry), and the US Army's microwave ranging system (SECOR)
1966 – 70 Photogrammetric position fixing from Balloons (Echos and Pageos)
1978 – 1995 GPS and GLONASS development
1995 – Operational GPS and GLONASS
2006 – Galileo development.

The following is a brief summary of the main features of the current GPS and GLONASS systems. Clearly these will have improved by the time of publication of this book. Galileo is not yet operative.

Navstar/Gps

(Navigational System with Time and Range/Global Positioning System) GPS provides real time navigation and positioning by *one-way microwave measurements* between satellites and GPS receivers.

At least four satellites above the horizon available everywhere 24 hours a day. In principle three dimensional coordinates can be derived from three distances measured from three satellites whose positions are known from ephemerides. The clocks of the satellites and the receiver have not been synchronised so a fourth satellite and distance is required.

Space segment

There are 24 satellites in six nearly circular orbits (i = 55° T = 12 hours sidereal). Altitude of 20 200 km. Lifetime 10 years.

Satellites are fitted with atomic clocks (2 rubidium 2 cesium per sat) freqency standards to 10^{-13}.

Fundamental operating frequency is 10.23 MHz: By multiplication by 154 to L1 1574.42 MHz (19.05 cm) wavelength and by 120 to L2 1227.60 MHz (24.45 cm) wavelength.

(Coarse Aquisition) C/A code modulated on L1 only, F 1.023 MHz (293 m) repetition rate of 1 ms.

Precise P Code mod on L1 and on L2 f 10.23 MHz (29.3 m). Repetition sequence of 266 days.

Also Navigation message on L1 and L2 (Sat ephemeris, clock correction to GPS time, coeffecients for ionosperic refraction model, a system status information).

Control Segment

Colorado Springs plus five other globally distributed stations. World Wide Geodetic System WGS 84

$$a = 6378137 \text{ m} \quad 1/f = 298.257\ 223\ 563$$

$$GM = 398600.4418 \times 10^{9} \text{ m}^{3}\text{s}^{-2}$$

$$\text{Angular velocity of Earth } 7.292115 \times 10^{-5} \text{rad s}^{-1}$$

Russian GLONASS (GLObal Navigation Satellite System)

Similar 24 satellites three orbits, $i = 64.8°$ $h = 19\ 100$ km.
Period T = 11 hours 15 min.

Appendix 10
References and Further Reading

Allan, A.L. 1997. *Practical Surveying and Computations*. ISBN 7506-3655-6. Butterworth-Heinemann. pp 571.

Allan, A.L. 2004. *Maths for Map Makers*. ISBN 1-870325-99-0. Whittles Publishing. pp 394.

Barrel, H. and Sears, J.E. 1969. The refraction and dispersion of air for the visible spectrum. *Phil. Trans. R. Soc.* A238.

Bomford, G. 1980. *Geodesy*. 4th Edition. Oxford. Clarendon Press. pp 855.

Bowring, B.A. and Vincenty, T. 1978. *Application of three-dimensional Geodesy to adjustments of horizontal networks*. Rockville Md USA. NOAA Technical Memorandum NOS-NGS-13.

Burnside, C.D. 1991. *Electromagnetic Distance Measurement*. Oxford. BSP Professional Books. pp 278.

Cooper, M.A.R. 1987. *Control Surveys in Civil Engineering*. London. William Collins Sons & Co Ltd. pp 381.

Cooper, M.A.R. and Cross, P.A. 1988. Statistical Concepts and their application in Photogrammetry and Surveying. *Photogrammetric Record* XIII(3), 645–678.

Cross, P.A. 1983. *Advanced Least Squares applied to Position Fixing*. Working Paper No. 6. University of East London. pp 205.

Froome, K.D. and Essen, L. 1969. *The velocity of light and radio waves*. London. Academic Press. pp 157.

Garland, G.D. 1997. *The Earth's Shape and Gravity*. ISBN 0-08-010822-9. Pergamon. pp 186.

RICS Guidance Note. 2003. *Guidelines for the use of GPS in Surveying and Mapping*. ISBN 1-84219-093-8. pp 98.

Iliffe, J.C. 2000. *Datums and Map Projections*. ISBN 1-870325-28-1. Whittles Publishing. pp 150.

Kendal, M.G. and Stuart, A. 1967. *The Advanced Theory of Statistics*. London. Griffin.

Leick, A. 2004. *GPS Satellite Surveying*. ISBN 0-471-05930-7. New York. John Wiley and Sons. pp 435.

Mikhail, E. 1976. *Observations and Least Squares*. ISBN 0-7002-2481-5. New York. Dun-Donnelley. pp 497.

Reuger, J.M. 1996. *Electronic Distance Measurement*. Berlin. Springer-Verlag. pp 265.

Reuger, J.M. 2003. *Electronic Surveying Instruments*. ISBN 0-7334-2083-4. Monograph 18, University of New South Wales. pp 156.

Snyder, J.P. 1987. *Map Projections: a working manual*. Washington D.C. US Geological Survey, US Government Printer. pp 256.

Star Almanac for Land Surveyors. The Stationery Office Annually. London. HMSO. pp 80.

Thomas, T.A. 1982. The Six Methods of Finding North using a suspended gyroscope. *Survey Review* 26(203) and 26(204) Jan and April 1982.

Thompson, E.H. 1969. *An Introduction to the Algebra of Matrices with some Applications*. London. Adam Hilger. pp 229.

Torge, W.T. 2001 *Geodesy*. ISBN 3-11-017072-8. de Gruter. pp 416.

Vanicek, P. and Krakiwsky, E.J. 1986. *Geodesy the Concepts*. Amsterdam. North-Holland. pp 697.

Journals and organizations

Survey Review
Cartographic Journal
Photogrammetric Record
Journal of Geodesy
Hydrographic Surveyor
GIM International
Geomatics World
GIS Professional
Ordnance Survey (see http://www.ordnancesurvey.co.uk)
Royal Institution of Chartered Surveyors (see http://www.rics.org/geo)
National Geodetic Survey (see http://www.ngs.noaa.gov/Geoid)

Websites for survey instrument manufacturers

http://www.spectraprecision.com
http://www.topconeurope.com
http://www.thalesnavigation.com
http://www.pentax.co.jp/piic/survey
http://www.leica-geosystems.com
http://www.leica.loyola.com
http://www.haff.com
http://www.riegl.com
http://www.kolidainstrument.com
http://www.trimble.com
http://www.ushikata.co.jp
http://www.sokkia.co.uk
http://www.nikon-trimble.com

Index